Advances in Intelligent Systems and Computing

Volume 412

Series editor

Janusz Kacprzyk, Polish Academy of Sciences, Warsaw, Poland
e-mail: kacprzyk@ibspan.waw.pl

About this Series

The series "Advances in Intelligent Systems and Computing" contains publications on theory, applications, and design methods of Intelligent Systems and Intelligent Computing. Virtually all disciplines such as engineering, natural sciences, computer and information science, ICT, economics, business, e-commerce, environment, healthcare, life science are covered. The list of topics spans all the areas of modern intelligent systems and computing.

The publications within "Advances in Intelligent Systems and Computing" are primarily textbooks and proceedings of important conferences, symposia and congresses. They cover significant recent developments in the field, both of a foundational and applicable character. An important characteristic feature of the series is the short publication time and world-wide distribution. This permits a rapid and broad dissemination of research results.

More information about this series at http://www.springer.com/series/11156

Muthukrishnan Senthilkumar
Vijayalakshmi Ramasamy
Shina Sheen · C. Veeramani
Anthony Bonato · Lynn Batten
Editors

Computational Intelligence, Cyber Security and Computational Models

Proceedings of ICC3 2015

Springer

Editors
Muthukrishnan Senthilkumar
Applied Mathematics and Computational Sciences
PSG College of Technology
Coimbatore, Tamil Nadu
India

Vijayalakshmi Ramasamy
Applied Mathematics and Computational Sciences
PSG College of Technology
Coimbatore, Tamil Nadu
India

Shina Sheen
Applied Mathematics and Computational Sciences
PSG College of Technology
Coimbatore, Tamil Nadu
India

C. Veeramani
Applied Mathematics and Computational Sciences
PSG College of Technology
Coimbatore, Tamil Nadu
India

Anthony Bonato
Department of Mathematics
Ryerson University
Toronto, ON
Canada

Lynn Batten
School of Information Technology
Deakin University
Melbourne
Australia

ISSN 2194-5357 ISSN 2194-5365 (electronic)
Advances in Intelligent Systems and Computing
ISBN 978-981-10-0250-2 ISBN 978-981-10-0251-9 (eBook)
DOI 10.1007/978-981-10-0251-9

Library of Congress Control Number: 2015958313

Printed on acid-free paper

This Springer imprint is published by SpringerNature
The registered company is Springer Science+Business Media Singapore Pte Ltd.

Preface

The most recent advancements in the dynamically expanding realm of Internet and networking technologies have provided a scope for research and development in computer science and its allied thrust areas. To provide a broad interdisciplinary research forum, the International Conference on Computational Intelligence, Cyber Security and Computational Models (ICC3 2015) is organized by the Department of Applied Mathematics and Computational Sciences of PSG College of Technology, during 17–19 December 2015.

The principal objective of this conference is to discuss the state-of-art scientific approaches, techniques and results to explore the cutting-edge ideas and to promote collaborative research in the areas of computational intelligence, cyber security and computational models to enable establishing research relations worldwide.

Computational intelligence (CI), a dynamic domain of modern information science, has been applied in many fields of engineering, data analytics, forecasting, biomedicine and others. CI systems use nature-inspired computational approaches and techniques to solve complex real-world problems. The widespread applications range from image and sound processing, signal processing, multi-dimensional data visualization, steering of objects to expert systems and many other potential practical implementations. CI systems have the capability to reconstruct behaviours observed in learning sequences and can form rules of inference and generalize knowledge in situations when they are expected to make prediction or to classify the object to one of the previously observed categories. CI track consists of the research articles which exhibit various potential practical applications.

Cyber security landscape is evolving rapidly, as the attacks are increasing in number and sophistication from a wider range of threat actors than ever before. The large-scale cyber attacks in various countries lead to the threat of information security which in turn could be a threat to national security and requires effective crisis management. Such information security risks are becoming more diversified, advanced and complex, and many conventional means of security fail to ensure information safety. Cyber security track in this conference aims to be a forum for the presentation of developments in computer security and for bringing together researchers and practitioners in the information security field to exchange practical ideas and experiences.

Computational experiments are inevitable in this era, because analytical solutions to many scientific problems may not be obtainable or be tedious to derive. Theory of computation, data analytics, high-performance computing, quantum computing, weather forecasting, flight simulation, Earth simulator, protein folding and so on need computational models like stochastic models, graph models and network models to make predictions about the performance of complicated systems. Solutions of numerous technical problems require extensive mathematical concepts to model the problem and to understand the behaviour of associated complex systems through computer simulations. With the advent of efficient computations, solutions can be found for various problems using computational modelling and research in the domain is gaining significance.

This is reflected in an increase in submissions to ICC3 2015 over the previous edition. We received 177 papers in total, and accepted 56 papers (31 %). Every submitted paper went through a rigorous review process. Where issues remained, additional reviews were commissioned.

The organizers of ICC3 2015 wholeheartedly appreciate the peer reviewers for their support and valuable comments for ensuring the quality of the proceedings. We also extend our warmest gratitude to Springer Publishers, for their continued support in bringing out the proceedings volume in time and for excellent production quality. We would like to thank all keynote speakers, international advisory committee members and the chair persons for their excellent contribution. We hope that all the participants of the conference would have been benefited academically and wish them success in their research career.

This ICC3 series traditionally results in new contacts between the participants and interdisciplinary communications realized often in new joint research. We believe that this tradition will continue in the future as well. The next ICC3 conference will be held in PSG College of Technology, Coimbatore in 2017.

Organization

Patron

Shri L. Gopalakrishnan
Managing Trustee
PSG & Sons Charities Trust
Coimbatore, India

Chairman

Dr. R. Rudramoorthy
Principal
PSG College of Technology
Coimbatore, India

Organizing Chair

Dr. R. Nadarajan
Professor and Head
Department of Applied Mathematics and Computational Sciences
PSG College of Technology
Coimbatore, India

Program Chair

Dr. Muthukrishnan Senthilkumar
Associate Professor
Department of Applied Mathematics and Computational Sciences

PSG College of Technology
Coimbatore, India

Computational Intelligence Track Chair

Dr. Vijayalakshmi Ramasamy
Associate Professor
Department of Applied Mathematics and Computational Sciences
PSG College of Technology
Coimbatore, India

Cyber Security Track Chair

Dr. Shina Sheen
Assistant Professor
Department of Applied Mathematics and Computational Sciences
PSG College of Technology
Coimbatore, India

Computational Models Track Chair

Dr. C. Veeramani
Assistant Professor
Department of Applied Mathematics and Computational Sciences
PSG College of Technology
Coimbatore, India

Advisory Committee Members

Prof. Anthony Bonato, Department of Mathematics, Ryerson University, Canada
Prof. Kiseon Kim, School of Information and Communication, Gwangju Institute of Science and Technology, Gwangju, South Korea
Prof. Yew-Soon Ong, Director, SIMTECH-NTU Joint Lab on Complex Systems, School of Computer Engineering, Nanyang Technological University
Prof. Ram Ramanathan, Business School, University of Bedfordshire, UK
Prof. Atilla Elci, Professor, University of Aksaray, Turkey

Contents

About the Editors

Muthukrishnan Senthilkumar is an Associate Professor in the Department of Applied Mathematics and Computational Sciences at PSG College of Technology. He has received Post Doctoral Fellowship from Department of Nanobio Materials and Electronics, (WCU), Gwangju Institute of Science and Technology (GIST), Gwangju, Republic of Korea. He is also a Visiting Professor to the School of Information and Communication, GIST, Republic of Korea. He has over 16 years of teaching and research experience. He has received his Ph.D. in retrial queueing models from Anna University, Chennai, India. His biography is recognized and included in the biographies of Marquis Who's Who in the World 2013 and 2016. His fields of interest include retrial queueing theory, data communication, reliability engineering and Epidemic Models. He has published several research articles in refereed reputed journals. Currently he is guiding four Ph.D. scholars.

Vijayalakshmi Ramasamy has over 19 years of academic experience and currently is an Associate Professor in the Department of Applied Mathematics and Computational Sciences at PSG College of Technology, India. She received her Ph. D. in graph mining algorithms from Anna University. Her research interest includes graph based data mining, especially complex network systems and social network analysis. Her research publications include 5 book chapters and about 26 peer-reviewed international journals and conference papers. She has been a reviewer for many reputed journals. She is currently actively associated with Cognitive Neuro-engineering Laboratory (CNEL), UniSA in the acquisition and analysis of brain wave data for cognitive modelling, pattern identification and visualization using graph-theoretic approaches. She has established a Computational Neuroscience Laboratory in 2012 at PSG College of Technology in association with Prof. Nanda Nandagopal, Professor of Defense from the University of South Australia, Australia.

Shina Sheen is an Assistant Professor (Selection Grade) in the Department of Applied Mathematics and Computational Sciences at PSG College of Technology. She holds a B.Sc. in Physics from Calicut University and an MCA and an M.Phil. in Computer Science from Bharathiyar University. Her Ph.D. thesis is in the area of malware detection using machine learning and data mining techniques. Her research focuses on security in smart devices, intrusion detection, cyber security and data mining. She has published articles in refereed reputed journals and conference proceedings. She holds the GIAC certified intrusion analyst (GCIA) security certification. She is a member of ACM, Cryptology Research Society of India and the Computer Society of India.

C. Veeramani is an Assistant Professor in the Department of Applied Mathematics and Computational Sciences at PSG College of Technology. He has received his Ph.D. degree in fuzzy optimization from Anna University, Chennai, India. He has over 10 years of teaching and research experience. His fields of interest include fuzzy optimization, fuzzy logic, soft computing and epidemic models. He has published his research articles in various reputed international journals and international conference proceedings. Currently, he is guiding four Ph.D. scholars in the area of fuzzy optimization and fuzzy epidemic models.

Anthony Bonato is Associate Dean, Students and Programs in the Yeates School of Graduate Studies at Ryerson University and Editor-in-Chief of the journal Internet Mathematics. He has authored over 90 publications with 40 co-authors on the topics of graph theory and complex networks. In 2009 and 2011, he was awarded Ryerson Faculty SRC Awards for excellence in research. In 2012, he was awarded an inaugural YSGS Outstanding Contribution to Graduate Education Award. In 2015, Bonato was appointed to the NSERC Discovery Mathematics and Statistics Evaluation Group for a 3-year term.

Lynn Batten is the Director of Information Security Group at Deakin University, Australia. Her current research interests include malicious software, information security and reliability, cryptology, computer forensics, coding theory and optimization techniques. She is a member of the editorial board of the Journal of Combinatorial Mathematics and Combinatorial Computing and of the Journal of the Australian Mathematical Society. She is on the Board of Directors of SECIA. She has published several books and research articles in refereed reputed journals.

Part I
Keynotes

The Game of Wall Cops and Robbers

Anthony Bonato and Fionn Mc Inerney

Abstract Wall Cops and Robbers is a new vertex pursuit game played on graphs, inspired by both the games of Cops and Robbers and Conway's Angel Problem. In the game, the cops are free to move to any vertex and build a wall; once a vertex contains a wall, the robber may not move there. Otherwise, the robber moves from vertex-to-vertex along edges. The cops capture the robber if the robber is surrounded by walls. The *wall capture time* of a graph G, written $W_{c_t}(G)$ is the least number of moves it takes for one cop to capture the robber in G. In the present note, we focus on the wall capture time of certain infinite grids. We give upper bounds on the wall capture time for Cartesian, strong, and triangular grids, while giving the exact value for hexagonal grids. We conclude with open problems.

1 Introduction

Wall Cops and Robbers is a new vertex pursuit game played on graphs, inspired by the games of Cops and Robbers and the Angel Problem. In Wall Cops and Robbers, there are two players: a cop and a robber. The game starts with the cop building a "wall" on a vertex which blocks off that vertex so that the robber cannot occupy it. After that, the robber selects a vertex. The cop can build a wall on any vertex on his turn except for the vertex that the robber currently occupies. Hence, we may think of the cop as playing off the graph. The robber, however, can only move along an edge to an adjacent vertex each turn. The robber is also allowed to skip his turn, but to avoid complexities, he may not move back to the vertex he occupied on his previous turn (that is, the robber cannot backtrack).

Supported by grants from NSERC and Ryerson University.

A. Bonato (✉) · F.M. Inerney
Ryerson University, Toronto, Canada
e-mail: abonato@ryerson.ca

M. Senthilkumar et al. (eds.), *Computational Intelligence, Cyber Security and Computational Models*, Advances in Intelligent Systems and Computing 412, DOI 10.1007/978-981-10-0251-9_1

The game of Cops and Robbers was first introduced by Quilliot [1] and independently by Nowakowski and Winkler [2]. For additional background on Cops and Robbers and its variants, see the book [3] and the surveys [4–6]. The Angel Problem was first introduced by Conway [7]. The Angel Problem is a turn-based game played on either an infinite chessboard or an infinite 3-dimensional chessboard. The game is played by an Angel and the Devil. The Angel has power k, where k is a positive integer, which allows the Angel to move k spaces in any direction. We can think of this as the Angel having k moves on his turn where he can only move to an adjacent space on each move. The Devil has the same power as the cop in Wall Cops and Robbers, except that he eats squares in the Angel Problem while the cop builds walls on vertices. The objective of the game is the Angel trying to elude capture by the Devil on an infinite chessboard, which corresponds to a Cartesian product of infinite two-way paths. For more information on the Angel Problem, see [8].

Firefighting on graphs also has some similar aspects to Wall Cops and Robbers. Firefighter was introduced by Hartnell [9], and it is graph process where firefighters try to contain a spreading fire. The fire starts at some vertex, and then the firefighters place themselves on vertices making them *protected*. Then on the next turn the fire spreads to all adjacent vertices that are not protected, and the firefighters protect another set of vertices. The game continues in this manner, and one of the goals for the firefighters save as many vertices from the fire as possible. See the survey [10] for various desired outcomes of Firefighter.

The objective of Wall Cops and Robbers is for the cop is to build walls to block off all adjacent vertices to the robber so that the robber can no longer move on his next turn. The objective of the game for the robber is to evade capture by the cop for as long as possible. The *wall capture time*, written W_{c_t}, of a graph is the least number of moves it takes for the cop to capture the robber on the given graph given that the cop and robber have both played their best strategies. Note that Wall Cops and Robbers is equivalent to the Angel problem, where the Angel has power $k = 1$ (although Angels can backtrack).

For an elementary example, consider the game played on the graph G in Fig. 1. We label the vertices 1, 2, 3, 4, 5, and 6 as in Fig. 1 and the cop builds a wall on vertex 3. If the robber chooses 1 or 2, then he will be stuck in the left triangle and will lose on the next turn. Any of the vertices in the right triangle would be a good choice for the robber, so he chooses 5. The cop builds a wall on 4. The robber can either move to 6 or skip his turn and remain at 5; since both give the same result of

Fig. 1 A labelled graph G

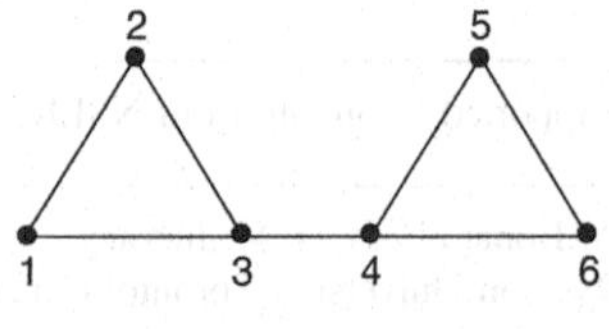

him losing next turn, let us say he moves to 6. The cop builds a wall on 5 and captures the robber as he can no longer move. Since it takes exactly three moves for the cop to capture the robber with both sides playing at their best, we have that $W_{c_t}(G) = 3$.

2 Grids

Given our limited space, we confine our discussion of the wall capture time to infinite grids and tilings of the plane in two dimensions. We study the wall capture time of infinite hexagonal grids, infinite Cartesian grids, infinite triangular grids, and infinite strong grids. Lastly, we study the wall capture time of n-layered infinite Cartesian grids which are certain subgraphs of three dimensional grids.

The infinite hexagonal grid, written H_∞, is a tiling of the plane by hexagons with vertices represented as vertices of the hexagons. We have the following result, whose proof is omitted for space considerations.

Theorem 1 $W_{c_t}(H_\infty) = 8$.

An infinite Cartesian grid, written $P_\infty \Box P_\infty$, is the Cartesian product of two infinite, two-way paths. A *trap* is two walls made by cops on a Cartesian grid such that they share the same x or y coordinate but not both, and are distance two apart. The vertex in between these two walls will be called the *middle vertex*. It is called a trap since if the robber moves onto the middle vertex, then the cop will close the trap by moving to the open vertex that is adjacent to the middle vertex that the robber did not just come from in the last move. The robber cannot move back to his previous vertex by the rules of the game, and then the cop will capture him by playing adjacent to him. Thus, moving into a trap guarantees that the robber will be captured in exactly two turns.

The proof of the following theorem, while elementary, involves the careful analysis of cases using traps and so is omitted here.

Theorem 2 $W_{c_t}(P_\infty \Box P_\infty) \leq 14$.

We next turn to the infinite triangular grid, written Δ_∞, which is a tiling of the plane by triangles with vertices represented as vertices of the triangles. We sketch the proof of this result below.

Theorem 3 $W_{c_t}(\Delta_\infty) \leq 138$.

Proof The cop will first trap the robber in a hexagon with lengths of sides 3, 21, 21, 21, 21, and 3. The cop will build the hexagon, taking the robber's vertex as the centre of this hexagon so that he is distance 11 from each of its walls. The robber will not move backwards or skip his turn while the hexagon is being built as this will allow the cop to build a smaller hexagon and thus, use fewer moves to capture the robber. We will describe the corners of the hexagon as follows: TLC, TRC, LC, RC, BLC, BRC with T standing for top, B standing for bottom, R standing for right,

L standing for left, and C standing for corner. The first 9 moves for the cop are as follows:

1. The first move of the cop is wasted as we are playing on an infinite grid, so the robber will just play so far away from the first cop that the wall he builds will be useless.
2. The cop plays one up and right of the LC.
3. The cop plays one down and right of the LC.
4. The cop plays one up and left of the RC.
5. The cop plays one down and left of the RC.
6. The cop plays on the TLC.
7. The cop plays on the TRC.
8. The cop plays on the BLC.
9. The cop plays on the BRC.

The robber must have moved towards one of the sides of the hexagon in these first 9 moves. If he moved up, then the cop's 10th move is to play on the vertex two down and left of the TRC. If he moved down, then the cop's 10th move is to play on the vertex two up and right of the BLC. If the robber went straight left or right without any diagonal movements then the cop would not have had to play there so the robber would not move like that. Otherwise, the cop could build a smaller hexagon. Then we know that the robber will move either up or down towards the TLC or TRC or the BLC or BRC. Now the robber is distance two away from a side (only one) and it is the cop's turn and thus, the cop can stop him getting on the sides of the hexagon.

As the robber runs along the sides of the hexagon, the cop will gain a move on the robber at the LC and RC due to the walls that were built at the start. The cop will use these two extra moves to build a wall on the vertex in between the TLC and TRC and a wall on the vertex in between the BLC and BRC in the order that the robber will approach these vertices.

The robber may be able to move toward the centre of the hexagon after most of the sides have been built by the cop in order to gain some extra moves. It is difficult to know exactly when this may happen and in which exact direction the robber would move at the start of the game and at this point in the game. Therefore, we will assume that the cop will build the entire hexagon even if the robber moves toward the centre of the hexagon early. Thus, the robber will force the cop to build a wall inside the hexagon on his 10th move as described above. We will assume also that the robber can be anywhere inside this hexagon after it is built. The robber will be in the centre of the hexagon as if he is near any of the sides, it will allow the cop to use those sides to trap him in a smaller subgraph in this next phase (Figs. 2 and 3).

The cop will then confine the robber to a parallelogram with diagonals of length 23 and the other two sides of length five. We will assume, without loss of generality, that the diagonals go from the top left side of the hexagon to the bottom right side of the hexagon. Since both the top left side and bottom right side of the hexagon are distance 11 from the centre, the diagonal between them is of length 23. The two diagonals of length 23 will be built distance five apart. One will be built

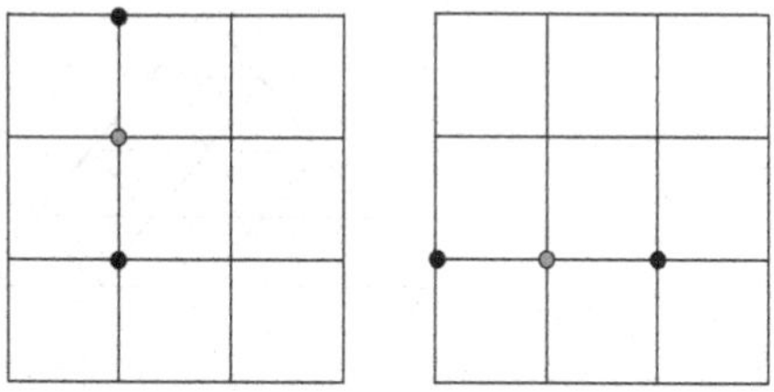

Fig. 2 The two possible traps the cop can build. The two walls of the trap are in *black* and the middle vertex is in *grey*

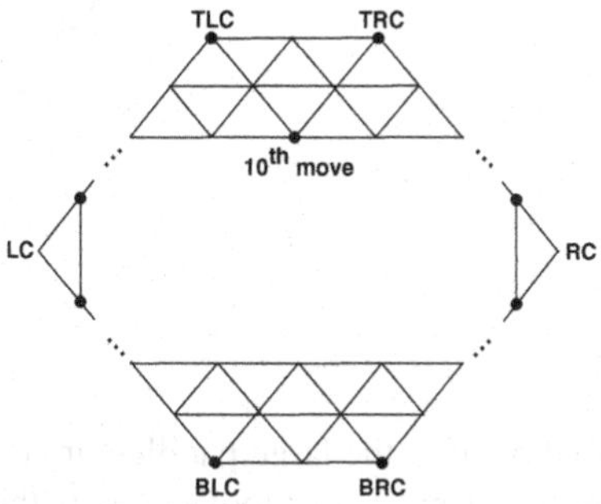

Fig. 3 The first 10 moves of the cop shown as *black vertices*, with the first move not shown as it is far away from the hexagon

distance two from the robber on his left and the other will be built distance three from the robber on his right. This guarantees that the robber is always at least distance two away from either diagonal with it being the cop's turn. The first wall built on these diagonals will be built on the left one since it is only distance two away from the robber. Thus, the robber will never be able to occupy a vertex on either diagonal or leave the parallelogram. The two sides of length five are already built as they are part of the sides of the hexagon. Once this parallelogram is completely built, we will assume that the robber is in the best position possible. Therefore, he will be in the centre of this parallelogram.

The cop will then be building two sides parallel to the sides of the hexagon of length three to trap the robber in a 5×4 parallelogram. The first of these two sides will be built distance one away from the robber, down and to his right, since it is possible to stop him bypassing the side as there are only three open vertices to cover. Thus, the cop will play adjacent to the robber on this side he will be building as seen in Fig. 4. The robber can move so that the side must actually be built as parallel to the sides of the parallelogram of length five. This does not matter as in either case the cop will trap the robber in a 5×4 parallelogram. From here, the other side will be built distance two away from the robber's starting position in the parallelogram on the opposite side of the first side built and parallel to the first side built. Now the robber is trapped in a 5×4 parallelogram. The robber will move

Fig. 4 The parallelogram

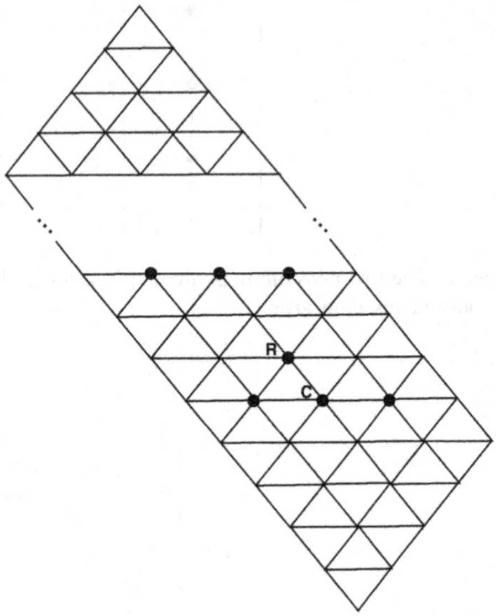

back to the vertex he started on after the large parallelogram's sides were built. The cop will capture the robber in at most four moves inside this small parallelogram, no matter where the robber moves inside it as there are only six open vertices and the cop can cut out one of two vertices that are distance three apart.

Now we will sum up all the moves of the cop. The cop took one wasted move at the start. He then takes $21 \times 4 + 3 + 3 + 1 - 6 = 85$ moves (subtract six for counting six vertices twice) to build the sides of the hexagon and a wall on the vertex on his 10th move. Followed by 42 moves to build the two diagonals of length 23 of the large parallelogram. Six moves to complete the sides of the 5×4 parallelogram. Then, four moves to capture the robber inside the 5×4 parallelogram. Thus, it took the cop $1 + 85 + 42 + 6 + 4 = 138$ moves to capture the robber. □

The infinite strong grid, written $P_\infty \boxtimes P_\infty$, is the strong product of two infinite, two-way paths. We will be labelling the vertices using Cartesian coordinates. We rely on the following fact from [7]. Note that in the Angel Problem, the Angel (that is robber in our case), unlike in Wall Cops and Robbers, is allowed to move to a vertex he occupied in the previous turn. However, such moves will only lengthen the play of the game. Hence, bounds on the length of the Angel Problem will also be bounds on the wall capture time played on the infinite strong grid.

Lemma 1 ([7]) *On the infinite strong grid, the cop can confine the robber to a* 35×35 *box by building three walls in each corner on his first 12 moves. The cop can confine the robber from there as he can always stop the robber reaching a side of this box if the robber is distance five from a side and it is the cop's turn.*

We now turn to the main result of this section.

Theorem 4 $W_{c_t}(P_\infty \boxtimes P_\infty) \leq 246$.

Proof Without loss of generality, we will assume that after the cop plays his first turn, that the robber will place himself sufficiently far away that this first wall will play no part in capturing the robber. Thus, we will count this move only at the end in the total number of moves but will not count it in our ordered turns for the cop below. The cop's plan is to capture the robber in a 35×32 box first and then capture the robber inside that box. The robber's best strategy has to be to move in one direction such as up or right, and or diagonally up and right so as to make it difficult for the cop to trap him. Otherwise, as moving back in the opposite direction while this box is being set up will just waste the robber's turn and possibly allow the cop to build a smaller box to contain the robber. If the robber skips his turn several times, then it will allow the cop to either encroach the sides of the box or if they are already built, then it will allow the cop to encroach the vertical or horizontal walls that enclose the robber after that. Therefore, it is not a good idea for the robber to skip his turn until he knows he will be captured and there are few turns remaining because even doing it once will allow the cop to be one turn ahead of the robber which may also result in fewer moves needed.

The cop will start by building the corners of a 35×35 box as in Lemma 1 and once a few moves have been made it will be clear what direction, if any, the robber has chosen to move in which will allow the cop to encroach one of the sides in three steps making it a 35×32 box (or encroach two of the sides in for a total of three steps, also making it a 35×32 box). The box must start as a 35×35 box since the cop requires 12 moves before the robber can be at a distance of five moves from any side, thus, making it necessary that the robber start at a distance of 17 moves from any side which a 35×35 box ensures.

We will assume that the robber is moving diagonally opposite of the corner in which the cop first plays as this makes it most difficult for the cop since he will not be able to encroach the sides of the box further in than just three steps. If the robber moves in any other fashion at the beginning, then the same strategy can be used by the cop. It will be shown that the robber cannot escape the box if he is at a distance of five moves from any side of the box, there have been three walls built in each corner of the box, and it is the cop's turn as in Lemma 1.

Let T, B, L, R, C represent top, bottom, left, right, and corner, respectively, of the 35×35 box. The way the moves of the cop will be described should be interpreted as follows: two up from the BLC means that the bottom left corner is (0, 0) and two up from that would be (0, 2). We will play the first cop in the BLC and thus, we assume the robber is always moving diagonally toward the TRC but of course the strategy of the cop can be easily modified for the robber moving diagonally toward any corner. The robber by moving in the opposite direction of the second cop stops the cop from encroaching one or two of the sides by more than a total of three steps.

1. The cop plays four up from the BLC.
2. The cop plays two down from the TRC.

3. The cop plays one right of the TLC.
4. The cop plays two right of the TLC.
5. The cop plays four right of the TLC.
6. The cop plays one down from the TRC.
7. The cop plays four down from the TRC.
8. It is now clear that the robber is moving to the TRC and thus, the cop can encroach the bottom side of the box in three steps to make it a 35 × 32 box by playing five up from the BLC.
9. The cop plays seven up from the BLC.
10. The cop plays three up and one left of the BRC.
11. The cop plays three up and two left of the BRC.
12. The cop plays three up and four left of the BRC.

Now the next move the robber makes will bring him to within five moves of one or two sides. In the present case, the robber has moved to the TRC. If the robber moves up in any direction, then the cop's next move is to play on the side of the box directly above the robber. If the robber moves diagonally down and to the right or directly to the right then the cop's next move is to play on the side of the box directly to the right of the robber. From there in either case, we have the scenario where it is the robber's turn and he is distance five away from a side with a wall on the side directly in his path if he were to go straight at the side. We know that the cop can block the robber along this side now no matter which move he makes by Lemma 1 and Fig. 5. With three walls built in each of the corners, the robber cannot run forever and must turn at the corners. We will assume the cop builds the entire box and that the robber can be anywhere inside that box afterwards. Therefore, it takes $35+35+30+30 = 130$ moves to build the walls of the box.

We assume the robber is in the centre of the 35 × 32 box. The cop will now confine the robber to a 32 × 12 box. The two sides of length 12 have already been built as they are part of the sides of the 35 × 32 box. The cop will build the two sides of length 32 that are distance 11 from each other. The robber can only ever be

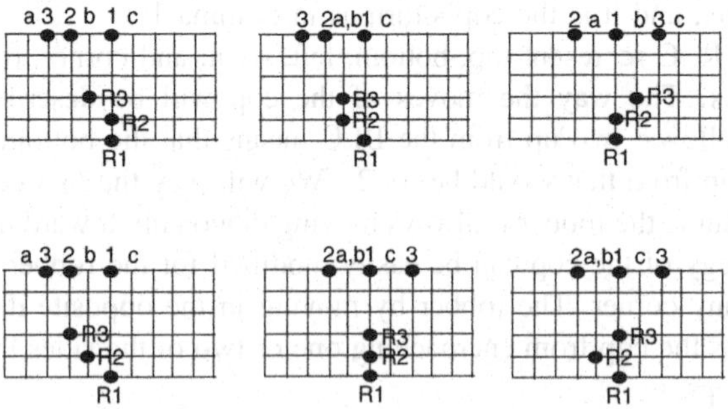

Fig. 5 The cop's strategy to stop robber reaching a side from a distance of five moves. In each of the figures, *a*, *b*, and *c* represent the fourth wall the cop would build if the robber moved diagonally up and to the *left* or straight up or diagonally up and to the *right*, respectively

distance five from either of these two sides but not both. This ensures the cop can stop the robber reaching either of these sides. It is difficult to know exactly what the best position the robber can attain while still making the cop complete the 32×12 box. Thus, we will say that once that box is completely built, we will let the robber be in the middle of this smaller box as that is best for him with the knowledge of the cop's strategy. Therefore, up to this point including the first move for the cop, there have been 131 moves for building the box and the first move and 60 moves for the two sides of length 32 in the smaller box for a total of 191 moves.

The robber is in the middle of the 32×12 box with the sides of this area completely covered with walls. The cop will now confine the robber to a 12×12 box. Two of the sides of length 12 have already been built as they are part of the sides of the 32×12 box. The cop will then build the other two sides distance 11 from each other with the robber directly in the middle so that he cannot be distance five from both of them at the same time. These two sides take 20 moves in total to build. Thus, the number of moves is now at 211.

We will again assume the robber to be anywhere in this new box which is 12×12 with the sides of this area completely covered with walls. The cop's strategy is to build a 6×6 box directly in the middle of the 12×12 box. This 6×6 box takes 20 moves to build. Now the robber can be inside the 6×6 box or outside it or he could have been on one of the vertices of the sides of the 6×6 box. Let these be cases A through C, respectively.

Case A: The robber is inside the 6×6 box.

The cop builds a 4×4 cross such that the intersection of the cross occurs in one of the middle vertices. The cop will build his first wall on the intersection vertex and it is guaranteed to be open as there are four possible intersection vertices in the 6×6 box. If the robber allows the cross to be built which means not being on one of those vertices while the cop needed to build a wall there, then he will be captured in at most three moves after that as a 2×2 area is the largest compartment left. This would total 10 moves for capturing him in the 6×6 box. If the robber occupies one of the vertices of the cross the cop needs to build a wall on, then the robber would occupy one of the vertices that is adjacent to the 2×2 compartment that will be left open. But then the cop just builds a wall adjacent to the robber in this 2×2 compartment. No matter what the robber does now, he will be captured in at most 10 moves. Therefore, Case A results in 10 moves.

Case B: The robber is outside the 6×6 box.

Since all the sides are symmetric we will consider the robber to be in the bottom 12×3 part of the 12×12 box. Then the cop can build two vertical walls or one horizontal and one vertical wall distance two and three away from the robber to confine him to a 7×4 box. We can say that it will take a maximum of 8 moves for the cop to capture the robber in that 7×4 area since there are only 10 open vertices in it as the cop can choose the two open vertices after the 8 moves such that they will not be adjacent. Therefore, Case B results in at most 12 moves.

Case C: The robber is on one of the vertices of one of the sides of the 6×6 box after all the other vertices of the sides of this box have a wall built on them.

If the robber is not on one of the corners of this 6×6 box, then the cop will build a wall on the outside of the 6×6 box on the adjacent vertex that is closest to a corner. The cop will keep playing adjacent to the robber on the outside of the 6×6 box until he cannot anymore or the robber moves. If the robber moves inside the 6×6 box then he will be captured in 10 moves by Case A. If he moves outside of the 6×6 box then he will be captured in 12 moves by Case B but the walls being built while he is on one of the sides of the box count towards these 12 moves. Therefore, if the robber does not occupy a corner vertex of the 6×6 box, then he will be captured in at most 13 moves by skipping his turn until he cannot move outside of the 6×6 box at which point he will move into the 6×6 box. This is due to there being exactly three adjacent vertices outside of the 6×6 box. Note that the move where the cop builds the wall to complete the 6×6 box was already counted before.

If the robber is on one of the corners of the 6×6 box then there are five adjacent vertices outside of the 6×6 box. The cop will build a wall on each of these vertices and once he has done so the robber will move into the 6×6 box for the same reasons as above. This results in 15 moves for the cop. Therefore, Case C results in at most 15 moves.

Since Case C results in the most moves for the cop, then that would be the robber's strategy. Hence, we have proven the desired result that the total moves to capture the robber is $211 + 20 + 15 = 246$. □

We also can say something about the wall capture time of layered Cartesian grids.

Theorem 5 *For n a positive integer, we have that*

$$W_{c_t}(P_\infty \square P_\infty \square P_n) \leq 44n^2 + 6n + 15.$$

3 Conclusion and Open Problems

The table below gives a summary of the bounds on the wall capture times for the various grids we considered.

Graph	W_{c_t}
H_∞	8
$P_\infty \square P_\infty$	≤ 14
$P_\infty \square P_\infty \square P_n$	$\leq 44n^2 + 6n + 15$
$P_\infty \square P_\infty \square P_\infty$	Open
Δ_∞	≤ 138
$P_\infty \boxtimes P_\infty$	≤ 246

We now present some open problems. For most of the graph classes we have studied, we only have an upper bound for the wall capture time. To improve on our results, one could find tight lower bounds.

For the three-dimensional infinite Cartesian grid, it is not known whether one cop can capture the robber on this graph. The same question is open for the infinite three-dimensional strong grid.

We have studied the Cartesian and strong graph products. There are 256 possible products, and some of the notable ones that we have not studied are the lexicographic, disjunction, and symmetric difference products; see [3].

References

1. Quilliot, A.: Jeux et pointes fixes sur les graphes. Thèse de 3ème cycle, pp. 131–145. Université de Paris VI (1978)
2. Nowakowski, R.J., Winkler, P.: Vertex-to-vertex pursuit in a graph. Discrete Math. **43**, 235–239 (1983)
3. Bonato, A., Nowakowski, R.J.: The game of Cops and Robbers on graphs. American Mathematical Society, Providence, Rhode Island (2011)
4. Baird, W., Bonato, A.: Meyniel's conjecture on the cop number: a survey. J. Comb. **3**, 225–238 (2012)
5. Bonato, A.: WHAT IS … Cop Number? Notices American Math. Soc. **59**, 1100–1101 (2012)
6. Bonato, A.: Catch me if you can: Cops and Robbers on graphs. In: Proceedings of the 6th International Conference on Mathematical and Computational Models (ICMCM'11) (2011)
7. Berlekamp, E.R., Conway, J.H., Guy, R.K.: Winning ways for your mathematical plays, vol. 2. Academic Press, New York (1982)
8. Conway, J.H.: The angel problem. In: Nowakowski R. (ed.) Games of No Chance, vol. 29, pp. 3–12. MSRI Publications (1996)
9. Hartnell, B.: Firefighter! An application of domination. Presentation at the 25th Manitoba Conference on Combinatorial Mathematics and Computing. University of Manitoba, Winnipeg, Canada (1995)
10. Finbow, S., MacGillivray, G.: The firefighter problem: a survey of results, directions and questions. Australasian J. Comb. **43**, 57–77 (2009)

Smartphone Applications, Malware and Data Theft

Lynn M. Batten, Veelasha Moonsamy and Moutaz Alazab

Abstract The growing popularity of smartphone devices has led to development of increasing numbers of applications which have subsequently become targets for malicious authors. Analysing applications in order to identify malicious ones is a current major concern in information security; an additional problem connected with smart-phone applications is that their many advertising libraries can lead to loss of personal information. In this paper, we relate the current methods of detecting malware on smartphone devices and discuss the problems caused by malware as well as advertising.

Keywords Android · Application · Malware · Advertising

1 Introduction

Although many operating systems offer applications (APPs), most are not open source. In this paper, we focus on Android as it is an open source system and therefore easier to test than a proprietary system. None-the-less, much of the discussion here applies to other operating systems too, including iOS, Blackberry and

L.M. Batten (✉)
School of IT, Deakin University, Melbourne, Australia
e-mail: lmbatten@deakin.edu.au

V. Moonsamy
Radboud University Nijmegen, Nijmegen, The Netherlands
e-mail: veelasha@cs.ru.nl

M. Alazab
Faculty of Information Technology, Isra University, Amman, Jordan
e-mail: moutaz.alazab@iu.edu.jo

M. Senthilkumar et al. (eds.), *Computational Intelligence, Cyber Security and Computational Models*, Advances in Intelligent Systems and Computing 412, DOI 10.1007/978-981-10-0251-9_2

Windows. APPs can be a great source of convenience, but they also bring problems connected with malicious code as well as data theft. We shall consider both of these aspects in this paper.

Android applications are available to users as a zipped file, identified as the Android application package, or APK file, which contains classes.dex, assets, library and resource files along with the AndroidManifest.xml which is a configuration file containing the list of permissions and the application's components including activities, services, intent receivers, and content provider, layout data and the application resources. The executable code for the APP resides in classes.dex and in the library file. In particular, the classes.dex file contains all the Java classes compiled to Dalvik Byte code. (For more details see for example http://developer.android.com/sdk/installing/studio-build.html).

The Android permission system is used to protect a smartphone's resources (see http://developer.android.com/guide/topics/security/permissions.html). Each application declares a list of permissions needed to protect access to resources (for example 'permission to access the Internet' is a common one). While the permission system helps prevent the intrusion of malware, it is not designed to detect malicious applications. In fact, several research papers have shown that some applications with no permissions at all can still access the operating system [1]. Brodeur [2] developed a no-permission application that gathers user information and forwards it to a pre-selected server, while the authors of [3] demonstrate a no-permission application that can reboot an Android device. Moreover, several papers (for example [4] and its references) consider the risk of over-privileged applications which request permissions that are not required for the application to execute. Hence, analysing such applications based on only the requested permissions can bias the analysis results.

Malware detection is an emerging topic in the study of the Android platform which relies heavily on its permission system to control access to restricted system resources and private information stored on the smartphone. However, there is no evidence providing a clear understanding of the key differences for permissions between clean and malicious applications.

Mobile phones provide tracking services for several reasons, both for convenience in assisting a person to find a location and in order to offer services such as information about a favourite restaurant which is nearby: so location identification can offer assistance with directions to a target destination and also in indicating facilities along the way. However, several publications (e.g. [5]) have been able to show that sensitive information, such as device ID and user location, is often leaked via advertising libraries.

In this paper, we present some of the recent work on malware detection on Android APPs and also on 'data leaks' from APPs; this latter refers to information taken from the smartphone without the knowledge of the user. Section 2 looks at malware while Sect. 3 considers leaky APPs. In Sect. 4, we describe what is generally required in setting up experiments and tests on APPs. This is followed by a brief summary and then references.

2 Malware

In addition to permissions, a second security mechanism for smartphones is traditional anti-malware analysis, which uses a pattern matching technique to identify malicious applications based on a byte-string that is unique to an application. Recent research work, for example [6], points out that such a signature is comprised of a package or file name that can be matched against an anti-malware database. Nevertheless, work by [7] and others, confirms that malicious applications can evade the current protection mechanisms by using one or more of the following techniques: null operation (NOP) insertion, arithmetic and branch insertion, Java reflection, Byte code encryption, Junk Code Insertion, payload encryption, native exploits or changing the package name. While the pattern matching technique is popular with anti-malware companies due to the high accuracy with which it identifies malicious applications in real time and with low run time, it has been shown to be ineffective in detecting sophisticated malware. According to a study conducted on ten anti-malware applications for Android [6], none of the evaluated applications is resistant against malware transformation techniques including poly-morphism, obfuscation and anti-reversing attacks.

A different approach to identifying malicious applications has been to leverage information from system calls dynamically. However, the authors of [8] argue that monitoring and intercepting system calls is inefficient in Android because system calls are basic interfaces provided by an operating system, and they are the only entrance to kernel mode from user mode; nevertheless, some malware do not necessarily make use of the system call interface and so can evade this analysis method [9]. A second problem is that it is difficult to identify behaviour with system calls. Thirdly, monitoring and intercepting system calls in real Android devices is not possible, as the kernel of such a device cannot use loadable kernel modules. In the literature, techniques used to identify malicious applications include permission-based detection, signature-based detection and system calls-based detection. There have even been attempts to apply machine learning algorithms on the smartphone devices. Some of these algorithms are mentioned below and more information about them can be found in [10].

Google deploys an automated system, known as "*Bouncer*" to test uploaded applications for malicious code. Once a developer has uploaded an application to the Google market, Bouncer compares uploaded applications with known malware and then runs the uploaded application in a virtual environment in order to identify any potential malicious behaviour. Nevertheless, some malicious applications targeting the Android market can evade Bouncer according to a report published by TrendMicro in 2012 [11].

A study by Amamra et al. [12] investigated the effectiveness of machine learning classifiers in detecting malware. The authors collected a dataset of 100 free applications from the Android market and 90 malicious applications from the Contagio mobile dump. They focused on leveraging information from system calls to identify malicious applications, and initially evaluated their framework using the following algorithms: Logistic Regression, support vector machine (SVM), artificial Neural Network and Naive Bayes. They achieved 92.5 % detection accuracy with an 8.5 % false positive rate using the SVM classifier.

Zhao et al. [13] developed a tool they call RobotDroid to detect smartphone malware during runtime. RobotDroid logs the intent issued and system resources accessed by applications; it then categorizes the logs and sorts them based on the timestamp. Their tool employs the SVM classifier to find those support vectors which are best able to identify malicious applications. The tool was tested using three malware families, lGeinimi, DroidDream and Plankton, and obtained 93.3, 90 and 90 % detection accuracy respectively.

Sahs and Khan in [14] collected a dataset of 2081 benign applications and 91 malicious applications and extracted permissions and Control Flow Graphs using Androguard to train a classification SVM. The authors mention (Section VII) that their system is limited to just permission and control flow graphs, and that other information-rich features can be extracted from the code itself, including constant declarations and method names.

Aung and Zawi [15] applied machine learning algorithms to the information retrieved from permissions and events. The authors applied information gain [16] on the given features in order to improve the detection accuracy and efficiency. In order to test their methodology, they included 3 machine learning algorithms: J48, Random Forest and CART, and obtained 89.36, 91.67 and 87.88 % accuracy respectively.

Amos et al., the authors of [17], developed a framework called STREAM to enable large-scale validation of mobile malware machine learning classifiers. They extract information using dynamic analysis about the battery, binder, memory, network, and permissions from a dataset of 408 benign applications and 1330 malicious applications. For the purpose of testing their framework, they include 6 machine learning algorithms: Random Forest, Naive Bayes, Multilayer Perceptron, Bayes net, Logistic and J48, and obtained 70.31, 78.91, 70.31, 81.25, 68.75 and 73.4 % detection accuracy respectively.

In [18], the authors are the first to examine behaviour in malicious applications using DroidBox. Using a dataset comprising samples that were collected from publicly available sources, each malicious application is executed for 60 s in a sandboxed environment and the log files generated are collected at the end of execution. Droidbox also generated two types of graphs (behaviour graphs and treemap graphs) for each sample. Both graphs helped the authors analyze the activities performed during run-time and also assisted in establishing patterns between variants from the same malware family. These graphs illustrate how some benign applications might leak data connected to short message service (SMS) texting and other features of the applications. The authors note that, while one would expect to see encrypted code in malware, not a single malicious application in their (small) sample set invoked a cryptographic activity, while several of the clean APPs did so.

Finally, it is worth noting that according to the authors of [19, 20], the detection accuracy and efficiency of any machine learning system are influenced by three main factors: the features used to represent the instances; the algorithm used to generate the classifier; and the parameter values of a classifier.

3 Applications Which Leak Data

3.1 Malware and APPs

Our 2012 paper [18] explains that the Android applications market has been infected by numerous malicious applications and that rogue developers are injecting malware into legitimate market applications which are then installed on open source sites accessible to consumers. We thus consider the situation of malware à propos APPs in this sub-section.

In [18], we demonstrated that Droidbox [21] can be a useful tool both in classifying malicious Android applications and in determining weaknesses in benign Android applications; weaknesses include the leaking of private data caused by such functionalities as location services and advertising and we consider these in the next sub-section. DroidBox can track sensitive data originating from the phone's database and add and modify output channels to detect leaks via outgoing SMS and to disclose full details of the network communication. Android applications can perform phone calls or send SMS to premium rate numbers that are declared by the attacker. DroidBox can disclose these operations, and is able to track sensitive data originating from the phone's database and detect leaks via outgoing SMS as well as disclose full details of the network communication.

Some malicious Android applications can evade anti-virus software by performing obfuscation and changing themselves during run-time [15]. DroidBox is designed to detect applications which attempt obfuscation by using cryptographic keys to encrypt or decrypt data; however, as noted at the end of the previous section, the use of cryptography is not necessarily an indication that an APP is malicious.

3.2 Tracking Services

We turn to a discussion of two types of tracking features for smartphones; these are: (i) Location Services and (ii) Advertising. For (i), smartphone owners can either turn on or turn off location tracking to prevent installed applications from discovering their physical locations. As for advertising, users are allowed to either turn on or 'limit' tracking by advertising libraries embedded in applications.

The Android permission-based model is used to restrict access to privileged system resources and to a user's private information. This is achieved by requiring the user to grant access to all permissions requested by the application in order for it to be successfully installed. Consequently, any advertising libraries embedded in an APP receive the same privileges as the APP that requested the permissions.

Pearce et al. [4] proposed a framework that can separate an advertising library from its main application. They introduced a new Application Programming Interface (API) as well as two additional permissions and applied a method known as privilege separation, which extracts the advertising component from the main functionality component of the application; this ensures that the advertising library does not inherit the same permissions assigned to its home APP. In [22], Shekhar et al. presented their method for separating applications and advertisements in the Android platform: a framework that can take as input an APP with embedded libraries and rewrite it so that the main functionality of the APP and the advertising libraries run as different processes. The authors also verified that, in the rewritten version of the APP, all the permissions requested by it were indeed required for the APP to function properly.

Stevens et al., in [23], performed a thorough analysis of third party advertising libraries to understand if they are unnecessarily accessing private information stored on users' smartphones. Additionally, the authors presented several vulnerabilities that attackers can exploit whilst being connected on the same network as the victim. Grace et al. [24] observed that some third-party advertising libraries employ unsafe mechanisms to retrieve and execute code from the Internet. Such behaviour renders a user's private information vulnerable to external attacks that can be carried out via the Internet.

The authors of [25] investigated tracking services on the Android and iOS smartphone platforms and described a simple and effective way to monitor traffic generated by tracking services to and from the smartphone and external servers. As part of the testing, they dynamically executed a set of Android and iOS applications, collected from their respective official markets. Their results indicate that even if the user disables or limits tracking services on the smartphone, applications can by-pass those settings and, consequently, leak private information to external parties. On the other hand, when testing the location 'on' setting, the authors notice that generally location is not tracked.

Two of the authors of [25] collaborated with additional researchers to investigate the same problem on other smartphone operating systems [26]. Using the experimental software platform Mallory, which was also used in [25], the authors investigated the 'tracking off' settings on the Blackberry 10 and Windows Phone 8 platforms in a manner similar to that used for Android and iOS and with similar results. The conclusion is that tracking settings on all four smartphone operating systems Android, iOS, Blackberry and Windows cannot be trusted to operate as proposed.

4 Experimental Work

In this section, we give a general explanation of how APP samples can be collected and tested for malware and for leaks. Any APP should first be tested to determine whether it is benign or malicious; this can be done using an existing free online service, as described in the next sub-section. The existence of this service also permits us to define malicious and clean APPs rigorously.

4.1 VirusTotal

VirusTotal (https://www.virustotal.com/), a subsidiary of Google, is a free online service that analyzes files and URLs enabling the identification of viruses, worms, trojans and other kinds of malicious content detected by antivirus engines and website scanners. It may also be used to detect false positives, that is, benign code detected as malicious by one or more scanners.

VirusTotal stores a list of identifiers (signatures) of known malware, which are contributed by many antivirus companies; this list is updated about every 15 min as signatures are being developed and distributed by antivirus companies. An APP can be uploaded onto their website where scanning is done via API queries to the approximately 50 different companies providing VirusTotal with information.

We define a *malicious APP* as one which is identified to be so by *at least one* of VirusTotal's antivirus products. We define a *clean* or *benign APP* to be one not identified as malicious by *any* of the VirusTotal antivirus products.

4.2 Setting up the Testing Environment

This section describes the experimental setup for most of the work done by the authors in the various publications [2, 7, 19–22, 26].

Smartphone malicious applications can be collected from several open source sites such as Contagion, Offensive and VXHeavens, while benign applications can be chosen from obvious sources such as: adobe_flash_player, official APP markets, antivirus, facebook, googlemaps, mobi and youtubedownloader. In all cases APPs should be checked by VirusTotal as described above.

For all of our testing, we used Windows supported by Linux and set up a virtual machine environment to separate the application from the network as mobile malware can be spread from mobile device to PC and vice versa.

In order to identify APPs and detect any changes during the testing, we used HashMyFiles installed in the Windows host to generate a unique identifying hash value. In doing dynamic testing (that is, executing the application to determine what it does), a decision about the running time has to be made. With a small set of samples, a tester may be able to allocate a longer time to each sample than available when using a large set. Whatever the time chosen, if the APP is designed to behave in a malicious way after the allocated run time has expired, this behaviour will not be detected in the testing.

4.3 Performance Evaluation

The standard measure of success in machine learning is the overall accuracy [27], which is defined as the percentage of all applications classified correctly. Research papers studying malware usually work with multi-classes in which each category has many more benign than malicious applications. In this (imbalanced) case, the accuracy measure may not be an adequate performance metric [28]. For example, if a classifier correctly identifies the entire dataset as benign, the classifier achieves high accuracy results while failing to detect the malicious applications.

In order to adequately reflect categorization performance, there are more accurate metrics that can be used in imbalanced cases, including 'recall' and 'precision', which can be combined into 'F-measure' [29, 30]. These metrics would normally be calculated on each class separately and then combined together to provide a weighted average.

5 Summary

Advertising has developed as the solution to the lack of a business model associated with the provision of free APPS for smartphones, as it allows application developers to offer free applications to the public while still earning revenue from in-application advertisements. Although location services are primarily used for purposes related to navigation, advertising companies tend to exploit this functionality in order to increase their revenue. Thus, advertising is unlikely to disappear from smartphones in the near future.

In this paper, we have explained how malware can be installed on APPs offered through the unofficial APP markets, and subsequently downloaded on to many smartphones with results such as theft of user identity and contacts, and fraudulent use of the smartphone for expensive calls. We have also explained how advertising can be used to capture and track user identity.

Based on these observations, the authors of [25] suggest the following (paraphrased) recommendations for three relevant parties:

1. *Novice Smartphone Users*. Download applications only from the official markets as they are less likely to be malicious than APPs from unofficial markets. While one cannot guarantee that all applications found on official markets are

clean, there is always a chance for any malicious applications to be deleted from the market when reported to the designated authorities.

2. *Device Manufacturers*. Smart-device manufacturers can provide users with pre-installed applications so that they have more control of their private information, instead of relying on the smartphone operating system.
3. *Academia/Industry*. Researchers from academia and industry within the field form an open-source research community to develop open-source applications that will help to compensate for the security vulnerabilities found in existing applications offered by the official application markets.

References

1. Moonsamy, V., Batten, L.M.: Zero permission android applications—attacks and defences. In: 3rd Applications and Technologies in Information Security, pp. 5–9. School of Information Systems, Deakin University Press, Australia (2012)
2. Brodeur, P.: Zero-permission android applications Part 2. Publication of Leviathan Security Group; accessed August 7, 2015, http://www.leviathansecurity.com/blog/zero-permission-android-applications-part-2/ (2012)
3. Lineberry, A., Richardson, D.L., Wyatt, T.: These aren't the permissions you're looking for, 2010. https://www.defcon.org/images/defcon-18/dc-18-presentations/Lineberry/DEFCON-18-Lineberry-Not-The-Permissions-You-Are-Looking-For.pdf
4. Pearce, P., Felt, A. P., Nunez, G., Wagner, D.: Android: privilege separation for applications and advertisers in android. In: 7th ACM Symposium on Information, Computer and Communications Security, pp. 71–72. ACM Digital Library, Arizona, USA (2012)
5. Moonsamy, V., Alazab, M., Batten, L.M.: Towards an understanding of the impact of advertising on data leak. Int. J. Secur. Networks, **7**(3), 181–193. Inderscience Publishers, London, England (2012)
6. Rastogi, V., Chen, Y., Jiang, X.: DroidChameleon: Evaluating android anti-malware against transformation attacks. In: 8th ACM Symposium on Information, Computer and Communications Security (ASIACCS 2013), pp. 329–334. ACM Digital Library, Arizona, USA (2013)
7. Park, Y., Lee, C. Lee, C., Lim, J., Han, S., Park, M. Cho, S.: RGBDroid: a novel response-based approach to android privilege escalation attacks. In: 5th USENIX conference on Large-Scale Exploits and Emergent Threats (LEET'12), 8 pp. Berkeley, California, USA (2012)
8. Peng, G., Shao, Y., Wang, T., Zhan, X., Zhang, H.: Research on android malware detection and interception based on behavior monitoring. Wuhan Univ. J. Nat. Sci. **17**, 421–427 (2012)
9. Egele, M., Scholte, T., Kirda E., Kruegel, C.: A survey on automated dynamic malware-analysis techniques and tools. ACM Comput. Surv. **44**(2), 6. Arizona, USA (2012)
10. Hall, M., Frank, E., Holmes, G., Pfahringer, B., Reutemann, P., Witten, I.H.: The WEKA data mining software: an update. SIGKDD Explorations, **11**(1). http://www.cs.waikato.ac.nz/ml/weka/ (2009)
11. TrendMicro, Repeating History. http://www.trendmicro.com/cloud-content/us/pdfs/security-intelligence/reports/rpt-repeating-history.pdf (2012)
12. Amamra, A., Talhi, C., Robert J.-M., Hamiche, M.: Enhancing Smartphone Malware Detection Performance by Applying Machine Learning Hybrid Classifiers. In: Kim, T.-H., Ramos, C., Kim, H.-K., Kiumi, A., Mohammed, S., Ślęzak, D. (eds.) Computer Applications

for Software Engineering, Disaster Recovery, and Business Continuity, CCIS, vol. 340, pp. 131–137. Springer, Heidelberg (2012)
13. Zhao, M., Zhang, T., Ge, F., Yuan, Z.: RobotDroid: A lightweight malware detection framework on smartphones. J. Networks **7**, 715–722 (2012)
14. Sahs, J., Khan, L.: A machine learning approach to android malware detection. In: Intelligence and Security Informatics Conference (EISIC), pp. 141–147. Odense, Denmark (2012)
15. Aung, Z., Zaw, W.: Permission-based android malware detection. Int. J. Sci. Technol. Res. **2**(3), 228–234 (2013)
16. Kent, J.T.: Information gain and a general measure of correlation. Biometrika **70**, 163–173 (1983)
17. Amos, B., Turner, H. White, J.: Applying machine learning classifiers to dynamic Android malware detection at scale. In: Wireless Communications and Mobile Computing, pp. 1666–1671 (2013)
18. Alazab, M., Moonsamy, V., Batten, L. M., Tian, R., Lantz, P.: Analysis of malicious and benign Android applications. In: 32nd International Conference on Distributed Computing Systems, pp. 608–616. IEEE, Los Alamitos, California, USA (2012)
19. Abawajy, J., Beliakov, G., Kelarev A., Yearwood, J.: Performance evaluation of multi-tier ensemble classifiers for phishing websites. In: 3rd Applications and Technologies in Information Security, pp. 11–16. School of Information Systems, Deakin University Press, Australia (2012)
20. Lu, Y., Din, S., Zheng, C., Gao, B.: Using multi-feature and classifier ensembles to improve malware detection. J. CCIT **39**, 57–72 (2010)
21. Lantz, P.: An android application sandbox for dynamic analysis. Master's thesis, Department of Electrical and Information Technology, Lund University, Lund, Sweden (2011)
22. Shekhar, S., Dietz, M., Wallach, D.: Adsplit: Separating smartphone advertising from applications. In: 20th USENIX Security Symposium, pp. 553–567. USENIX, Bellevue, USA (2012)
23. Stevens, R., Gibler, C., Crussell, J., Erickson, J., Chen. H.: Investigating user privacy in android adlibraries. In: IEEE Mobile Security Technologies (MoST 2012), 10 pp. California, USA (2012)
24. Grace, M.C., Zhou, W., Jiang, X., Sadeghi A.: Unsafe exposure analysis of mobile in-app advertisements. In: 5th ACM Conference on Security and Privacy in Wireless and Mobile Networks, pp. 101–112. ACM, Arizona, USA (2012)
25. Moonsamy, V., Batten, L.M., Shore, M.: Can smartphone users turn off tracking service settings?' In: MoMM, 9 pp. ACM Digital Library, Arizona, USA (2013)
26. Rahulamathavan, Y., Moonsamy, V., Batten, L.M., Shunliang, S., Rajarajan, M.: An analysis of tracking settings in Blackberry 10 and Windows Phone 8 Smartphones. In: ACISP 2014, LNCS vol. 8544, pp. 430–437. Springer, Heidelberg (2014)
27. Chong, I.-G., Jun, C.-H.: Performance of some variable selection methods when multicollinearity is present. Chemometr. Intell. Lab. Syst. **78**, 103–112 (2005)
28. Barber, B., Hamilton, H.: Parametric algorithms for mining share frequent itemsets. J. Intell. Inf. Syst. **16**(3), 277–293 (2001)
29. Christen, P.: Data matching: concepts and techniques for record linkage, entity resolution, and duplicate detection. Springer, Heidelberg (2012)
30. Mani, I. Zhang, I.: kNN approach to unbalanced data distributions: a case study involving information extraction. In: ICML'03 Workshop on Learning from Imbalanced Data Sets, 7 pp. Washington, DC, USA (2003)

Towards Evolutionary Multitasking: A New Paradigm in Evolutionary Computation

Yew-Soon Ong

Abstract The design of population-based search algorithms of evolutionary computation (EC) has traditionally been focused on efficiently solving a single optimization task at a time. It is only very recently that a new paradigm in EC, namely, multifactorial optimization (MFO), has been introduced to explore the potential of evolutionary multitasking (Gupta A et al., IEEE Trans Evol Comput [1]). The nomenclature signifies a multitasking search involving multiple optimization tasks at once, with each task contributing a unique factor influencing the evolution of a single population of individuals. MFO is found to leverage the scope for implicit genetic transfer offered by the population in a simple and elegant manner, thereby opening doors to a plethora of new research opportunities in EC, dealing, in particular, with the exploitation of underlying synergies between seemingly unrelated tasks. A strong practical motivation for the paradigm is derived from the rapidly expanding popularity of cloud computing (CC) services. It is noted that CC characteristically provides an environment in which multiple jobs can be received from multiple users at the same time. Thus, assuming each job to correspond to some kind of optimization task, as may be the case in a cloud-based on-demand optimization service, the CC environment is expected to lend itself nicely to the unique features of MFO. In this talk, the formalization of the concept of MFO is first introduced. A fitness landscape-based approach towards understanding what is truly meant by there being underlying synergies (or what we term as genetic complementarities) between optimization tasks is then discussed. Accordingly, a synergy metric capable of quantifying the complementarily, which shall later be shown to act as a "qualitative" predictor of the success of multitasking is also presented (Gupta A et al., A study of genetic complementarity in evolutionary multitasking [2]). With the above in mind, a novel evolutionary algorithm (EA) for MFO is proposed, one that is inspired by bio-cultural models of multi-factorial inheritance, so as to best harness the genetic complementarity between tasks. The salient feature of the algorithm is that it incorporates a unified

Y.-S. Ong (✉)
School of Computer Engineering, Nanyang Technological University, Singapore 639798, Singapore
e-mail: ASYSOng@ntu.edu.sg

M. Senthilkumar et al. (eds.), *Computational Intelligence, Cyber Security and Computational Models*, Advances in Intelligent Systems and Computing 412, DOI 10.1007/978-981-10-0251-9_3

solution representation scheme which, to a large extent, unites the fields of continuous and discrete optimization. The efficacy of the proposed algorithm and the concept of MFO in general, shall finally be substantiated via a variety of computation experiments in intra and inter-domain evolutionary multitasking.

Keywords Multi-factorial optimization · Evolutionary computation · Genetic complementarity

References

1. Gupta, A., Ong, Y.S., Feng, L.: Multifactorial evolution: towards evolutionary multitasking, Accepted IEEE Trans. Evol. Comput. doi:10.1109/TEVC.2015.2458037
2. Gupta, A., Ong, Y.S., Da, B., Feng, L., Handoko, D.: A study of genetic complementarity in evolutionary multitasking. To be released soon

Generating a Standardized Upper Ontology for Security of Information and Networks

Atilla Elçi

Abstract A usable functional interface between ontology and security integrating related information is needed for security engineering as well as creating secure systems. That in turn necessitates ontologizing security of information and networks to start with and then standardization. Having involved in the fields of semantic technology and information assurance, I have strived to facilitate establishing an interface between them and for this reason SIN Conference Series I created included all interest areas of semantics, metadata and ontology aspects. In a keynote talk and its proceedings paper in SIN 2014, I took up this subject and drove to the point that generic ontology for security of information and networks is timely, and it should better be standardized. In the present paper I investigate through examples where available to drive the point that the standard upper ontology for security may be developed through community sourcing and then standardized through competent agencies.

Keywords Information security · Semantic · Ontology · Secure ontology · Standards

1 Introduction

Realization of security-critical systems involves quality security engineering from analysis to design to development as well as implementation, testing, deployment and maintenance. Sphere of interest is quite large expanding to computers, network, internet and cloud, confidentiality, integrity, availability of data and applications, information flow, access control, privacy, trust, algorithms and protocols, cryptology and so on. The body of knowledge involved is beyond grasp of any individual,

A. Elçi (✉)
Department of Electrical and Electronics Engineering,
Aksaray University, Aksaray 68100, Turkey
e-mail: aelci@acm.org

M. Senthilkumar et al. (eds.), *Computational Intelligence, Cyber Security and Computational Models*, Advances in Intelligent Systems and Computing 412, DOI 10.1007/978-981-10-0251-9_4

extremely detailed and highly parametric, and certainly very complex to keep current in all aspects even for experts. It is thought that semantic technology with existing theory, standardized languages and associated practical tools together with properly configured domain ontologies can provide highly appreciable services in alleviating associated issues. In order to realize such synergy, a functional interface integrating related information between ontology and security is imperative which in turn necessitates ontologizing the domain of security of information and networks. Eventually such ontologies would be standardized.

Having worked in the fields of semantic technology and information assurance, I have strived to facilitate establishing an interface between security and ontology and for this reason SIN Conference Series [1] I created included all pertinent research interest areas of security, semantics, metadata and ontology aspects. In a keynote talk and in the associated paper in the proceedings of SIN 2014, I took up this subject [2]. The contents of that paper was such that the wide span of the interest area was highlighted in a long list of specialized research topics initially, then surveyed the few studies on ontologizing security related works, and standards on security and ontologies. It was shown that although numerous standards existed on either but there were none on the joint topic of information security ontology. It was highlighted that a generic ontology for security of information and networks is needed, earlier the better. Furthermore, it was concluded that a dire need existed for standardization of high-level information security ontology, perhaps an upper ontology for all others to link up to so that a forest of linked ontologies allowing all concerned to link their big data can evolve in time.

In this paper I inquire into how to realize an upper ontology for security domain and eventually standardize it. Next section introduces a vision to develop the security upper ontology. Standardization issue is taken up in section three and conclusions follow in section four.

2 Generating Security Ontology

As reviewed in [2], few studies exist in relation to ontologizing security and standardization. All, inclusive of those by this author, are minor and isolated initiatives studying certain aspects of ontology and security. At the World Wide Web Consortium (W3C), respected standardization body for Web matters, there has been little work with respect to information security ontology; and, hardly any results exist in producing standardized security ontology [3].

The Security Activity at W3C is organized through its subgroups variously named as Web Security Interest, Web Cryptography Working, Web Application Security, Privacy Interest, Technical Architecture Group (TAG), Web Payments Interest and XML Security Working Group. These groups provide a facilitating forum for discussions on how to improve standards, related implementations and extending existing standards in order to further Web security. The WebCrypto group has announced a draft Web Cryptography API [4] featuring message confidentiality and authentication services as building blocks for improved Web security. The Web Application Security Group "aims to enable secure mash-ups, address click-jacking, and to create a more robust Web security environment through light-weight policy expression that meshes with HTML5's built-in security policies." The XML Security Working Group, XMLSec for short [5], has already produced three W3C Recommendations on XML signature, encryption and signature properties. While TAG is "responsible for the security, sanity, and layering of the overall web platform", Web Security Interest, Privacy Interest and Web Payments Interest Groups appear as advisory.

It is to be concluded that the W3C Security Activity has so far been concerned with instigating secure versions of existing Web standards and applications through extensions. This modus operandi has been the norm for quite some time now in the industry as there have been isolated research studies on aspects of ontology and security. For example as early as 2002, Jürgens [6] proposed extending UML to UMLsec for integrating security related information in UML diagram specifications in the form of UML profiles. Such information provided through standard UML extension mechanism can then be used for model-based security engineering and verification through formal semantics of UML. As would be known, UML is fairly high up in the spectrum of ontology formalism above concept model, RDF/S but just below OWL (see: Fig. 7.5: The ontology spectrum: Weak to strong semantics in [7, 10]). Similarly, Fatemi et al. [8] took up securing ontologies to protect company private information. This work proposed creating a secure flavor of OWL in order to remedy this issue through one of two approaches. In the first, a Secure Web Ontology Language (SOWL) is proposed with extensions that may be non-compliant to OWL Recommendation thus requiring a new version of OWL. In the next, a scheme called OWL + S is proposed whereby OWL is enhanced by implicit attachment of security ontology. Considering the extent and maturity of OWL recommendations, it would not be reasonable to propose extensions requiring modification of its syntax. Doubtless there could be other similar attempts at investigating security of semantic systems, for example references in [9].

In the end, rather than modifying well-established and in-widespread-use existing ontology languages unreasonably, it would be advisable to leave them intact but to develop security ontologies. In that, rather than attempting isolated individual domain ontologies for subareas of security field, it would be pertinent to go for an upper level ontology for all the expanse of information security. (Here again, by information security I mean to cover information technology and information assurance including as well as all associated flavors.) Upper-level ontology, sometimes called variously as top level as well, normally includes domain concepts as generic common knowledge broad and abstract terms and relations in a hierarchical and networked structure. The Suggested Upper Merged Ontology (SUMO), the Descriptive Ontology for Linguistic and Cognitive Engineering (DOLCE), PROTON, UMBEL, Cyc and Basic Formal Ontology (BFO) are examples of upper-level ontology [10]. Upper level ontology is abstract or conceptual; it is meant to alleviate integration of middle-level operational and low-level data base schema ontologies developed by disparate individuals.

Consensus building nature of upper ontology is to be kept in view while attempting to develop one. All stake-holders, those with an interest in using it, should be involved. The crowdsourcing modus operandi suits well. "Crowdsourcing" is defined as the "practice of obtaining needed services, ideas, or content by soliciting contributions from a large group of people and especially from the online community rather than from traditional employees or suppliers" by Merriam-Webster Dictionary [11]. "By definition, crowdsourcing combines the efforts of numerous self-identified volunteers or part-time workers, where each contributor, acting on their own initiative, adds a small contribution that combines with those of others to achieve a greater result [12]". Crowdsourcing should be employed with a twist: considering the specific expertise requirement constricting the likely public audience, the crowd is "the community" composed of concerned, related and professional people of the extended domain. Hence, rather than outsourcing to or commissioning from a specific named group, "community sourcing" and "community curating" leading to collaborative ontology development must be employed. With use of proper tools, such as Collaborative Protegé [13] allowing and managing contributions of multiple contributing editors this scheme is feasible. A recommender system and trust scores among parties would serve here for social harvesting and sourcing up front. At the well advanced stage with reasonably large set of ontology definitions in hand, the touch up then may be affected by a commissioned committee work as it's been done in W3C.

3 Standardizing Security Ontology

International Standards Organization defines a standard as a formal document providing requirements, specifications, guidelines or characteristics of materials, products, processes and services fit for the purpose if used consistently. Standards "ensure that products and services are safe, reliable and of good quality. For business, they are strategic tools that reduce costs by minimizing waste and errors, and increasing productivity. They help companies to access new markets, level the playing field for developing countries and facilitate free and fair global trade" [14]. Standards often elicit compliance. They come in various flavors but in this paper we are interested in "technical standards" or "industry standards" in information security and ontologies domains. A technical standard establishes uniform engineering or technical criteria, methods, processes and practices in reference to technical systems.

Technical standards may be developed unilaterally but their acceptance and enforcement in practice would depend on the pulling power of the entity developed it. A community standard however developed through consensus would carry willing acceptance of its stake-holders thus it becomes a voluntary de facto standard.

Thus we would need a standardization process that secures the formal consensus of the security and ontology communities. Ontolog Community is a good example of an organization of loosely coupled people of common interests in ontology area. Its weekly virtual working meetings establish a platform to present individual points of views, discussion, consensus building and evolution of resolutions. The case in point is the Ontology-Based Standards Mini Series of virtual meetings [15].

As soon as a reasonably mature draft standard ontology evolves in the process, the product should eventually have to get the consensus of technical experts. Consequently, the development of the standard upper ontology for security would better continue through the joint concerted effort of such standardization organizations operative in the interest area as W3C, IEEE Standards Association, NIST, and ISO. These organizations have practical competence in the processes of standardization. W3C has been active in producing voluntary standards called "recommendation" in Web realm [16]. IEEE SA has produced the Standard Ontologies for Robotics and Automation [17] and for Learning Technology [18].

It is only through such a process of security community harvesting, cooperation and collaboration of field experts, finalization by standardization organizations, standard upper ontology for security will be rendered open, international and authoritative in the field.

4 Conclusions

I will conclude with the same sentence as in [2]: "It should be clear that generic ontology for security of information and networks is needed and that earlier the better." Certainly the outcome will be much more usable if the security ontology gets standardized. Furthermore, for a standardized ontology it is best if it is of high-level, perhaps an information security upper ontology for all others to link up to. Provided all concerned link their big data, this then should come handy in evolving a forest of linked ontologies, eventually helping as well in unifying the terminology.

This paper proposes a social harvesting approach involving community sourcing and volunteer-driven standardization for evolving a standardized upper ontology for security of information and networks.

References

1. International Conference on Security of Information and Networks. www.sinconf.org. Accessed 29 Sept 2015
2. Elci, A.: (Keynote Talk) Isn't the time ripe for a standard ontology on security of information and networks? In: Poet, R., Elci, A., Gaur, M.S., Orgun, M.A., Makarevich, O. (eds.) Proceedings of SIN 2014, the Seventh International Conference on Security of Information and Networks, pp. 1–3 (SIN 2014), 9–11 Sept 2014. Glasgow, UK. Published by ACM, ISBN 978-1-4503-3033-6. doi:10.1145/2659651.2664291 (2014)

3. World Wide Web Consortium. Security Activity. www.w3.org/Security/. Accessed 29 Sept 2015
4. W3C Web Cryptography Working Group. Web Cryptography API. www.w3.org/TR/WebCryptoAPI/. Accessed 29 Sept 2015
5. W3C XML Security Working Group. www.w3.org/2008/xmlsec/. Accessed 29 Sept 2015
6. Jürjens, J.: UMLsec: Extending UML for secure systems development. In: Jezequel, J.-M., Hussmann, H., Cook, S. (eds.) UML 2002—The Unified Modeling Language, pp. 412–425. Dresden, Springer-Verlag. And, for a brief summary of the proposal. https://en.wikipedia.org/wiki/UMLsec (2002)
7. Daconta, M.C., Obrst, L.J., Smith, K.T.: The Semantic Web: A Guide to the Future of XML, Web Services, and Knowledge Management. John Wiley & Sons. ISBN:0471432571 (2003)
8. Fatemi, M.R., Elci, A., Bayram, Z.: A proposal to ontology security standards. In: Proceedings of the 2008 International Conference on Semantic Web and Web Services (SWWS'08), of the 2008 World Congress in Computer Science, Computer Engineering, and Applied Computing (WORLDCOMP'08), pp. 183–186. Las Vegas, USA, July 14–17
9. Sicilia, M.-A., García-Barriocanal, E., Bermejo-Higuera, J., Sánchez-Alonso, S.: What are Information Security Ontologies Useful for? In: Garoufallou, E,. Hartley, E., Richard, J., Gaitanou, P. (eds.) Metadata and Semantics Research, vol. 544, pp. 51–61. Communications in Computer and Information Science, doi:10.1007/978-3-319-24129-6_5. Springer International Publishing. ISBN: 978-3-319-24128-9 (2015)
10. Open Semantic Framework. A Basic Guide to Ontologies. http://wiki.opensemanticframework.org/index.php/A_Basic_Guide_to_Ontologies. Accessed 29 Sept 2015
11. http://www.merriam-webster.com/dictionary/crowdsourcing. Accessed 29 Sept 2015
12. https://en.wikipedia.org/wiki/Crowdsourcing. Accessed 29 Sept 2015
13. Tudorache, T.: Collaborative protege. Available: http://protegewiki.stanford.edu/wiki/CollaborativeProtege. Accessed 29 Sep 2015 (2011)
14. http://www.iso.org/iso/home/standards.htm
15. Ontolog Community. Ontology-based standards. http://ontolog.cim3.net/cgi-bin/wiki.pl?ConferenceCall_2013_11_07. Accessed 29 Sept 2015
16. http://www.w3.org/standards/. Accessed 29 Sept 2015
17. IEEE SA—1872-2015—IEEE Standard Ontologies for Robotics and Automation. http://standards.ieee.org/findstds/standard/1872-2015.html. Accessed 29 Sept 2015
18. IEEE Standard for Learning Technology. http://standards.ieee.org/findstds/standard/1484.13.1-2012.html. Accessed 29 Sept 2015

Part II
Computational Intelligence

Analysis of Throat Microphone Using MFCC Features for Speaker Recognition

R. Visalakshi, P. Dhanalakshmi and S. Palanivel

Abstract In this paper, a visual aid system has been developed for helping people with sight loss (visually impaired) to help them in distinguishing among several speakers. We have analyzed the performance of a speaker recognition system based on features extracted from the speech recorded using a throat microphone in clean and noisy environment. In general, clean speech performs better for speaker recognition system. Speaker recognition in noisy environment, using a transducer held at the throat results in a signal that is clean even in noisy. The characteristics are extracted by means of Mel-Frequency Cepstral Coefficients (MFCC). Radial Basis function neural network (RBFNN) and Auto associative neural network (AANN) are two modeling techniques used to capture the features and in order to identify the speakers from clean and noisy environment. RBFNN and AANN model is used to reduce the mean square error among the feature vectors. The proposed work also compares the performance of RBFNN with AANN. By comparing the results of the two models, AANN performs well and produces better results than RBFNN using MFCC features in terms of accuracy.

Keywords Radial basic function neural network · Autoassociative neural network · Mel-frequency cepstral coefficients · Speaker recognition · Throat microphone · Visually impaired · Access control

1 Introduction

The natural and fundamental way of communication is speech for humans. Every human voice has various attributes to communicate the information such as emotion, gender, attitude, health and identity. The aim of Speaker Recognition (SR) by machine is the task of recognizing a person automatically based on the information

---

R. Visalakshi (✉) · P. Dhanalakshmi · S. Palanivel
Department of CSE, Annamalai University, Annamalai Nagar, India
e-mail: visalakshi_au@yahoo.in

M. Senthilkumar et al. (eds.), *Computational Intelligence, Cyber Security and Computational Models*, Advances in Intelligent Systems and Computing 412, DOI 10.1007/978-981-10-0251-9_5

received from his/ her speech signal [1–3]. No two individual's sound are identical because their vocal tract shapes, larynx sizes and other parts of their voice production organs are different and each voice has its own characteristic manner of speaking namely rhythm, intonation style, pronunciation, vocabulary etc., [4]. There is a variation between Speech Recognition (what is being said) and Speaker Recognition (who is speaking). SR is categorized into Speaker Identification and Speaker Verification.

- Speaker identification: It is the process of determining, to which of the speakers, the given utterance belongs to.
- Speaker verification: It is the process of accepting or rejecting the identity claim of the speaker.

Depending on the spoken text, SR methods are further divided into text-dependent and text-independent cases. Text-dependent SR systems require the speaker to produce speech for the same text in both training and testing whereas in text-independent, SR system has no restriction on the sentences or phrase to be spoken.

1.1 Outline of the Work

In order to identify the speech data from clean and noisy environment, the features are extracted using MFCC are discussed in Sect. 3. The three layer RBFNN model and AANN model are used to capture the features and in order to identify the speakers from clean and noisy environment. The performance of RBFNN is compared to AANN feed forward neural network. Experimental results show that the accuracy of AANN with MFCC can provide a better result are discussed in Sect. 4. Speaker recognition system may be viewed as working in four stages namely Analysis, Features Extraction, Modeling and Testing [5]. Figure 1 illustrates the block diagram of speaker recognition.

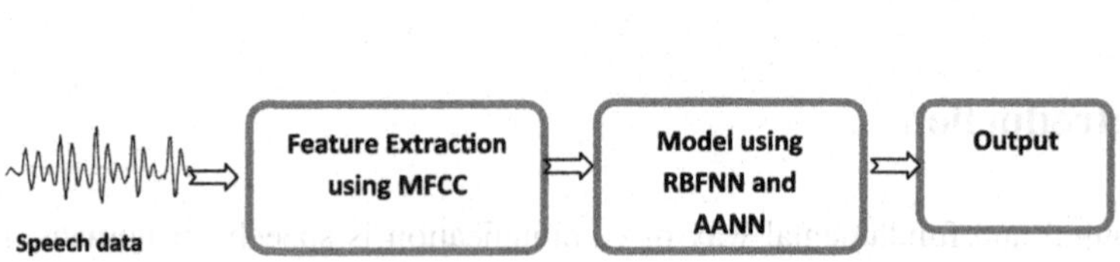

Fig. 1 Block diagram of speaker recognition

2 Related Work

Many works have been carried out in the literature about speaker recognition in clean and noisy environment using Normal and Throat microphone. Zhu and Yang [5] used WLPCC and VQ method. WLPCC shows the high accuracy when it is used as the parameter of the speaker recognition system. Mubeen [6] used MFCC and AANN model achieved 91 % (obtained using NM features alone) to 94 % (NM and TM combined). Krishnamoorthy [7] used MFCC and GMM-UBM method provides the performance of 78.20 % using only limited data and 80 % using both limited and noisy data. Sadic [8] the CVA is newly used for text-independent speaker recognition in which performance of CVA is compared with those of FLDA and GMM. Erzin [9] to investigate throat microphone in high noise environments, such as airplane, motorcycle, military field, factory and street crowd environments. From the literature study, some of the researcher use GMM, SVM, HMM and AANN method and acoustic features such as VQ, LPCC, WLPCC, MFCC and new other features also used. RBFNN method is generally used in speech recognition, classification of audio signal, etc., but RBFNN method is not much used for speaker recognition. Hence this work compares RBFNN method with the most used AANN method for recognizing the speaker even in noisy environment using Throat microphone.

3 Acoustic Feature Extraction Techniques

In this proposed method feature extraction based on MFCC are used for speaker recognition.

3.1 Pre-processing

Preprocessing of the speaker signal must be done in order to extract the acoustic features from the speech signals and it is divided into successive analysis frame. In this work sampling rate of 8 kHz, pulse code modulation(PCM) format, 16 bit monophonic is deployed.

3.2 Mel Frequency Cepstral Coefficients

In speaker recognition the Mel-Frequency Cepstral Coefficients (MFCC) features is the most normally used characteristics. It units the benefits of the cepstrum analysis

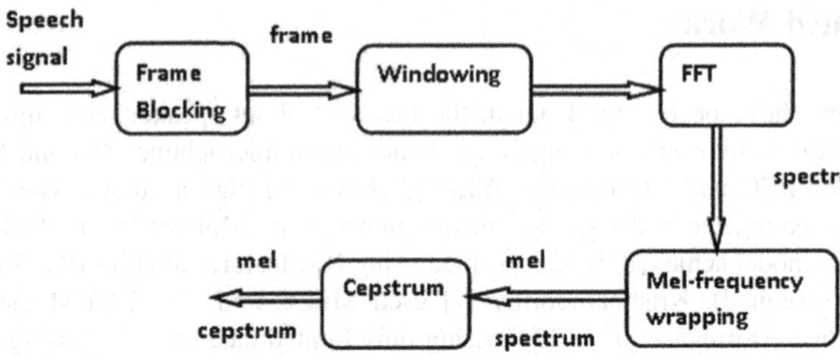

Fig. 2 Block diagram of Mel cepstral coefficients

with a perceptual frequency scale based on critical bands. The steps for calculating the MFCC from the speech signal are as follows in Fig. 2.

Following fragmenting the speech signal into overlapping frames, the frequency response of each frame is calculated by Discrete Fourier Transform (DFT). Next the spectrogram of the speech signal is acquired.

4 Experimental Results

4.1 Datasets

The experimental evaluation of the proposed speaker recognition system is carried out using a database collected from Throat microphone in clean and noisy environment. The corpus of speech are collected in the department of linguistics in annamalai university for clean environment. For noisy environment, the corpus of speech data are collected in class room and laboratories from 25 students for duration ranging from 1 to 2 h. Speech signals are recorded in with sampling rate of 8 kHz, 16 bits with mono channel. Both training and testing files are independent of text. The total no of utterances is 1000. We used 750 for training utterances and remaining 250 is used for testing utterances.

4.2 Modeling Using RBFNN and AANN

In the initial process of RBFNN training 39 MFCC features are extracted from the available speaker dataset. The extracted features are fed as input to RBFNN model. It is a feed forward neural network consists of an input layer, hidden layer and an output layer [10]. The RBF centre value is located by using the k-means algorithm. The complexity in determining the weights are reduced by using the least squares algorithm and the number of means is chosen to be 1, 2 and 5 for determining the number of nodes in hidden layer. Analysis shows that the system performance is optimal, when the value of k is 5. After completing the training process, testing for speaker recognition takes place. In the testing phase, feature vectors of a specified speaker is fed to the input units, by using the functions compute the confidence score value as output. The highest among the obtained confidence score value decides the class to which the speaker belongs. The performance analysis by varying the k value is shown in Fig. 3.

The AANN model is used to capture of distributed of 39 MFCC features respectively. The distribution is usually different for different speaker. The feature vectors are given as input and compared with the output to calculate the error. In this experiment the network is trained for 500 epochs. The confidence score is calculated from the normalized squared error and the speaker is decided based on highest confidence score. The network structures *39L 78N 4N 78N 39L* for MFCC gives a good performance and this structure is obtained after some trial and error.

The performance of speaker recognition is obtained by varying the expansion layer and third layer compression layer of AANN model. Speech signals are divided into frames of length 20 ms with 8000 samples per second, with shift of 10 ms is used for training and testing the designed AANN model. The performance of the system is evaluated, and the method achieves about 97.0 %. Figures 4 and 5 shows that the performance of RBFNN and AANN for speaker recognition in clean

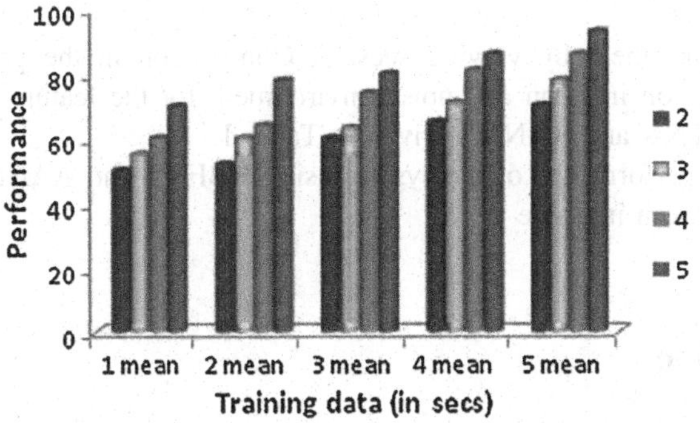

Fig. 3 Performance of RBFNN for different mean

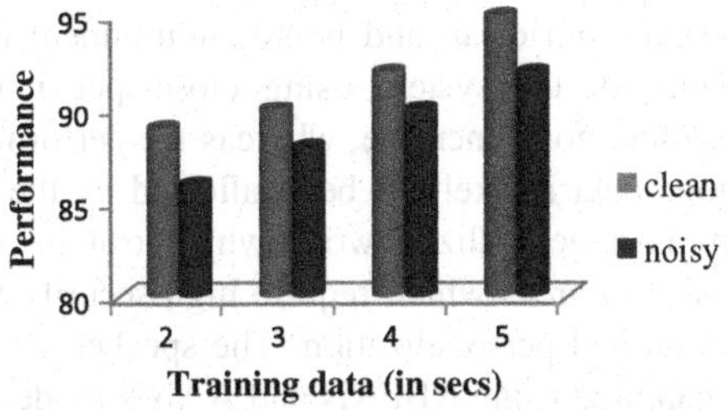

Fig. 4 Performance of RBFNN for speaker recognition in clean noisy environment

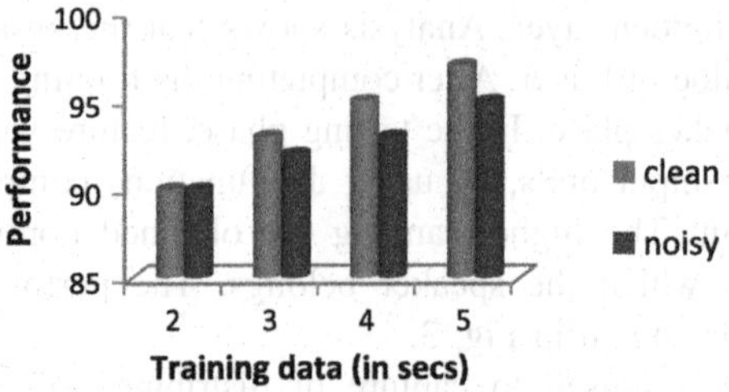

Fig. 5 Performance of AANN for speaker recognition in clean and noisy enviroment

Table 1 Speaker recognition performance using MFCC for AANN and RBFNN

Category	RBFNN (%)	AANN (%)
Clean	95.0	97.0
Noisy	90.0	95.0

Table 2 Speaker recognition performance using RBFNN and AANN

RBFNN (%)	AANN (%)
94.0	97.0

and noisy environment by varying secs. A Comparison in the performance of speaker recognition in clean and noisy environment for the feature vectors using MFCC for RBFNN and AANN is given in Table 1.

The overall performance of the system using RBFNN and AANN for speaker recognition is given in Table 2.

5 Conclusion

In this paper, we have proposed a speaker recognition system using RBFNN and AANN. MFCC features are extracted from the voice signal that can later be used to represent each speaker. Experimental results show that, the characteristics of the speech data are collected from clean and noisy environment using throat microphone. The performance of the system using close-speaking microphone data degrades as the background noise increase, whereas the performance of the system using throat microphone data is likely to be unaffected by the background noise. The proposed system is a generalized work which can be useful for visually impaired people and also in areas which require high security like access control, transaction authentication and personalization. The speaker information present in the source features is captured using RBFNN and AANN model. By comparing the results of the two models, AANN performs well and produces better results than

RBFNN using MFCC features in terms of accuracy. In future, throat microphone can be used to analyze the performance of speech impaired people. Various acoustic features can be analyzed and the performance of different pattern recognition techniques can be studied.

References

1. Yujin, Y., Peihua, Z., Qun, Z.: Research of speaker recognition based on combination of LPCC and MFCC. In: IEEE (2010)
2. Tomi, K., Li H.: An overview of text independent speaker recognition: from features to supervectors. ScienceDirect, Speech Communication (2010)
3. Jin, Q., Stan Jou, S.C., Schultz, T.: Whispering speaker identification (2007)
4. Nigade, A.S., Chitode, J.S.: Throat microphone signals for isolated word recognition using LPC. Int. J. Adv. Res. Comput. Sci. Soft. Eng. (IJARCSSE), **2**(8) (2012)
5. Zhu, L., Yang, Q.: Speaker recognition system based on weighted feature parameter. In: International Conference on Solid State Devices and Materials Science, pp. 1515–1522. Macao (2012)
6. Mubeen, N., Shahina, A., Nayeemulla Khan, A., Vinoth, G.: Combining spectral features of standard and throat microphones for speaker identification. In: IEEE ICRTIT, pp. 119–122. Chennai, Tamil Nadu (2012)
7. Krishnamoorthy, P., Jayanna, H.S., Prasanna, S.R.M.: Speaker recognition under limited data condition by noise addition. In: Expert System with Applications, pp. 13487–13490 (2011)
8. Sadic, S., Bilginer Gulmezoglu, M.: Common vector approach and its combination with GMM for text independent speaker recognition. In: Expert System with Applications, pp. 11394–11400 (2011)
9. Erzin, E.: Improved throat microphone speech recognition by joint analysis of throat and acoustic microphone recordings. IEEE **17**(7), 1558–7916 (2009)
10. Wali, S.S., Hatture, S.M., Nandyal, S.: MFCC based text-dependent speaker identification using BPNN. Int. J. Signal Process. Syst. **3**(1) (2015)

Single-Pixel Based Double Random-Phase Encoding Technique

Nitin Rawat

Abstract A new encryption technique based on single-pixel compressive sensing along with a Double Random-Phase encoding (DRPE) is proposed. As compared with the conventional way of image compression where the image information is firstly capture and then compress, the single-pixel compressive sensing collects only a few large coefficients of the data information and throws out the remaining which gives scrambled effect on the image. Further, to enhance the complexity of the image data, the double random phase encoding along with a fractional Fourier transform is implemented to re-encrypt it. The single-pixel based compressive sensing, DRPE and fractional Fourier transform act as a secret keys. At the receiver end, the original image data is reconstructed by applying the inverse of double random phase process and an l_1-minimization approach. The peak-to-peak signal-to-noise ratio and the minimum number of compressive sensing measurements to reconstruct the image are used to analyze the quality of the decryption image. The numerical results demonstrate the system to be highly complex, robust and secure.

Keywords Single-pixel · Compressed sensing · Double Random-Phase encoding · Fractional fourier transform · Key · l_1-norm

1 Introduction

Several encryption techniques have been applied to secure the data information from attackers [1–3]. A well-known Double Random-Phase encoding (DRPE) proves to be more secure and robust encryption technique [4, 5]. In DRPE, the image is scrambled by two independent phase functions with a Fourier transform

N. Rawat (✉)
School of Mechatronics, Gwangju Institute of Science and Technology,
Gwangju 500-712, South Korea
e-mail: nitincad4@gmail.com

M. Senthilkumar et al. (eds.), *Computational Intelligence, Cyber Security and Computational Models*, Advances in Intelligent Systems and Computing 412, DOI 10.1007/978-981-10-0251-9_6

(FT). However, more complex algorithms are used such as jigsaw transform [6], Henon chaotic [7], and Arnold transform [8] for pixel scrambling in DRPE as well. These methods enhance the security system of the data by scrambling the content which can be unlocked only by the right decrypted key. The fractional Fourier transform (FRT) is introduced for further improvements in the DRPE approach [9, 10]. FRT provides an extra degree of freedom and enlarges the key space resulting in higher security of data as compared to the FT approach [5]. In encryption, data compression plays an important role and have applied in various encryption based methods [11, 12]; although the encryption scheme cannot achieve perfect compression with high security, it is still significant owing to the high computational complexity of cracking.

Compressive sensing (CS) is a new sampling paradigm which extracts only few essential features and throws out the remaining from an image [13]. CS states that the image contains many non-zero coefficients when transformed into an appropriate basis and is possible to measure a sparsely represented N-dimensional image from $M < N$ incoherent measurements using an optimization algorithm [14, 15]. In compression based encryption methods, the camera captures all the information which means the data is first collected and then compressed [11, 12].

In this paper, we propose an encryption method based on single-pixel compressive sensing along with a double random phase encoding technique. The method not only captures the image information but also compresses while capturing. The single-pixel compressive sensing method itself is encrypts the image into a complex form. Furthermore, the double random phase encoding process along with a fractional Fourier transform enhances the complexity by re-encrypting the image data into multiple times. Numerical results show our system to be more robust, complex, and secure and can sustain various attacks with high dimensional reduction capability.

2 Fundamental Knowledge

2.1 Fractional Fourier Transform (FRT)

The FRT is a generalization of Fourier transform (FT). Since we always look for ways to enhance the security of the data, the FRT involves an extra parameter of the transform order 0 to 1 [9, 10]. In case of FRT, the parameters, such as fractional orders and the scaling factors along with the x and y-axis [16] make it complex to decode as they serve as additional keys for image decryption. Furthermore, the FRT mixes the signal by rotating it through any arbitrary angle in frequency-space domain. Here $(\alpha, \beta) = p\pi/2$ are the angles at which FRT can be calculated.

In two-dimensional case, the αth order FRT F^{α} of a function $f(x, y)$ can be expressed as

$$F^{\alpha}\{f(x,y)\} = \int\int_{-\infty}^{+\infty} K^{\alpha}(x,y;u,v)f(x,y)dxdy \tag{1}$$

where K^{α} is the transform kernel given by

$$K^{\alpha}(x,y;u,v) = A_{\varphi}\exp[i\pi(x^2+y^2+u^2+v^2)\cot\varphi_{\alpha} - 2(xu+yv)\cot\varphi_{\alpha}] \tag{2}$$

and

$$A_{\varphi} = \frac{\exp[i\pi\mathrm{sgn}(\sin\varphi_{\alpha})/2 + i\varphi_{\alpha}]}{|\sin\varphi_{\alpha}|} \tag{3}$$

2.2 Double Random Phase Encoding (DRPE) Using FRT Technique

Figure 1 shows a conventional double random phase encoding technique. Let an image $A(x,y)$ is multiplied by an independent random phase functions $\exp[i\phi_{1\&2}(x,y)]$ and is transformed through the FRT order. The FRT order of (α, β)

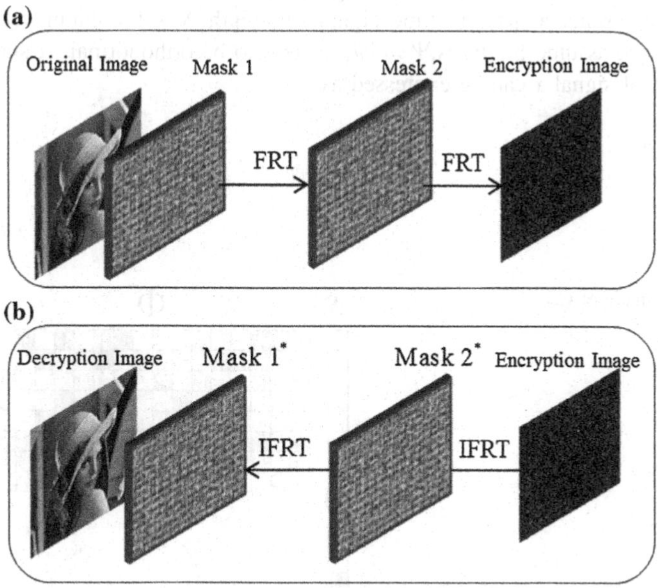

Fig. 1 Conventional double random phase encoding process. **a** Encoding and; **b** decoding process '*FRT*' and '*IFRT*' are fractional Fourier transform and inverse fractional Fourier transform

is performed where the range of the order is considered between $(-2$ to $2)$. The encryption process can be expressed

$$I(x,y) = F_\alpha\{F_\beta\{[A(x,y)\exp[i\phi_1(x,y)]\exp[i\phi_2(x,y)]]\}\} \tag{4}$$

where $F_{\alpha,\beta}$ represents the order of FRT. The image $A(x,y)$ is multiplied by the random phases $\exp[i\phi_{1\&2}(x,y)]$ where an FRT of αth and bth order give the encrypted image in the fractional domain. The FRT parameters and the random phases act as a secret key to the encoding process.

Similarly the decryption procedure is expressed as

$$A'(x,y) = F_\beta^{-1}\{\{F_\alpha^{-1}[I(x,y)\exp[i\phi_2(x,y)^*]\exp[i\phi_1(x,y)]^*\}\} \tag{5}$$

where $F_{\alpha,\beta}^{-1}$ represents the inverse FRT through an order of (α,β) and $\exp[i\phi_{2\&1}(x,y)]^*$ are the complex conjugates. The inverse process gives the decryption image.

2.3 Compressive Sensing (CS)

CS relies on two principles: sparsity and incoherence. A small collection of non-adaptive linear measurements of a compressible image contain enough information for perfect reconstruction [14, 17–19]. Figure 2 shows a schematic of CS approach where x represents a discrete time signal of length $N \times 1$ column vector. For simplicity, we assume the basis, $\Psi = [\psi_1, \ldots \psi_N]$ to be orthonormal. A sparse representation of signal x can be expressed as

$$x = \sum_{n=1}^{N} s_n \psi_n \tag{6}$$

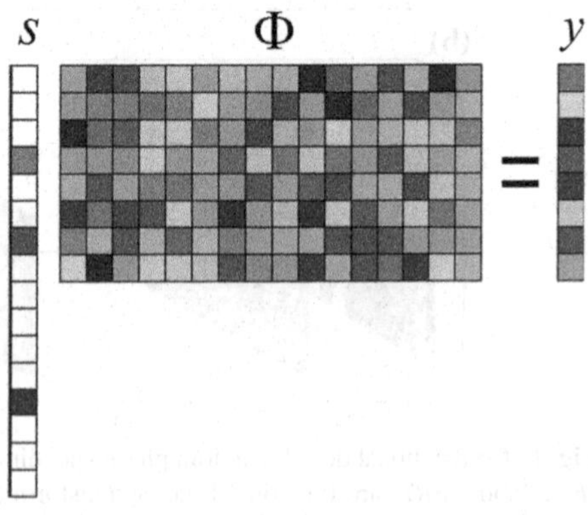

Fig. 2 Schematic of CS approach

where x is the linear combination of K chosen from Ψ. A signal x is K-sparse if it is a linear combination of only K basis vectors, which means that only K of the coefficients s_n in Eq. (6) are nonzero, and the remaining are zeros. The CS imaging system can be re-written in a matrix form and is expressed as:

$$y = \Phi x = \Phi \Psi s \tag{7}$$

where y represents an $M \times 1$ column vector, x is an image ordered in an $N \times 1$ vector and the measurement matrix as $M \times N$. Ψ is an image representation in some sparsifying basis and s denotes the non-zero coefficients of an image. The incoherence between the sensing matrix Φ and the sparsifying Ψ operator can be expressed as:

$$\mu(\Phi, \Psi) = \sqrt{n} \cdot \max_{1 \leq k,j \leq n} \left| \langle \varphi_k, \psi_j \rangle \right| \tag{8}$$

where φ_k, ψ_j denotes the column vector of Φ and Ψ respectively and n is the length of the column vector. If Φ and Ψ are highly correlated, the coherence is large, else it is small. It follows $\mu(\Phi, \Psi) \in [1, \sqrt{n}]$. In CS, we are interested in $M \ll N$ case where the goal is to acquire less number of measurements than the number of pixels in the image.

If both Ψ and Φ of them are uncorrelated, then scan be well recovered from $n = O(d \log N)$ measurements under the *Restricted Isometry Property* (RIP) [13]. Once the above conditions satisfy the CS theory, a recovery can be possible by using l_1-norm minimization; the proposed reconstruction is given by $y = \Phi s$, where s is the solution to the convex optimization program $\left(\|x\|_{l_1} := \sum_i \|\alpha\|_1 \right)$.

2.4 Digital Micro-Mirror Device

DMD is an array of individual pixels consists of an array of electrostistically actuated micro-mirrors of size 1024 × 768 and an example of a transmissive spatial light modulator (SLM) that either passes or blocks parts of the beam [20]. The array dimensions being determined by the resolution of the particular DMD, where in our case it is of 1024 × 768. DMD memory is loaded by row. An entire row must be loaded even if only one pixel in the row needs to be changed. The light falling on the DMD can be reflected in two directions depending on the orientation of the mirrors rotate in one of two states (+10 degrees and −10 degrees from horizontal). Light will be collected by the subsequent lens if the mirror is in the +10 degrees state. The positive state is tilted toward the illumination and is referred to as the state whereas the negative state is tilted away from the illumination and is referred to as the off state. Thus, the DMD can pass or throws away the pixel information by choice.

2.5 Single-Pixel Compressive Sensing

The quintessential example of CS is the single-pixel CS based camera originally developed at Rice University [18] and is used in various applications such as terahertz imaging [21], compressive holography [22], and in active illumination based techniques [23]. The main goal is to directly compute random linear measurements by collecting a few large coefficients and neglecting the rest of the object image. Figure 3 shows a schematic of single-pixel base CS approach where the object information is obtained via DMD with 1024 × 768 micromirrors, two lenses, a single-photon detector, and an A/D converter [24]. The lenses used are Bi-convex (AR Coating: 350–700 nm) of size 1 inch diameter.

The DMD directly measures the set of largest coefficients and throws out the remaining. After repeating the process until M values are acquired, a reconstruction can be obtained by using l_1-minimization.

In single-pixel CS based camera, the light source illuminates the object information and projects it to the DMD. Each mirror of the DMD corresponds to a particular pixel and can be independently passes towards Lens 2 (corresponding to a 1 at that particular pixel) or away from Lens 2 (corresponding to a 0 at that particular pixel). The light reflected through lens 2 onto the single-photon detector while producing an output voltage. The output voltage is totally depends on the DMD modulation pattern that corresponds to one measurement. The single-photon detector integrates the product of $s[n]\phi_m[n]$ to compute the measurement $y[m] = \langle s, \phi_m \rangle$. Finally a perfect signal or an image can be recovered by applying l_1-magic.

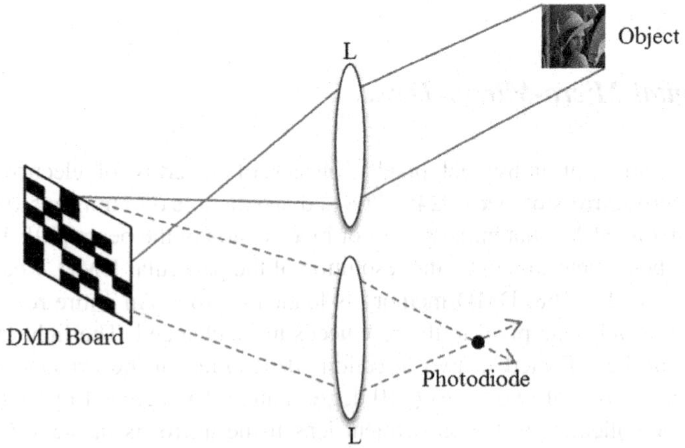

Fig. 3 Schematic of the single pixel camera approach

3 Digital Image Encryption Process

3.1 Encryption and Decryption Procedure

Figure 4 shows the flow chart of the single-pixel CS based encryption technique where the original image is passed onto a single-pixel through lenses and DMD mirror arrays. The image information is compressed while capturing, by directly estimating the set of few large coefficients and is used as secret key 1. Further, the image information is re-encrypted by a well-know DRPE technique with two independent phase masks $\exp[i\phi_{1\&2}(x,y)]$ which serve as secret key 2 and secret key 3. To enhance the security of the data, the FRT approach is introduced which gives an extra degree of freedom.

At the decryption side, the original information is extracted by reversing the DRPE and FRT process. A secret key 4 is introduced on the decryption side where the reconstruction is only possible by using an l_1-magic.

3.2 Numerical Experiments

In order to verify the results of the proposed system, the Peak-to-peak Signal-to-Noise Ratio (PSNR) between the original image and reconstructed image is used and is expressed as:

$$\mathrm{PSNR} = 10\log\frac{255^2}{\left(\frac{1}{MN}\right)\sum_{i=1}^{N}\sum_{j=1}^{M}\left[A(i,j)-A'(i,j)\right]^2} \tag{9}$$

where $A(i,j)$ and $A'(i,j)$ are the reconstructed and original image. The PSNR of decrypted image calculated is 27.012 dB which indicates a good quality of an image.

The encoding process is shown in Fig. 5 where a gray scale image of size 256 × 256 (Fig. 5a) is used. In Fig. 5b, the size of encryption image has been reduced because of being sampled through DMD which consists of an array of

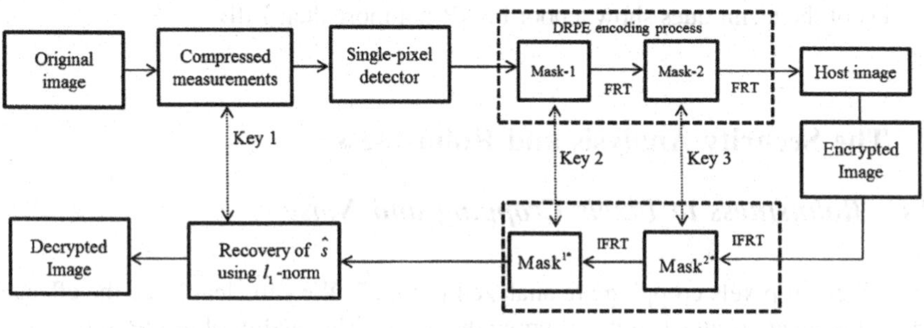

Fig. 4 Flow chart of the proposed encoding system

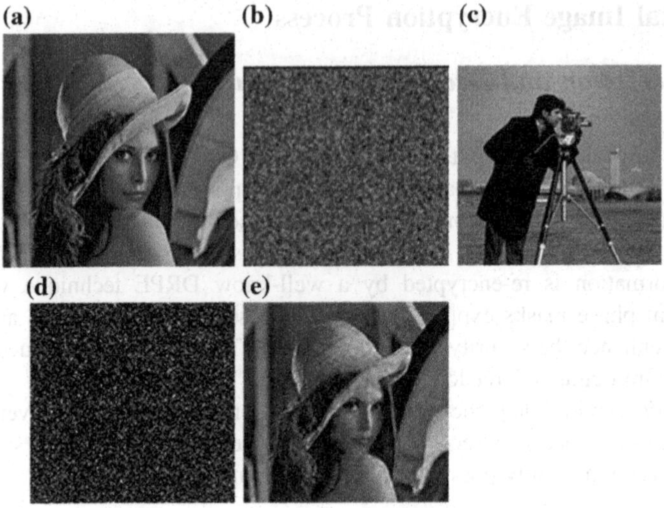

Fig. 5 Results of encoding process. **a** Original image; **b** encrypted image after DMD device; **c** host image; **d** re-encrypted image after DRPE and FRT technique; **e** reconstructed image after using l_1-norm

N tiny mirrors. Each mirror of DMD corresponds to a particular pixel and can be independently oriented either towards or away from lens 2.

The DMD reflects the generated pattern onto a single-pixel detector via lens 2. Further, in Fig. 5d, the measurements onto a single-pixel detector are re-encrypted by DRPE technique. An extra security enhancement is provided by using FRT approach in DRPE. The decryption process is shown where a random matrix (Fig. 5d), Φ of $M \times N$ is used which is highly incoherent with any fixed basis Ψ. Finally, in Fig. 5e, the sparse coefficient s can be recovered by using l_1-magic. In Fig. 6a, the number of CS measurements versus the PSNR is shown where the image is well reconstructed with only few measurements. The PSNR improves as the number of CS measurements increases. We have compared our proposed method to two other well-known techniques such as DRPE in Fourier domain and DRPE in FRT domain. Only 30 % of the measurements are used for all the three techniques and PSNR is measured. It can be clearly seen that our proposed method shows a remarkable PSNR up to 25 dB whereas the rest of the techniques show a poor PSNR not more than 7 dB.

4 The Security Analysis and Robustness

4.1 Robustness to Pixels Cropping and Noise

The effects of pixels cropping are analyzed in Fig. 7. We can clearly see the effects where the image is attacked by cropping the pixels. The quality of the reconstructed

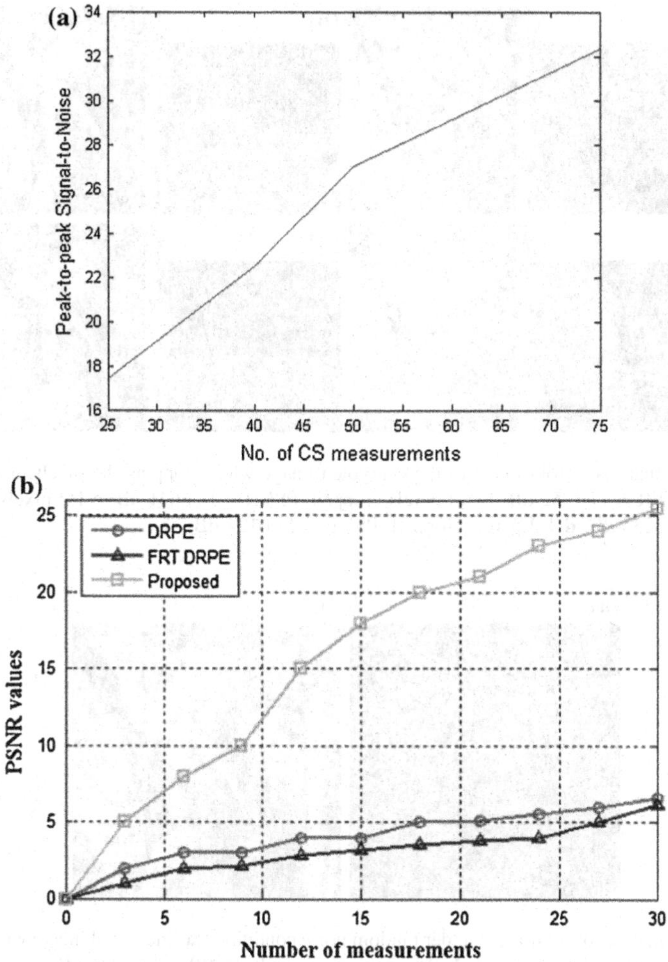

Fig. 6 CS measurements versus PSNR values. **a** As the number of CS measurements (M) increases, the PSNR increase as well which indicates the reconstruction quality. **b** Comparison of our proposed method with two well-known conventional DRPE method

image will degrade, resulting in smaller PSNR. We analyze 12.5–75 % pixel cropping which shows that the more the pixels are cropped, the smaller the PSNR is.

Figure 8 shows the reconstruction results of encryption information when using random noise effect. A Gaussian random noise is introduced where the encrypted information is added to random noise with value ranged from 0 to 1. The PSNR calculated is 23.256 dB. The image is well recognized after applying random noise.

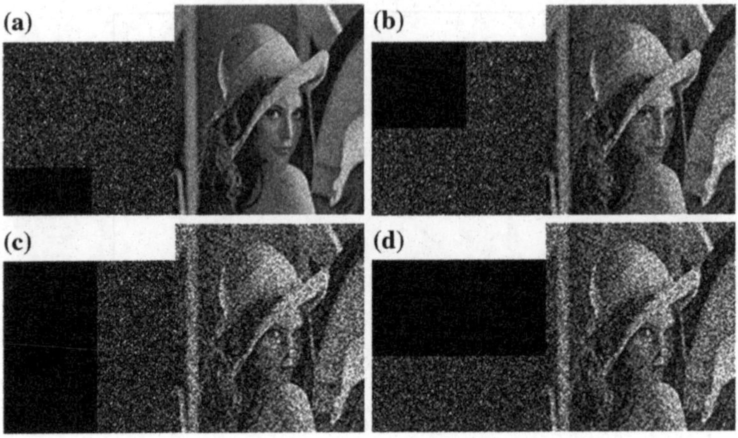

Fig. 7 The analysis of robustness of the encrypted image when cropping the pixels. **a** 1/8 pixels cropped, PSNR = 17.1452 dB; **b** 1/4 pixels cropped, PSNR = 14.5641 dB; **c** 1/2 pixels cropped, PSNR = 11.8134 dB; **d** 1/2 pixels cropped, PSNR = 11.0124 dB

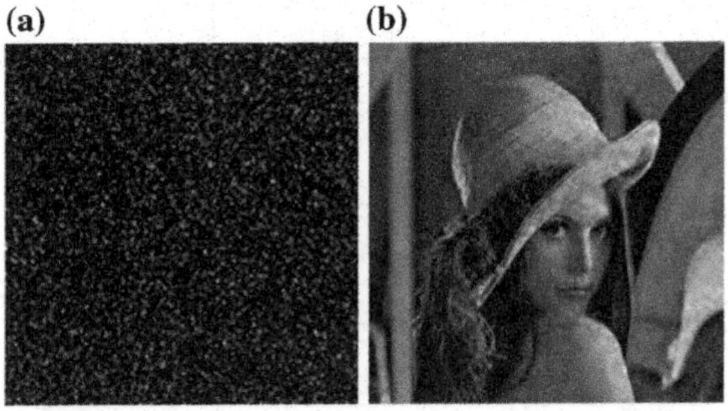

Fig. 8 The analysis of robustness under random noise conditions. **a** Encrypted image with random noise of value ranged from 0 to 1. **b** Reconstruction result, PSNR = 22.1075 dB

4.2 Robustness to Using Wrong Keys

By introducing a single-pixel based CS along with a DRPE technique, the proposed image encryption method has a high security and robustness. As shown in Fig. 9, we analyze the robustness when using wrong keys, where the decryption images are very sensitive to the correct keys. Figure 9a shows the reconstructed image when using wrong secret key-1, giving a PSNR of 3.254 dB. Similarly in Fig. 9b, c using the wrong random phases with FRT orders which act as secret keys- 2, and 3 give a PSNR of 2.4301 and 3.825 dB.

If an only if all the keys are available, the correct decrypted image can be obtained. Our proposed system shows the system to be more complex and robust.

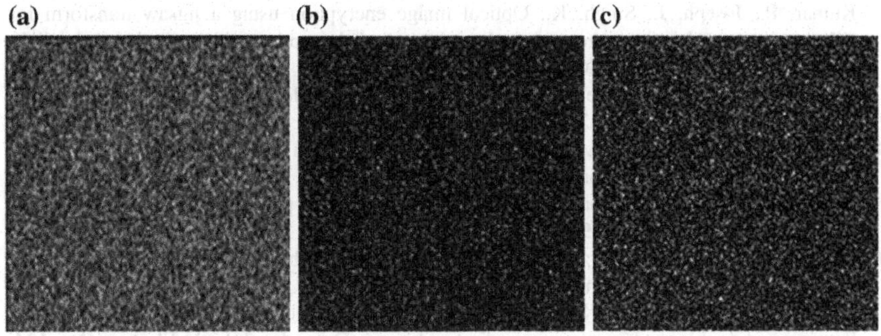

Fig. 9 The analysis of robustness to the wrong keys. **a** The reconstructed image with wrong key-1, PSNR = 4.0425 dB. **b** The reconstructed image with wrong key-2, PSNR = 2.4301 dB. **c** The reconstructed image with wrong key-3, PSNR = 3.825 dB

5 Conclusion

In this paper, we have proposed a novel image encryption method based on single-pixel CS along with a well-known DRPE technique. Our system reduces the dimensional size of image by directly capturing only a few large coefficients from an image and throws out the remaining. In order to improve the information security, the DRPE technique enhances the security by introducing two independent phase functions along with FRT orders.

Numerical experiments show the effectiveness of introducing single-pixel CS based encryption technique along with DRPE approach. It is evident from the simulations that improper selection of even using a single key during decryption misleads the attacker from the actual information. The proposed system not only have multiple encrypting but also have high dimensional reduction capability.

References

1. Fridrich, J.: Image encryption based on chaotic maps. In: 1997 IEEE International Conference on System, Man and Cybernatics Computational Cybernatics and Simulation, vol. 2, pp. 1105–1110 (1997)
2. Izmerly, O., Mor, T.: Chosen ciphertext attacks on lattice-based public key encryption and modern (non-quantum) cryptography in a quantum environment. Theor. Comput. Sci. **367**, 308–323 (2006)
3. Seyedzadeh, S.M., Mirzakuchaki, S.: A fast color image encryption algorithm based on coupled two-dimensional piecewise chaotic map. Signal Process. **92**, 1202–1215 (2012)
4. Javidi, B., Horner, J.L.: An optical pattern recognition system for validation & security verification. Opt. Eng. **33**, 1752–1756 (1994)
5. Refregier, P.: Optical image encryption based on input plane and Fourier plane random encoding. Opt. Lett. **20**, 767–769 (1995)

6. Kumar, P., Joseph, J., Singh, K.: Optical image encryption using a jigsaw transform for silhouette removal in interference-based methods and decryption with a single spatial light modulator. Appl. Opt. **50**, 1805–1811 (2011)
7. Wei-bin, C., Xin, Z.: Image encryption algorithm based on Henon chaotic system. Int. Conf. Image Anal. Signal Process. **2009**, 94–97 (2009)
8. Chen, W., Quan, C., Tay, C.J.: Optical color image encryption based on Arnold transform and interference method. Opt. Commun. **282**, 3680–3685 (2009)
9. Joshi, M., Singh, K.: Color image encryption and decryption using fractional Fourier transform. Opt. Commun. **279**, 35–42 (2007)
10. Hennelly, B., Sheridan, J.T.: Fractional Fourier transform-based image encryption: phase retrieval algorithm. Opt. Commun. **226**, 61–80 (2003)
11. Lu, P., Xu, Z., Lu, X., Liu, X.: Digital image information encryption based on Compressive Sensing and double random-phase encoding technique. Opt.—Int. J. Light Electron Opt. **124**, 2514–2518 (2013)
12. Liu, X., Cao, Y., Lu, P., Lu, X., Li, Y.: Optical image encryption technique based on compressed sensing and Arnold transformation. Opt.—Int. J. Light Electron Opt. **124**, 6590–6593 (2013)
13. Wakin, M.B.: An Introduction to compressive sampling. IEEE Sign. Proc. Mag. **25**, 21–30 (2008)
14. Candès, E.J., Romberg, J., Tao, T.: Robust uncertainty principles: exact signal frequency information. IEEE Trans. Inf. Theory **52**, 489–509 (2006)
15. Donoho, D.L.: Compressed sensing. IEEE Trans. Inf. Theory **52**, 1289–1306 (2006)
16. Unnikrishnan, G., Joseph, J., Singh, K.: Optical encryption by double-random phase encoding in the fractional Fourier domain. Opt. Lett. **25**, 887–889 (2000)
17. Candès, E.J., Plan, Y., Theory, A.A.R.: A probabilistic and RIPless theory of compressed sensing. IEEE Trans. Inf. Theory **57**, 7235–7254 (2011)
18. Takhar, D., et al.: A new cmpressive imaging camera architecture using optical-domain compression. SPIE Opt. + Photonics 6065 (2006)
19. Candes, E.J., Tao, T.: Near-optimal signal recovery from random projections: universal encoding strategies? IEEE Trans. Inf. Theory **52**, 5406–5425 (2006)
20. Feather, G.A. Monk, D.W.: The digital micromirror device for projection display. In: Proceedings of the Seventh Annual IEEE International Conference, pp. 43–51 (1995)
21. Chan, W.L., Charan, K., Takhar, D., Kelly, K.F., Baraniuk, R.G., Mittleman, D.M.: A single-pixel terahertz imaging system based on compressed sensing. Appl. Phys. Lett. **93**, 121105 (2008)
22. Clemente, P., Durán, V., Tajahuerce, E., Andrés, P., Climent, V., Lancis, J.: Compressive holography with a single-pixel detector. Opt. Express **38**, 2524–2527 (2013)
23. Magalhães, F., Araújo, F.M., Correia, M.V., Abolbashari, M., Farahi, F.: Active illumination single-pixel camera based on compressive sensing. Opt. Soc. Am. A **50**, 405–414 (2011)
24. Davenport, M.A., Takhar, D., Laska, J.N., Sun, T., Kelly, K.F., Baraniuk, R.G.: Single-pixel imaging via compressive sampling. IEEE Sign. Proc. Mag. **25**, 83–91 (2008)

Kernel Online Multi-task Learning

S. Sumitra and A. Aravindh

Abstract Many real world learning problems can be divided into a number of dependent subtasks. The conventional machine learning strategy considers each learning problem as a single unit and does not incorporate information associated with the tasks that are closely related with it. Such anomalies have been rectified in the Multi-Task learning (MTL) paradigm, where the model learns a modeling problem, by making use of its associated tasks. Such modeling strategies improve the generalization capacity of the model. In this paper we proposed a mathematical framework for multi-task learning using kernel online learning approach. We applied the proposed algorithm on a synthetic dataset as well as real time data set and the results were found to be promising.

Keywords Online Multi-task learning · Kernel methods · Multi-class classification

1 Introduction

The multi-task learning (MTL) generates a model for each of the tasks associated with a problem under consideration by making use of the knowledge transfer between subtasks. Hence if a learning problem can be divided into T different related tasks, MTL algorithms generate T models for each of its tasks, by incorporating the useful information from other tasks along with its own. On the other hand, conventional machine learning approach considers T related tasks as inde-

S. Sumitra (✉) · A. Aravindh
Department of Mathematics, Indian Institute of Space Science and Technology, Trivandrum, India
e-mail: sumitra@iist.ac.in
URL: http://www.iist.ac.in/

A. Aravindh
e-mail: aravindh.sc13m058@iist.ac.in

M. Senthilkumar et al. (eds.), *Computational Intelligence, Cyber Security and Computational Models*, Advances in Intelligent Systems and Computing 412, DOI 10.1007/978-981-10-0251-9_7

pendent tasks and would not consider other tasks information for developing a model for a subtask. The MTL uses all data points from different related classification problems for training and build the classifier for each problem simultaneously. The knowledge transfer between the sub problems helps to increase the generalization capacity of MTL models. The MTL learning frame work has been successfully applied on many real world problems like spam filtering, object recognition and computer vision.

The classical multi-task learning is often studied in batch learning set-up which assumes that the training data of all tasks are available. This assumption is not suitable for many practical applications where the data samples arrive sequentially one at a time. Also, batch multi-task learning algorithms involve huge computational cost. The main contribution of this paper is the mathematical formulation of MTL for online learning, by making use of kernel theory concepts.

2 Related Work

This paper focuses on multi-task learning in online paradigm. Unlike batch learning method which assumes training data samples are available, online learning method builds the model from the stream of data samples with simple and fast model updates. Thus these algorithms are naturally inclined to solve large number of data samples which arrive sequentially. Several online learning algorithms are presented in [1–6]. [2] derived algorithms for online binary and multi-class classification based on analytical solutions and simple constrained optimization problems. The online perceptron algorithm [7] updates the model weight based on the misclassified sample. [3] formulated online classification algorithms for binary classification problems as multi-class problems.

The multi-task learning approach has been extensively studied and its performance has been found to exceed the conventional learning methods. Pontil et al. [8] developed an approach to multi-task learning method for batch learning, based on the assumption that the solution vector corresponding to each task can be split into a global vector and task specific vector (noise vector). The noise vectors are small when the tasks are closely related to each other. The algorithm then estimates global and local parameter of all tasks simultaneously by solving an optimization problem analogous to SVMs used for single task learning. Guangxia et al. [9] proposed a collaborative approach to online multi-task learning where the global model is developed using the training samples of all the tasks. This global model is incorporated into the task specific model to improve the performance of learning model. The various multi-task batch learning techniques are discussed in [10–14]. The method described in this paper incorporated the definition of global and task specific function model as formulated in [8], into online kernel learning strategy [15]. We applied the proposed algorithm on various data sets for analyzing its efficiency and our experimental results were found to be promising.

The rest of the paper is organized as follows. The next section describes the concept of kernel theory and multi-tasking. Section 4 presents the proposed algorithm. The experimental results are given in Sects. 5 and 6 concludes the work with future extension.

3 Methods

In this section we describe the basic methods we used in our work.

3.1 Kernel Theory

Let $\mathcal{D} = \{(x_i, y_i) \,|i = 1, \ldots, N\}$ be the training data points, where $x_i \in \mathcal{X} \subseteq \mathbb{R}^n$ and $y_i \in \mathbb{R}$. Let f is the function that generates the data. Let f belongs to a reproducing kernel Hilbert space ($\mathcal{F}$). By definition, RKHS is a space of functions in which every point evaluations are bounded linear functionals. Therefore if we define $L_{x_i}(f) : \mathcal{F} \rightarrow \mathbb{R}$, such that $L_{x_i}(f) = f(x_i), x_i \in \mathcal{X}, f \in \mathcal{F}$, then by definition of RKHS, $L_{x_i}, i = 1, 2, \ldots$ are bounded linear functionals and hence by Riesz representation theorem, $\exists k_{x_i} \in \mathcal{F}$ such that

$$L_{x_i}(f) = f(x_i) = \langle f, k_{x_i} \rangle, f \in \mathcal{F} \tag{1}$$

where k_{x_i} unique; depends only on L_{x_i} that is x_i. Hence we can define a function $k : \mathcal{X} \times \mathcal{X} \rightarrow \mathbb{R}$ such that

$$k(x_i, x_j) = \langle k_{x_i}, k_{x_j} \rangle, x_i, x_j \in \mathcal{X}$$

k is called the reproducing kernel of $\mathcal{F}$. Corresponding to every RKHS there exists a unique reproducing kernel and vice versa. $\mathcal{F}$ is the closure of the span of $\{k_{x_i}, x_i \in \mathcal{X}\}$ and hence every $f \in \mathcal{F}$ can be written as $f = \sum_i \alpha_i k_{x_i}, \alpha_i \in \mathbb{R}$. The inner product $\langle ., . \rangle_{\mathcal{F}}$ induces a norm on $f \in \mathcal{F} : \| f \|_{\mathcal{F}}^2 = \langle f, f \rangle_{\mathcal{F}} = \sum_i \sum_j \alpha_i \alpha_j k(x_i, x_j)$.

The kernel learning methods use regularized risk functional defined as

$$R(f) = \sum_{i=1}^{N} l(y_i, f(x_i)) + \frac{\lambda}{2} \| f \|^2 \tag{2}$$

where $\lambda > 0$ is the regularization parameter and $l(y_i, f(x_i))$ is empirical loss term. Using representor theorem, the $f \in \mathcal{F}$ that minimizes (2) can be represented as $f = \sum_{i=1}^{N} \alpha_i k_{x_i}, x_i \in \mathcal{X}$.

3.2 Kernel Online Learning

The description of the kernel online frame work by choosing the empirical loss term as the least square function is given below [15]. Let the data sample (x_i, y_i) arrives at time instant *i*. Let f_i be the model developed at time i. The learning algorithm produces sequence of hypothesis $(f_1, f_2, \ldots, f_{m+1})$, where f_1 is arbitrarily initialized hypothesis and f_i for $i > 1$ is the hypothesis chosen after $(i-1)$th data point.

$$f_i = \sum_{i=1}^{t} \alpha_i k_{x_i}, \quad \text{where} \quad \alpha_i \in R, \quad k_{x_i} \in \mathcal{F}.$$

The instantaneous regularized risk function on a single sample (x_{i+1}, y_{i+1}) at time $i+1$ is defined as

$$\hat{J}_{i+1} = \frac{1}{2} \| f(x_{i+1}) - y_{i+1} \|^2 + \frac{\lambda}{2} \| f \|^2 \tag{3}$$

$$f_{i+1} = \arg\min_{f \in \mathcal{F}} \hat{J}_{i+1}(f) \tag{4}$$

The procedure for applying gradient descent method for solving (4) is given below.

Given initial approximation, $f_1 \in \mathcal{F}$

$$f_{i+1} = f_i - \eta_{i+1} (\nabla \hat{J}_{i+1}(f))_{f=f_i} \tag{5}$$

where $\eta_{i+1} > 0$ is the learning rate, which is generally taken as constant $\eta_i = \eta$.

Now, $f_i(x_{i+1}) = \langle f_i, k_{x_{i+1}} \rangle, k_{x_{i+1}} \in \mathcal{F}$. Therefore

$$\nabla \hat{J}_{i+1}(f_i) = [f(x_{i+1}) - y_{i+1}] k_{x_{i+1}} + \lambda f \tag{6}$$

Using Eq. (5), the update rule is derived as

$$f_{i+1} = (1 - \eta_{i+1} \lambda) \sum_{j=1}^{t} \alpha_j k_{x_j} - \eta_{i+1} (f_i(x_{i+1}) - y_{i+1}) k_{x_{i+1}} \tag{7}$$

Hence, at step $i+1$, the coefficients are updated as

$$\begin{aligned} \alpha_{i+1} &:= -\eta_{i+1} (f_i(x_{i+1}) - y_{i+1}) \\ \alpha_j &:= (1 - \eta_{i+1} \lambda) \alpha_j, \; j \leq i \end{aligned} \tag{8}$$

4 Kernel Online Multi-task Learning Algorithm

This section describes the proposed method for online multi-task learning using kernel methods. The multi-task learning set up comprises training data samples from T number of tasks and learns all the models simultaneously. In online environment, each data instance comes sequentially and the models are learnt. Let (x^t_{i+1}, y^t_{i+1}) be the incoming data instance of task t at $(i+1)$th iteration, where $t \in 1, \ldots, T$. The function corresponding to task t is defined as [8]

$$f_t = g_0 + g_t \tag{9}$$

where $g_0, g_t \in \mathcal{F}$ and g_0 is the global model which is common to all tasks and g_t are individual functions for each task. Hence,

$$f_t(x) = \sum_{j=1}^{i} \beta_j k_{x_j}(x) + \sum_{j=1}^{i_t} \alpha_j k_{x_j}(x) \tag{10}$$

where i is the total number of data samples arrived, i_t is the number of data samples arrived till iteration i for the task t, $\alpha_j, j = 1, 2, \ldots i_t \in \mathbb{R}$ and $\beta_j, j = 1, 2, \cdots \in \mathbb{R}$.

Analogous to online kernel learning for independent task, we can construct the updation rule for the global and task specific model parameters using gradient descent technique whose regularized objective function with least square error loss is given by

$$\hat{J}_{i+1}(f) = \tfrac{1}{2} \parallel \langle g_0, k_{x_{i+1}} \rangle + \langle g_t, k_{x_{i+1}} \rangle - y_{i+1} \parallel^2 + \tfrac{\lambda}{2} \parallel g_0 \parallel^2 + \tfrac{\lambda}{2} \parallel g_t \parallel^2 \tag{11}$$

$\lambda > 0$ is the regularization parameter.

The model function f_t for the task t at time $i+1$ can be found by minimizing the objective function (11). We used stochastic gradient descent to find the minimum.

Differentiating (11) w.r.t g_0 and g_t respectively, we get

$$\nabla \hat{J}_{i+1}(f)/_{g_0 = g_0^i} = [\langle g_0, k_{x_{i+1}} \rangle + \langle g_t, k_{x_{i+1}} \rangle - y_{i+1}] k_{x_{i+1}} + \lambda g_0 \tag{12}$$

$$\nabla \hat{J}_{i+1}(f)/_{g_t = g_t^i} = [\langle g_0, k_{x_{i+1}} \rangle + \langle g_t, k_{x_{i+1}} \rangle - y_{i+1}] k_{x_{i+1}} + \lambda g_t \tag{13}$$

From (5), (12) and (13) the update rule for the global and task specific model can be obtained as,

$$g_0^{i+1} = g_0 - \eta\big[\big(\langle g_0, k_{x_{i+1}} \rangle + \langle g_t, k_{x_{i+1}} \rangle - y_{i+1}\big) k_{x_{i+1}} + \lambda g_0\big] \tag{14}$$

$$g_t^{i+1} = g_t - \eta\big[\big(\langle g_0, k_{x_{i+1}} \rangle + \langle g_t, k_{x_{i+1}} \rangle - y_{i+1}\big) k_{x_{i+1}} + \lambda g_t\big] \tag{15}$$

Using (10), (14) and (15), the update rule for global and task specific coefficients, β and α respectively can be obtained as follows:

$$\begin{aligned}\beta_{i+1} &:= -\eta(g_0(x_{i+1}) + g_t(x_{i+1}) - y_{i+1})\\ \beta_j &:= (1-\eta\lambda)\beta_j,\ j \le i\end{aligned} \tag{16}$$

$$\begin{aligned}\alpha_{i_t+1} &:= -\eta(g_0(x_{i+1}) + g_t(x_{i+1}) - y_{i+1})\\ \alpha_j &:= (1-\eta\lambda)\alpha_j,\ j \le i_t\end{aligned} \tag{17}$$

where,

$$g_0(x_{i+1}) = \sum_{j=1}^{i} \beta_j k_{x_j}(x_{i+1})$$
$$g_t(x_{i+1}) = \sum_{j=1}^{i_t} \alpha_j k_{x_j}(x_{i+1})$$

The global parameter is updated for each data arrival but the task specific parameter is updated only when the data generated from the particular task arrives.

5 Experiments

The algorithm was applied on synthetic data as well as real world data for classification and regression problems. For all the examples, the RKHS space generated by Gaussian kernel $k(x,x') = exp\left(-\frac{\|x-x'\|^2}{2\sigma^2}\right)$ was taken as the hypothesis space. The hyper parameters, kernel width σ and regularization parameter λ were determined using cross validation.

The performance of the algorithms were assessed using cross validation techniques. The performance measure used for regression was mean-squared error (MSE). MSE is given by

$$MSE = \frac{1}{N_t}\sum_{i=1}^{N_t}(f(x_i) - y_i)^2 \tag{18}$$

where N_t is the number of testing points.

For classification, the performance measure used was accuracy where

$$Accuracy = \frac{TP + TN}{N_t} \tag{19}$$

where *TP* is the number of true positives and *TN* is the number of true negatives.

The norm difference is calculated using

$$\| f - g \|^2 = \| f \|^2 + \| g \|^2 - 2\langle f, g \rangle \tag{20}$$

where $f = \sum_i a_i k_{x_i}$ and $g = \sum_i b_i k_{x_i}$.

5.1 Regression Analysis on Synthetic DataSet

Figure 1 compares the MSE obtained from the Online Kernel learning model (OKL) with that of the proposed Kernel Online Multi-Task Regression (KOMTR) for the synthetic dataset generated from normal distribution with certain mean and variance. The graph is plotted between MSE and the training size for the three synthetic datasets. Figure 2 shows the comparison between OKL and OMTR for the three synthetic datasets generated using uniform distribution with dimension n = 30 i.e. $x_i = uniform(0, 1)$, $i = 1, \cdots, 30$. The choice of parameters through cross-validation are RBF $\sigma = 1$, $\lambda = 0.1$. It is seen that the MSE of OMTR is relatively small when compared to online independent learning learning.

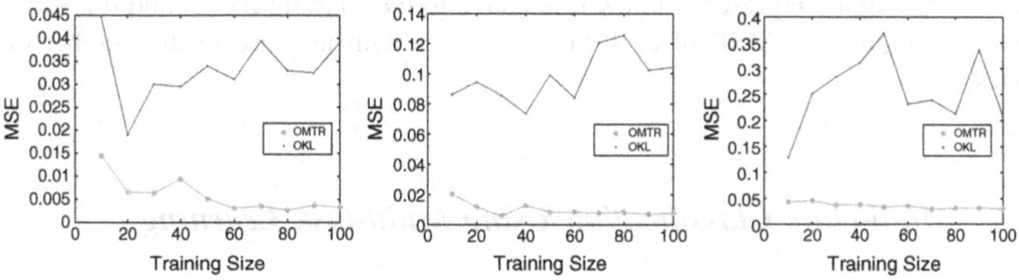

Fig. 1 Synthetic dataset-normal distribution: MSE comparison between OKL and OMTR

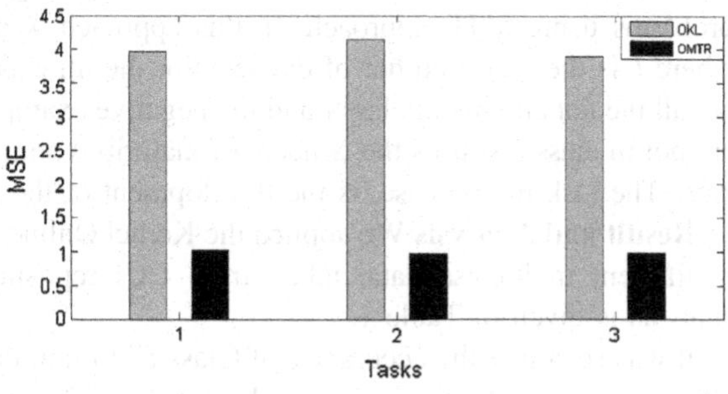

Fig. 2 Synthetic dataset-uniform distribution: MSE comparison between OKL and OMTR

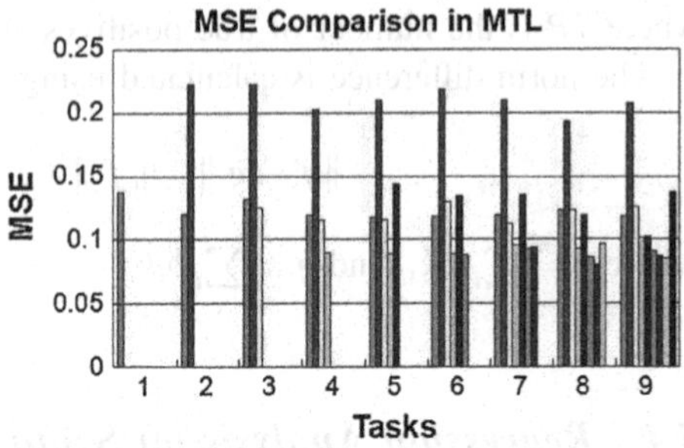

Fig. 3 Comparison of MSE of school dataset with varying number of tasks

5.2 *Regression Analysis on Real Time DataSet*

The algorithm is applied to the dataset obtained from the Inner London Education Authority (http://www.public.asu.edu/~jye02/Software/MALSAR). This data consists of examination of 15,362 students from 139 secondary schools. The goal is to predict the exam scores of the students based on the following inputs: year of exam, gender, verbal reasoning band, ethnic group, gender of the school and school denomination. The total number of inputs for each of the students in each school is 27. Each school dataset is considered as one task and hence we have 139 tasks.

Figure 3 is plotted between varying number of tasks and mean square error. This is to study the variation of MSE with increasing number of tasks. Each distinct color in the figure represents a new task added to the learning model and it is clear from the figure that MSE of each task decreases with increase in the number of tasks.

5.3 *Multi-class Classification Using Multi-task Learning*

We modeled the multi-class classification problem as multi-task learning problem. We applied the cost function (11) on multi-class classification problems. The use of regularized cost function in classification problems has been studied in [16].

We adopted the One-Vs-All method for analyzing the multi-class classification problems using MTL approach. In this approach, k different classifiers are built, where k is the total number of classes. For the ith classifier, the positive examples are all the data points in class i and the negative examples are the data points which are not in class i. It uses the concept of majority voting for classifying an unknown data. The task in this case, is the development of the classifiers $i, i = 1, 2, \ldots k$.

Result and Analysis We applied the Kernel Online Multi-Task Classification on 6 different multiclass data, taken from UCI repository. The description of the analysis is given in Table 1.

It was seen that the accuracies of Glass and Ecoli dataset increased significantly when compared to the single task learning. The reason for the same can be

Table 1 Comparison of accuracies: single task learning (STL) and online multi task learning (OMTL)

Dataset	STL	OMTL
Iris	95.8518	95.444
Wine	98.1746	96.7785
Glass	75	78.254
Ecoli	78.1095	81.9403
Dermatology	97.1233	96.5
Image segmentation	80.9524	79

Table 2 Norm difference: Ecoli data set

Solution	f1	f2	f3	f4	f5
f1	0	196.5676	178.1972	178.0624	181.8905
f2		0	175.1147	178.9596	97.9565
f3			0	4.9255	81.4696
f4				0	85.7386
f5					0

Table 3 Norm difference: Iris data set

Solution	f1	f2	f3
f1	0	1214	6835
f2		0	8021
f3			0

described in terms of norm difference between the solution vectors of each of the classifiers. The lesser the norm distance between solution vectors of the tasks, more closely the tasks are related. Table 2 gives the norm difference between the solution vectors of the 5 different tasks of Ecoli data set while Table 3 gives that of 3 different tasks of Iris data set.

Comparing the results of Tables 2 and 3, it is clear that the distance between the solution vectors of Ecoli DataSet is less than that of Iris DataSet and hence the tasks of Ecoli data set are more related than Iris. Due to the closely related classes in Ecoli and Glass data, multi tasking classification showed an improved performance when compared with that of single task learning.

6 Conclusion

A kernel approach to online multi-tasking has been described in this paper. We compared the performance of the developed algorithm with that of single task cases in the case of regression and the results were promising. The developed frame work was applied on multi-class classification problems also. Our analysis showed that the multi-tasking concepts is a good approach for solving multi-class problems, if the classes are closely related. The work can be extended to weighted multi-tasking, where the information flow for studying the function for a specific task should be more from the tasks that is closely related to it than other tasks.

References

1. Cesa-Bianchi, N., Conconi, A., Gentile, C.: On the generalization ability of on-line learning algorithms. IEEE Trans. Inform. Theory **50**(9), 2050–2057 (2004)
2. Crammer, K., Dekel, O., Keshet, J., Shalev-Shwartz, S., Singer, Y.: Online passive-aggressive algorithms. J. Mach. Learn. Res. **7**, 551–585 (2006)
3. Crammer, K., Singer, Y.: Ultraconservative online algorithms for multiclass problems. J. Mach. Learn. Res. **3**, 951–991 (2003)
4. Yang, L., Jin, R., Ye, J.: Online learning by ellipsoid method. In: Proceedings of the 26th Annual International Conference on Machine Learning, ICML-09, pp. 145. Montreal, QC, Canada (2009)
5. Zhao, P., Hoi, S.C.H., Jin, R.: DUOL: a double updating approach for online learning. In: Advances in Neural Information Processing Systems (Proceedings of NIPS-2009), pp. 2259–2267
6. Li, Y., Long, P.M.: The relaxed online maximum margin algorithm. In: Advances in Neural Information Processing Systems (Proceedings of NIPS-1999), pp. 498–504
7. Rosenblatt, F.: The perceptron: a probabilistic model for information storage and organization in the brain. Psychol. Rev. **65**(6), 386–408 (1958)
8. Evgeniou, T., Pontil, M.: Regularized multitask learning. In: KDD-04, Aug. 2004
9. Li, G., Hoi, S.C.H., Chang, K., Liu, W., Jain, R.: Collaborative online multitask learning. IEEE Trans. Knowl. Data Eng. **26**(8) (2014)
10. Pong, T.K., Tseng, P., Ji, S., Ye, J.: Trace norm regularization: reformulations, algorithms, and multi-task learning. SIAM J. Optim. **20**(6), 3465–3489 (2010)
11. Tibshirani, R.: Regression shrinkage and selection via the lasso. J. Roy. Stat. Soc.: Ser. B, **58** (1), 267–288 (1996)
12. Argyriou, A., Evgeniou, T., Pontil, M.: Convex multi-task feature learning. Mach. Learn. **73** (3), 243–272 (2008)
13. Jalali, A., Sanghavi, S., Ruan, C., Ravikumar, P.K.: A dirty model for multi-task learning. Adv. Neural Inf. Process. Syst. (NIPS 2010) **23** (2010)
14. Gong, P., Ye, J., Zhang, C.: Robust multi-task feature learning. In: International Conference on Knowledge Data and Data Mining (KDD'12), Aug. 12–16. Beijing, China (2012)
15. Kivinen, J., Smola, A.J., Williamson, R.C.: Online learning with kernels. IEEE Trans. Signal Process. **52**(8) (2004)
16. Zhang, P., Peng, J.: SVM vs regularized least squares classification. In: Proceedings of the 17th International Conference on Pattern Recognition (ICPR-04), pp. 1051–4651 (2004)

Performance Evaluation of Sentiment Classification Using Query Strategies in a Pool Based Active Learning Scenario

K. Lakshmi Devi, P. Subathra and P.N. Kumar

Abstract In order to perform Sentiment Classification in scenarios where there is availability of huge amounts of unlabelled data (as in Tweets and other big data applications), human annotators are required to label the data, which is very expensive and time consuming. This aspect is resolved by adopting the Active Learning approach to create labelled data from the available unlabelled data by actively choosing the most appropriate or most informative instances in a greedy manner, and then submitting to human annotator for annotation. Active learning (AL) thus reduces the time, cost and effort to label huge amount of unlabelled data. The AL provides improved performance over passive learning by reducing the amount of data to be used for learning; producing higher quality labelled data; reducing the running time of the classification process; and improving the predictive accuracy. Different Query Strategies have been proposed for choosing the most informative instances out of the unlabelled data. In this work, we have performed a comparative performance evaluation of Sentiment Classification in a Pool based Active Learning scenario adopting the query strategies—Entropy Sampling Query Strategy in Uncertainty Sampling, Kullback-Leibler divergence and Vote Entropy in Query By Committee using the evaluation metrics Accuracy, Weighted Precision, Weighted Recall, Weighted F-measure, Root Mean Square Error, Weighted True Positive Rate and Weighted False Positive Rate. We have also calculated different time measures in an Active Learning process viz. Accumulative Iteration time, Iteration time, Training time, Instances selection time and Test time. The empirical results reveal that Uncertainty Sampling query strategy showed better overall performance than Query By Committee in the Sentiment Classification of movie reviews dataset.

K. Lakshmi Devi (✉) · P. Subathra · P.N. Kumar
Department of CSE, Amrita School of Engineering, Amrita Vishwa Vidyapeetham, Coimbatore 641112, Tamil Nadu, India
e-mail: laksdevi115@gmail.com

P. Subathra
e-mail: p_subathra@cb.amrita.edu

P.N. Kumar
e-mail: pn_kumar@cb.amrita.edu

M. Senthilkumar et al. (eds.), *Computational Intelligence, Cyber Security and Computational Models*, Advances in Intelligent Systems and Computing 412, DOI 10.1007/978-981-10-0251-9_8

Keywords Sentiment classification · Active learning · Uncertainty sampling · Query by committee · Kullback-Leibler divergence · Vote entropy

1 Introduction

Sentiment Classification (SC) refers to the task of determining the semantic orientation or polarity of any given text towards a particular aspect or topic. SC has been extensively studied and explored in varied fields like NLP [1, 2], machine learning and data mining [3], political debates [4], news articles [5] etc. Many supervised and unsupervised approaches which have been proposed to perform SC are mainly focussed on learning models that are trained using labelled data. However, in scenarios where there is availability of huge amounts of unlabelled data, it is a tedious and expensive job to label whole amount of data. For instance, it is easy to collect millions of web pages without much cost and effort, but human annotators are required to label all these web pages which are very expensive and time consuming. Here comes the role of Active Learning (AL), a supervised machine leaning approach which is used to create labelled data from the available unlabelled data by actively choosing the most appropriate or most informative instances and then submitting to human annotator (also called as oracle) for annotation.

AL thus reduces the time, cost and effort to label huge amount of unlabelled data. AL plays a vital role in reducing the annotation cost by requesting labels for a few carefully chosen instances thereby producing an accurate classifier.

This finds applications in real time situations where it is difficult to get labelled data. The input to the AL process is a set of available labelled instances and a huge set of unlabelled instances. The major goal of AL process is to produce a classifier with maximum possible accuracy by avoiding the need for supplying labelled data than necessary. The learning phase asks advice from the oracle only when the training utility of the outcome of a query is high, thereby attempting to keep human effort as minimum as possible [6]. There are three scenarios in which an Active Learner can pose queries:

- Pool-based sampling
- Membership query synthesis
- Stream-based selective sampling

Different query strategies are used in each of these three settings in order to choose the instances which are most informative. In this work, we have performed a comparative analysis of the performance of query selection frameworks viz. Uncertainty Sampling (US) versus Query By Committee (QBC) in a Pool-based active AL process. Query strategies are required in an AL process to select most informative instances to be labelled by the oracle. The query strategies that are investigated in this work are Entropy Sampling Query Strategy in US, Kullback-Leibler (KL) divergence and Vote Entropy in QBC.

2 Related Work

SC using machine learning (ML) are broadly classified into supervised and unsupervised SC which differs according to the presence or absence of labelled data respectively [1, 7]. The supervised approaches are considered to provide better accuracy than unsupervised approaches because of their better predictive power and hence are widely being adopted [8]. However in real time situations, it is difficult and expensive to obtain large set of labelled data in contrast to unlabelled data which will be plentiful and easy to obtain and synthesize. AL comes into aid in such situations where we want to handle large unlabelled datasets. A system that incorporates AL is designed to provide better performance with less training where in the learner actively chooses the instances using which it learns [9].

The first and foremost AL scenario which is to be examined is learning using Membership queries [10]. In this particular setting, the learner may ask for labels for any unlabelled example present in the input space which also includes the queries which are generated by the learner de novo instead of sampling from underlying distribution [11]. The major limitation associated with this setting is that the Active Learner may end up querying random instances devoid of meaning [11]. In order to address the limitations of the above approach, two schemes viz. Stream based selective sampling and Pool-based selective sampling have been introduced. Selective sampling works based on the assumption that collecting and synthesizing an unlabelled instance incurs no cost, hence can be initially sampled from the actual underlying distribution and then subsequently the learner decides whether or not to ask for its label. In Stream based selective sampling, every unlabelled instance is drawn one at a time from a stream of instances and then the learner chooses to query or discard it [9]. The learner's decision to query or discard is based on different approaches like defining some utility measures [11, 12], computing region of uncertainty [13], defining version space [14] etc. This setting is mainly adopted in mobile and embedded devices where the memory and processing power are limited. Pool-based selective sampling works well in scenarios where huge amount of unlabelled data can be collected at once and stored in a pool. Queries are chosen from the pool in a greedy manner and only those instances which satisfy the utility criteria are selected. As against Stream based selective sampling which makes query decision for each instance individually, Pool-based selective sampling initially evaluates the whole data and then ranks instances according to the utility measure and then selects the best query. This setting is mainly adopted in text classification, information extraction etc.

3 Proposed System

The proposed system focuses on Pool-based sampling scenario where in there exists a small set of labelled data L which are annotated by an oracle and a sufficiently large pool of unlabelled data U. The selection strategy assigns each instance of U, a value that signifies how important the labelling of that instance is to an AL process and then ranks the instances in decreasing order of *informativeness*. The top ones are submitted to the oracle for annotation according to the batch size *b*. The data points are chosen from the pool in a greedy manner in such a way that only those data points are selected which have high *informativeness* measure. After annotation, the labelled instances are then added to the already available labelled set L and subsequently the *informativeness* value corresponding to each unlabeled instance is updated. The process iterates until a stopping criteria is met (such as, when instances lack in sufficient *informativeness*)—thereafter the oracle stops labelling. The resulting labelled instances can be used to train the classifier. The proposed system architecture of the Sentiment Classification using US and QBC AL process is given in Fig. 1.

3.1 Initial Training Set Construction

As Pool based scenario initially starts with a small set of labelled set, in order to choose this set, many techniques are applied which seed the AL process. One of the techniques adopts random sampling which is the most common and widely used approach [15]. Clustering techniques like K-means which assumes that most representative instances are likely to be present at the centres of clusters [16], k-medoid [17], special clustering [18] etc. can be used for building the initial set of labelled examples.

3.2 Query Strategy Frameworks

In this work, we mainly focus on the query strategy framework using an AL process. The query strategy frameworks that are investigated in this work are:

- Uncertainty Sampling(US)
- Query-By-Committee(QBC)

3.3 Uncertainty Sampling(US)

Uncertainty Sampling (US) is the simplest and most widely used active learning query strategy framework in text classification [19]. In this, the unlabelled instances which the Active Learner finds least certain to label are queried iteratively. This sampling technique relies on the probability of a particular data point belonging to one of the labels. The probability of an instance represents the informativeness measure of that instance. The major aim of this sampling scheme is how to calculate

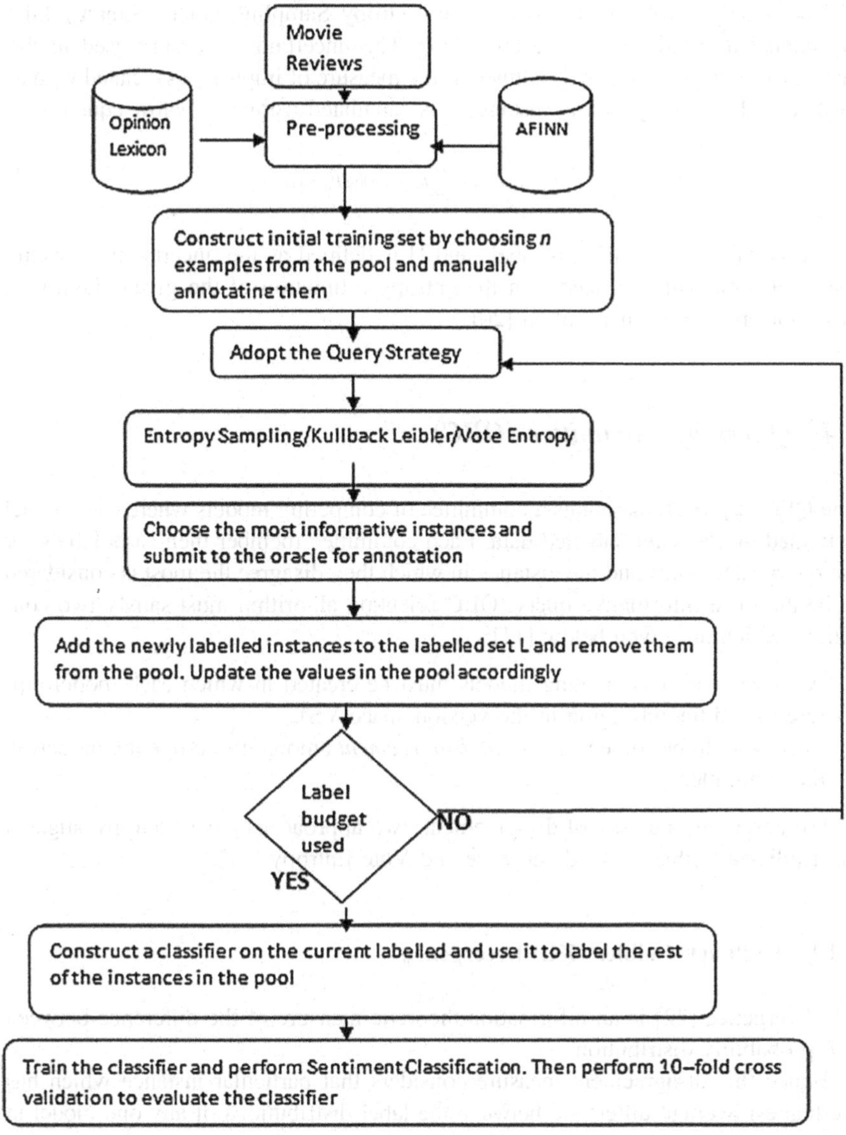

Fig. 1 Sentiment classification using active learning

the uncertainty associated with any unlabelled instance; higher the uncertainty of an instance, higher the informativeness associated with it [20].

A classifier (typically Support Vector Machine (SVM) or Naive Bayes (NB)) is fed with instances labelled so far by the oracle and subsequently this classifier is employed to classify unlabelled instances; the instances which are least certain are chosen for submitting to oracle for annotation.

The widely employed US strategy is Entropy Sampling Query Strategy (also considered as traditional or baseline US). The uncertainty measure used in this strategy is entropy. Entropy is defined as the measure of impurity associated with an instance. The entropy of any instance x is calculated using the below equation:

$$x^*_H = argmaz_z - \sum_i P_\theta(y_i|x) \log P_\theta(y_i|x) \quad (1)$$

where y_i ranges over all class labels and H is defined as the uncertainty measurement function which is based on the entropy estimation of the given classifier's posterior probability distribution [20].

3.4 *Query by Committee (QBC)*

The QBC approach maintains a committee of competing models where each model is trained on the same labelled data. Each committee member then casts labels for the query candidates and the instance in which they disagree the most is considered to be the most informative query. QBC selection algorithm must satisfy two conditions which are given below [11]:

- A committee of competing models must be created in which each model represents a different region in the version space(VS).
- There should be some *measure of disagreement* among the competing models in the committee.

For enforcing measure of disagreement, two approaches have been investigated viz Kullback-Leibler (KL) divergence and Vote Entropy [21].

3.4.1 Kullback-Leibler (KL) Divergence

KL divergence [22] is an information-theoretic measure of the difference between two probability distributions.

Hence this disagreement measure considers that particular instance which has the highest average difference between the label distributions of any one model in the committee and the consensus [22].

$$X^*_{KL} = argmax_x \frac{1}{C} \sum_{c=1}^{C} D(P_\theta \| P_c) \quad (2)$$

where D $(P_\theta \| P_c) = \sum_i P_\theta(y_i|x) \log \frac{P_\theta(y_i|x)}{P_c(y_i|x)}$

Given Θ model in the committee, C as the committee size, then $P_c(y_i|x)$ is the "consensus" probability that y_i is the correct label.

$$P_c(y_i|x) = \frac{1}{c}\sum_{c=1}^{c} P_\theta(y_i|x) \tag{3}$$

3.4.2 Vote Entropy

The Vote entropy query strategy is defined by the equation below:

$$x^*_{VE} = argmax_x - \sum_i \frac{V(y_i)}{C}\log\frac{V(y_i)}{C} \tag{4}$$

where y_i ranges over all labels and V (y_i) is given as the number of votes that a particular label obtains from the predictions of the committee members where C is given by the committee size.

3.5 Supervised Active Learners

In this work, we have used NB as the base classifier for Entropy Query Sampling Strategy, which is simple, efficient and widely adopted in text classification. The main advantage of using NB in active learning is that in order to collect the term frequencies (TF) information, only a single pass is required. For KL Divergence and Vote Entropy Query Strategies, we employed a committee of NB and SVM classifiers.

4 Experiments

The dataset used in this work is sentence polarity dataset v1.0 [23]. The dataset consists of 5331 positive movie reviews and 5331 negative movie reviews taken from popular movie review website "rotten tomatoes". The data is cleaned by tokenising, removing unwanted punctuations, stop words etc. and subsequently performing lemmatisation and finally N-grams are built which are taken as features. A lexical resource called as AFINN is used which is a list of English words with their corresponding valence score [24]. The polarities of the words in the dataset were retrieved from AFINN and scores were calculated; frequency of positive, negative, very positive and very negative words were also taken as features using positive and negative opinion lexicons. We have used Java Class Library for Active Learning (JCLAL) for implementing the Active Learning system. For the initial training set construction, a 10-fold cross validation evaluation was adopted in which 5 % of the dataset was chosen to build the labelled set and the remaining instances constituted the unlabeled set.

5 Results and Analysis

The query strategies used in this work are Entropy Query Sampling Strategy which is an US technique, KL divergence and Vote Entropy Query Strategies which are QBC techniques. These query strategies were employed by the AL process which iteratively selected the most informative instances out of unlabelled set. The labelling of these chosen instances was done by the oracle and these labelled instances were added to the labelled set and unlabelled set was updated accordingly. The base classifier used for Entropy Query Sampling Strategy was NB and the QBC used a committee of NB and SVM.

Table 1 shows the performance of US versus QBC using the evaluation metrics which match closely. Results demonstrate that both US and QBC techniques have produced comparable results in terms of Accuracy, Weighted Precision, Weighted Recall, Weighted F-measure, Root Mean Square Error(RMSE), Weighted True positive Rate and Weighted False Positive Rate.

Figure 2 demonstrates the different metrics where in US and QBC differ. Various timings depicted on Y axis is measured in milliseconds. The metrics that we have investigated here are given below:

- *Accumulative Iteration Time* The results reveal that the QBC sampling approach requires less Accumulative Iteration Time than US. In addition, among the QBC approaches, Vote Entropy strategy requires less Accumulative Iteration time than KL divergence scheme.
- *Iteration Time* The time required to complete one iteration is less for US than QBC strategy.

Table 1 Performance of selected metrics—uncertainty sampling (US) versus query by committee (QBC)

Active learning			
	Entropy sampling	Kullback Leibler	Vote entropy
Correctly classified instances	63.73414231	63.62382791	63.62382791
Weighted precision	0.638458655	0.637518021	0.637518021
Root mean squared error	0.487338177	0.511894014	0.511394014
Weighted true positive rate	0.637341423	0.636238279	0.636238279
Weighted F measure	0.636669213	0.635456036	0.635456036
Weighted recall	0.637341423	0.636238279	0.636238279
Weighted false positive rate	0.362515444	0.363607642	0.363607642

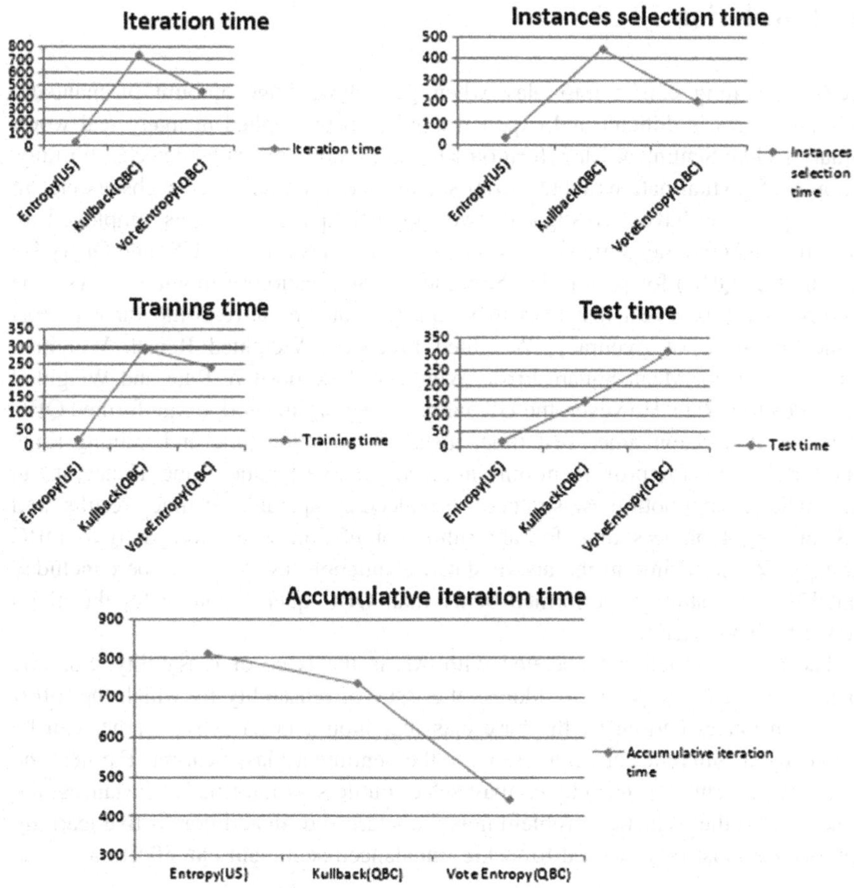

Fig. 2 Salient metrics: performance of US versus QBC (Y axis-time in milli secs)

- *Training Time* This measure indicates the total amount of time required to complete training of the classifier. The results reveal that the US outperformed QBC in terms of training time.
- *Instances Selection Time* As the QBC techniques use a committee of classifiers, its selection time of instances is much more greater than US.
- *Test Time* The test time is also greater for QBC techniques than US strategy.

Our analysis is that the US query strategy framework outperformed QBC in terms of Iteration time, Test time, Instances selection time and training time. The QBC showed improvement only in Accumulative Iteration time. Both these strategies produced comparable accuracy results though. It was therefore concluded that US query strategy outperformed the QBC query strategy for the given movie reviews dataset.

6 Conclusion

Active Learning comes into play when generating huge amount of manually annotated data is difficult and expensive. AL can be applied in many real world situations like Sentiment Classification where in sentiment can be detected for huge amount of textual data extracted from social media, e-commerce websites etc. In this paper, we have investigated two popular query strategies employed in Pool-based Active Learning scenario viz. Uncertainty Sampling (US) and Query By Committee (QBC) for performing Sentiment Classification of movie reviews. The empirical results reveal that both these query strategies have comparable performance in terms of Accuracy, Weighted Precision, Weighted Recall, Weighted F-measure, Root Mean Square Error, Weighted True positive Rate and Weighted False Positive Rate. However, the US query strategy framework outperformed QBC in terms of Iteration time, Test time, Instances selection time and training time. The QBC showed improvement only in Accumulative Iteration time. Hence, it can be concluded that both these strategies produced comparable accuracy results, and US strategy took less time for algorithm completion when compared to QBC strategy. While taking all the above stated evaluation results, it can be concluded that US query strategy performed better than QBC query strategy for the given movie reviews dataset.

The major limitation associated with AL is the issue of reusability. Suitable methods must be proposed to address the issue of reusability by which the informative instances chosen by the base classifier inbuilt in an active learner can be reused by a different classifier used for the sentiment classification. Further, the Uncertainty Sampling techniques may select outliers as informative instances and hence this outlier selection problem must be addressed. In addition, active learning approaches must be proposed to tackle imbalanced sentiment classification.

References

1. Pang, B., Lee, L., Vaithyanathan, S.: Thumbs up?: sentiment classification using machine learning techniques. In: Proceedings of the ACL-02 Conference on Empirical Methods in Natural Language processing-Volume 10. Association for Computational Linguistics, (2002)
2. Turney, P.D.: Thumbs up or thumbs down?: semantic orientation applied to unsupervised classification of reviews. In: Proceedings of the 40th Annual Meeting on Association for Computational Linguistics. Association for Computational Linguistics, 2002
3. Pang, B., Lee, L.: Opinion mining and sentiment analysis. Found. Trends Inf. Retrieval **2**(1–2):1–135 (2008)
4. Maks, Isa, Vossen, Piek: A lexicon model for deep sentiment analysis and opinion mining applications. Decis. Support Syst. **53**(4), 680–688 (2012)
5. Xu, Tao, Peng, Qinke, Cheng, Yinzhao: Identifying the semantic orientation of terms using S-HAL for sentiment analysis. Knowl.-Based Syst. **35**, 279–289 (2012)
6. Olsson, F.: A literature survey of active machine learning in the context of natural language processing. (2009)

7. Rothfels, J., Tibshirani, J.: Unsupervised sentiment classification of English movie reviews using automatic selection of positive and negative sentiment items. CS224N-Final Project (2010)
8. Fersini, E., Messina, E., Pozzi, F.A.: Sentiment analysis: Bayesian Ensemble Learning. Decis. Support Syst. **68**, 26–38 (2014)
9. Settles, B.: Active learning. Synth. Lect. Artif. Intell. Mach. Learn. **6**(1), 1–114 (2012)
10. Angluin, D.: Queries and concept learning. Mach. Learn. **2**(4), 319–342 (1988)
11. Settles, B.: Active learning literature survey. University of Wisconsin, Madison **52**(55–66):11 (2010)
12. Dagan, Engelson, S.: Committee-based sampling for training probabilistic classifiers. In: Proceedings of the International Conference on Machine Learning (ICML), pp. 150–157. Morgan Kaufmann (1995) (Cited on page(s) 8, 28)
13. Atlas, L.C., Ladner, R.: Improving generalization with active learning. Mach. Learn. **15**(2), 201–221 (1994). doi:10.1007/BF00993277. Cited on page(s) 7, 8, 24, 58, 62
14. Mitchell.: Generalization as search. Artif. Intell. **18**:203–226 (1982). doi:10.1016/0004-3702(82)90040-6 (Cited on page(s) 8, 21)
15. Lee, M.S., Rhee, J.K., Kim, B.H., Zhang, B.T.: AESNB: Active example selection with naive Bayes classifier for learning from imbalanced biomedical data. In: Proceedings of the 2009 Ninth IEEE International Conference on Bioinformatics and Bioengineering, **38**: 15–21 (2009)
16. Kang, J., Ryu, K., Kwon, H.: Using cluster-based sampling to select initial training set for active learning in text classification. Adv. Knowl. Discov. Data Min. **3056**: 384–388. 39, 96, 135, 139, 200 (2004)
17. Nguyen, H.T., Smeulders, A.: Active learning using pre-clustering. In: Proceedings of the 21st International Conference on Machine Learning, 623–630, 39, 52, 118 (2004)
18. Dasgupta, S., Ng, V.: Mine the easy, classify the hard: A semisupervised approach to automatic sentiment classification. In: Proceedings of the Joint Conference of the 47th Annual Meeting of the ACL and the 4th International Joint Conference on Natural Language, **2**(701–709), Suntec, Singapore. **39**:45 (2009)
19. Lewis, D.D., Gale, W.A.: A sequential algorithm for training text classifiers. In Proceedings of the 17th annual International ACM SIGIR conference on Research and Development in Information Retrieval, Springer, New York. **36**(3–12): 40 (1994)
20. Zhu, J., et al.: Active learning with sampling by uncertainty and density for word sense disambiguation and text classification. In: Proceedings of the 22nd International Conference on Computational Linguistics-Volume 1. Association for Computational Linguistics (2008)
21. Dagan, I., Engelson, S.P.: Committee-based sampling for training probabilistic classifiers. In: Proceedings of 1995 International Conference on Machine Learning, **42**:150–157 (1995)
22. Kullback, Leibler, R.A.: On information and sufficiency. Ann. Math. Stat. **22**:79–86 (1951)
23. https://www.cs.cornell.edu/people/pabo/movie-review-data
24. http://neuro.imm.dtu.dk/wiki/AFINN

An Enhanced Image Watermarking Scheme Using Blocks with Homogeneous Feature Distribution

Kavitha Chandranbabu and Devaraj Ponnaian

Abstract This paper proposes a gray scale image watermarking scheme based on Discrete Cosine Transform (DCT). In this scheme, the host image is divided into non-overlapping blocks. The pixel values within the blocks are permuted and DCT is applied for each block separately. The permutation makes the distribution of the DCT coefficient uniform within the block and hence embedding watermark in any region of the block produces similar result. Since, the embedding is done in blocks, multiple embedding is also possible. Experimental results also show that the performance of the scheme is good against various image processing operations and cropping attack.

Keywords Discrete cosine transform · Block permutation · Frequency domain · Watermark

1 Introduction

Nowadays, digital information is largely used due to the ease of storage and transmission. Digitally stored information such as images, audio and video can be copied, edited and shared easily which makes it difficult to identify the ownership of the content. Digital watermarking techniques are developed to address the copyright issue for digital information. In digital image watermarking, an image that represents the owner identity is embedded in the host image. Embedding can be done either in spatial domain or on frequency domain of the host image. The watermark embedding in the spatial domain is considered to be weak against

K. Chandranbabu (✉) · D. Ponnaian
Department of Mathematics, College of Engineering, Anna University Chennai, Guindy, Chennai 600025, India
e-mail: kavistha@gmail.com

D. Ponnaian
e-mail: devaraj@annauniv.edu

M. Senthilkumar et al. (eds.), *Computational Intelligence, Cyber Security and Computational Models*, Advances in Intelligent Systems and Computing 412, DOI 10.1007/978-981-10-0251-9_9

various image processing operations and hence, most of the algorithms are designed to embed the watermark in the frequency domain which is obtained by using transforms such as discrete cosine transforms (DCT) [1–12], discrete Fourier transforms (DFT), and discrete wavelet transforms (DWT). Frequency domain embedding identifies the region of embedding based on the sensitivity of Human Visual System (HVS). The DCT reserves most of the signal information in the low frequency components. Hence, algorithms [9–12] using DCT, embeds the watermark in the middle frequency band to overcome the tradeoff between the robustness and imperceptibility.

In [9], the watermarking algorithm is designed for image content authentication using DCT where the watermark generated from the original image is embedded in middle frequency range. In [10], the DCT coefficients are modified to embed watermark based on the inter-block DCT coefficient correlation. The modification parameter is computed using the DC coefficient and some low frequency coefficient. The scheme proposed in [11], determines the DCT of each 8×8 block and computes the complexity of each block by counting the number of nonzero entries in each block. Based on the complexity the blocks are selected and watermark is embedded in the low frequency band of the selected blocks. An algorithm based on Chinese remainder theorem and DCT was proposed in [13]. In [12], sub sampling technique along with Just Noticeable Distortion (JND) function is used for watermarking where the watermark is embedded in middle frequency domain. Some schemes [14, 15] also use the DC coefficients for marking the watermark to improve the robustness.

In [16], using sub sampling four sub images are formed for which DCT is applied. A permutated binary watermark image is embedded into the DCT coefficients of two of the sub images which are selected according to a random coordinates sequence generated by a secret key. This paper proposes a modified version of the embedding technique discussed in [16] and embeds the watermark into non-overlapping blocks of the host image which have permuted pixel values. When DCT is applied to the permuted blocks, the features within the block are uniformly distributed and hence, the watermark can be embedded in any region of the block. This makes the scheme overcomes the restriction of embedding the watermark in the middle frequency band alone. Moreover, based on the size of watermark the watermark can also be embedded more than once.

2 Block Permutation

Most of the watermarking algorithms use the concept of creating blocks from the host image and embedding watermark in it. Here, along with the concept of creating blocks, block permutation is also done before embedding watermark. The embedding algorithm in this paper divides the host image into non overlapping block, permutes the pixel values within blocks and applies DCT for each block separately. The DCT has the property of concentrating the low frequency

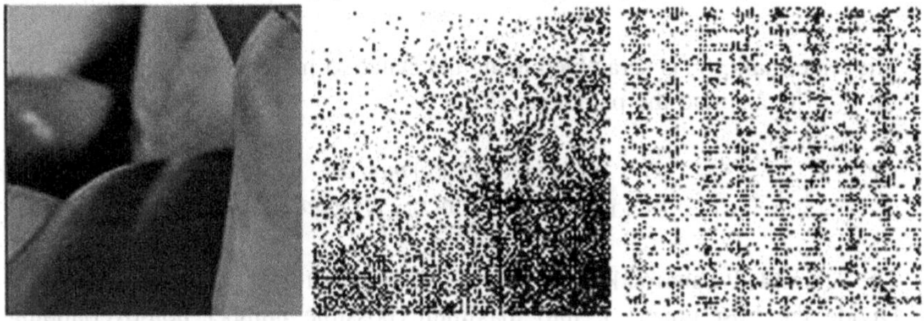

Fig. 1 **a** Original image. **b** DCT of (**a**). **c** DCT of permuted image (**a**)

components in the top left corner which varies diagonally till bottom right corner. The top left most component is the DC coefficient. The remaining components are called AC coefficients. Due to this property of the DCT, embedding the watermark in the low frequency component may result in degradation of the original image whereas the high frequency components may be lost while applying compression functions. Hence, always the middle band serves as the region of embedding. To overcome this restriction, the pixel values within the blocks are permuted and then DCT is applied, which makes the frequency distribution uniform throughout the block. Hence, embedding the watermark in any region has equal chance of being marked in low or high frequency component. Due to this technique the quality of the original image is also maintained up to the desired level as well as the embedding is made robust. The region of embedding is also increased. The permutation technique used in this scheme is first permutating all rows of the block using the function $t = ((p * s) mod\, D) + 1$, where D is the number of rows in the block, p is the prime number, s and t are the actual and destination row locations, respectively. The row permuted image is again permuted columnwise using the same function with different prime number. Figure 1 shows the distribution of the DCT block with and without permutation.

3 Proposed Scheme

Consider a gray scale image, I, of size $H \times W$, where H and W are height and width of the image, respectively. Let the binary watermark, w be of size $H_1 \times W_1$, where H_1 and W_1 represents the height and width of the watermark. Since, the watermark is usually much smaller than the host image, it can be embedded using few blocks of the host image and this embedding algorithm requires the host image to be of size $H \geq 4.H_1$ and $W \geq 4.W_1$, which is sufficient to create blocks.

3.1 Watermark Embedding

1. The watermark, w, is scrambled using the standard map (1) with the secret key x_0, $y_0 \in [0, 2\pi)$ and the parameter $|K| > 0$.

$$\left.\begin{aligned} x_{n+1} &= (x_n + y_n)\, mod\, 2\pi \\ y_{n+1} &= (y_n + K.\sin(x_{n+1}))\, mod\, 2\pi \end{aligned}\right\} \tag{1}$$

2. The scrambled watermark is transformed into a vector, w' of size $H_1 \times W_1$.
3. Divide the host image, I, into maximum possible number non-overlapping blocks of sub images $B_{i,j}$, of size $BD \times BD$ where $BD = \min(2.H_1, 2.W_1)$, $1 \le i \le \lfloor \frac{H}{BD} \rfloor$ and $1 \le j \le \lfloor \frac{W}{BD} \rfloor$. The part of the image that cannot be divided into blocks of the required size are not used for embedding and are retained as such in the watermarked image.
4. The pixels values within the blocks are permuted using the permutation process discussed in Sect. 2 to make block features uniformly distributed.
5. The permuted blocks rearranged to form a rectangular matrix

$$\begin{aligned} Q = [B_{1,1}, B_{2,1}, \ldots .B_{i,j}, B_{i+1,j}, B_{i,j+1}, B_{i+1,j+1} \ldots], \\ i = 1, 3, \ldots, \frac{H}{BD} - 1, j = 1, 3, \ldots, \frac{W}{BD} - 1 \end{aligned} \tag{2}$$

6. In the embedding process, 1/4th of the watermark vector w' is embedded in each block. Hence, for a single embedding of the watermark vector w' four blocks from Q are necessary.
7. A random sequence $Z = (u,\ v)$ with $1 \le u, v \le 4$ of size $H_1 \times W_1 \times \lfloor (size\ of\ Q)/4 \rfloor$ is generated. This sequence is used as a coefficient selector during the embedding process.

The steps involved in embedding the watermark vector w' is as follows:

a. Choose first four blocks from the matrix Q defined in (2).
b. Apply the DCT for each block separately to obtain the four DCT coefficient sets, D_1, D_2, D_3 and D_4.
c. The watermark is embedded in the DCT coefficient set in the zigzag order leaving the DC coefficient and some of the rows diagonally below the DC coefficient. The watermark bits are embedded starting from the diagonally rth row where $r = \lceil (BD/t) \rceil$, $t \in (1, BD - 1)$.
d. For the mth bit of the watermark vector w', i.e. w'_m, there are four DCT coefficients in the sets D_1, D_2, D_3 and D_4 out of which a pair is selected using the coefficient selector, $Z_m = (u, v)$. Within the set D_u and D_v the mth DCT coefficient in the zigzag order starting from the diagonally rth row is used for embedding. Let it be denoted by V_u and V_v.

e. Now, compare V_u and V_v. If the bit $w'_m = 1$ and $V_u < V_v$, then swap u and v in Z_m. Similarly, if the bit $w'_m = 0$ and $V_u > V_v$, then swap u and v in Z_m.
f. Using the swapped Z_m, reselect V_u and V_v and compute the following to embed the watermark bit.

$$V_a = \frac{|V_u| + |V_v|}{2}, \; D = \frac{V_u - V_v}{2}$$

$$\textit{If } \left|\frac{D}{V_a}\right| \leq \beta$$

$$V'_u = V_u + \alpha(2w'_m - 1)|V_a|; \quad V'_v = V_v - \alpha(2w'_m - 1)|V_a|$$

$$\textit{else}$$

$$V'_u = V_u; \quad V'_v = V_v$$

where $\alpha \in [0.05, 0.1)$, $\beta = (.1 - a) \times 10$ and α is the watermark embedding strength and β is the threshold.
g. Apply the inverse DCT to the altered set D_1, D_2, D_3 and D_4 and update the modified blocks into the matrix Q.
h. Select the next four blocks Q and apply step (b)–step (g), until the number of blocks remaining in Q are less than four or all blocks exhausted.
i. The embedded blocks in the rectangular matrix are again replaced to their corresponding positions in the host image.

4 Watermark Detection

Let $\bar{I}$ be the watermarked image from which the watermark is to be detected. From the image $\bar{I}$ the rectangular matrix Q is created as in embedding process. The swapped coefficient selector $Z = (u, v)$ obtained during embedding process is used for watermark detection.

a. Choose first four blocks from the matrix Q formed from $\bar{I}$.
b. Apply the DCT for each block separately to obtain the four DCT coefficient sets, $\bar{D}_1, \bar{D}_2, \bar{D}_3$ and $\bar{D}_4$.
c. The watermark bits are extracted from the DCT coefficient set in the zigzag same as in embedding process. The mth bit of the extraction vector $\bar{w}'$, i.e. $\bar{w}'_m$, is determined using the DCT coefficients V_u and V_v selected using the swapped coefficient selector $Z_m = (u, v)$.

$$\bar{w}'_m = \begin{cases} 1, & if\ V_u \geq V_v, \\ 0, & if\ V_u < V_v. \end{cases}$$

d. The extraction vector is converted into a two dimensional array of size $H_1 \times W_1$, for which inverse permutation is applied using the standard map with same key. The image produced is the watermark $\bar{w}$, that is embedded the set of blocks.
e. The Normalised Correlation (NC) of the extracted watermark is computed using formula (3) given below.

$$NC = \frac{\sum_i \sum_j w(i,j) \cdot \bar{w}(i,j)}{\sum_i \sum_j [w(i,j)]^2} \quad (3)$$

where $w(i,\ j)$ and $\bar{w}(i,j)$ represents the $(i,\ j)$th pixel value in the original watermark and extracted watermark, respectively.
f. Select the next four blocks Q and apply step (b)–step (e), until the number of blocks remaining in Q are less than four or all blocks exhausted.
g. Finally, using $\bar{w}$ and their corresponding NC from each block the extracted watermark w_e is obtained by taking weighted average as given in (4).

$$w_e = \sum_i \frac{NC(i)}{Total} * \bar{w}_i \quad (4)$$

where $NC(i)$ is the normalized correlation of ith extracted watermark, *Total* is the sum of $NC's$ of all extracted watermark and $\bar{w}_i$ is the ith extracted watermark from the blocks.

5 Experimental Results

The proposed scheme is tested using a gray scale image of size 512 × 512 and a binary watermark of size 64 × 64. The performance of the scheme is analysed using two types of measure: peak signal to noise ratio (PSNR) and normalized correlation (NC). PSNR is used to test the quality of the watermarked image against the original image. For the 8 bit gray scale image the PSNR is computed using formula (5):

$$PSNR(dB) = 10.\log_{10}\left(\frac{255^2}{MSE}\right) \quad (5)$$

where $MSE = \frac{1}{D \times D} \sum_i \sum_j [I(i,j) - \bar{I}(i,j)]^2$, I and $\bar{I}$ are the original and watermarked image. Figure 2 shows the original image, marked image and extracted watermark.

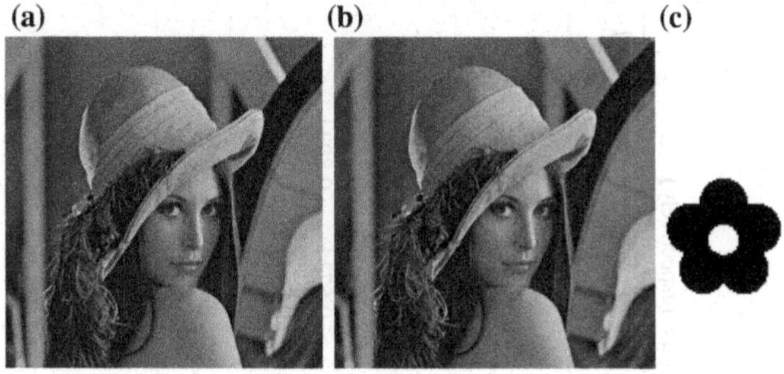

Fig. 2 **a** Original image. **b** Watermarked image with $t = 1.5$ (PSNR = 56.1787 dB). **c** Extracted watermark (NC = 1)

The PSNR value of the marked image is 55.8328 dB for $\alpha = .09$ and $t = 1.5$ which is sufficiently more than the desired value of 30 dB to make watermark imperceptible. The NC gives the correlation between the original and extracted watermark and it is 1 for Fig. 2c. Hence, the extracted watermark is same as the original one.

The comparison of the proposed scheme with and without block permutation is given in Table 1. From the table, it is clear that the detection of the watermark under various attack is equally good irrespective of the region of embedding when embedded in permuted blocks whereas without permutation the performance of the scheme varies when embedded in different region. Moreover, when watermark is embedded without permuting values in the block, the PSNR and NC values in the table shows, the quality of the watermark extracted from middle frequency region is low when compared to low frequency region. Table 1 also shows the comparison of the scheme with [16]. From the NCs of extracted watermark without permutation ($t = 127$) and that of [16], it can be seen that multiple embedding and weighted average improves the performance of the extraction process. From the table, it is also clear that the proposed scheme preforms better than [16] under most of the attacks and in some cases it is equally good.

The algorithm is tested against various attacks such as jpeg compression, cropping, and by applying various noises and filters. The results are compared with [10] in Table 2, which shows that the performance is good against attacks such as cropping and jpeg compression whereas for other attacks the results show it is resistant against these attack. Figure 3 shows the results of the cropping attack where it can be seen that the extracted watermark is same as original watermark. The watermark extracted after distorting the watermarked image under various attacks is given in Fig. 4.

Table 1 Comparison of the proposed scheme with [16] (t = 127 and t = 3 represents the low and middle frequency regions, respectively)

	With permutation				Without permutation				Reference [16]
	t = 127		t = 3		t = 127		t = 3		
PSNR of watermarked image	55.7029 dB		50.0422 dB		52.2830 dB		83.2314 dB		
Attacks	PSNR	NC	PSNR	NC	PSNR	NC	PSNR	NC	NC
JPEG 80	59.1356	0.9182	59.4832	0.9350	68.4566	0.9898	67.0116	0.9844	0.9312
JPEG 40	56.8508	0.8682	57.2301	0.8730	61.2878	0.9495	61.0322	0.9416	0.8278
JPEG 20	55.9678	0.8321	55.8975	0.8478	58.2775	0.8977	57.5986	0.8851	0.7521
Gaussian low pass filter (7 × 7)	67.6268	0.9880	67.0116	0.9898	81.2441	0.9988	81.2441	1	0.7883
Average filter (5 × 5)	55.1375	0.7954	55.4178	0.8261	52.8368	0.6330	52.1272	0.5734	0.8034
Median filter (5 × 5)	57.7708	0.8863	57.8100	0.8911	58.5959	0.8989	58.0739	0.8941	0.8540
Gaussian noise (0, 0.01)	53.1284	0.7443	53.2645	0.7136	54.2849	0.8087	53.9041	0.7924	0.7804
Gaussian noise (0, 0.02)	52.8809	0.6721	52.8211	0.6703	53.6854	0.7768	53.6061	0.7232	0.5797
Salt and Pepper noise (0.01)	54.2327	0.8087	54.1472	0.8039	56.5459	0.8598	55.9937	0.8448	0.6458

Table 2 Comparison of the proposed scheme (t = 127) with [10] under various attacks

Attacks	Proposed scheme	[10]	Attacks	Proposed scheme	[10]
JPEG compression (CR = 7.92)	0.9681	1	Cropping (top-left corner)	1	0.9954
JPEG compression (CR = 12.2)	0.9561	0.9810	Cropping (top-right corner)	0.9994	0.9973
JPEG compression (CR = 15.54)	0.9278	0.9280	Cropping (bottom-left corner)	1	0.9924
JPEG compression (CR = 18.67)	0.9140	0.8847	Cropping (bottom-right corner)	0.9994	0.9981
Salt and pepper noise (0.01)	0.8087	0.8122	Median filtering 3 × 3	0.9368	0.9118
Gaussian noise, mean = 0, variance = 0.001	0.8670	0.8816	Histogram equalization	0.9537	0.9253

Fig. 3 The cropped images and the corresponding extracted watermark

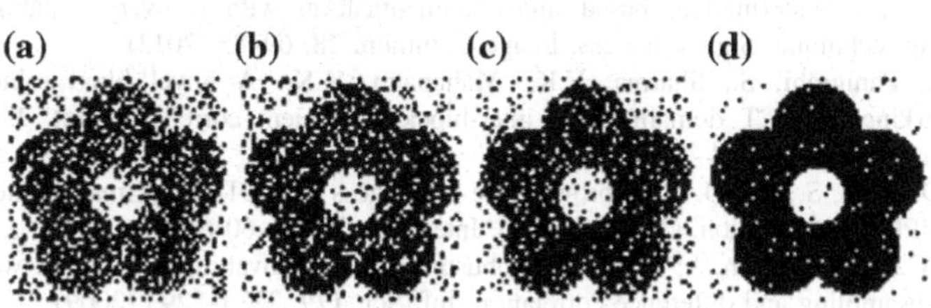

Fig. 4 The extracted watermark after **a** Salt and Pepper noise (0.01). **b** White Gaussian noise, mean = 0, variance = 0.001. **c** Median filter (3 × 3). **d** Histogram equalization

6 Conclusion

This paper introduces the techniques of embedding the watermark in the DCT of the permuted block. A watermarking scheme is proposed based on this technique. The simulation results show that embedding the watermark after permutation performs alike irrespective of the region of embedding. This overcome the restriction in selecting the region of embedding watermark and increases the region of embedding. Moreover, the performance of the scheme against JPEG compression, cropping, Gaussian noise, salt and pepper noise and median filter is equally good when compared with the existing schemes.

References

1. Chu, W.C.: DCT-based imagewatermarking using subsampling. IEEE Trans. Multimedia **5**(1), 34–38 (2003)
2. Eyadat, M., Vasikarla, S.: Performance evaluation of an incorporated DCT block-based watermarking algorithm with human visual system model. Pattern Recogn. Lett. **26**, 1405–1411 (2005)
3. Al-Haj, A.: Combined DWT-DCT digital image watermarking. J. Comput. Sci. **3**(9), 740–746 (2007)
4. Jiansheng, M., Sukang, L., Xiaomei, T.: A digital watermarking algorithm based on DCT and DWT. In: Proceedings of the International Symposium on Web Information Systems and Applications (WISA'09), pp. 104–107 (2009)
5. Zhu, G., Zhang, S.: Research and implementation of DCT-based image digital watermarking algorithm. In: IEEE Proceedings of the International Symposium on Electronic Commerce and Security, pp. 195–198 (2009)
6. Wei, Z, Ngan, K.N.: Spatio-temporal just noticeable distortion profile for grey scale image/video in DCT domain. IEEE Trans. Circuits Syst. Video Technol. **19**(3), 337–346 (2009)
7. Xu, Z.J., Wang, Z.Z., Lu, Q.: Research on image watermarking algorithm based on DCT. Procedia Environ. Sci. **10**, 1129–1135 (2011)
8. Yesilyurt, M., Yalman, Y., Ozcerit, A.T.: A new DCT based watermarking method using luminance component. Elektronika ir Elektrotechnika **19**(4), 47–52 (2013)
9. Rosales-Roldan, L., Cedillo-Hernandez, M., Nakano-Miyatake, M., Perez-Meana, H., Kurkoski, B.: Watermarking-based image authentication with recovery capability using halftoning technique. Signal Process. Image Commun. **28**, 69–83 (2013)
10. Das, C., Panigrahi, S., Sharma, V.K., Mahapatra, K.K.: A novel blind robust image watermarking in DCT domain using inter-block coefficient correlation. Int. J. Electron. Commun. (AEU) **68**, 244– 253 (2014)
11. Lin, S.D., Shie, S., Guo J.Y.: Improving the robustness of DCT-based image watermarking against JPEG compression. Comput. Stand. Interfaces **32**, 54–60 (2010)
12. Chang, C., Lin, P., Yeh, J.: Preserving robustness and removability for digital watermarks using subsampling and difference correlation. Inf. Sci. **179**, 2283–2293 (2009)
13. Patra, J.C., Phua, J.E., Bornand, C.: A novel DCT domain CRT-based watermarking scheme for image authentication surviving JPEG compression. Digital Signal Process. **20**, 1597–1611 (2010)

14. Taherinia, A.H., Jamzad, M.: A robust image watermarking using two level DCT and wavelet packets denoising. In: IEEE Proceedings of the International Conference on Availability, Reliability and Security, pp. 150–157 (2009)
15. Ali, M., Ahna, C.W., Pant, M.: A robust image watermarking technique using SVD and differential evolution in DCT domain. Optik **125**, 428–434 (2014)
16. Lu, W., Lu, H., Chung, F.: Robust digital image watermarking based on subsampling. Appl. Math. Comput. **181**, 886–893 (2006)

Performance Analysis of ApEn as a Feature Extraction Technique and Time Delay Neural Networks, Multi Layer Perceptron as Post Classifiers for the Classification of Epilepsy Risk Levels from EEG Signals

Sunil Kumar Prabhakar and Harikumar Rajaguru

Abstract Epilepsy being a very common and chronic neurological disorder has a pathetic effect on the lives of human beings. The seizures in epilepsy are due to the unexpected and transient electrical disturbances in the cortical regions of the brain. Analysis of the Electroencephalography (EEG) Signals helps to understand the detection of epilepsy risk levels in a better perspective. This paper deals with the Approximate Entropy (ApEn) as a Feature Extraction Technique followed by the possible usage of Time Delay Neural Network (TDNN) and Multi Layer Perceptron (MLP) as post classifiers for the classification of epilepsy risk levels from EEG signals. The analysis is done in terms of bench mark parameters such as Performance Index (PI), Quality Values (QV), Sensitivity, Specificity, Time Delay and Accuracy.

Keywords EEG · TDNN · MLP · PI · QV

1 Introduction

To measure the electrical activity of the brain, EEG is used [1]. It has several important advantages when compared to the other methods. For the direct measurement of the electrical activities of the brain, EEG is highly utilized [2]. Generally the temporal resolution of the EEG Signals is very high [3]. Certain difficulties are present with the EEG analysis, diagnosis and interpretation. With the

---

S.K. Prabhakar (✉) · H. Rajaguru
Department of ECE, Bannari Amman Institute of Technology, Sathyamangalam, India
e-mail: sunilprabhakar22@gmail.com

H. Rajaguru
e-mail: harikumarrajaguru@gmail.com

M. Senthilkumar et al. (eds.), *Computational Intelligence, Cyber Security and Computational Models*, Advances in Intelligent Systems and Computing 412, DOI 10.1007/978-981-10-0251-9_10

help of strip charts, EEG signals were analyzed by an encephalographer which is quite time consuming and it is hectic in nature. So some sort of automated systems for the detection and classification of epilepsy risk levels from EEG signals are required [4]. So in this paper, ApEn is employed as a feature extraction technique and the features extracted from it are given as inputs to the Neural Network Classifiers. This paper is organized as follows: In Sect. 2, the methods and materials are discussed followed by the analysis of Approximate Entropy (ApEn) as a feature extraction technique in Sect. 3. In the literature, various feature extraction techniques are available like Time Frequency Distribution (TFD) [5], Fast Fourier Transform (FFT) [6], Eigenvector Methods (EM) [7], Wavelet Transform (WT) [8] and Auto Regressive Methods (ARM) [9]. The main purpose of engaging ApEn as feature extraction technique here is because of its ability to quantify the total amount of regularities and the unpredictable fluctuations present in a series of data. In Sect. 4, the post classifiers such as Time Delay Neural Network (TDNN) and Multi Layer Perceptron (MLP) Network in brief are discussed. When various other types of neural networks are available such as Radial Basis Functions (RBF), Support Vector Machines (SVM), Extreme Learning Machine (ELM) in literature [10], the TDNN is implemented here because the total number of connection parameters and invariance under shifts in time which are independent can be easily reduced. Also the shift invariance aids greatly to the successful recognition even when the patterns misalign with respect to time. MLP is also opted here because of its ability to generalize and classify an unknown pattern with a very well known pattern which shares the same characteristic features. MLP Neural Network is highly fault tolerant also. In MLP, the relearning process even after damage can be relatively quick and easier and thus MLP has an edge over other neural network classifiers. Finally it followed by the results and discussion in Sect. 5.

2 Materials and Methods

For the performance analysis of the epilepsy risk levels using Approximate Entropy (ApEn) as a feature extraction technique followed by TDNN and MLP as a Post Classifier, the raw EEG data of 20 epileptic patients who were under treatment in the Neurology Department of Sri Ramakrishna Hospital, Coimbatore in European Data Format (EDF) are taken for study. The pre processing stage of the EEG signals is given more attention because it is vital to use the best available technique in literature to extract all the useful information embedded in the non-stationary biomedical signals [11]. The EEG records which were obtained were continuous for about 30 s and each of them was divided into epochs of two second duration. The channel used to acquire the EEG data is a 16 channel bipolar measurement type. Totally 22 electrodes in 10-20 format are used in this experiment. Generally a two second epoch is long enough to avoid unnecessary redundancy in the signal and it is long enough to detect any significant changes in activity and to detect the presence of artifacts in the signal [11]. For each and every patient, the total number of

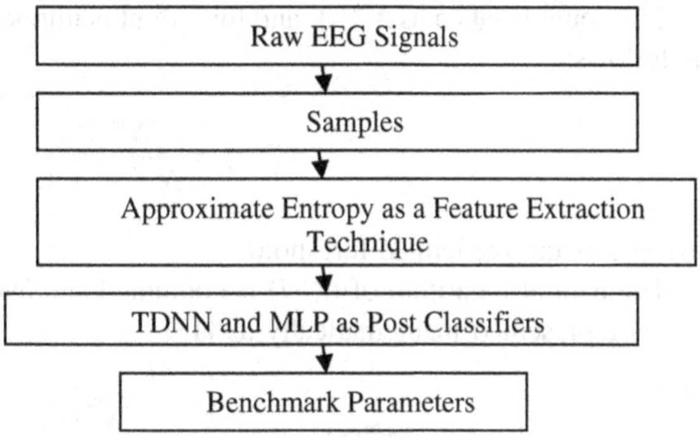

Fig. 1 Block diagram of the procedure

channels is 16 and it is over three epochs. In this paper the exhaustive analysis and results are shown only for a single epoch. The frequency is considered to be 50 Hz and the sampling frequency is considered to be about 200 Hz. Each and every sample corresponds to the instantaneous amplitude values of the signal which totals to 400 values for an epoch [11]. The total number of artifacts present in the data is four. Chewing artifact, motion artifact, eye blink and electromyography (EMG) are the four number of artifacts present and approximately the percentage of data which are artifacts is 1 %. No attempts were made to select certain number of artifacts which are of more specific nature. The main objective to include artifacts is to differentiate the spike categories of waveforms from non spike categories. Figure 1 shows the block diagram of the procedure.

The block diagram of the procedure is given in Fig. 1. Initially the raw EEG Signals are taken and it is sampled. ApEn is used here as a Feature Extraction Technique and later TDNN and MLP are used as Post Classifiers for the perfect classification of epilepsy risk levels from EEG Signals and then the bench mark parameters are analyzed in terms of performance index and quality values.

3 Approximate Entropy as Feature Extraction Technique

In this paper, Approximate Entropy is used as a feature extraction technique. A sequence $L = l(1), l(2), l(3)\ldots l(N)$ is considered, where the N data denotes the total number of data [12]. The subsequence of L should also be represented and is expressed mathematically as follows

$$l(i) = [l(i), l(i+1), \ldots l(i+m-1)$$

Between two vectors l_i and l_j, the distance is defined as follows

$$d(l_i, l_j) = \max[L(i+k-1) - L(j+k-1)]$$

The count is taken as $N^m(i)$, and this count is important to compute the ratio [12] as follows:

$$C_r^m(i) = \frac{N^m(i)}{N - m + 1}$$

where r is the predefined threshold.

The natural logarithm of $C_r^m(i)$ is computed and averaged for all the values of i and is represented mathematically as [13]

$$\phi^m(r) = \frac{1}{N - m + 1} \sum_{i=1}^{N-m+1} \ln C_r^m(i)$$

The iteration is done and ApEn is calculated [12] as follows

$$ApEn = \phi^m(r) - \phi^{m+1}(r)$$

4 Neural Networks as Post Classifiers

In this paper, the Time Delay Neural Network (TDNN) and the Multi Layer Perceptron (MLP) Neural Networks are used.

4.1 Time Delay Neural Network (TDNN)

It is an Artificial Neural architecture, where the main intention of it is to work on data which is sequential in manner and it is feed forward in nature. The TDNN units easily recognize the features which are highly independent of time shift [14]. Its application is higher and forms an integral part in the pattern recognition system. Augmentation of the input signal is done initially and other input is represented as delayed copies. Since there are no internal states present here, the Neural Network is generally assumed to be time-shift invariant.

4.2 Multi Layer Perceptron (MLP)

Multilayer Perceptron is actually a feed forward Neural Network. It is trained with the help of standard Back Propagation Algorithm. They generally come under the classification of supervised networks. For the sake of training, a desired response is always required. Transformation of the input data into a particular desired response data can be easily done with the help of MLP and hence it is widely used for pattern recognition and classification purposes. Most of the Neural Network application generally involves MLP's because it is considered as a very powerful and versatile classifier [11]. Any input-output map can be virtually approximated even with one or two hidden layers here. Even in the most difficult problems, MLPs performance is good as that of any other optimal statistical classifiers. The LM algorithm is used as the standard Training algorithm which helps to minimize the MSE criteria. It has good convergence properties and it is rapid and robust.

The steepest descent rule is engaged and the weight is updated according to the following equation

$$W(k+1) = W(k) + \alpha \frac{\partial E(k)}{\partial W(k)} + \mu \Delta W(k)$$

where $W(k)$ is the weight at the kth iteration, α is the learning rate, (k) is the difference between Neural Network output and the expected output. $\Delta W(k)$ is the weighted difference between the kth and $(k-1)$th iteration (this item is optimal), and μ is the momentum constant. For adaptive algorithms, it varies with time, but this requires many iterations and leads to a high computational burden [11]. The application of the non-linear least squares Gauss-Newton has been used to solve many supervised NN training problem.

$$W(k+1) = W(k) + \Delta W(k),\ k = 0, 1, \ldots$$

where W(k) denotes the NN weight vector at the kth iteration and ΔW(k) is the changed weight.

5 Results and Discussion

For ApEn as a feature extraction technique and TDNN and MLP as Post Classifier, based on the Quality values, Time Delay and Accuracy the results are computed and tabulated in Tables 1 and 2. The formulae for the Performance Index (PI), Sensitivity, Specificity and Accuracy are given as follows

$$PI = \frac{PC - MC - FA}{PC} \times 100$$

where PC—Perfect Classification, MC—Missed Classification, FA—False Alarm, The Sensitivity, Specificity and Accuracy measures are stated by the following

Table 1 Average values for ApEn with TDNN network

Parameter	Epoch 1	Epoch 2	Epoch 3
PC (%)	97.5	95.20	94.37
MC (%)	1.66	3.75	4.16
FA (%)	0.833	1.04	1.45
PI (%)	97.35	94.83	93.81
Sensitivity (%)	99.16	98.95	98.54
Specificity (%)	98.33	96.25	95.83
Time delay (s)	2.05	2.12	2.13
Quality values	23.57	22.51	22.10
Accuracy (%)	98.75	97.60	97.18

Table 2 Average values for ApEn with MLP network

Parameter	Epoch 1	Epoch 2	Epoch 3
PC (%)	98.12	96.25	94.79
MC (%)	1.04	2.5	3.33
FA (%)	0.83	1.25	1.87
PI (%)	98.02	96.04	94.35
Sensitivity (%)	99.16	98.75	98.125
Specificity (%)	98.95	97.5	96.66
Time delay (s)	2.02	2.07	2.09
Quality values	23.85	22.85	22.09
Accuracy (%)	99.06	98.125	97.39

$$Sensitivity = \frac{PC}{PC + FA} \times 100$$

$$Specificity = \frac{PC}{PC + MC} \times 100$$

$$Accuracy = \frac{Sensitivity\ +\ Specificity}{2} \times 100$$

The Specificity and Sensitivity Analysis for the application of ApEn as a feature extraction technique followed by TDNN and MLP as Post Classifier is shown in Fig. 2. The Time Delay and Quality Value Analysis for the application of ApEn as a feature extraction technique followed by TDNN and MLP as Post Classifier is shown in Fig. 3. Similarly the Performance Index and Accuracy Analysis for the application of ApEn as a feature extraction technique followed by TDNN and MLP as Post Classifier is shown in Fig. 4.

It is inferred from Fig. 2 that the specificity and sensitivity measures are not constant throughout at all. It is found that when ApEn is engaged with TDNN, an average specificity rate of about 96.8 % is found and if ApEn is engaged with MLP, an average specificity rate of about 97.7 % is found. Also if the sensitivity measures are considered, it is found as of 98.88 % when ApEn is performed with TDNN and 98.67 % is found when ApEn is performed with MLP.

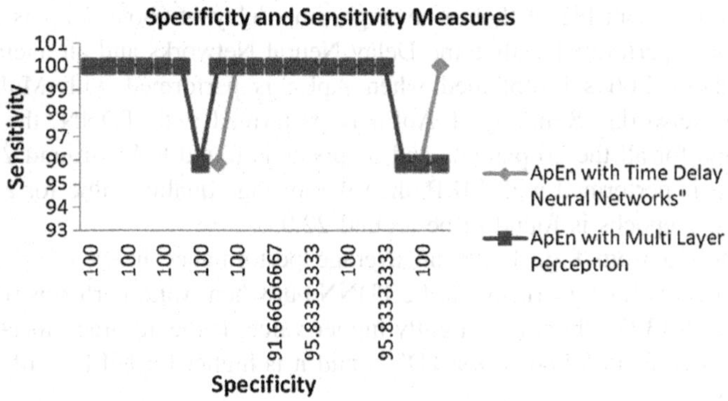

Fig. 2 Specificity and sensitivity measures

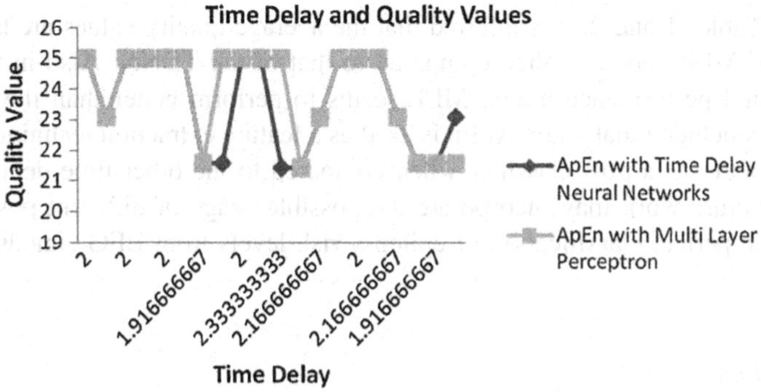

Fig. 3 Time delay and quality value measures

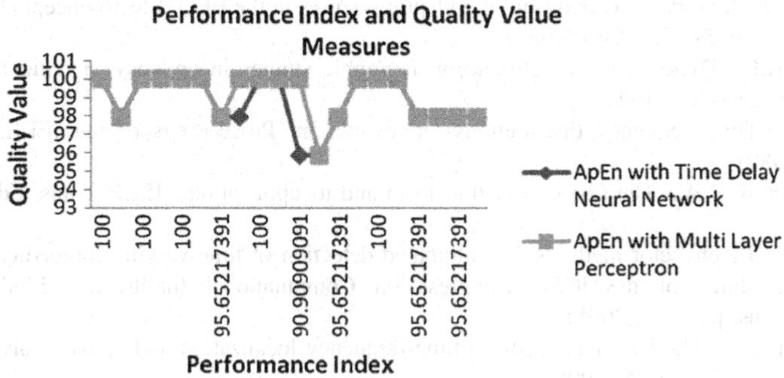

Fig. 4 Performance index and quality value measures

It is inferred from Fig. 3 that an average time delay of about 2.1 s is obtained when ApEn is performed with Time Delay Neural Networks and an average time delay of about 2.06 s is obtained when ApEn is performed with Multi Layer Perceptron Networks. Similarly if ApEn is performed with TDNN, the average quality value for all the 20 patients and 3 epochs is found to be around 22.72 but when ApEn is performed with MLP, then the average quality value for all the 20 patients and 3 epochs is found to be around 22.93.

It is inferred from Fig. 4 that an average performance index of 95.33 % is obtained when ApEn performs with the TDNN but when ApEn performs with MLP it is around 96.13 % which is a slightly higher value. If the accuracy measures are considered it is about 97.84 % for TDNN and it is higher for MLP as of 98.19 % respectively.

6 Conclusion

From the Tables 1 and 2, it is inferred that the average quality values are higher in the case of MLP network when compared to that of the TDNN. Also in terms of accuracy and performance index, MLP seems to perform better than the TDNN. Thus it is concluded that when ApEn is used as a feature extraction technique, MLP performs better as a post classifier when compared to the other time delay neural network. Future work may incorporate the possible usage of different post classifiers for the perfect classification of epilepsy risk levels from EEG signals.

References

1. Hazarika, N., Chen, J.Z., Tsoi, A.C., Sergejew, A.: Classification of EEG signals using the wavelet transform. Sig. Process. **59**(1), 61–72 (1997)
2. Harikumar, R., Vijayakumar, T.: Wavelets and morphological operators based classification of epilepsy risk levels. Math. Probl. Eng. **2014**(813197), 13 pp. (2014). doi:10.1155/2014/813197
3. Gotman, J.: Automatic recognition of epileptic seizures in the EEG. Electroencephalogr. Clin. Neurophysiol. **54**, 530–540 (1982)
4. Finley, K.H., Dynes, J.B.: Electroencephalographic studies in epilepsy: a critical analysis. Brain **65**, 256–265 (1942)
5. Leon C.: Time-frequency distributions—a review. In: Proceedings of the IEEE, vol. 77, no. 7 (1989)
6. Cooley et al., J.W.: The fast fourier transform and its applications. IEEE Trans. Educ. **12**(1) (1969)
7. Guler, U.: Eigenvector methods for automated detection of time-varying biomedical signals. In: Proceedings of the ICSC Congress on Computational Intelligence Methods and Applications, pp. 1–6 (2005)
8. Daubechies, I.: The wavelet transform time-frequency localization and signal analysis. IEEE Trans. Inf. Theory **36**:5 (1990)

9. Zuhairi, B., Nidal, K.: Autoregressive method in short term load forecast. In: Proceedings of the 2nd IEEE International Conference on Power and Energy, December 1–3, (2008), Johor Bahru, Malaysia
10. Guoqiang, P.Z.: Neural networks for classification : a survey. IEEE Trans. Syst. Man Cybern. —C: Appl. Rev. **30**(4), 451–462, November (2000)
11. Harikumar, R., Sukanesh R.: A comparison of GA and neural network (MLP) in patient specific classification of epilepsy risk levels from EEG signals. Eng. Lett. **14**(1) (2006)
12. Reddy, S., Kulkarni P.K.: EEG signal classification for epilepsy seizure detection using improved approximate entropy. Int. J. Public Health Sci. **2**(1), 23–32 (2013)
13. George, J.A., Jiang, Fan, S.-Z., Abood, M.F., et al. (2015). Sample entropy analysis of eeg signals via artificial neural networks to model patients. In: Consciousness Based on Anesthesiologists Experience, BioMed Research International, vol. 2015, Article ID: 343478, 8 pp., 2015, doi:10.1155/2015/343478
14. Lang, K.J. et al.: A time—delay neural network architecture for isolated word recognition. Neural Netw. **3**, 23–43 (1990)

Suspicious Human Activity Detection in Classroom Examination

T. Senthilkumar and G. Narmatha

Abstract The proposed work aims in developing a system that analyze and detect the suspicious activity that are often occurring in a classroom environment. Video Analytics provides an optimal solution for this as it helps in pointing out an event and retrieves the relevant information from the video recorded. The system framework consists of three parts to monitor the student activity during examination. Firstly, the face region of the students is detected and monitored using Haar feature Extraction. Secondly, the hand contact detection is analyzed when two students exchange papers or any other foreign objects between them by grid formation. Thirdly, the hand signaling of the student using convex hull is recognized and the alert is given to the invigilator. The system is built using C/C++ and OpenCV library that shows the better performance in the real-time video frames.

Keywords Video analytics · Face detection · Hand movement detection · Haar casacade

1 Introduction

Video Analytics is used for various applications like motion detection, human activity prediction, people counting and suspicious activity recognition etc. The human face plays a crucial role in our social interaction, conveying people's

T. Senthilkumar (✉)
Department of Computer Science and Engineering, Amrita School of Engineering, Amrita Vishwa Vidyapeetham, Coimbatore, India
e-mail: t_senthilkumar@cb.amrita.edu

G. Narmatha
M.Tech (Computer Vision and Image Processing) Department of Computer Science and Engineering, Amrita School of Engineering, Amrita Vishwa Vidyapeetham, Coimbatore, India
e-mail: narmatha.gopal@gmail.com

M. Senthilkumar et al. (eds.), *Computational Intelligence, Cyber Security and Computational Models*, Advances in Intelligent Systems and Computing 412, DOI 10.1007/978-981-10-0251-9_11

identity [1]. Face recognition is done for automated person identification from which the orientation of human head is predicted. Additionally, the hand contact detection and hand signaling are recognized to determine the suspicious action carried out in examination hall using better classification algorithms [2]. When considering the different biometric techniques, facial recognition is not of much reliable but the key advantage is that it does not require the passer-by being aware of system. It just recognizes the individuals among the crowd even without their knowledge whereas other biometrics like fingerprint, iris scan and speech recognition cannot perform these kind of mass identification. Manual monitoring of examination hall may highly prone to error [3]. So a system is proposed that automatically detects the suspicious action that are carried out in examination hall. PCA for classification will reduce dimensionality and improve the results [4].

This system automatically detects head orientation and hand gesture recognition based on which an alert is given to the invigilator. Faces exhibt emotions that are complex to predict [5]. The approaches exists to detect face based on skin features may produce more false positives since there occurs problem such as illumination condition, camera characteristics and ethnicity [6]. Head Orientation detection is done to know is there any abnormal rotation of head using Haar features. Hand Gesture Recognition is done to monitor whether the person is showing hand signaling to the neighboring person or not. It is done by drawing a convex hull along the contour region of hand where a minimum points inside or on the hull maintains the convexity property [7]. Similarly hand contact detection is also detected using grid formation in the fixed environment and the alert is given based on the combination of above three results [8].

2 Literature Survey

The human behavior analysis involves stages such as motion detection with the help of background modeling and foreground segmentation, object classification, motion tracking and activity recognition. For person identification from a surveillance video, there are several methods employed in early system. The proposed system uses face recognition technique to identify students present in the examination hall and to locate features such as eyes, ears, nose, and mouth etc., from the detected face region image [9].

Turk and Pentland discovered that while using the Eigen faces techniques, the residual error could be used to detect faces in images, a discovery that enabled reliable real-time automated face recognition systems. PCA removes the correlations among the different input dimensions and significantly reduce the data dimensions. But PCA is sensitive to scale and hence it should be applied only on data that have approximately the same change [10, 11]. It may lose important information for discrimination between different classes.

Tsong-Yi Chen, Chao-Ho Chen, Da-Jinn Wang, and Yi-Li Kuo proposed skin color based face detection approach that is based on NCC color space where the initial face candidate is obtained by detecting the skin color region and then the face feature of the candidate is analyzed to determine whether the candidate is real face or not [7]. It allows fast processing and robust to resolution changes but have higher probability of false positive detection.

Kjeldsen and Kendersi devised a technique for doing skin-tone segmentation in HSV space, based on the premise that skin tone in images occupies a connected volume in HSV space. They further developed a system that used a backpropagation neural network to recognize gestures from the segmented hand images.

Haar features approach introduced by Viola et Jones can be used for face recognition. The strength of this methodology is that it gives low false positive rate. Hence it has a very high accuracy while detecting faces, it needs to be coupled with a classification algorithm like AdaBoost to give the best performance and hence has an extra overhead attached with it [12, 13]. It has a high detection accuracy and minimizes the computational time. The approach was used to construct a face detection system that is approximately 15 faster than any previously mentioned approaches [14].

3 Proposed Method

The proposed method identifies the suspicious human activity detection in an examination hall based on three categories (Fig. 1).

The first category is to detect the head motion detection by extracting Haar features from the input frame and the second is to detect Hand contact detection based on grid formation since the classroom is set as a fixed environment [15]. The Third category is to detect and recognize the hand signaling for a single person (Table 1).

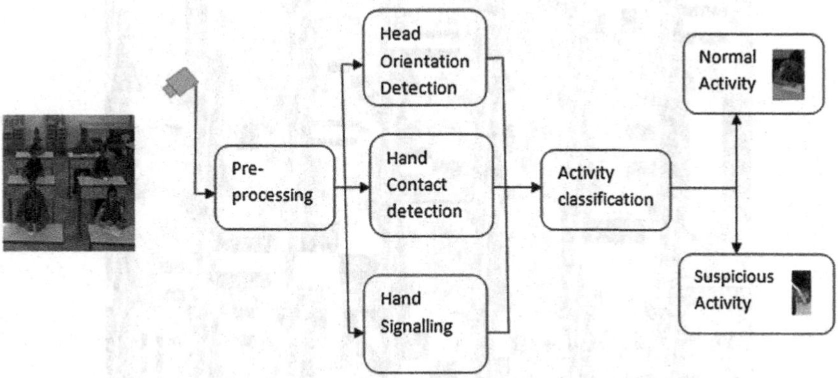

Fig. 1 Architecture diagram for suspicious human activity detection

Table 1 Malpractices constraints and algorithm used

Malpractices	Constraints	Position	Algorithm
Face detection	Head motion detection	Head rotation to the right	Viola Jones algorithm
		Head rotation to the left	
		Head rotation top and bottom	
Hand contact detection	Hand contact detection between two persons	Hand contact sidewise detection	Background subtraction, erosion, dilation, grid formation
		Hand contact backward detection	
	Hand signaling of the person in first row	Does not support for cluttered background and multiple hand detection	Skin color detection, convex hull

3.1 Head Motion Detection

When a frame is captured from a video sequence, it is pre-processed and a face detection method is first performed to determine the position of the face. Next the face regions local features are extracted to examine their activity where it is normal or suspicious activity [16, 17]. The face region is detected using the extracted Haar feature from the give image and their behavior is monitored. Haar-like features are digital image features used in object recognition. Each feature is the difference calculation of the white area minus the dark area. Simple Haar features provide much faster implementation and requires less data (Fig. 2).

Haar can detect face only if the face orientation does not exceed above 30° in all other directions. The AdaBoost classifier is used to produce a better result in a less computation time.

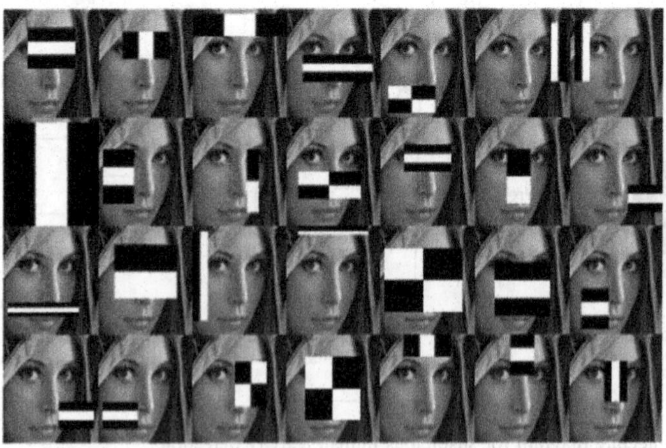

Fig. 2 Face detection using Haar features

3.2 Hand Movement Detection

Hand recognition system uses hand gestures to communicate with the computer system. This algorithm divided into three parts: pre processing, segmentation and classification. Background subtraction is a major pre-processing step where the foreground is extracted from the static background. Followed by, the hand movement from a given video frame is detected by constructing a grid for every student in an examination hall and a group vector line segmenting each rows and columns [18]. The value of the vector lines are summed up from the binary image and stored in an array. When there is a movement of hand occurs,the value exceeds the threshold and an alert is given to the invigilator (Fig. 3).

3.3 Hand Signaling

Hand signaling is detected by using hand gesture analysis that involves a sequence of image processing steps. This phase contains two steps, hand detection and gesture recognition [9]. The hand is segmented from the background by analyzing the skin color over the skin color range [14, 19]. RGB color space is most native to many video capture devices. Based on NCC color space, the skin region of hand is analyzed by determining the color feature from the input image (Figs. 4 and 5).

Fig. 3 Grid formation for hand contact detection

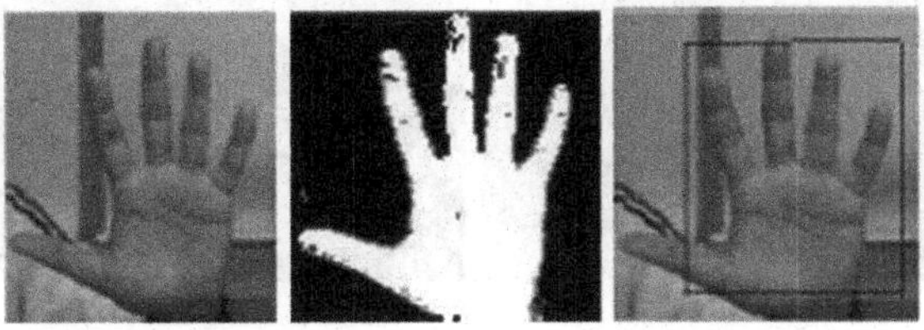

Fig. 4 Hand region segmentation

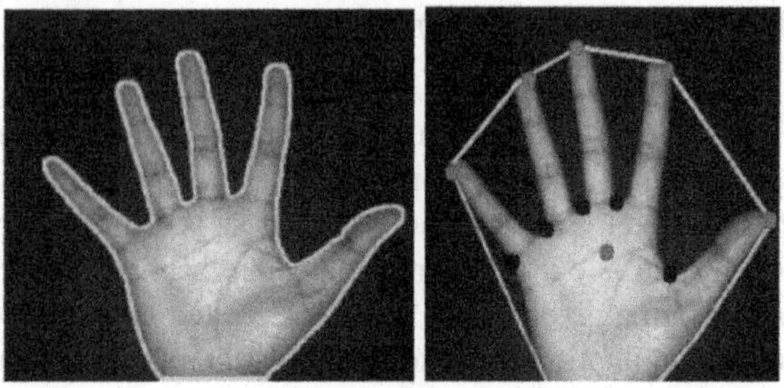

Fig. 5 Hand signaling detection

After the hand gets detected multiple erosion is made followed by dilation such that palm region is only exposed. A Convex hull is drawn around the hand contour that uses minimum points in the contour region to form the hull and maintains the convexity property [7].

4 Result Analysis

A real-time video of a classroom is taken from a static camera and is preprocessed so as to get a clear view of each student in the frame. First the head movement is detected and the hand contact between the students is monitored.

4.1 Face Detected Using Haar Cascade

The proposed system detects the multiple face from the real-time video stream. The results below given are formed by running the Haar cascade in an input the video stream classroom dataset (Fig. 6).

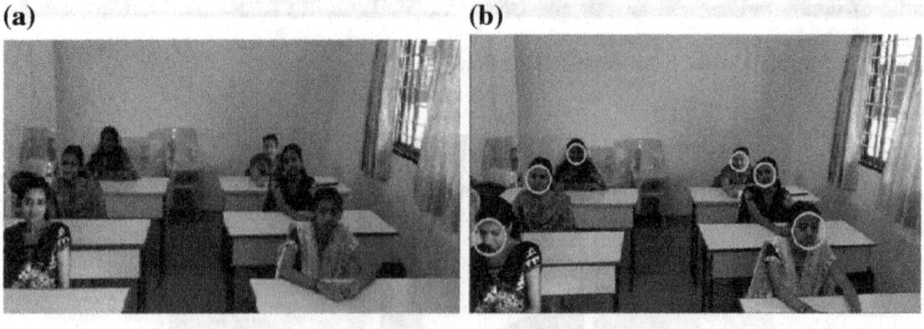

Fig. 6 **a** Input image. **b** Face detected image using Haar cascade

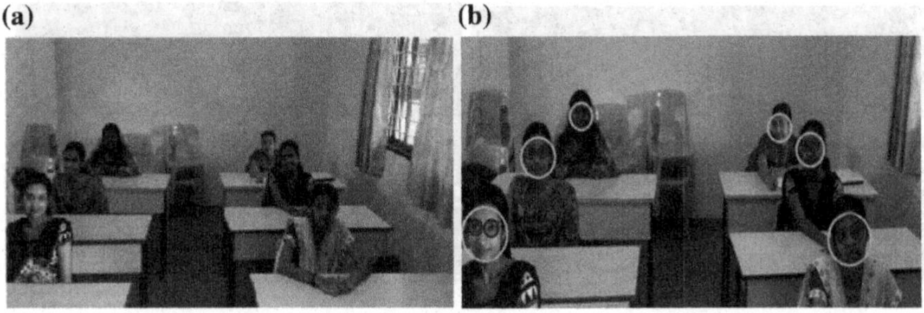

Fig. 7 **a** Input image. **b** Face and eye detected image

When there is a head pose variation beyond the threshold value then the face is detected separately and its activity is monitored. The eye region based on Haar feature can also be detected. But the disadvantage is that the people who are far away from the camera position, their eye cannot be detected and monitored by this method (Fig. 7).

4.2 Hand Movement Detection

Hand Contact Detection can be monitored by undergoing three stages: Background Subtraction, Segmentation and Classification. Background Subtraction is the major pre-processing step that extracts the moving foreground from static background. The background subtraction calculates the foreground mask performing a subtraction between the current frame and a background model, containing the static part of the scene (Fig. 8).

Dilation and Erosion

The two most basic operations in mathematical morphology are erosion and dilation. Dilation adds pixels to the boundaries of objects in an image, while erosion removes pixels on object boundaries. The number of pixels added or removed from

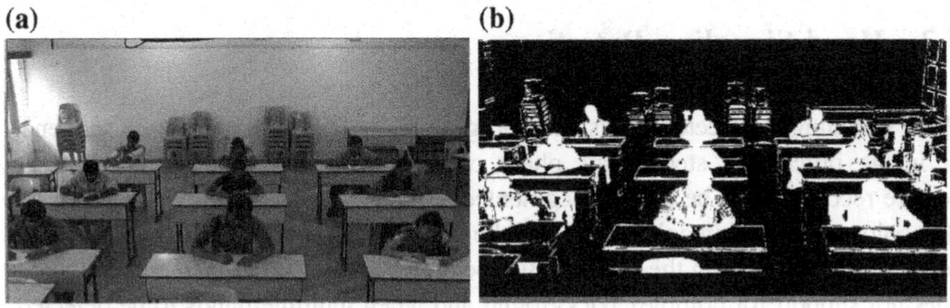

Fig. 8 **a** Input image. **b** Background subtracted image

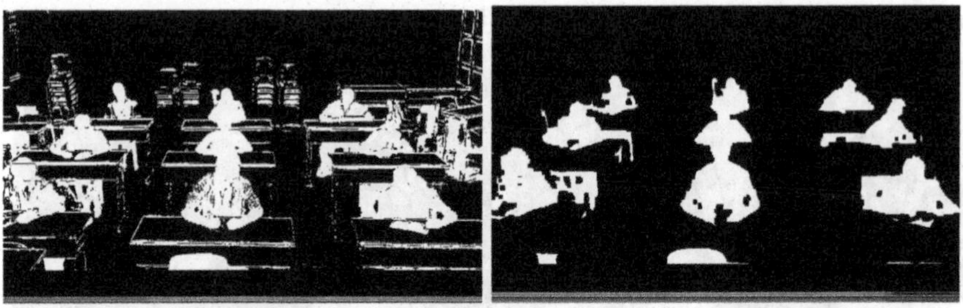

Fig. 9 Binary image before and after morphological operation

(a) (b)

Fig. 10 **a** Input image and, **b** suspicious hand movement activity detected

the objects in an image depends on the size and shape of the structuring element used to process the image (here it is 3 × 3 square structuring element). Hence after the morphological processing the noise effects in the frame are removed highlighting the persons in the dataset as shown in the Fig. 9.

The input real-time video of a classroom is taken where the background is static. Hence manually the region of each student is segmented and monitored individually by drawing a vector line between each columns and the value of their sum is stored in an array. When the students try to exchange paper with other, the value stored exceeds the threshold, and an alert is given to the invigilator as shown in Fig. 10b.

4.3 Hand Signaling Detection

Hand signaling is detected by using hand gesture analysis that involves a sequence of image processing steps. This phase contains two steps, hand detection and gesture recognition. The hand is segmented from the background by analyzing the skin color over the skin color range and the numbers of fingers are detected by using convex hull as shown in the Fig. 11. But the drawback of using this method is that it does not support for complex background and when multiple hands to be detected.

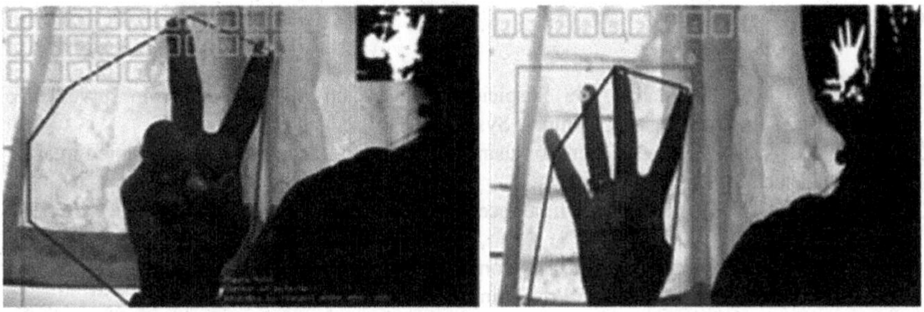

Fig. 11 Hand Signaling detection

Table 2 Performance of the detectors on datasets

Malpractice constraints		True positive (%)	False negative (%)	PPR
Face detection		99	5	0.987
Hand Contact Detection	Sidewise	85	20	0.848
	Backward	74	10	0.732
Hand signaling (for single person)		80	24	0.789

4.4 *Performance Analysis*

The parameter that are used for performance analysis are

- TP: true positives, i.e., number of correct matches.
- FN: false negatives, matches that were not correctly detected.

Positive Predictive Value (PPV) (Table 2),

$$\mathrm{PPV} = \frac{\mathrm{TP}}{\mathrm{TP} + \mathrm{FP}}$$

5 Conclusion

Thus our proposed system detects the suspicious human activity carried out in an examination hall by using three criteria that are head motion, hand contact detection and hand signaling recognition. This method gives better results and reduces the number of false positives when compared to other approaches. Now a days, having human invigilators in the examination halls requires lot of man power and is inconvenient at times. To avoid this we have developed a system that has the vision and ability to automatically invigilate in an examination hall. The developed system gives the high accuracy rate and minimizes the computation time. The system is currently being extended using convolution Neural Network under Deep learning with cascading [20].

References

1. Gowsikhaa, D., Manjunath, A.S.: Suspicious human activity detection from surveillance videos. Int. J. Internet Distrib. Comput. Syst. **2**(2), (2012)
2. Toygar, O., Acan, A.: Face recognition using pca, lda and ica approaches on colored images. J. Electr. Electron. Eng. **3**(1) (2003)
3. Viola, P., Jones, M.J.: Robust real-time face detection. Int. J. Comput. Vision **57**(2), 137–154 (2004)
4. Chellappa, R., Vaswani, : Principal components space analysis for image and video classification. IEEE Trans. Image Processing **15**(7), 1816–1830 (2006)
5. Jafri, R., Arabnia, H.R.: A survey of face recognition techniques. J. Inf. Process. Syst. **5**(2) (2009)
6. AL-Mohair, H.K., Mohamad-Saleh, J., Suandi, S.A.: A human skin color detection: a review on neural network perspective. Int. J. Innovative Comput. Inf. Control **8**(12) (2012)
7. Nayana, P.B., Kubakaddi, S.: Implementation of hand gesture recognition technique for HCI using open CV. Int. J. Recent Dev. Eng. Technol. **2**(5) (2014), ISSN 2347-6435
8. Fan, X.A., Lee, T.H., Xiang, C.: Face recognition using recursive fisher linear discriminent. IEEE Trans. Image processing **15**(8), 2097–2105 (2006)
9. Ojala, T., Pietikainen, M., Harwood, D.: A comparative study of texture measures with classification based on Feature distributions. Pattern Recogn. **29**, 51–59 (1996)
10. Rodriguez, Y.: Face detection and verification using local binary patterns. Ph.D. Thesis, École Polytechnique Fédérale de Lausanne, (2006)
11. Schneiderman, H., Kanade, T.: A statistical method for 3D object detection applied to faces and cars. In: IEEE Conference on Computer Vision and Pattern Recognition 2000 Proceedings, vol. 1. Digital Object Identifier 10.1109/CVPR.2000.855895
12. Yan, Z., Yang, F., Wang, J., Shi, Y., Li, C., Sun, M., Face orientation detection in video stream based on harr-like feature and LQV classifier for civil video surveillance. International Communications Satellite Systems Conferences (2013)
13. Borges, P.V.K., Conci, N., Cavallaro, A., Video-based human behavior understanding: a survey. IEEE Trans. Circ. Syst. Video Technol. **23**(11) (2013)
14. Hadid, A., Heikkil¨a, J.Y., Silven, O., Pietikinen, M.: Face and eye detection for person authentication in mobile phones. In: *Machine Vision Group IEEE Transactions*, 244-1354-0/07
15. Chen,T.-Y., Chen, C.-H., Wang, D.-J., Kuo, Y.-L.: A people counting system based on face-detection using skin color detection technique. In: Fourth International Conference on Genetic and Evolutionary Computing (2010)
16. Ko, T.: A survey on behaviour analysis in video surveillance applications. In: Applied Imagery Pattern Recognition Workshop '08. 37th IEEE, pp. 1–8, (2008)
17. Goo, G., Lee, K., Ka, H., Kim, B.S., Kim, W.Y., Yoon, J.Y., Kim, J.J.: Analysis of crowded scenes in surveillance videos. Can. J. Image Process. Comput. Vision **1**(1): 52–75, (2010)
18. Maurin, B., Masoud, O., apanikolopoulos, N.: Camera surveillance of crowded traffic scenes. IEEE Comput. Soc. Press **22**(4), 16–44 (2010)
19. Yang, M.-H., Kriegman, D.J., Ahuja, N.: Detecting faces in images: a survey. IEEE Trans. Pattern Anal. Mach. Intell. **24**, 34–58 (2002)
20. Tang, X., Ou, Z., Su, T., Zhao, P.: Cascade adaboost classifiers with stage features optimization for cellular phone embedded face detection system. In: ICNC 2005, LNCS 3612, pp. 688–697, (2005)

H_∞ State Estimation of Discrete Time Delayed Neural Networks with Multiple Missing Measurements Using Second Order Reciprocal Convex Approach

K. Maheswari

Abstract This paper focuses on H_∞ state estimation for a general class of discrete-time nonlinear systems of the neural network with time-varying delays and multiple missing measurements which is described by the unified model. The H_∞ performance for the systems described by the unified model is analyzed by using sector decomposition approach together with the Lyapunov stability theory. By constructing triple Lyapunov-Krasovskii functional, a new sufficient condition is established to ensure the asymptotic mean square stability of discrete-time delayed neural networks. Second order convex reciprocal technique is incorporated to deal with partitioned double summation terms and the conservatism of conditions for the state estimator synthesis is reduced efficiently. Finally, a numerical example is given to demonstrate the effectiveness of the proposed design method.

Keywords H_∞ state estimator · Sector decomposition approach · Time varying delays · Second order reciprocal convex approach

1 Introduction

In the past few decades, state estimation problem have found successful applications in many areas, such as system modeling, signal processing, and control engineering [1–3]. One often needs to estimate the neuron state through measured network output, and then utilize the estimated neuron state to achieve certain practical performances by making use of relatively large scale neural networks. The problems of filtering involve estimating the state of a system using the output measurement. In this sense, H_∞ state estimation approach is closely related to many robustness problems such as stabilization and sensitivity minimization of uncertain

K. Maheswari (✉)
Department of Mathematics, Kumaraguru College of Technology, Coimbatore 641049, Tamilnadu, India
e-mail: maheswari25krish@gmail.com

M. Senthilkumar et al. (eds.), *Computational Intelligence, Cyber Security and Computational Models*, Advances in Intelligent Systems and Computing 412, DOI 10.1007/978-981-10-0251-9_12

systems, and has therefore gained more attention. Therefore, the key idea is to minimize the estimation error such that the L_2 gain of the controlled system can reach the desired level.

The majority of the existing research results were developed in the context of continuous systems and deterministic neural networks with or without delays. In implementing the applications of neural networks, discrete-time neural networks play a vital role than their continuous time counterparts in today's digital world. The application of discrete-time systems is witnessed in various fields such as secure communication, biological neural network, gene regulatory networks, digital communication etc. Most existing literature concerning state estimation problems implicitly assumes that the network output measurements always contain information about the current state of neurons. However, this is not always true in practice since network measurements are often delayed and interrupted due to finite speed of the signal transmission [4].

Time delay is one of the instability sources that arises naturally in connection with the system process and information flow for dynamical systems. Moreover, it is a common phenomenon in many engineering and industrial systems such as those in communication networks, manufacturing etc. [5]. Therefore, a state estimator procedure has been designed to enhance the tolerance capacity of the state estimator for the time delay, which is done through the sector decomposition method [6].

Sector decomposition approach divides the sector into several parts and is proposed to increase the upper bounds of the time-varying delays with their lower bounds fixed. This approach is more general and less conservative where the non-linear systems satisfy the globally Lipschitz conditions with time-varying delays. Probabilistic missing measurements is introduced in the synthesis of the state estimator to avoid the partial data loss in the system output measurement.

Based on the above discussions, the aim of this paper is to study the asymptotic mean square stability for a class of discrete-time neural networks. In order to reduce the conservatism, a novel method known as the sector-decomposition approach is introduced for the state estimator synthesis. By utilizing some most updated techniques like the second order reciprocal convex approach on the triple Lyapunov-Krasovskii functional, the explicit expression of the desired estimator gains is obtained and are established in terms of LMIs. The feasibility of the derived criteria can easily be checked by resorting to Matlab LMI Toolbox.

2 Problem Description and Preliminaries

Consider the discrete time unified model described by

$$\begin{aligned} x(k+1) &= Ax(k) + A_d x(k-\tau(k)) + C_p\psi(\xi(k)) + C_w w(k) \\ \xi(k) &= M_q x(k) + M_{qd} x(k-\tau(k)) + L_p\psi(\xi(k)) + L_w w(k) \end{aligned} \tag{1}$$

with the initial condition function $x(k) = \varpi(k)\,\forall k \in [-\tau(k), 0]$, where $x(k) \in \Re^n$ is the system state, $\tau(k) \in [\tau_1, \tau_2]$, $A \in \Re^{n\times n}$, $A_d \in \Re^{n\times n}$, $C_p \in \Re^{n\times L}$, $C_w \in \Re^{n\times m}$, $A \in \Re^{n\times n}$, $M_q \in \Re^{L\times n}$, $M_{qd} \in \Re^{L\times n}$, $L_p \in \Re^{L\times L}$, $L_w \in \Re^{L\times m}$, $\xi \in \Re^L$ is the nonlinear function ψ; $\psi \in C(\Re^L; \Re^L)$ is the nonlinear function satisfying $\psi(0) = 0$; $w(k) \in \Re^m$ is the disturbance input that belongs to $l_2 \in [0, \infty)$, $L \in N$ is the number of nonlinear functions. The time varying delay is $\tau(k) \in \Re$, ϖ is the given function in $[-\tau(k), 0]$. From the measured output of the system (1) is chosen to be as:

$$\begin{aligned} y(k) &= \mho M_y x(k) + L_{yw} w(k) \\ &= \sum_{i=1}^{l} \rho_i M_{yi} x(k) + L_{yw} w(k) \end{aligned} \tag{2}$$

where, $M_y \in \Re^{l\times n}$, $L_{yw} \in \Re^{l\times s}$ are known matrices,

$M_{yi} = \mathrm{diag}\{\underbrace{0, \ldots, 0}_{i-1}, 1\,\underbrace{0, \ldots, 0}_{l-i}\} M_y (i = 1, \ldots, l), \mho = \mathrm{diag}\{\rho_1, \ldots, \rho_l\}.\rho_i(i = 1, \ldots, l)$, are l individual random variables. The signal here is the combination of the system states given by $z(k) = M_z x(k)$ where $z(k) \in \Re^r$ is the immeasurable state that is to be estimated from the measured output $y(k)$ with probabilistic time delays and $M_z \in \Re^{r\times n}$ is a constant matrix. Construct the filter for $z(k)$ and by using the augmented vector the system is reduced to

$$\begin{aligned} e(k+1) &= \bar{A}e(k) + Ne(k) + \bar{A}_d e(k - \tau(k)) + \bar{C}_p g(\bar{\xi}(k)) + \bar{C}_w w(k) \\ \bar{\xi}(k) &= \bar{M}_q e(k) + \bar{M}_{qd} e(k - \tau(k)) + \bar{L}_p g(\bar{\xi}(k) + L_w w(k) \end{aligned} \tag{3}$$

where,

$$g(\bar{\xi}(k)) = \begin{bmatrix} \psi(\xi(k)) \\ f(\hat{\xi}(k)) \end{bmatrix}, \bar{A} = \begin{bmatrix} A & 0 \\ 0 & \tilde{A} \end{bmatrix}, \bar{A}_d = \begin{bmatrix} A_d & 0 \\ 0 & A_d \end{bmatrix},$$

$$N = \begin{bmatrix} 0 & 0 \\ -K(\mho - U)M_y & 0 \end{bmatrix}, \bar{C}_p = \begin{bmatrix} C_p & 0 \\ 0 & C_p \end{bmatrix}, \bar{C}_w = \begin{bmatrix} C_w \\ \tilde{C}_w \end{bmatrix},$$

$$\bar{M}_q = \begin{bmatrix} M_q & 0 \\ 0 & M_q \end{bmatrix}, \bar{M}_{qd} = \begin{bmatrix} M_{qd} & 0 \\ 0 & M_{qd} \end{bmatrix}, \bar{L}_p = \begin{bmatrix} L_p & 0 \\ 0 & L_p \end{bmatrix}.$$

The aim of this paper is to design a filter such that the closed-loop system (3) is asymptotically stable in the mean square for all possible multiple missing measurements (2) and there exists a positive scalar γ that satisfies the following inequality:

$$J = \sum_{k=0}^{\infty} \{\tilde{z}^T(k)\tilde{z}(k) - \gamma^2 w^T(k) w(k)\} < 0 \tag{4}$$

under the zero conditions for all nonzero $w(k) \in l_2[0,\infty)$ and all allowed time varying delays $\tau(k)$, where γ is called the upper bound of the L_2 gain for the system (3). The optimal filter can further be obtained by finding the minimal value of γ satisfying (4).

Assumption 1 The time varying delay $\tau(k)$, satisfies the condition $\tau_1 \le \tau(k) \le \tau_2$ in which τ_1, τ_2 are known non-negative integers. Here, we denote $\tau_{12} = \tau_2 - \tau_1$ and $\tau^* = (\tau_2 - \tau_1)^2/2$.

Assumption 2 The nonlinear functions in (1) are globally Lipschitz. Therefore, the following inequality holds good: $l_i^- \le \frac{\psi_i(\rho)-\psi_i(\upsilon)}{\rho-\upsilon} \le l_i^+, \forall \rho, \upsilon \in \Re, i = 1, \ldots.., L$. where $l_i^+ > l_i^- \ge 0$.

Definition 1 The error system (1) is said to be stable in the mean square if for any $\varepsilon > 0$, there is a $\gamma(\varepsilon) > 0$ such that $\mathcal{E}\{\| r(t) \|^2\} < \varepsilon$, $t > 0$, when $\mathcal{E}\{\| r(0) \|^2\} < \gamma(\varepsilon)$. In addition, if $\lim_{t\to\infty} \mathcal{E}\{\| r(t) \|^2\} = 0$, for any initial conditions, then the error system (1) is asymptotically stable in the mean square.

3 Main Results

In this section, the author deals with the H_∞ state estimation problem. First, the asymptotic mean square stability of the augmented system (3) is investigated using the second order reciprocal approach.

Theorem 1 *Under the Assumption 2, the augmented system (3) with known diagonal positive semi-definite matrices U, σ and Λ, given non-negative integers $0 \le \tau_1 \le \tau_2$ without external disturbances, satisfying the eigenvalue problem, minimize γ subject to $[\Xi]_{10\times 10}$ is asymptotically mean square stable, and the upper bound of the L_2 gain for the system is minimal, if there exists symmetric positive definite matrices $P > 0$, $Q_1 > 0$, $Q_2 > 0$, $R_1 > 0$, $R_2 > 0$, $Z > 0$, $G_k, k = 1,2,3$, any matrix S, a positive scalar γ and Π_1 with appropriate dimension, such that the following LMIs hold for l = 1, 2:*

$$\begin{bmatrix} \Phi_l & \Pi_1^T \\ * & -Z \end{bmatrix} < 0, \tag{5}$$

$$\begin{bmatrix} 2Z & 0 & G_1 & 0 \\ * & Z & 0 & G_2 \\ * & * & 2Z & 0 \\ * & * & * & Z \end{bmatrix} > 0, \begin{bmatrix} Q_2 & G_3 \\ * & Q_2 + \tau^* Z \end{bmatrix} > 0. \tag{6}$$

where

$$\phi_l = \begin{bmatrix} \Xi & 2\Gamma_{1l}\hat{Z} \\ * & -2\hat{Z} \end{bmatrix} + \begin{bmatrix} \Gamma_{1l}^T \\ 0_n \end{bmatrix}\Pi_1 + \Pi_1^T \begin{bmatrix} \Gamma_{1l}^T \\ 0_n \end{bmatrix}^T,$$

$$\bar{Z} = \begin{bmatrix} 3Z & G_1 + G_2 \\ * & 3Z \end{bmatrix} \text{ and } \Xi = [\Xi]_{10\times 10}, \Gamma_{1l}, \hat{Z}$$

are defined as

$$\begin{aligned}
\Xi_{1,1} &= \bar{A}^T P\bar{A} + \sum_{i=1}^{l} \sigma_i^2 \bar{N}_i^T P\bar{N}_i - P + R_1 + \tau_1^2 Q_1 - \tau_{12}^2 Q_2 - \frac{\tau_{12}^3(\tau_2 - \tau_1 + 1)}{4} Z_1 \\
&\quad + \bar{M}_z, \Xi_{1,5} = \bar{A}^T P\bar{A}_d, \Xi_{1,9} = \bar{A}^T P\bar{C}_p + \bar{C}_q^T F\Pi, \Xi_{1,10} = \bar{A}^T P\bar{C}_w, \Xi_{2,2} = \tau_1^2 Q_2 \\
&\quad - \frac{(\tau_{12})^3(\tau_2 - \tau_1 + 1)}{4} Z, \Xi_{3,3} = -R_1 + R_2, \Xi_{3,4} = -Q_2 - \tau^* Z, \Xi_{3,5} = G_3, \\
\Xi_{4,4} &= -R_2 - G_3, \Xi_{4,5} = -Q_2 - G_3^T + Q_2^T + \tau^* Z, \Xi_{5,5} = \bar{A}_d P\bar{A}_d + Q_2 - G_3, \\
\Xi_{5,9} &= \bar{A}_d^T P\bar{C}_p + \bar{M}_{qd}^T F\Pi, \Xi_{5,10} = \bar{A}_d^T P\bar{C}_w, \Xi_{6,6} = -Q_1, \Xi_{7,7} = -Z, \\
\Xi_{7,8} &= -G_1^T - G_2^T, \Xi_{8,8} = -Z, \Xi_{9,9} = \bar{C}_p^T P\bar{C}_p - 2\Pi + \bar{L}_p^T F\Pi + \Pi F\bar{L}_p, \\
\Xi_{9,10} &= \bar{C}_p^T P\bar{C}_w + \Pi F\bar{L}_w, \Xi_{10,10} = \bar{C}_w^T P\bar{C}_w - \gamma^2 I, \\
\Gamma_{11} &= \begin{bmatrix} 0_n & 0_n & 0_n & 0_n & 0_n & 0_n & -I_n & 0_n & 0_n & 0_n \\ 0_n & 0_n & 0_n & 0_n & \tau_{12}I_n & 0_n & 0_n & -I_n & 0_n & 0_n \end{bmatrix}, \\
\Gamma_{12} &= \begin{bmatrix} 0_n & 0_n & \tau_{12}I_n & 0_n & 0_n & 0_n & -I_n & 0_n & 0_n & 0_n \\ 0_n & 0_n & 0_n & 0_n & I_n & 0_n & 0_n & -I_n & 0_n & 0_n \end{bmatrix}, \\
\hat{Z} &= \begin{bmatrix} Z & 0 \\ 0 & Z \end{bmatrix}, \bar{M}_z = \begin{bmatrix} 0 & 0 \\ 0 & M_z^T M_z \end{bmatrix}, F = \begin{bmatrix} H & 0 \\ 0 & H \end{bmatrix}, \Pi = \begin{bmatrix} \Sigma & 0 \\ 0 & \Lambda \end{bmatrix}, \\
\bar{N}_i &= \begin{bmatrix} 0 & 0 \\ -KM_{yi} & 0 \end{bmatrix}, (i = 1, .., l), H = \text{diag}\{h_1, \ldots., h_L\} \geq 0, \Sigma = \text{diag}\{\varepsilon_1, \ldots., \varepsilon_L\},
\end{aligned}$$

Proof Consider the system (3) with $w(k) = 0$. At least one equilibrium is located at the origin, since $e(k) = 0$ and $\bar{\xi}(k) = 0$, which means that $e_{eq} = 0$ and $\bar{\xi}_{eq}(k) = 0$.

Construct the Lyapunov-Krasovskii functional candidate as:

$$V(k) = V_1(k) + V_2(k) + V_3(k) + V_4(k) + V_5(k) \text{ where}$$

$$V_1(k) = e^T(k)Pe(k),$$

$$V_2(k) = \sum_{i=k-\tau_1}^{k-1} e^T(i)R_1e(i) + \sum_{i=k-\tau_2}^{k-\tau_1} e^T(i)R_2e(i),$$

$$V_3(k) = \tau_1 \sum_{i=-\tau_1}^{-1} \sum_{j=k+i}^{k-1} e^T(j)Q_1e(j)$$

$$V_4(k) = \tau_{12} \sum_{i=-\tau_2}^{-\tau_1-1} \sum_{j=k+i}^{k-1} \eta(j)Q_2\eta(j),$$

$$V_5(k) = \tau^* \sum_{i=-\tau_2}^{-\tau_1-1} \sum_{j=i}^{-\tau_1-1} \sum_{l=k+j}^{k-1} \eta^T(l)Z\eta(l),$$

$P > 0$, $R_1 > 0$, $R_2 > 0$, $Q_1 > 0$, $Q_2 > 0$, $Z > 0$, then $V(k) > 0, \forall\ e(k) \neq 0,\ \forall\ \bar{\xi}(k) \neq 0$, and $V(k) = 0$ if and only if $e(k) = 0$ and $\bar{\xi}(k) = 0$.

Setting $\eta(k) = e(k+1) - e(k)$, $\alpha = (\frac{\tau_2-\tau(k)}{\tau_2-\tau_1})$, $\beta = (\frac{\tau(k)-\tau_1}{\tau_2-\tau_1})$.

Calculating the difference of $V_i(k), (i = 1, 2, 3, 4, 5)$ gives

$$\begin{aligned} E\{\Delta V_1(k)\} &= e^T(k)\bar{A}^TP\bar{A}e(k) + e^T(k)\sum_{i=1}^{l}\sigma_i^2\bar{N}_i^TP\bar{N}_ie(k) \\ &\quad + e^T(k-\tau(k))\bar{A}_d^TP\bar{A}_de^T(k-\tau(k)) + g^T(\bar{\xi}(k))\bar{C}_p^TP\bar{C}_pg(\bar{\xi}(k)) \\ &\quad + 2e^T(k)\bar{A}^TP\bar{A}_de(k-\tau(k)) + 2e^T(k)\bar{A}^TP\bar{C}_pg(\bar{\xi}(k)) \\ &\quad + 2e^T(k-\tau(k))\bar{A}_d^TP\bar{C}_pg(\bar{\xi}(k)) - e^T(k)Pe(k), \end{aligned} \tag{7}$$

$$E\{\Delta V_2(k)\} = e^T(k)R_1e(k) - e^T(k-\tau_2)R_2e(k-\tau_2), \tag{8}$$

$$\begin{aligned} E\{\Delta V_3(k)\} &= \tau_1 \sum_{i=-\tau_1}^{-1}\left[\sum_{j=k+1+i}^{k} e^T(j)Q_1e(j) - \sum_{j=k+i}^{k-1} e(j)Q_1e(j)\right] \\ &= \tau_1^2e^T(k)Q_1e(k) - \left[\sum_{i=k-\tau_1}^{k-1} e(i)\right]^T Q_1 \left[\sum_{i=k-\tau_1}^{k-1} e(i)\right], \end{aligned} \tag{9}$$

$$E\{\Delta V_4(k)\} = -\tau_{12}\left[\sum_{i=-\tau_2}^{-\tau_1} \eta^T(k)Q_2\eta(k) - \sum_{i=k-\tau_2}^{k-\tau_1-1} \eta^T(i)Q_2\eta(i)\right] \tag{10}$$

By assumption 2, we note that $\tau_1 \leq \tau(k) \leq \tau_2$ and from convex reciprocal lemma [7],

$$E\{\Delta V_4(k)\} \leq \tau_{12}^2 \eta^T(k) Q_2 \eta(k) - \frac{1}{\alpha}\left[\sum_{i=k-\tau_2}^{k-\tau(k)-1} \eta(k)\right]^T Q_2 \left[\sum_{i=k-\tau_2}^{k-\tau(k)-1} \eta(k)\right] - \frac{1}{\beta}\left[\sum_{i=k-\tau(k)}^{k-\tau_1-1} \eta(k)\right]^T Q_2 \left[\sum_{i=k-\tau(k)}^{k-\tau_1-1} \eta(k)\right], \tag{11}$$

$$E\{\Delta V_5(k)\} = \frac{\tau_{12}^3(\tau_2 - \tau_1 + 1)}{4} \eta^T(k) Z \eta(k) - \tau^* \sum_{i=k-\tau_2}^{k-\tau(k)-1} \sum_{j=k-\tau_2}^{k-\tau(k)-1} \eta^T(k) Z \eta(k) - \tau^* \sum_{i=k-\tau_2}^{k-\tau(k)-1} \sum_{j=k-\tau(k)}^{k-\tau_1-1} \eta^T(k) Z \eta(k) - \tau^* \sum_{i=k-\tau(k)}^{k-\tau_1-1} \sum_{j=i}^{k-\tau_1-1} \eta(k)^T Z \eta(k),$$

By applying second order convex lemma [8], we get

$$\begin{aligned} E\{\Delta V_5(k)\} \leq & \frac{\tau_{12}^3(\tau_2 - \tau_1 + 1)}{4} \eta^T(k) Z \eta(k) \\ & - \frac{1}{\alpha^2}\left[\sum_{i=k-\tau_2}^{k-\tau(k)-1} \sum_{j=i}^{k-\tau(k)-1} \eta(j)\right]^T Z \left[\sum_{i=k-\tau_2}^{k-\tau(k)-1} \sum_{j=i}^{k-\tau(k)-1} \eta(j)\right] \\ & - \frac{(\tau_2 - \tau_1)^2}{2} \frac{\alpha}{\beta}\left[\sum_{i=k-\tau_2}^{k-\tau(k)-1} \sum_{j=i}^{k-\tau(k)-1} \eta(j)\right]^T Z \left[\sum_{i=k-\tau_2}^{k-\tau(k)-1} \sum_{j=i}^{k-\tau(k)-1} \eta(j)\right] \\ & - \frac{1}{\beta^2}\left[\sum_{i=k-\tau(k)}^{k-\tau_1-1} \sum_{j=i}^{k-\tau_1-1} \eta(j)\right]^T Z \left[\sum_{i=k-\tau(k)}^{k-\tau_1-1} \sum_{j=i}^{k-\tau_1-1} \eta(j)\right]. \end{aligned} \tag{12}$$

Let us define

$$\chi(k) = \left[e^T(k)\, e^T(k+1)\, e^T(k-\tau_1)\, e^T(k-\tau_2)\, e^T(k-\tau(k)) \sum_{i=k-\tau_1}^{k-1} e^T(i) \sum_{i=k-\tau_2}^{k-\tau(k)-1} e^T(i) \sum_{i=k-\tau(k)}^{k-\tau_1-1} e^T(i)\; g^T(\bar{\xi}(k))\right]^T,$$

Using (7)–(12), we derive

$$\Delta V(k) \leq \chi^T(k)\left\{\Omega - \Gamma_1^T(k)\begin{bmatrix} Z_1 & G_1 + G2 \\ * & Z \end{bmatrix}\Gamma_1(k)\right\}\chi(k),$$

where

$$\Gamma_1(k) = \begin{bmatrix} 0_n & 0_n & (\tau(k)-\tau_1)I_n & 0_n & 0_n & 0_n & -I_n & 0_n & 0_n \\ 0_n & 0_n & 0_n & 0_n & (\tau_2-\tau(k))I_n & 0_n & 0_n & -I_n & 0_n \end{bmatrix},$$

Furthermore, if M denotes the matrix with $w(k)=0$ in the matrix Ξ then

$$\Delta V(k) \le \chi^T M \chi \text{ where } M = \left\{ \Xi - \Gamma_1^T(k) \begin{bmatrix} Z & G_1 + G2 \\ * & Z \end{bmatrix} \Gamma_1(k) \right\}$$

Under the zero initial conditions system (3) and J in (4) is equivalent to,

$$\begin{aligned} J &= \sum_{k=0}^{\infty} E\left[\tilde{z}^T(k)\tilde{z}(k) - \gamma^2 w^T(k)w(k)\right] \\ &= \sum_{k=0}^{\infty} E\{\chi^T M \chi - \gamma^2 w^T(k)w(k) + e^T(k)\bar{M}_z e(k)\} \\ &= \Upsilon^T \Xi \Upsilon \end{aligned} \tag{13}$$

where $\Upsilon(k) = [\chi^T(k) \quad w^T(k)]^T$. Since $\Xi < 0$, $J < 0$ holds for any $\Upsilon^T \neq 0$, and for $w(k) \in l_2[0,\infty)$. Meanwhile, $M < 0$ is also derived from $\Xi < 0$ since M is the principal minor of the matrix Ξ, which means that the system without external disturbances is asymptotically mean square stable for all possible multiple missing measurements. This completes the proof.

Theorem 2 *For the given non-negative integers $\tau_2 > \tau_1 \ge 0$, a positive integer q, and two known diagonal semi positive definite matrices U and σ, the guaranteed H_∞ performance state estimation problem is optimal and the upper bound of the L_2 gain of the system is minimal, if there exist positive definite matrices $P, R_1, R_2, Q_1, Q_2, Z, G_k, k = 1,2,3,$ of appropriate dimensions, diagonal positive semi-definite matrices Λ and Σ, any matrix S, a positive scalar γ that satisfy the eigenvalue problem, minimize γ subject to $\bar{\Theta} = [\Theta]_{(21+i)\times(21+i)}$, $(i = 1,\ldots l)$. Moreover, the feedback gain is obtained by*

$$K = P_2^{-1} S \tag{14}$$

Proof Take

$$P = \begin{bmatrix} P_1 & 0 \\ 0 & P_2 \end{bmatrix}, R_1 = \begin{bmatrix} R_{11} & R_{12} \\ * & R_{13} \end{bmatrix}, R_2 = \begin{bmatrix} R_{21} & R_{22} \\ * & R_{23} \end{bmatrix},$$

$$Q_1 = \begin{bmatrix} Q_{11} & Q_{12} \\ * & Q_{13} \end{bmatrix}, Q_2 = \begin{bmatrix} Q_{21} & Q_{22} \\ * & Q_{23} \end{bmatrix}, Z = \begin{bmatrix} Z_1 & Z_2 \\ * & Z_3 \end{bmatrix}$$

and constitute the above matrices Ξ into $\bar{\Theta} \le 0$. Defining

$$P_2K = S \tag{15}$$

Then by virtue of Schur Complement [9] the matrix $\bar{\Theta} \leq 0$ is equal to Ξ. This completes the proof.

4 Numerical Example

In this section, a numerical example is presented to show the effectiveness of the proposed estimator design.

Consider the following discrete time-varying delayed BAM neural network.

$$\begin{aligned} x(k+1) &= E_1x(k) + F_1f(y(k)) + G_1f(y(k-d(k))) \\ y(k+1) &= E_2x(k) + F_2f(y(k)) + G_2f(y(k-d(k))) \\ z_1(k) &= H_1x(k) \\ z_2(k) &= H_2y(k) \end{aligned} \tag{16}$$

where $x(k) = [x_1(k), x_2(k)]^T, y(k) = [y_1(k), y_2(k)]$ where $E_1, E_2, F_1, F_2, G_1, G_2, H_1, H_2$ are the given matrices.

Theorem 1 [2]	85
Theorem 1 (q = 1)	38.08
Theorem 1 (q = 2)	59.056

The Table shows that the upper bound is greatly enlarged with the increase of sector decomposition q, which means that conservativeness of the conditions is reduced. With the above parameters, we obtain the estimator gain as follows;

$$K = \begin{bmatrix} -0.2387 & -0.2457 & -0.1092 & -0.3833 \\ 0.2246 & 0.8032 & 0.0724 & 0.1573 \\ 0.0802 & 0.1709 & -0.2361 & -0.1094 \\ -0.0170 & -0.0958 & 0.0714 & -0.0394 \end{bmatrix}$$

The estimator error performances of two subsystems are depicted in the Fig. 1, which shows that the system states are well estimated since the estimator errors in both of the cases converge to zero finally.

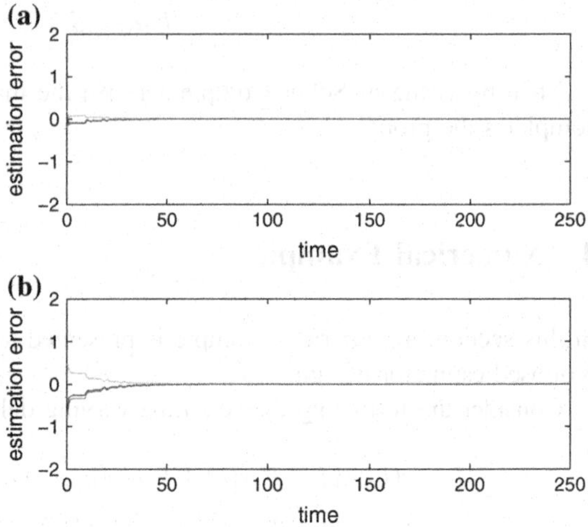

Fig. 1 The state trajectories of system (16)

5 Conclusion

This paper presents the H_∞ estimation problem for discrete time nonlinear systems with time varying delays and multiple missing measurements based on the unified model which consists of a linear dynamic system and a static non-linear operator. In order to increase the tolerance capacity of the controller for the time varying delays, a novel method called the sector decomposition method is introduced. A linear combination of positive functions weighted by the inverses of the squared convex parameters based on the reciprocally convex approach is utilized to handle the function combinations arising from the triple summation terms in the derivation of LMI conditions that further reduces the conservatism for the state estimator design. The methodologies and techniques developed in this paper are expected to be extended into the state estimations and control issues of general switched systems.

References

1. Liu, M., Zhang, S., Fan, Z., Qiu, M.: H_∞ state estimation for discrete-time choatic systems based on a unified model. IEEE Trans. Syst. Man Cybern.-Part B. **42**, 1053–1063 (2012)
2. Arunkumar, A., et al.: Robust state estimation for discrete-time BAM neural networks with time-varying delays. Neurocomputing **131**, 171–178 (2014)
3. Balasubramaniam, P., Jarina, B.: Robust state estimation for discrete-time genetic regulatory networks with random delays. Neurocomputing **122**, 349–369 (2013)
4. Zidong, W., Yurong, L.: Robust state estimation for discrete-time stochastic neural networks with probabilistic measurement delays. Neurocomputing **74**, 256–264 (2010)
5. Balasubramaniam, P., Krishnasamy, R., Rakkiyappan, R.: Delay-dependent stability of neutral systems with time varying delays using delay-decomposition approach. Appl. Math. Model. **36**, 2253–2261 (2012)

6. Kwon, O.M., et al.: New criteria on delay dependent stability for discrete time neural networks with time varying delays. Neurocomputing **121**, 185–194 (2013)
7. Park, P., Ko, J.W., Jeong, C.: Reciprocally convex approach to stability of systems with time-varying delays. Automatica **47**, 235–238 (2011)
8. Lee, W.I., Park, P.: Second order reciprocally convex approach to stability of systems with interval time varying delays. Appl. Math. Comput. **229**, 245–253 (2014)
9. Boyd, S., Ghaoui, L., Feron, E., Balakrishnan, V.: Linear matrix inequalities in system and control theory. SIAM, Philadelphia (1994)

A Fuzzy Methodology for Clustering Text Documents with Uncertain Spatial References

V.R. Kanagavalli and K. Raja

Abstract Fuzzy ERC (Extraction, Resolving and Clustering) architecture is proposed for handling the uncertain information that can be either queried explicitly by the user and the system can also cluster the documents based on the spatial keyword present in them. This research work applies fuzzy logic techniques along with information retrieval methods in resolving the spatial uncertainty in text and also finds the spatial similarity between two documents, in other words, the degree to which two or more documents talk about the same spatial location. An experimental analysis is performed with Reuter's Data set. The results obtained from the experiment are based on the empirical evidence of the document clustering based on the spatial references present in them. It is concluded that the proposed work will provide users a new way in retrieving documents that have similar spatial references in them.

Keywords Information retrieval · Text clustering · Uncertain spatial reference · Fuzzy logic

1 Introduction

The information and knowledge sharing era is exploding with information that people are continuously sharing over various sources across the globe. All this information is mostly presented transferred and shared using natural language, since it provides flexibility and spontaneity to the users. Along with the spontaneity

V.R. Kanagavalli (✉)
Department of Computer Sciences and Applications, Faculty of Science & Humanities, Sathyabama University, Chennai, India
e-mail: Kanagavalli.teacher@gmail.com

V.R. Kanagavalli
Department of Computer Applications, Sri Sai Ram Engineering College, Chennai, India

K. Raja
Alpha College of Engineering, Chennai, India
e-mail: raja_koth@yahoo.co.in

M. Senthilkumar et al. (eds.), *Computational Intelligence, Cyber Security and Computational Models*, Advances in Intelligent Systems and Computing 412, DOI 10.1007/978-981-10-0251-9_13

comes the issue of vagueness and ambiguity. There are document classification systems that classifies and groups the documents that are speaking about the same concept. But the same type of classification is not successfully handled if it happens to be based on spatial keywords. This is due to the inherent ambiguity and uncertainty that is associated with the spatial terms found in natural language descriptions. Most of the text documents contain spatial references in them and the user's queries are often associated with a spatial location. At present it is very difficult to retrieve documents that discuss about the same geographic location using different terms. The source of the difficulty is the level of uncertainty and fuzziness associated with natural language. Thus this research work proposes algorithms for clustering the text documents based on the crisp and uncertain spatial references present in the text document.

2 Related Work

2.1 Uncertain Spatial Referencing in Natural Language

Natural language is prone to ambiguity and there is a vast amount of literature handling ambiguity in natural language. There are various applications and domains upon which it is tested and experimented. Li et al. [1] present a hybrid approach to add on to the efficiency of InfoXtract. The work concentrates on identifying the sense in which a location name appears in the text document. It combines the three different methodologies to devise a hybrid approach. In the graph based approach it employs Prim's algorithm which is followed by pattern matching technique. The results are finally fine tuned by default sense heuristics to predict the sense in which a location name appears in a text document.

Durupinar et al. [2] shows the application of natural language processing to extract the spatial relationships between objects in a crime scene description. A dependency parser is used for extracting the syntactic relationship between the objects present in the textual descriptions. Once the spatial relationship is identified, a text-to-scene system is used for reconstructing the description into a 3D scene. The authors of the work have used intuitive relevancy scores for finding the relationship between the spatial relations. The retrieval of the photographs of the crime-scenes is facilitated by construction of an index after the information retrieval process is over. Bordogna et al. [3] deal with the uncertainty in location based queries using possibility distribution. The candidate locations are extracted in the initial filtering phase and then the results are checked for match with the query posed in the refinement phase. The drawback of their method is that the system performance hugely relies on the underlying Geographic Information System (GIS).

Zhou and Yao [4] evaluate the performance of the information retrieval based on the satisfaction of the user preferences. Bitters [5] has worked on providing better geospatial reasoning in a Natural Language Processing (NLP) by maintaining a geographic neighborhood. The ambiguity of vague, implicit and incomplete geographic name references is resolved by maintaining the hierarchy of the spatial locations discussed in a discourse. Users may need information levels at various levels which deal with granulation of information. In their work, the authors Mulkar-Mehta et al. [6, 7] propose a methodology to deal with the granularity of objects, locations and actions that are embedded in an event description.

The index terms are judged and weighed for their importance by applying a set of fuzzy rules with the tf-idf value, ambiguity level and the relation with other index terms as factors, by Ropero et al. [8]. Homonyms in the geographic names lead to uncertainty which is handled by the authors in the work [9]. An annotated corpus is used as a training data set for machine learning methods for the recognition and classification of spatial arguments present in the spatial expressions in a given sentence [10].

2.2 Fuzzy Information Retreival

Masrah and Trevor [11] discuss the application of fuzzy logic in computing the similarity between words for grouping the documents. They have proposed an asymmetric word similarity algorithm for resolving the synonymy problem and prove the efficiency of their algorithm compared to the traditional tf-idf method. Kyoomarsi et al. [12] proposes a fuzzy logic based artificial intelligence method for generating summary from a given text document. The parameters considered by the fuzzy logic method are mean Term Frequency–Inverse Sentence Frequency (TF-ISF), length and position of the sentence, similarity of the sentence to the title, keywords, and cohesion of sentence to sentence, sentence to centroid. These parameters are supplied to the Mamdani based fuzzy logic along with the fuzzy rules that selects the candidate sentences for the summary from the given text document. Das et al. [13] discuss a fuzzy based method of extracting n gram and bi gram keywords which are rigidly collocated, from a web based document using fuzzy sets and fuzzy membership functions. The presence of corpus is not necessary for the systems.

2.3 Clustering Text Documents

Clustering is used in information retrieval for finding relevant documents in a quick manner and to provide different levels of details. The searching of relevant information in a cluster can be either top down or bottom up. The clustering algorithm can be implemented using multiple techniques like Matrix-based, Heuristic, Simulated Annealing, Genetic algorithm or Scatter/Gather technique. Each of these techniques has their advantages and limitations and is suitable for specific applications. The results of clustering algorithms depend upon various parameters like similarity measures, linking methods, similarity threshold and indexing methods.

Cutting et al. [14] Scatter/Gather algorithm is a clustering approach wherein a group of documents are selected from various document clusters and then they are regrouped or re-clustered depending on the requirement of the user. The disadvantage of using a scatter/gather algorithm is the difficulty in choosing a judicious value of k- the initial number of clusters and the inability to include the documents dynamically into the clusters. Steinbach et al. [15] presents a comparative study of document clustering techniques. The authors present the reasoning for better performance of bisecting k-means over other methods. The experiment results by the author portray the drawback of agglomerative hierarchical clustering algorithms. Chen [16] discusses a possibility theory based Self organizing maps for generating a classification of text documents or text corpus. The first step described is to produce a cluster of documents using self organizing maps which is then analyzed using association rule mining. The categories obtained are then refined using possibility measures of necessity and possibility to obtain a more focused theme categorization of the given document set. Liu et al. [17] propose a new feature selection method for text clustering where the weight of the terms are taken into account for and the clustering and featuring are done iteratively.

Sahoo et al. [18] proposes the incremental hierarchical clustering algorithm to the text document set and uses katz's distribution to the word occurrence data. This work modifies the existing COBWEB/CLASSIT algorithm to suit the text data. But the disadvantage of the work is that the selection of keywords is not clear and presently it takes into consideration all the keywords except the stop words. Also the spatial nature of the document is not taken into consideration in this work. Aggarwal et al. [19] discusses an algorithm that uses the content and auxiliary attributes that is present along with the text for clustering the documents. An initial set of clusters is created based on the content of the text documents which is then refined based on the metadata that is present along with the text document. The work also classifies the document set based on the content and auxiliary attribute based clusters.

3 Proposed System

The proposed design preprocesses the text document, identifies the uncertain spatial terms from the text phrases, assigns fuzzy spatial relevance score to these terms, clusters the documents based on the fuzzy spatial relevance scores. The whole process consists of the Preprocessing phase, Extraction phase, Resolving phase and Clustering phase. The conceptual framework for fuzzy ERC approach is shown in Fig. 1.

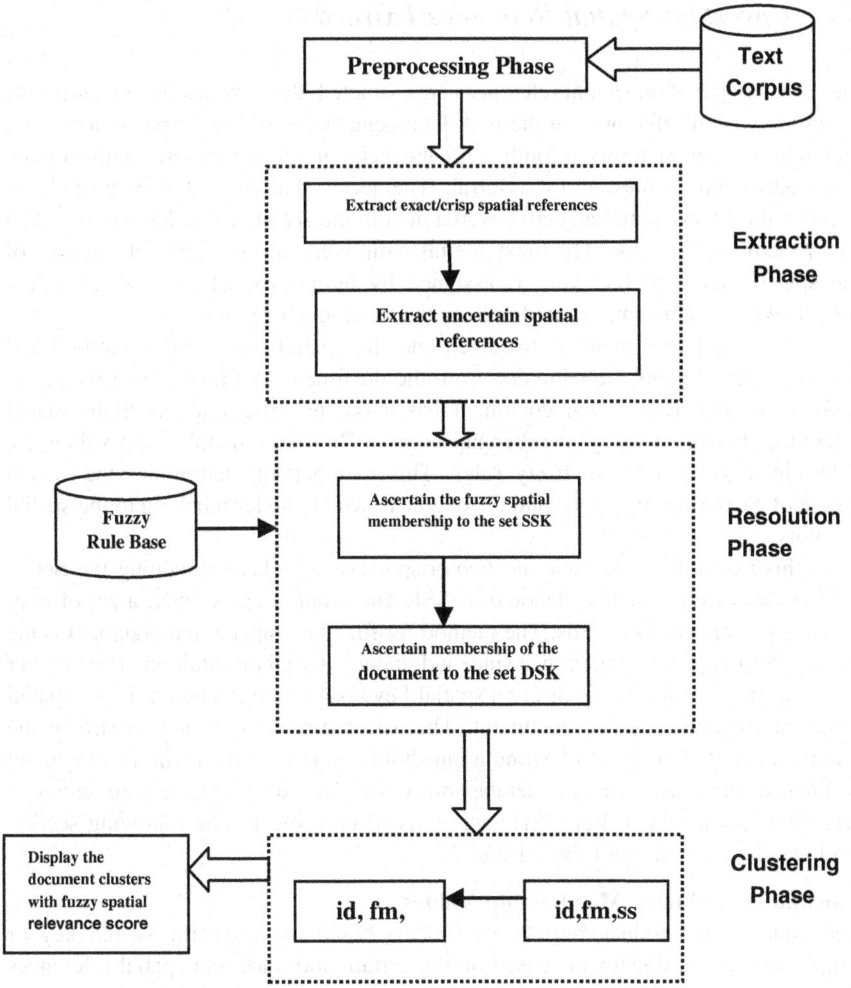

Fig. 1 Conceptual framework for Fuzzy ERM approach

3.1 Preprocessing Phase

The real data for this study were obtained from the Reuter's Data Set. The language considered is English. The text documents that are extracted from the Reuter's data set are converted into XML documents. These documents are then preprocessed for identifying the probable document vectors. The standard crisp spatial terms are identified using GeoNames.org

3.2 *Uncertain Spatial Reference Extraction*

Once the crisp certain spatial references are extracted, the next task is to identify the uncertain spatial references in the text document. A list of fuzzy spatial adjectives and relative spatial terms is built with the help of human experts with domain knowledge and is present for referral. The terms that immediately precede or succeed the fuzzy spatial adjective is assigned to the set of spatial keywords with a fuzzy membership value. The fuzzy membership value or the degree of certainty of the spatial nature of the word is assigned by human experts and stored in the predefined list. This can be modified as per the domain requirement.

A given text document is tokenized and the spatial tokens are identified and distinct spatial tokens are tabulated from the document. A fuzzy spatial relevance score is generated for each document based on the crisp and uncertain spatial references based on fuzzy membership values. The fuzzy membership values are determined using a set of fuzzy rules. This membership determines the spatial nature of each token extracted and the degree to which the term belong to the spatial category.

In this research work, there are two crisp classes, a class containing the Set of Spatial Keywords which is denoted as SSK and another class DSK, a set of documents with spatial keywords. The method for fuzzy membership assignment is the method of intuition and inference since judgment is based on intuition. The number of occurrences of the fuzzy or crisp spatial keyword in the document is the spatial keyword frequency in the document. The assumption is that an increase in the number of spatial keywords beyond a threshold reduces its impact in the document and hence carries a less fuzzy membership value. The fuzzy membership values of crisp and spatial search keys (M1) are as listed in Table 1. The following section explains the contents of Tables 1 and 2.

Calculation of Fuzzy Membership Values

The system can produce retrieve documents based on a specific search key or simply cluster the documents based on the certain and uncertain spatial references found in the text document. In case of a crisp spatial search key is presented to the

Table 1 Fuzzy membership values of crisp and fuzzy spatial search key

Frequency (%) of crisp spatial search key K (M2 = 1)	Fuzzy membership value M0 to SSK	Frequency % of Fuzzy spatial adjectives with spatial search key K (M2 < 1)	Fuzzy membership value M1 to SSK
0.1–0.5	0.6	0.1–0.5	0.6
0.6–0.10	0.8	0.6–0.10	0.8
1.1–1.5	0.7	1.1–1.5	0.7
1.6–2.5	0.6	1.6–2.5	0.6
2.6–4.0	0.4	2.6–4.0	0.4
>4.0	0.3	>4.0	0.3

Table 2 Fuzzy membership values of sample fuzzy spatial adjective

Fuzzy spatial adjective	Fuzzy membership value—M2
Near to/near the	0.4
To the left/right/front/back of	0.6
Adjacent to /close to/close proximity to/behind the/	0.7
Suburbs of/downtown/neighborhood of/	0.7

system, then the Table 1 is used for assigning the fuzzy membership values. If a spatial search key is found in the online gazetteer then the spatial nature of the search key is made for sure and hence the membership value M2 to the set of spatial keywords is assigned as one. If the spatial search key is uncertain and ambiguous, if found along with fuzzy spatial adjectives, then, M2 is assigned a value less than one depending upon the fuzzy spatial adjectives. The value of M2 for the fuzzy spatial adjectives is listed in Table 2. Now the next issue is to find the importance of the spatial keyword in the document. The frequency of occurrence is used to assign a fuzzy spatial relevance score. The crisp spatial search key and the fuzzy spatial adjectives are now analysed to find their eligibility to belong to the set of candidate spatial keywords, SSK. The frequency of occurrence is calculated per 1000 words on average and then converted as a percentage for convenience. If the number of occurrences of crisp or fuzzy spatial search key is less than 5 per 1000 words, then it is moderately important and hence the fuzzy membership assigned is 0.6. If it is between 6 and 10 per 1000 words, the importance is high, hence thus, a membership of 0.8 to the set of the candidate spatial search key. Similarly the other scores are assigned and are listed in the Table 1. Also, a fuzzy spatial keyword may also be manually added to the system by the user to incorporate the knowledge about the colloquial names associated with the spatial locations. This type of manual addition to the knowledge base will allow the identification of documents containing informal descriptions of the events happening in the corresponding locations. A sample list of fuzzy spatial adjective is given in Table 2.

The M2 value for the crisp and fuzzy spatial locations is calculated first and then fuzzy membership values are assigned. These fuzzy memberships to set of candidate spatial keywords (SSK) will then be used to identify the fuzzy membership value to the set of documents with spatial keywords (DSK). The Cumulative fuzzy membership value M for the document for the spatial search key value K_M is defined as Fuzzy sum of the values M0 and M1.

3.3 Uncertainty Identification

Uncertainty creeps in when there is less evidence for certainty. If the text document contains crisp spatial references, then the spatial relevance of the document may be easily identified. But in cases of vague spatial references, the certainty of the spatial relevance of the document is less.

The improved FSS score generator algorithm of fuzzy ERC is formulated by incorporating membership values to the spatial adjectives present in the document. Let X = {x1, x2, x3 …, xn} be the set of documents and S = {s1, s2, s3 …, sn} be the set of spatial locations present in the gazetteer. Also, for each document, the fuzzy spatial quantifiers like, near to, in front of, in the suburbs of, adjacent to etc. should be identified and should be marked in a map interface.

An ordered list of document with the fuzzy spatial relevance of the documents with respect to a spatial query can be generated from the fuzzy membership values of the spatial terms present in the document. Fuzzy rules are used here to ascertain the spatial nature of each content word extracted from the document.

The FSS Score Generator Algorithm

Step 1: For each document in the corpus do the steps from step 2 through 6.
Step 2: Identify the fuzzy spatial keywords and fuzzy spatial adjectives.
Step 3: Calculate the fuzzy membership value of the keywords to the set of spatial keywords (SSK) based on the presence of fuzzy spatial keywords and fuzzy spatial adjectives.
Step 4: Find the frequency of fuzzy spatial keyword/quantifiers along with the fuzzy spatial qualifiers in an array. The fuzzy membership value of the document to the set DSK is assigned in descending order of the frequency count in the array. This would give us the similar text documents based on the crisp or fuzzy spatial keyword criteria. A.
Step 5: If a Crisp Spatial Keyword (CSK) is found in the gazetteer entries, then
    Step 5.1: Create index entry (doc_id, CSK, doc_addr) in a table
    Step 5.2: If the spatial keyword is fuzzy (FSK) and has a fuzzy
        Membership value M2 > 0.6 then
      Check (if the doc_id already exists in the index)
        a. If the doc_id exists in the index, then
Add (g(fsk), where g(fsk) is the fuzzy membership value of the spatial keyword to the index entry)
b. If the doc_id does not exist in the index, then
i. Create index entry (doc_id, g(fsk), doc_addr) in index table
Step 6: In the index maintain fuzzy spatial similarity score for each of the document id.
Step 7: Apply fuzzy clustering for the documents with initial cluster centers as c_i
Step 8: Refine the clusters
Step 9: Display the result

3.4 *Clustering Phase*

The uncertain spatial references are extracted and resolved in the earlier phases. The controlling parameter for the clustering algorithm is the fuzzy spatial similarity threshold value for including documents in a cluster. There is no maximum or minimum number or size of clusters set in this work since it depends on the size and number of the documents chosen for the study. Also the degree of overlap between clusters is not applied as a controlling parameter since fuzzy similarity scores are used which would result in a document belonging to two different clusters with varying degrees of membership. The clustered file organization matches a keyword against a set of centroids, the average representative of a group of documents. This system uses predefined linguistic terms like In front/rear side of, to the left/right of, Close/far to/from, In the neighbourhood/proximity/vicinity of, Downtown/In the suburbs of etc., to be associated with the spatial references.

Fuzzy ERC Algorithm

Step 1: Scan the index generated for the document set
Step 2: Find the CSK present in the index.

Step 2.1: For each unique CSK, generate a cluster centre.
Step 2.2: Identify documents with similar CSK.

Step 3: Form clusters with CSK as cluster centers.
Step 4: Scan the index for FSK.
Step 5: Associate the FSK to all the nearest CSK s by the use of fuzzy spatial adjectives.
Step 6: If the degree of fuzzy spatial similarity score is similar cluster the documents into the same cluster by assigning the doc_id to the clusters identified by the CSKs with varying degrees of membership.
Step 7: If a document is similar to more than one document with varying degrees of similarities then assign it to more than one cluster.
Step 8: Repeat the process till either the index entries are exhausted of CSK and FSK or there are no more new clusters generated by the algorithm.

4 Implementation and Results

The Reuters Data set is chosen since it is generic and has data multiple domains. Figure 2 shows the user interface which allows the addition and modification of the fuzzy relative spatial terms.

Figure 3 shows the addition of fuzzy spatial terms or local spatial names to the system so that the local knowledge can be incorporated into the system which may not be available in the standard gazetteer used. A fuzzy membership value is added to the fuzzy spatial term denoting the degree of certainty that they may be spatial terms or non-spatial terms.

The following Fig. 4 shows the spatial information present in a single document selected by the user from the corpus. It shows the spatial terms identified in the document by the system and also the fuzzy scores of the spatial references found in the specified document.

Figure 5 shows the cluster of document along with the document ids and the number of occurrences of the given spatial term.

The difference between the existing system and our proposed system is that the existing systems cluster the documents based on the concept discussed in them based on the key terms extracted from the document. It does not incorporate the uncertain spatial references found in the document.

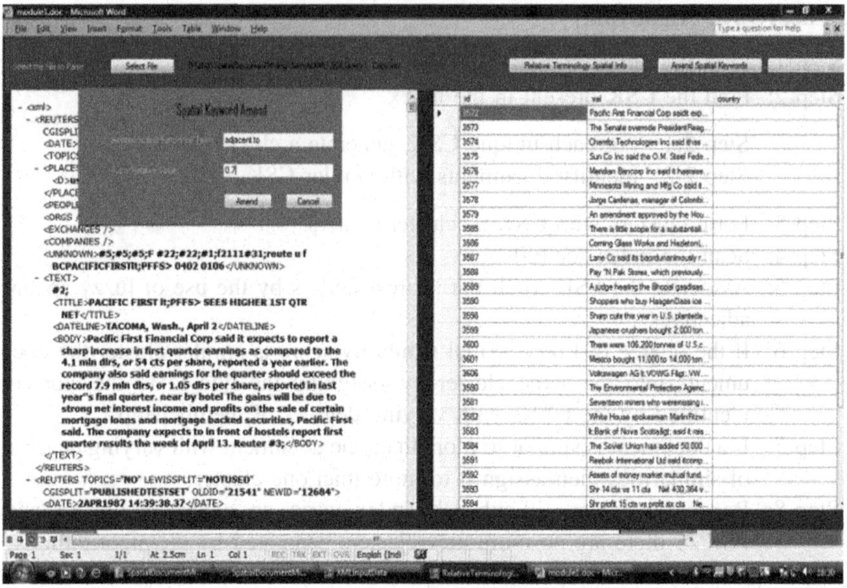

Fig. 2 Adding relative spatial term with fuzzy membership value

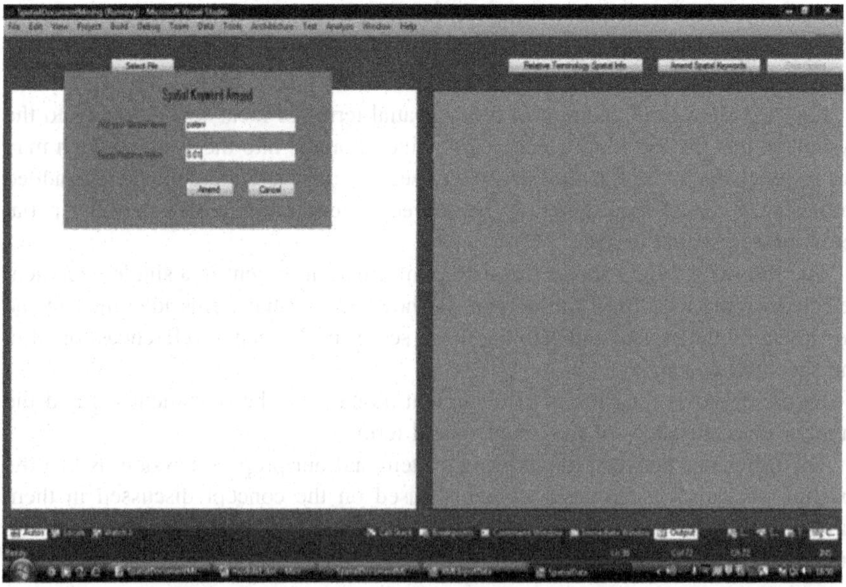

Fig. 3 Adding fuzzy spatial term with fuzzy membership value

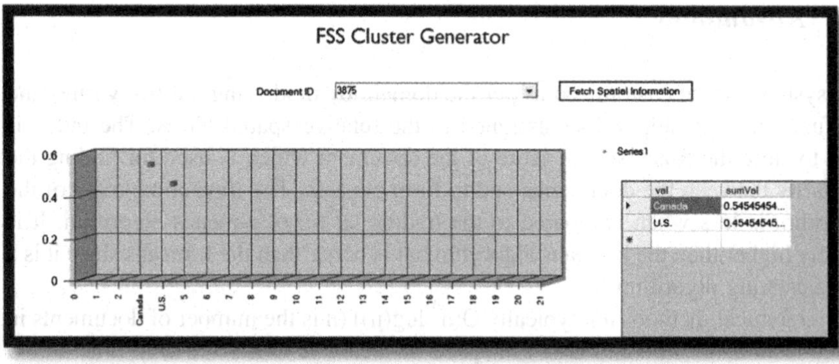

Fig. 4 Fuzzy scores of spatial references found in a specific document

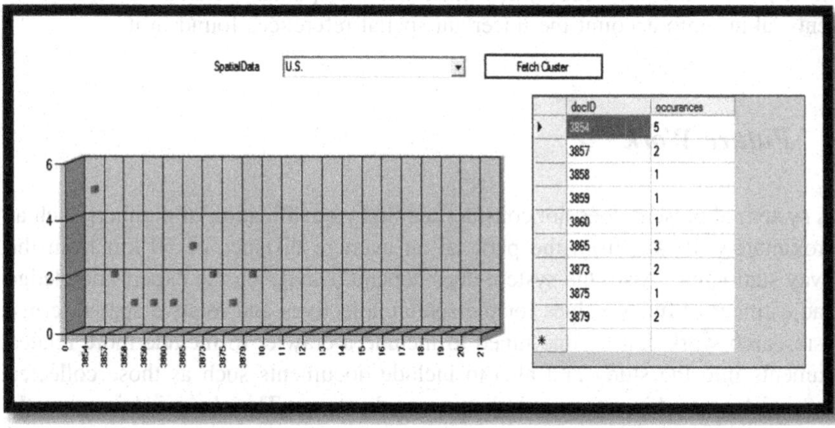

Fig. 5 Fetching clusters for a given spatial location with FSS

5 Conclusion and Future Work

The research problem is concerned with discovering spatial terms in text document data sets. The fuzzy ERC approach with FSS score generator algorithm implements the spatial adjectives to decrease the uncertainty associated with the spatial terms present in the text document. This fuzzy modeling technique is used to classify the text documents according to the spatial terms present in it. This approach is tested on Reuter's data set. The existing hard clustering systems cluster the documents based on the concepts or themes present in the document. It does not take into account the fuzzy and uncertain spatial references found in the text documents.

5.1 *Advantages*

The system can be customized as per the domain by modifying the fuzzy rules and the fuzzy membership values assigned to the relative spatial terms. The index is used to store the fuzzy spatial score of the document which is used for finding the similarity between the documents and to form clusters. The time complexity of the algorithm is less when compared to the traditional fuzzy c-means algorithm. It is slightly higher than the k-means algorithm but is better than the k-means since it is a soft clustering algorithm.

Hierarchical methods are typically $O(n^2 \log(n))$ (n is the number of documents in the collection) in time and $O(n^2)$ in space. K-means is O(k x i x n) in time and O (k + n) in space, where k is the number of clusters and i the number of iterations, but the quality of the clustering results is quite dependent of the initial random cluster seeding. But the proposed system is independent of the cluster seed value and takes O(nlogn). Thus the proposed system identifies the spatially related cluster of documents taking into account the uncertain spatial references found in it.

5.2 *Future Work*

This system at present does not consider the fuzzy qualification of numbers such as approximately 10 km from the park, at an average distance of 50 km from the railway station etc. Also, the system does depend heavily on the expert knowledge for adjustment of fuzzy values for the spatial adjectives and relative spatial terms. The research work can be fine tuned to include a crawler to include more related documents into the study and also to include documents such as those collected from social networking sites and chatting applications. The clustering is currently based on only spatial categories which can be fine tuned to include a theme based sub-category. Also, the fuzzy values for the spatial adjectives and relative spatial terms can be determined by using statistical methods or by using inputs from knowledgebase built by experts.

References

1. Li, H., Srihari, R.K., Niu, C., Li, W.: InfoXtract location normalization: a hybrid approach to geographic references in information extraction. Proceedings of the HLT-NAACL 2003 Workshop on Analysis of Geographic References, pp. 39–44. Alberta, Canada (2003)
2. Durupinar, F., Kahramankaptan, U., Cicekli, I.: Intelligent indexing, querying and reconstruction of crime scene photographs. In: Proceedings of TAINN2004, Izmir, Turkey, pp. 297–306 (2004)
3. Bordogna, G., Pagani, M., Pasi, G., Psaila, G.: Managing uncertainty in location-based queries. Fuzzy Sets Syst. **160**, 2241–2252 (2009). doi:10.1016/j.fss.2009.02.016

4. Zhou, B., Yao, Y.: Evaluating information retrieval system performance based on user preference. JIIS **34**, 227–248 (2010)
5. Bitters B.: Geospatial reasoning in a natural language processing (NLP) environment. In: Proceedings of the 25th International Cartographic Conference, CO-253 (2011)
6. Mulkar-Mehta, R., Hobbs, J.R., Hovy, E.: Granularity in natural language discourse. In: International Conference on Computational Semantics, Oxford, UK, pp. 360–364 (2011)
7. Mulkar-Mehta, R., Hobbs, J.R., Hovy, E.: Applications and discovery of granularity structures in natural language discourse. In: Logical Formalizations of Commonsense Reasoning—Papers from the AAAI 2011 Spring Symposium (SS-11-06) (2011)
8. Ropero, J., et al.: Term weighting for information retrieval using fuzzy logic, www.intechopen.com. ISBN 978-953-51-0393-6 (2012)
9. Leetaru, K.H.: Fulltext geocoding versus spatial metadata for large text archives: towards a geographically enriched wikipedia. D-Lib Magazine **18**(9/10) (2012). doi:10.1045/september2012-leetaru
10. Kordjamshidi, P., Van Otterlo, M., Moens, M.-F.: Spatial role labeling annotation scheme. Handbook of Linguistic Annotation, Edited by James Pustejovsky, Nancy Ide, 01/2014: chapter Spatial Role Labeling Annotation Scheme: pp. 28; Springer
11. Azmi Murad, M.A., Martin, T.: Word similarity for document grouping using soft computing. IJCSNS Int. J. Comput. Sci. Netw. Secur. **7**(8) (2007)
12. Kyoomarsi, F., Khosravi, H., Eslami, E., Dehkordy, P.K.: Optimizing machine learning approach based on fuzzy logic in text summarization. Int. J. Hybrid Inf. Technol. **2**(2) (2009)
13. Das, B., Pal, S., Mondal, S.Kr., Dalui, D., Shome, S.K.: Automatic keyword extraction from any text document using N-gram rigid collocation. Int. J. Soft Comput. Eng. (IJSCE) **3**(2), 238–242 (2013), ISSN: 2231–2307
14. Cutting, D.R., Karger, D.R., Pedersen, P., Tukey, J.W.: Scatter/gather: a cluster based approach to browsing large document collections. In: Proceedings of the Fifteenth Annual International ACM SIGIR Conference on Research and Development in Information Retrieval, Interface Design and Display, pp. 318–32927 (1992)
15. Steinbach, M., Karypis, G., Kumar, V.: A comparison of document clustering techniques. In: Proceedings of the Text Mining Workshop KDD, pp. 109–110 (2000)
16. Chen, Phoebe, Y.-P.: A hybrid framework using SOM and fuzzy theory for textual classification in data mining. Modelling with Words, pp. 153–167. Springer, Berlin Heidelberg (2003)
17. Liu, T., Liu, S., Chen, Z., Ma, W.Y.: An evaluation of feature selection for text clustering. In: Proceedings of the ICML Conference, Washington, DC, USA, pp. 488–495 (2003)
18. Sahoo, N., Callan, J., Krishnan, R., Duncan, G., Padman, R.: Incremental hierarchical clustering of text documents. In: Proceedings of the 15th ACM International Conference on Information and Knowledge Management, 06–11 Nov 2006, Arlington, Virginia, USA (2006). doi:10.1145/1183614.1183667
19. Aggarwal, C.C., Zhao, Y., Yu, P.S.: On the use of side information for mining text data. IEEE Trans. Knowl. Data Eng. **26**(6), 1415–1429 (2014)

A Novel Feature Extraction Algorithm from Fingerprint Image in Wavelet Domain

K. Sasirekha and K. Thangavel

Abstract The robustness of a fingerprint authentication system depends on the quality of the features extracted from the fingerprint image. For extracting good quality features, the quality of the image is to be improved through denoising and enhancement. In this paper, a set of invariant moment features are extracted from the approximation coefficient in the wavelet domain. Initially the fingerprint image is denoised using Stationary Wavelet Transform (SWT), a threshold based on Golden Ratio and weighted median. Then the denoised image is enhanced using Short Time Fourier Transform (STFT). A unique core point is then detected from the enhanced image by using complex filters to determine a Region of Interest (ROI), which is centered at the enhanced image. Then the ROI is decomposed using SWT at level one of Daubechies wavelet filter for extracting efficient features. The decomposed image is partitioned into four sub-images to reduce the effects of noise and nonlinear distortions. Finally a total of four sets of seven invariant moment features are extracted from four partitioned sub-images of an ROI of the approximation coefficient as it will contain low frequency components. To measure the similarity between feature vectors of an input fingerprint with the template stored in the database, the Euclidean Distance is employed for FVC2002 dataset. Using a simpler distance measure can substantially reduce the computational complexity of the system.

Keywords Fingerprint · Denoising · STFT · ROI · SWT · Moment features

K. Sasirekha (✉) · K. Thangavel
Department of Computer Science, Periyar University, Salem, India
e-mail: ksasirekha7@gmail.com

K. Thangavel
e-mail: drktvelu@yahoo.com

M. Senthilkumar et al. (eds.), *Computational Intelligence, Cyber Security and Computational Models*, Advances in Intelligent Systems and Computing 412, DOI 10.1007/978-981-10-0251-9_14

1 Introduction

Biometric authentication refers to verifying individuals based on their physiological and behavioral characteristics. Biometric methods provide a higher level of security and are more convenient for the user than the traditional methods of personal authentication such as passwords and tokens [1]. Among all the biometrics, fingerprint is a great source for identification of individuals [2, 3].

Fingerprint images are contaminated with noise during its acquisition and transmission. So far several techniques have been developed for reducing the noises in fingerprint image in spatial, frequency and wavelet domain [4]. Denoising by thresholding in the wavelet domain has been developed principally by Donoho [5]. Enhancement of the denoised image is necessary to improve the quality of the fingerprint. The most popular and widely used technique is STFT [6].

The accuracy of the fingerprint authentication system is greatly influenced by the quality of features extracted from the enhanced fingerprint image. Two main categories of the fingerprint verification system are minutiae based methods [7] and image based methods [8]. The minutiae based method may not use the rich discriminatory information available in the fingerprints and it is complex. Gabor filter based method [8] gives the best accuracy than other image based methods but it requires a larger storage space and a significantly high processing time. Various transform based methods using Discrete Wavelet Transform (DWT) and Discrete Cosine Transform (DCT) [9] were also proposed. However, these methods have not considered the invariance to an affine transform. Moment invariants have been widely applied to image pattern recognition in a variety of applications due to its invariant features on image translation, scaling and rotation. Moment invariants are first introduced by Hu [10].

In this paper, an efficient feature extraction method in wavelet domain using SWT is proposed. Initially, the noise is removed from the acquired fingerprint image using our previously proposed modified universal threshold in wavelet domain. The denoised image is enhanced using STFT and the ROI is extracted by identifying the core point using complex filters. Then the ROI is decomposed using SWT at level one of Daubechies wavelet filter. Finally, the proposed method extracts the Hu's invariant moments from the approximation coefficient as it is noise free subband.

The paper is organized as follows. In Sect. 2, the fingerprint image is denoised, enhanced and then ROI is extracted. In Sect. 3, the feature extraction in wavelet domain is proposed. Section 4 provides experimental results of the proposed method on fingerprint images. Finally this paper concludes with some perspectives in Sect. 5.

2 Fingerprint Image Denoising and Enhancement

2.1 Denoising Using Modified Universal Threshold

Denoising of the fingerprint image is indispensable to get a noise free fingerprint image. The SWT is studied over the conventional DWT as it is not a time-invariant transform [11]. Figure 1 shows the process of fingerprint denoising.

The modified universal threshold based on golden ratio and weighted median is given as [11]:

$$T = \sigma\left(\sqrt{1.618 * Log(N)}\right) \tag{1}$$

where (σ) is the noise variance and N is the number of elements in the image. Weighted median is used to compute the median of the high pass portion of the image instead of the conventional method. The following classical weight function (W) [12] is adopted for computing the weighted coefficient of the diagonal subband (HH) which is given by (2),

$$W(x, y) = \frac{1}{e^{|HH(x,y)|}} \tag{2}$$

where x and y are the coordinates in the HH subband. The weight W will be multiplied with HH to get weighted diagonal subband as given in (3),

$$HH_1(x, y) = W(x, y) * HH(x, y) \tag{3}$$

The noise variance (σ) is then calculated from the weighted diagonal subband (HH1) as given in (4),

$$\sigma = \frac{Median(HH_1)}{0.6745} \tag{4}$$

After computing the noise variance, the modified universal threshold is applied to the noisy fingerprint image and reconstructs using ISWT to get noise free image.

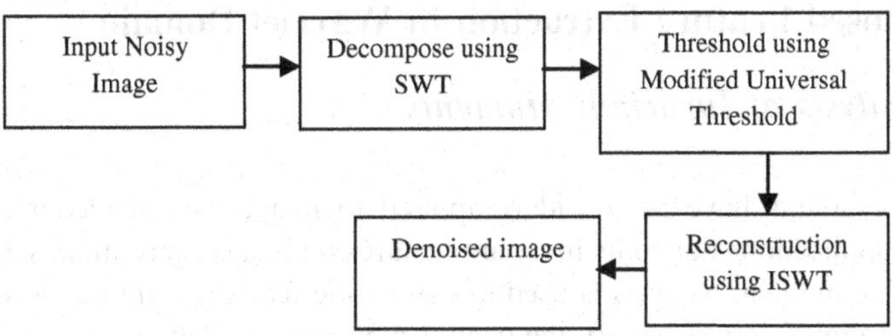

Fig. 1 Block diagram for denoising in wavelet domain

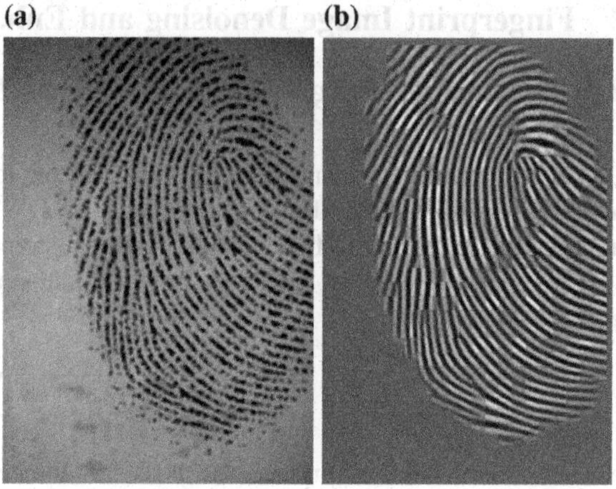

Fig. 2 Fingerprint enhancement using STFT. **a** Original image. **b** Enhanced image

2.2 *Fingerprint Image Enhancement and ROI Detection*

2.2.1 Image Enhancement Using STFT

The denoised fingerprint image is enhanced using contextual filtering in the Fourier domain [13]. Figure 2 depicts the enhanced image using STFT.

2.2.2 ROI Detection

The core point is determined from the enhanced image for detecting the ROI. The reliable detection of the position of a reference point can be accomplished by detecting the maximum curvature using complex filtering methods [14]. The orientation of the reference point is determined by using LMS orientation algorithm. The ROI around the core point is determined for feature extraction. The size of the ROI can be experimentally determined. In this paper, a size of 64 × 64 ROI is used as in [14].

3 Proposed Feature Extraction in Wavelet Domain

3.1 *Analysis of Invariant Moments*

Moment invariants have been widely applied to image pattern recognition in a variety of applications due to its invariant features on image translation, scaling and rotation. The moment analysis is used to extract the features from the decomposed ROI. For a digital image I(x, y), the central moments are defined as

$$\mu_{pq} = \sum_x \sum_y (x - \bar{x})^p (y - \bar{y})^q I(x, y) dxdy \tag{5}$$

The normalized central moment is defined as

$$\eta_{pq} = \frac{\mu_{pq}}{\mu_{00}^{\gamma}} \text{ where } \gamma = \frac{p+q}{2} + 1 \tag{6}$$

For p + q = 2, 3, …

A set of seven invariant moments can be derived from the second and third moments proposed by Hu [10]. Hu described M_1–M_6 as absolute orthogonal invariants and M_7 as a skew orthogonal invariant. The Hu set of invariant moments are given below:

$$\begin{aligned}
M_1 &= \eta_{20} + \eta_{02} \\
M_2 &= (\eta_{20} - \eta_{02})^2 + 4\eta_{11}^2 \\
M_3 &= (\eta_{30} - 3\eta_{12})^2 + (3\eta_{21} - \eta_{03})^2 \\
M_4 &= (\eta_{30} + \eta_{12})^2 + (\eta_{21} + \eta_{03})^2 \\
M_5 &= (\eta_{30} - 3\eta_{12})(\eta_{30} + \eta_{12})[(\eta_{30} + \eta_{12})^2 - 3(\eta_{21} + \eta_{03})^2] \\
&\quad + (3\eta_{21} - \eta_{03})(\eta_{21} + \eta_{03})[3(\eta_{30} + \eta_{12})^2 - (\eta_{21} + \eta_{03})^2] \\
M_6 &= (\eta_{20} - \eta_{02})[(\eta_{30} + \eta_{12})^2 - (\eta_{21} + \eta_{03})^2] + 4\eta_{11}(\eta_{30} + \eta_{12})(\eta_{21} + \eta_{03}) \\
M_7 &= 3(\eta_{21} - \eta_{03})(\eta_{30} + \eta_{12})[(\eta_{30} + \eta_{12})^2 - 3(\eta_{21} + \eta_{03})^2] \\
&\quad - (\eta_{30} - 3\eta_{12})(\eta_{21} + \eta_{03})[3(\eta_{30} + \eta_{12})^2 - (\eta_{21} + \eta_{03})^2]
\end{aligned}$$

3.2 *Feature Extraction in Wavelet Domain*

Wavelets are mathematical functions that analyze data according to scale or resolution. Figure 3 shows the block diagram of the proposed method. The extracted ROI is decomposed using SWT at level one. The daubechies filter is used as a wavelet filter. Figure 4 shows the decomposition of ROI into four subbands.

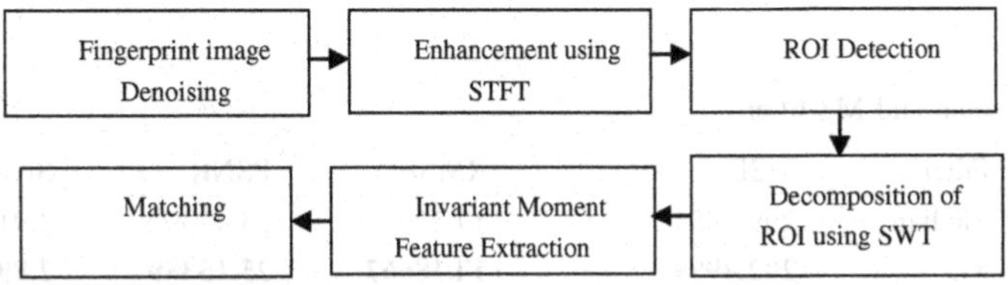

Fig. 3 Block diagram of the proposed method

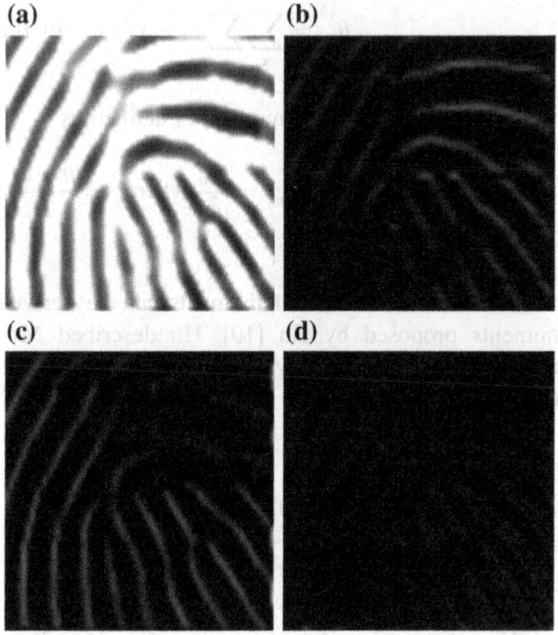

Fig. 4 Decomposition of ROI. **a** Approximation. **b** Horizontal. **c** Vertical. **d** Diagonal

4 Experimental Results

4.1 Denoising Fingerprint Image

The fingerprint database used in this work is FVC2002 [15]. The performance results in spatial domain using median filter and M3 filter for Gaussian noise is given in Table 1 for level 0.003.

The performance results of traditional universal threshold with Gaussian noise based on hard thresholding is given in Table 2 for noise level 0.003. The performance results of proposed modified universal threshold based on GR and weighted median with Gaussian noise of level as 0.003 based on thresholding is given in Table 3.

From the above tables and from Fig. 5 it clear that the performance of proposed modified universal threshold based on GR and weighted median with hard thresholding is improved than median filter, M3 filter and traditional universal threshold.

Table 1 Median and M3 filter

S. no	Filter	MSE		RMSE	PSNR	SNR
1	Median	299.1854		17.22912	23.47392	2.917146
2	**M3**	**202.4989**		**14.18667**	**25.15389**	**2.808041**

Table 2 Traditional universal threshold with hard thresholding

S. no	Filter	MSE	RMSE	PSNR	SNR
1	coif2	105.1573	10.20314	28.03343	0.688197
2	coif4	112.4269	10.55088	27.74158	0.667238
3	sym2	100.9692	9.992777	28.2189	0.805166
4	sym4	104.6807	10.18133	28.05088	0.690382
5	db1	103.7121	10.10698	28.13845	0.891723
6	**db2**	**100.833**	**9.98619**	**28.22458**	**0.806413**
7	db4	106.1137	10.25298	27.98811	0.705205

Table 3 Proposed modified universal threshold based on GR and weighted median with hard thresholding

S. no	Filter	MSE	RMSE	PSNR	SNR
1	coif2	0.094375	0.306977	58.42915	0.001189
2	coif4	0.096344	0.310167	58.33925	0.00113
3	sym2	0.089106	0.298275	58.67923	0.001111
4	sym4	0.096615	0.310585	58.32809	0.001235
5	db1	0.104979	0.322996	58.0118	0.000606
6	**db2**	**0.088247**	**0.296831**	**58.72143**	**0.001019**
7	db4	58.14849	58.14849	58.14849	58.14849

4.2 Invariant Moment Feature Extraction

In order to reduce the effects of noise and nonlinear distortions, the decomposed ROI is partitioned into four smaller sub-images (4 quadrants) in an anticlockwise sequence as depicted in Fig. 6. So each sub-image has a size of 32 × 32.

For each sub-image, a set of seven invariant moments is computed, so four sets of invariant moments of the sub-images are extracted as features to represent a

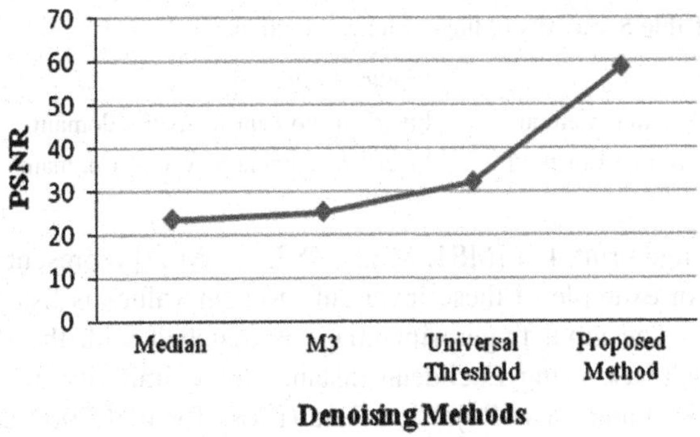

Fig. 5 Comparison of PSNR for different denoising methods

Fig. 6 ROI partition into four sub-image

Table 4 Seven moments with the different sub-images of the fingerprint image (m = invariant moment, i = sub-images)

	I_1	I_2	I_3	I_4
M_1	6.659714	6.653635	6.662685	6.661865
M_2	20.20554	22.17764	19.95373	19.95373
M_3	25.54887	28.61435	27.35306	27.35306
M_4	30.12149	29.69914	28.87533	28.87533
M_5	58.0675	60.01646	57.39719	57.39719
M_6	40.60153	40.83938	39.97735	39.97735
M_7	59.20406	59.04041	57.90297	57.90297

Table 5 Results of fingerprint recognition

	Feature extraction	RR (%)	FRR (%)
Existing method	Invariant moment in spatial domain	96.5	3.5
Proposed method	Invariant moment in wavelet domain	98.6	1.4

fingerprint. Let [MS1, MS2, MS3, …, MS7] represents a set of invariant moments. An example of these invariant moment values is listed as in Table 4.

The input fingerprint image is matched with the template image stored in the database using Euclidean distance as a similarity measure. The Recognition Rate (RR) and False Rejection Rate (FRR) for FVC2002 dataset is given in Table 5.

5 Conclusion

In this paper, a set of invariant moment features are extracted from the approximation coefficient of the decomposed ROI in the wavelet domain. Initially the fingerprint image is denoised using Stationary Wavelet Transform (SWT), a threshold based on Golden Ratio and weighted median. Short Time Fourier Transform is used to enhance the denoised image. The complex filter is used to locate the core point. The ROI is extracted based on the core point which is centered at the enhanced image. Then the ROI is decomposed using SWT at level one of daubechies wavelet filter for efficient feature extraction. A set of invariant moment features are extracted from partitioned sub-images of an ROI of the approximation coefficient as it is noise free. This method is efficient even for low quality fingerprint image and the modified universal threshold is simple in computation and yet it is effective in denoising. The proposed method has better performance in matching and got accuracy of 98.6 % using Euclidean Distance as a similarity measure for FVC2002 dataset.

References

1. Sutcu, Y., Tabassi, E., Sencar, H.T., Memon, N.: What is biometric information and how to measure it?. IEEE Int. Conf. Technol. Homel. Secur. (HST) 12–14, (2013)
2. Sasirekha, K., Thangavel, K.: A comparative analysis on fingerprint binarization techniques. Int. J. Comput. Intell. Inf. **4**(3) (2014)
3. Sasirekha, K., Thangavel, K., Saranya, K.: Cryptographic key generation from multiple fingerprints. Int. J. Comput. Int. Inf. **2**(4) (2013)
4. Kanagalakshmi, K., Chandra, E.: Performance evaluation of filters in noise removal of fingerprint image. Int. Conf. Electron. Comput. Technol. (ICECT) **1**, 117–121 (2011)
5. Donoho, David L.: Denoising by soft-thresholding. IEEE Trans. Inf. Theory **41**(3), 613–627 (1995)
6. Chikkerur, S., Cartwright, A.N., Govindaraju, V.: Fingerprint enhancement using STFT analysis. Pattern Recogn. **40**(1), 198–211 (2007)
7. Jang, X., Yau, W.Y.: Fingerprint minutiae matching based on the local and global structures. Proc. Int. Conf. Pattern Recognit. **2**, 1024–1045, (2000)
8. Jain, A.K., Prabhakar, S., Hong, L., Pankanti, S.: Filterbank-based fingerprint matching. IEEE Trans. Image Process. **9**, 846–859 (2000)
9. Amornraksa, T., Tachaphetpiboon, S.: Fingerprint recognition using DCT features. Electron. Lett. **42**(9), 522–523 (2006)
10. Hu, M.K.: Visual pattern recognition by moment invariants, IRE Trans. Inform. Theory IT-8, pp. 179–187, (1962)
11. Sasirekha, K., Thangavel, K.: A novel wavelet based thresholding for denoising fingerprint image. IEEE Int. Conf. Electron. Commun. Comput. Eng. 119–124, (2014)
12. Wang, C., Li, L., Yang, F., Gong, H.: A new kind of adaptive weighted median filter algorithm. IEEE Int. Conf. Comput. Appl. Syst. Model. **11**, 667–671 (2010)
13. Yang, J.C., Park, D.S.: Fingerprint verification based on invariant moment features and nonlinear BPNN. Int. J. Control, Autom. Syst. **6**(6):800–808 (2008)
14. Nilsson, K., Bigun, J.: Localization of corresponding points in fingerprints by complex filtering. Pattern Recogn. Lett. **24**, 2135–2144 (2003)
15. http://bias.csr.unibo.it/fvc2002/download.asp

Motor Imagery Classification Based on Variable Precision Multigranulation Rough Set

K. Renuga Devi and H. Hannah Inbarani

Abstract In this work classification based on Variable Precision Multigranulation Rough Set for motor imagery dataset is proposed. The accurate design of BCI (Brain Computer Interface) depends upon efficient classification of motor imagery movements of patients. In the first phase pre-processing is carried out with Chebyshev type2 filter in order to remove the noises that may exist in signal during acquisition. The daubechies wavelet is used to extract features from EEG Signal. Finally classification is done with Variable Precision Multigranulation Rough Set. An experimental result depicts higher accuracy according to variation of alpha and beta values in Variable Precision multigranulation rough set.

Keywords Chebyshev type 2 filter · Daubechies wavelet · Variable precision multigranulation rough set

1 Introduction

A brain-computer interface (BCI) provides a direct functional interaction between the human brain and the external device. Alternative ways of communication and control is needed for people with severe motor disabilities spinal cord injury (SCI), amyotrophic lateral sclerosis (ALS) to do their everyday activities [1]. To help such people BCI is developed which is used for assisting sensory motor function. To design BCI, EEG signals are recorded when the patients perform motor imagery tasks. Motor imagery involves imagination of various body parts resulting in sensorimotor cortex activation which modulates sensorimotor oscillation in EEG [2, 3]. The frequency of the sensorimotor rhythm is in the range of 13–15 Hz [1]. Some of the applications of BCI are helping paralyzed, video games and virtual reality, creative expression, neural prosthetics, wheelchairs [1]. An efficient algorithm for

K. Renuga Devi (✉) · H. Hannah Inbarani
Department of Computer Science, Periyar University, Salem, Tamilnadu, India
e-mail: renuga.star@yahoo.co.in

M. Senthilkumar et al. (eds.), *Computational Intelligence, Cyber Security and Computational Models*, Advances in Intelligent Systems and Computing 412, DOI 10.1007/978-981-10-0251-9_15

classifying different user commands is an important part of a brain-computer interface. The goal of this paper is to classify different motor imagery tasks, left hand, right hand, both feet, or tongue. The dataset are collected from BCI Competition III and IV, which is an open competition concerning BCI-related data classification, then Variable precision multigranulation rough set is applied and classification accuracy is evaluated. We test the algorithms with different parameters on the training set using cross-validation and then verify the best four of them on the final test set. This result shows that it is possible to use a variable precision multigranulation algorithm for EEG data classification and obtain good results compared to existing techniques. Section 2 discusses the preliminaries, Sect. 3 depicts proposed algorithm, Sect. 4 depicts results and discussion and Sect. 5 concludes the proposed work.

2 Preliminaries

2.1 *Variable Precision Multigranulation Rough Set*

The extended rough set model Variable Precision Rough Set is the generalization of standard set inclusion relation [4–7]. The extended notion should be able to allow for some degree of misclassification in largely correct classification [4–7]. The Multigranulation Rough Set (MGRS) is constructed on the basis of a family of binary relations instead of a single indiscernibility relation, where the set approximations are defined by using multiple equivalence relation [8]. The form of decision rules in MGRS is "OR" unlike the "AND" rules from Pawlak's rough set model [8]. VPRS measurement is based on ratios of elements contained in various sets [9–11]. The positive region of the set X contains elements of U which can be classified into X with classification error not greater than β [9]. The β Boundary region of X consists of all elements which cannot be classified either into X or into ~X with the classification error not greater than β [9]. The β Negative region of X consists of all elements which can be classified into complement of X or into ~X with the classification error not greater than β [9]

$$Pos_p^{\beta}(z) = \bigcup_{pr(z/x_i) \geq \beta} \{Xi\varepsilon E(P)\} \text{ with } P \subseteq C \quad [9] \tag{1}$$

$$Bnd_p^{\beta}(z) = \bigcup_{1-\beta < pr(z/x_i) < \beta} \{Xi\varepsilon E(P)\} \text{ with } P \subseteq C \quad [9] \tag{2}$$

$$Neg_p^{\beta}(z) = \bigcup_{pr(z/x_i) \leq 1-\beta} \{Xi\varepsilon E(P)\} \text{ with } P \subseteq C \quad [9] \tag{3}$$

In Eq (1), (2) and (3) $z \subseteq U$. Variable Precision Multigranulation Rough Set is general notions of set approximations. It allows for a controlled degree of misclassification [12]. In the variable precision multi-granulation rough set model, the requirement of accuracy on each granulation is determined by means of parameter $\alpha_{(\text{threshold})}$, $\beta_{(\text{threshold})}$, w_i^{α} and $\mu_{pi}^{X}(x)$ [12]. Let $S = (U; A)$ be an information system, U is a non-empty and finite set of objects, called a universe, and A is a non-empty and finite set of attributes, $X \subseteq U$ and $P = \{P_i \subseteq A | P_i \cap P_j = \phi(i \neq j), i, j \leq l\}$ [12]. l is the number of partitions. $\mu_{pi}^{X}(x)$, w_i^{α} are the supporting characteristic function. $\mu_{pi}^{X}(x)$ is computed by considering equivalence relation of decision and partition [12]. The parameter α determines the precision of every granulation which are used to approximate the target concept [12]. $\sim$ denotes complementary operation of the set. These partition contains different elements that have same equivalence class. Similarily for each decision equivalence class are found. Then lower and upper approximation sets of X with respect to P are [12]

$$\underline{VP}(x)_{\beta}^{\alpha} = \{x \varepsilon \cup | \sum_{i=1}^{l} w_i^{\alpha} \mu_{pi}^{x}(x) \geq \beta\} \quad [10] \tag{4}$$

$$\overline{VP}(x)_{\beta}^{\alpha} = \sim \underline{P}(\sim X)_{\beta}^{\alpha} \quad [10] \tag{5}$$

$$wi^{\alpha} = \begin{cases} \frac{1}{l} \alpha \leq \mu_{pi}^{X}(x) \leq l \\ 0 \ \mu_{pi}^{X}(x) < \alpha \end{cases} \quad [10] \tag{6}$$

3 Proposed Methodology

In the first step BCI Competition IV data set 1 and BCI Competition III data set 3a are acquired from website http://www.bbci.de/competition/download. Next step involves preprocessing using cheby2 filter to remove low and high frequency that exist in signal. Then daubechies wavelet is applied to extract features from signal. Feature Extraction mainly involves extraction of sensorimotor rhythm from the EEG signal. Then finally classification is done with variable precision multigranulation rough set and classification accuracy is evaluated using accuracy measures. Then it is compared with existing techniques.

3.1 Acquistion of Dataset

Motor Imagery data are collected from BCI Competition IV dataset 1. These data sets were recorded from healthy subjects. For each subject two classes of motor imagery were selected from the three classes left hand, right hand, and foot. The recording was made using Brain Amp MR plus amplifiers and an Ag/AgCl electrode cap. Signals from 59 EEG positions were measured that were most densely distributed over sensorimotor areas. Another Motor imagery dataset is collected from BCI Competition III dataset 3a which is Multiclass motor imagery data set. EEG was sampled with 250 Hz, it was filtered between 1 and 50 Hz. The task was to perform imagery left hand, right hand, foot or tongue movements according to a cue.

3.2 Chebyshev Type 2 Filter for Preprocessing

Chebyshev filters are used to separate one band of frequencies from another that has a flat pass band, a moderate group delay, and an equiripple stop band. It performs faster than other filter since they are carried out by recursion rather than convolution. In the BCI competition data set, our goal is to classify sensorimotor rhythm. Sensorimotor rhythm occurs in the range of mu and beta rhythm [1–3] and therefore the frequency band range is set between 4 and 40 Hz and it is low-pass filtered using the Chebyshev Type II filter of order 3 with a cut-off frequency of 0.7 Hz and Pass-band (ripple) attenuation 0.5 dB [13].

3.3 Daubechies Wavelet for Feature Extraction

The Discrete Wavelet Transform (DWT) with the Multi-Resolution Analysis (MRA) is applied to decompose EEG signal at resolution levels of the components of the EEG signal (δ, θ, α, β and γ) [2, 13]. The number of decomposition levels is chosen based on the dominant frequency components of the signal. In this work, Daubechies 4 (db4) is selected because its smoothing feature was suitable for detecting changes of the EEG signals [13]. The frequency band [fm/2: fm] of each detail scale of the DWT is directly related to the sampling rate of the original signal, which is given by fm = fs/2l + 1, where fs is the sampling frequency, and l is the level of decomposition [13]. In this study, the sampling frequency is 100 Hz of the EEG signal. The highest frequency that the signal could contain, from Nyquist' theorem, would be fs/2 [13]. The signals were decomposed into details D1–D5 and one final approximation A5. The frequency range from 25 to 50 contains Gamma Signal, 12.5–25 contains beta Signal, 6.25–12.5 contains Alpha and Mu Signal, 3.12–6.25 contains theta signal, 0–3.12 contains delta. Among this decomposition levels Mu and Beta contains sensorimotor rhythm which is passed as input.

3.4 Classification Based on VPMGRS

Algorithm for Variable Precision multigranulation Rough set is as follows. Input are decomposed frequency bands Mu, Beta of EEG Signal.

Algorithm:

Step 1: Find partitions for each attribute including decision using equivalence class based on multigranulation Rough set.

(i) The intersection of any two different cells is empty and the union of all the cells equals the original set. These cells are formally called equivalence classes.

(ii) Let us consider the Sample data U = $\{x_1, x_2, x_3, x_4, x_5, x_6, x_7, x_8, x_9, x_{10}\}$ and X denotes equivalence class of decision variable X = $\{x_1, x_2, x_6, x_8, x_9\}$. Let p_1, p_2 be the two partitions based on conditional attributes.

$$U/p_1 = \{\{x_1, x_7\}\{x_2, x_3, x_4, x_6, x_8, x_9\}\{x_5\}\{x_{10}\}\}$$
$$U/P_2 = \{\{x_1\}\{x_2, x_3, x_6, x_8, x_9\}\{x_4, x_5\}\{x_7\}\{x_{10}\}\}$$

Step 2: Find the Supporting Characteristic Function $\mu_{pi}^X(x)$ based on equivalence class of partition and decision.

(i) Equivalence class of decision variable X = $\{x_1, x_2, x_6, x_8, x_9\}$. The First equivalence class of partition p1 is$\{x_1, x_7\}$. Here Total Number of elements are 2. Only x_1 are in X so Supporting Characteristic Function for X_1 and X_7 are 1/2, $\mu_{P1}^X(x_1) = 0.5$, $\mu_{P1}^X(x_7) = 0.5$.

(ii) The second equivalence class of partition p1 is $\{x_2, x_3, x_4, x_6, x_8, x_9\}$. Here Total Number of elements are 6, X = $\{x_1, x_2, x_6, x_8, x_9\}$ only x_2, x_6, x_8, x_9 are in X so Supporting Characteristic Function is 4/6, $\mu_{P1}^X(x_2) = 0.6$, $\mu_{P1}^X(x_3) = 0.6$, $\mu_{P1}^X(x_4) = 0.6$, $\mu_{P1}^X(x_6) = 0.6$, $\mu_{P1}^X(x_8) = 0.6$, $\mu_{P1}^X(x_9) = 0.6$ Partition p_1 and p_2 supporting characteristic function values of each element are

$$\begin{array}{lll} \mu_{P1}^X(x_1) = 0.5 & \mu_{P1}^X(x_2) = 0.6 & \mu_{P1}^X(x_3) = 0.6 \\ \mu_{P1}^X(x_4) = 0.6 & \mu_{P1}^X(x_5) = 0 & \mu_{P1}^X(x_6) = 0.6 \\ \mu_{P1}^X(x_7) = 0.5 & \mu_{P1}^X(x_8) = 0.6 & \mu_{P1}^X(x_9) = 0.6 \\ \mu_{P1}^X(x_{10}) = 0 & \mu_{P2}^X(x_1) = 1 & \mu_{P2}^X(x_2) = 0.6 \\ \mu_{P2}^X(x_3) = 0.6 & \mu_{P2}^X(x_4) = 0 & \mu_{P2}^X(x_5) = 0 \\ \mu_{P2}^X(x_6) = 0.6 & \mu_{P2}^X(x_7) = 0 & \mu_{P2}^X(x_8) = 0.6 \\ \mu_{P2}^X(x_9) = 0.6 & \mu_{P2}^X(x_{10}) = 0 & \end{array}$$

Step 3: Compute Supporting Characteristic Function wi$^\alpha$ based on $\mu_{pi}^X(x)$

$$wi^\alpha = \begin{cases} \frac{1}{l} \alpha \le \mu_{pi}^X(x) \le l \\ 0 \; \mu_{pi}^X(x) < \alpha \end{cases} \quad (7)$$

wi$^\alpha$ varies according to the $\mu_P^X(x_i)$. If $\mu_P^X(x_i)$ is greater than α then wi$^\alpha$ is 1/l, where l is number of partitions. If $\mu_P^X(x_i)$ is less than α then wi$^\alpha$ is 0.

Step 4: If the Product of wi$^\alpha$ and $\mu_{pi}^X(x)$ is greater than or equal to β Then it is added to lower approximation. The Complementary Operation of lower approximation gives upper approximation.

$$\underline{VP}(x)_\beta^\alpha = \{x\varepsilon\cup| \sum_{i=1}^{l} w_i^\alpha \mu_{pi}^x(x) \ge \beta\} \quad (8)$$

$$\overline{VP}(x)_\beta^\alpha = \sim \underline{P}(\sim X)_\beta^\alpha \quad (9)$$

For Example

Let us consider $\alpha = 0.3$, $\beta = 0.3$, $\mu_{P1}^X(x_1) = 0.5$, $\mu_{P2}^X(x_1) = 1$

(i) According to Eq. (7) Since $\alpha < \mu_{P1}^X(x_1)$ wi$^\alpha$ = 1/l = 1/2 = 0.5, Where l is number of partitions

(ii) Since $\alpha < \mu_{P2}^X(x_1)$, wi$^\alpha$ = 1/l = 1/2 = 0.5.

(iii) According to Eq. (8) Product of wi$^\alpha$, $\mu_{P1}^X(x_1)$ + wi$^\alpha$, $\mu_{P2}^X(x_1)$ = 0.5*0.5 + 0.5*1 = 0.75

(iv) Since the Product 0.75 > β so x_1 is added to lower approximation

$$\underline{VP}(x)_{0.3}^{0.3} = \{\mathbf{x_1, x_2, x_3, x_4, x_6, x_8, x_9}\}$$
$$\underline{VP}(x)_{0.6}^{0.6} = \{\mathbf{x_2, x_3, x_6, x_8, x_9}\}$$
$$\underline{VP}(x)_{0.3}^{0.7} = \{\mathbf{x_1}\}$$

Step 5: Rules are generated based on lower approximation.

Step 6: Classification is performed using VPMGRS and accuracy are evaluated using accuracy Measure.

4 Experimental Results and Discussion

4.1 *Experimental Results of Proposed Methodology*

In Variable precision multigranulation rough set accuracy varies according to the alpha and beta values. Classification accuracy are evaluated using accuracy measures. Accuracy is the proportion of true results (both true positives and true

Table 1 Accuracy calculation of BCI competition IV data set 1

Beta values	Alpha values	Precision		Classification accuracy
		class 1 hand	class 2 foot	
0.1	0.1–0.6	1	1	100
	0.7	0.9615	0.8529	90
	0.8	0.7500	0.6667	70
0.2	0.1–0.5	1	1	100
	0.6	0.9231	0.8235	86.66
	0.7	0.7619	0.6410	68.33
0.3	0.1–0.2	1	1	100
	0.3	0.9091	1	95
	0.4	0.8824	1	93.33
	0.5	0.8571	0.8125	83.33
	0.6	0.7368	0.6098	65

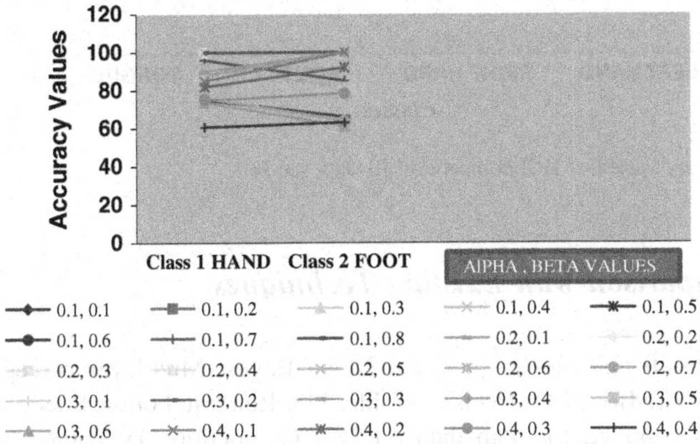

Fig. 1 Accuracy values of BCI competition IV data set 1

negatives) among the total number of cases examined. Table 1 and Fig. 1 depicts the accuracy values according to variation of alpha and beta values of BCI Competition IV data set 1. Table 2 and Fig. 2 depicts the accuracy values according to variation of alpha and beta values of BCI Competition III data set 3a.

$$Accuracy = \frac{\text{Number of Correct Predictions}}{\text{Number of Correct Prediction} + \text{Number of incorrect Prediction}} \quad (10)$$

Table 2 Accuracy for BCI competition III data set 3a

Beta values	Alpha values	Precision				Classification accuracy
		Class-1 left hand	Class-2 right hand	Class-3 foot	Class-4 tongue	
0.1	0.1–0.2	1	1	1	1	100
	0.3	0.8333	0.5833	1	0.6364	67.79
0.2	0.1	0.9333	0.9286	0.9286	0.8750	91.52
	0.2	0.7143	0.4444	0.6842	0.8667	66.17

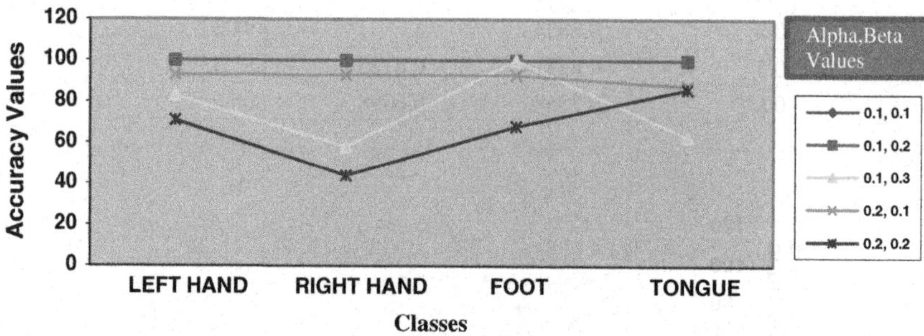

Fig. 2 Accuracy values of BCI competition III data set 3a

4.2 *Comparison with Existing Techniques*

Existing Classification techniques like Naïve Bayes, Multilayer perceptron, IBk, Kstar, LWL, AdaBoostM1, Decision Table, J48, Random Forest, Random Tree and Rep Tree are applied for both data set BCI Competition IV dataset 1 and BCI Competition III data set 3a and classification accuracies are analyzed. It is clear from the Table 3 and Fig. 3 depicts existing techniques classification accuracies are less than 55 %. Considering classification accuracy for BCI Competition IV Dataset 1 maximum accuracy of 50.83 is given by IBk which is less than proposed technique in this work. Similarily considering classification accuracy for BCI Competition III data set 3a maximum accuracy of 55.03 is given by J48 less than proposed work.

Table 3 Classification accuracy with existing techniques

Classification technique	Classification accuracy for BCI competition IV dataset 1	Classification accuracy for BCI competition III dataset 3a
KNN	48.33	20.33
Naïve Bayes	33.50	50.90
Multilayer Perceptron	41.17	48.83
IBk	50.83	52.67
Kstar	49.17	53.23
LWL	46.33	36.70
AdaBoostM1	38.67	24.13
Decision table	49.17	26.80
J48	45.33	55.03
Random forest	47.50	50.63
Random tree	49.67	47.57
Rep tree	47.50	37.50

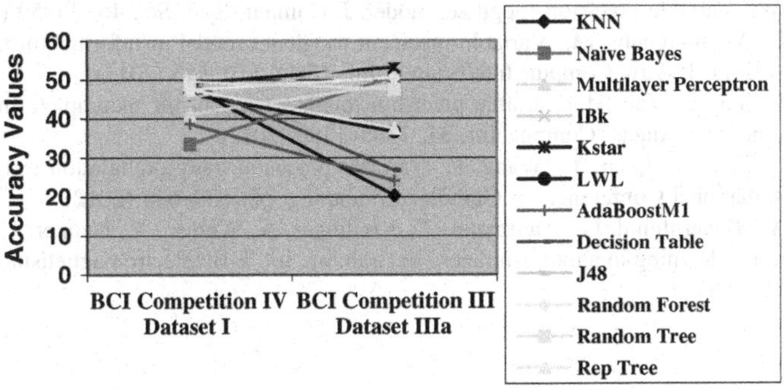

Fig. 3 Accuracy values of existing classification algorithm

5 Conclusion

Thus the BCI Competition IV dataset 1, BCI Competition III data set 3a shows higher accuracy when compared with existing techniques. BCI Competition IV dataset 1 is a two class problem (Right Hand, Foot). BCI Competition III data set 3a is a four class problem left hand, right hand, both feet, and tongue. Variable precision multigranulation rough set involves partition by multigranulation rough set and then based on partition μ values are computed and supporting charactersitic function are computed based on partition. Both Supporting Characteristic Function w_i^{α}, μ, α, β values determine classification accuracy. The variation of α, β values in both BCI Competition Data set depicts accuracy of more than 90 % and is disscused in detail in experimental results and discussion.

References

1. Alexander, A.F., Husek, D., Snasel, V., Bobrov, P., Mokienko, O., Tintera, J., Rydlo, J.: Brain-computer interface based on motor imagery: the most relevant sources of electrical brain activity. In: Proceedings of Soft computing in Industrial Applications, pp. 1–10 (2012)
2. Dokare, I., Kant, N.: Classification of EEG signal for imagined left and right hand movement for brain computer interface applications, biomedical engineering: application, basics and communication, p. 26 (2014)
3. Niazi, I.K., Jiang, N., Jochumsen, M., Nielsen, J.F., Dremstrup, K., Farina, D.: Detection of movement-related cortical potentials based on subject-independent training, Med. Biol. Eng. Comput. **51**, 507–512 (2013)
4. Beynon, M.J., Peel, M.J.: Variable precision rough set theory and data discretisation: an application to corporate failure prediction. Omega **29**, 561–576 (2001)
5. Inuiguchi, M.: Attribute reduction in variable precision rough set model. Int. J. Uncertainty Fuzziness Knowl. Based Syst. **14**(4), pp. 61–479 (2006)
6. Inuiguchi, M.: Structure-based attribute reduction in variable precision rough set models. J. Adv. Comput. Intell. Intell. Inform. **10**(5), 657–665 (2006)
7. Ningler, M., Stockmanns, G., Schneider, G., Kochs, H.-D., Kochs, E.: Adapted variable precision rough set approach for EEG analysis. Artif. Intell. Med. **47**, 239–261 (2009)
8. Qian, Y., Liang, J., Yao, Y., Dang, C.: MGRS: a multigranulation rough set, information sciences, pp. 1–22 (2009)
9. Ziarko, W.: Variable precision rough set model. J. Comput. Syst. Sci. **46**, 39–59 (1993)
10. Kusunoki, Y., Inuiguchi, M.: Variable precision rough set model in information tables with missing values. J. Adv. Comput. Intell. Intell. Inf. **15**(1), 110–116 (2011)
11. Gong, Z., Shi, Z., Yao, H.: Variable precision rough set model for incomplete information systems and its –reducts. Comput. Inf. **31**, 1385–1399 (2012)
12. Wei, W., Liang, J., Qian, Y., Wang, F.: Variable precision multi-granulation rough set. In: IEEE International Conference on Granular Computing, pp. 639–643 (2012)
13. Kaiser, V., Bauernfeind, G., Kaufmann, T., Kreilinger, A., Kübler, A., Neuper, C.: Cortical effects of user learning in a motor-imagery bci training. Int. J. Bioelectromagnetism **13**(2), 60–61 (2011)

Fault Tolerant and Energy Efficient Signal Processing on FPGA Using Evolutionary Techniques

Deepa Jose and Roshini Tamilselvan

Abstract In this paper, an energy efficient approach using field-programmable gate array (FPGA) partial dynamic reconfiguration (PDR) is presented to realize autonomous fault recovery in mission-critical (space/defence) signal processing applications at runtime. A genetic algorithm (GA) based on adaptive search space pruning is implemented, for reducing repair time thus increasing availability. The proposed method utilizes dynamic fitness function evaluation, which reduces the test patterns for fitness evaluation. Hence, the scalability issue and large recovery time associated with refurbishment of larger circuits is addressed and improved. Experiments with case study circuits, prove successful repair in minimum number of generations, when compared to conventional GA. In addition, an autonomous self-healing system for FPGA based signal processing system is proposed using the presented pruning based GA for intrinsic evolution with the goal of reduced power consumption and faster recovery time.

Keywords Signal processing · Genetic algorithms · Intrinsic evolution · Reconfigurable hardware · Intrinsic evolution · Power consumption · FPGA

1 Introduction

Evolvable hardware (EH) merges the efficient search capability of evolutionary algorithms (EA) with the flexibility of reprogrammable devices, thereby offering a natural framework for reconfiguration. EAs are commonly used as a method for

D. Jose (✉)
Department of ECE, KCG College of Technology, Karapakkam, Chennai 600097, India
e-mail: deepa.ece@kcgcollege.com

R. Tamilselvan
Department of Electronics and Communication, College of Engineering, Anna University, Guindy, Chennai 600025, India

M. Senthilkumar et al. (eds.), *Computational Intelligence, Cyber Security and Computational Models*, Advances in Intelligent Systems and Computing 412, DOI 10.1007/978-981-10-0251-9_16

finding an optimized solution to solve algorithmic problems which have complex or incomplete formulations. EH utilizes EA for the automatic design of circuits to map tasks into hardware [1, 2]. EH methods are used for self-repair, in order to restore lost functionality of digital logic circuits implemented in reprogrammable logic devices such as Field Programmable Gate Arrays (FPGA). This framework has gained significance in using EH for fault-tolerant systems since reconfiguration is proved to heal hardware faults. EH techniques aim to provide fault recovery capability without incurring the penalties of degraded system performance in the final hardware.

Recent research has explored the possibility of using GA techniques to increase FPGA reliability and autonomy [3–7]. In extrinsic evolvable hardware, circuit evolution is performed off-line using simulators in powerful computers to find appropriate solutions. The solution obtained is finally implemented in hardware. In intrinsic evolvable hardware the EA is included in the final system and every candidate solution is evaluated in hardware. With increasing size of configurations/applications to be regenerated on the FPGA, the search space increases resulting in an intractable problem for the GA to solve. As VLSI systems are becoming more complex and large, it is essential to modify already existing GAs for self recovery of such systems. The aim of this work is to extend the feasibility of using EH techniques by dynamically reducing the search space during run-time. The fault location information is used to dynamically prune the search space while attaining complete quality of recovery. Thus fault recovery time is reduced resulting in increased availability of the system. In contrast to the previous approaches, the proposed approach deals with look-up-tables (LUT). Hence, this approach offers fine grain recovery granularity. The main characteristic of the existing evolutionary techniques is their dependence upon either exhaustive functional fitness evaluation or exhaustive resource testing during regeneration. Exhaustive fitness function evaluation requires test pattern application to all the inputs. It is proved that, exhaustive evaluation increases computational complexity, fault recovery time and power consumption [8, 9]. To avoid the same, the present approach makes use of repeatability of the processing architecture to reduce the test pattern application for fault detection and fitness evaluation thereby reducing power consumption as well as fitness evaluation time.

2 Related Work

GA-based fault-handling schemes in FPGAs have been successfully employed as a reconfiguration mechanism for autonomous self-adaptive EH systems. These approaches utilize exhaustive testing for fault isolation and offline regeneration mechanisms [10]. A combinational approach of TMR with GAs and other n-plex spatial voting techniques deliver real-time fault-handling. But these techniques increase power consumption and hardware overhead n-fold during fault-free operation Moreover, these techniques require exhaustive evaluation of the function being refurbished. STARS is a model of exhaustive evaluation of a resource-oriented diagnostic that executes built-in self-test (BIST) on sub-sections of FPGA [11]. Segments of the FPGA are repeatedly taken offline in sequence for testing. At the same time, the functionality is transferred to a new location within the reprogrammable fabric. The shortcomings of this approach are fault detection latency and increase in power consumption. A competitive runtime (CRR) fitness assessment scheme based on pairwise discrepancy detection between competing configurations is presented in [12]. This approach does not require additional test vectors for exhaustive fitness calculation. However, fitness evaluation without exhaustive testing leads to undesirable emergent behaviors over a period of time. In case of CRR, one potential risk is functional drift occurring over a period of time. During this time period repeated repairs accumulate on long missions encountering multiple failures. This approach also requires larger number of generations for healing smaller circuits. In [13], a quadrature decoder having 16 LUTs implemented in Xilinx Virtex FPGA is successfully evolved in the presence of a stuck-at fault. A genetic representation for evolutionary fault recovery in Xilinx FPGAs is also presented. An approach to reduce the overall evolution runtime by operating directly on the bitstream used to configure the reconfigurable device is presented in [14]. Self recovery of a 4 bit × 4 bit adder on a Xilinx Virtex-II Pro FPGA device is achieved in as low as 0:4 s. A complete implementation of an EH system composed of a 2D array of 16 processing elements (PEs) on a Xilinx Virtex-5 FPGA chip is demonstrated in [15]. Here the evolutionary algorithm has the ability to reconfigure the functionality of PEs. The EA is run on the embedded microprocessor with the ability to internally reconfigure the PEs through ICAP [15].

3 Modified GA for Fault Repair

3.1 Search Space Pruning

The proposed GA is presented in Fig. 1. The GA used here is a variation of the elitist algorithm. Unlike the pre-existing GA, the present work uses heuristics to reduce the number of LUTs involved in the mutation and crossover operations. These heuristics are also used to decide the cross over operation. The algorithm requires backtracking the connection of outputs to the LUTs. For the pruning technique, two different techniques are used based on whether one or multiple outputs of the circuit are at fault. If one output is at fault, then all LUTs which are connected to non-faulty outputs are not marked. This will drastically reduce the search space since there is high probability that the LUT which is connected to a correct output will not be the source of the fault. If there are multiple faults, then the LUTs are marked in iterations. For example, in case of three faulty outputs, all LUTs connected to these faulty outputs are marked. Subsequently, only those LUTs connected to two faulty outputs are marked. Finally, the LUTs connected to two

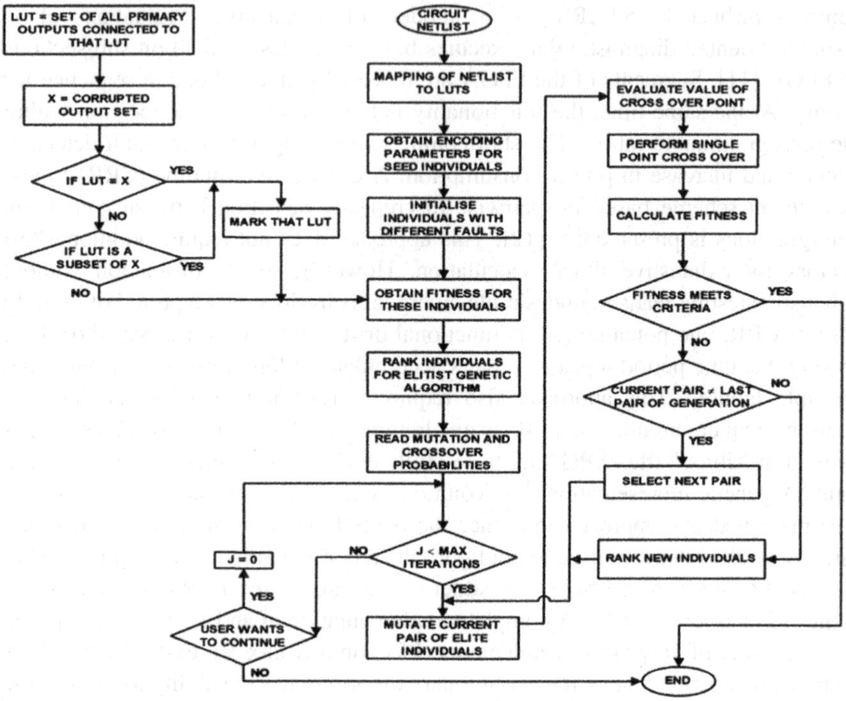

Fig. 1 Flowchart for the proposed GA

faulty outputs are marked. A modified heuristic is used if the above mentioned heuristic is unable to provide a good solution. In the modified heuristic, all LUTs that are connected to at least one of the faulty outputs are marked. But this simple heuristic provides less pruning. This heuristic is easier to apply, but reduction in search space is very low. With the modified heuristic crossover should be done randomly and is less dependent on marked LUTs. Both these heuristics are applied together to get a better solution. The procedure for marking of LUTs is depicted in Fig. 1.

Unlike other EH approaches, the current approach can reduce downtime, increase availability or speed up the fault regeneration process. In the present work, 10 individuals are initially taken as parents. The parents and the offsprings formed will serve to produce the next generation and the same process is repeated until the objective is met. The fitness of each individual is found by applying pseudo-exhaustive test patterns as explained in the following section. This GA produces faster results since we are trying to remove faulty LUTs form the new generations as much as possible. The probability of mutation-pm, and the probability of crossover-pc is fixed adaptively. These probabilities denote the number of individuals in each generation that will take part in the mutation and crossover operations. For function mutation, the mutation operation will be performed only on marked LUTs. For connection mutation, a connection can be mutated only if it is connected to marked LUTs, faulty outputs or faulty inputs. The mutation operator randomly changes the LUT's functionality or reconnects one input of the LUT to a new output. For the crossover operation, the point of crossover-xi needs to be fixed. Here, LUTs are arranged in ascending order. The numbering starts from (n + 1) and go up to (n + L). This serves as the identification for the LUTs. Fitness value is assigned to an individual via bit-by-bit functional output comparison with the known correct output. The correct output is obtained from the isolated fault-free module. A bit by bit comparison of fitness evaluation scheme is adopted due to its simplicity in hardware implementation. The fitness value is incremented by 1 for each matching output. GAs need a method to assign positive real values to individuals of the population, for the selection mechanism to work. The elite individual with the highest fitness value in the current population is retained for the future generations to assure monotonically increasing performance. The maximum value of fitness for an individual, achieved via exhaustive evaluation is 2n × m, where n and m are the number of primary inputs (PI) and primary outputs (PO) of the circuit, respectively [17].

4 Proposed Framework for GA

The proposed GA is used to recover 8-point FFT blocks. The FFT processor consists of iterative MSA modules. Online checkers are provided to detect faults in the FFT modules. The FFT block along with online checkers is connected in duplex mode in order to provide fault masking during refurbishment mode. As shown in Fig. 2, each FFT module is configured into partial reconfigurable modules (PRMs). Once fault is detected in any of the two FFT modules, only the faulty MSA module in FFT is refurbished using GA. The faulty FFT block is taken offline for recovery, while the fault free module continues to provide functional output in uniplex mode. In contrast to previous approaches, only the faulty MSA module is refurbished instead of entire faulty FFT. During refurbishment, the backtracking heuristic is applied to mark the LUTs to reduce search space. While refurbishment process is going on, faultless FFT block output is diverted to the system output using the multiplexer. This increases the availability of the FFT system. As shown in Fig. 2, the reconfiguration manager is connected to GA engine during the refurbishment phase. The fitness scores are reported to the GA engine. Based on information from the reconfiguration manager, GA engine and online checkers, the output actuator selects one of the outputs of the FFT blocks. The system remains in duplex mode or in uniplex mode, as required by mission objectives. Faulty modules can be reconfigured dynamically via FPGA partial reconfiguration. Test patterns are applied to MSAs during the refurbishment phase for fitness evaluation. The FFT module consists of identical rows of MSAs. Hence test pattern application to non-faulty rows is enough to evaluate the fitness of the faulty row. The input to the MSA rows of the FFT module is selected from the functional input, or test pattern

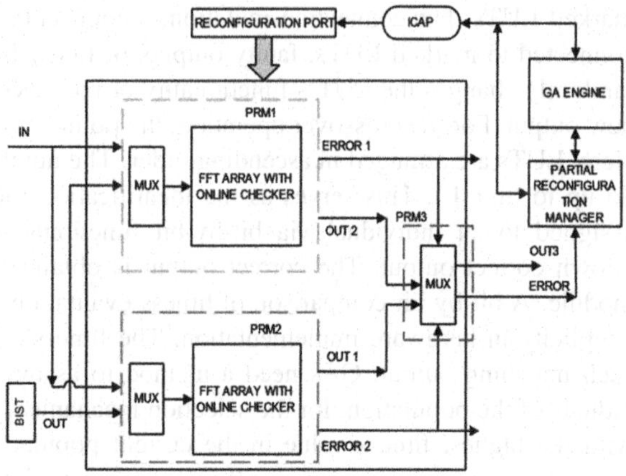

Fig. 2 Fault tolerant reconfigurable framework on FPGA using GA

generator's output. In previous approaches for fitness evaluation, test patterns are applied, either to all three modules in triple modular redundancy (TMR) mode, or to two modules in duplex mode. In the present work, test patterns need to be applied only to faulty FFT block. For previous approaches 2^{24} test patterns are required for the FFT blocks. However in the current approach, only $2^3 \times 4$ test patterns are required. This results in drastic reduction in fault refurbishment time as well as extra power consumption due to test pattern application. Thus, the fault free module is not used for fitness evaluation. This alleviates the requirement of exhaustive fitness evaluation method as opposed to other schemes and is referred to as the module-free fitness evaluation [16, 17]. If all 'n' PIs of the circuit are required for fitness evaluation, the conventional GA approach requires 2^n tests. Therefore, the test patterns increase exponentially for large systems. However, as discussed before the present approach alleviates this problem.

5 Experimental Setup and Results

As shown in Fig. 3, the simulator in the form of C++ based console application consists of three main components: dynamic pruning procedure, fitness report and GA. The GA is implemented using an object oriented architecture that contains classes which model the FPGA resources with flexible geometries. In the designed FPGA model, a Configurable Logic Block (CLB) is composed of four LUTs. The LUTs used here are three input LUTs. The number of PIs is identified based on the repeated structure of the architecture for test pattern application during the recovery process.

The simulator includes an adaptive pseudo-exhaustive test pattern generator that generates test patterns of varying width. When this simulator is run in the pruned GA mode, the test pattern generator (TPG) component applies test patterns and

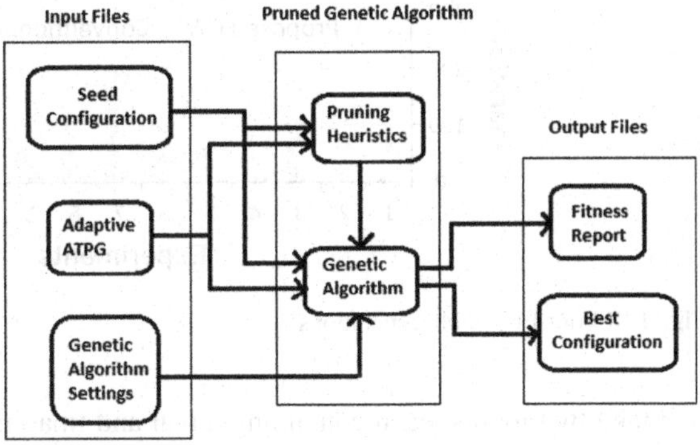

Fig. 3 Files used by the GA simulator

obtains resource performance information. This resource performance information forms the input to the GA. The GA reads the seed configuration file and performs GA based evolutionary repair using the pruned search space for faster refurbishment. Initially, a 1-bit and 3-bit multiplier, add and subtract module (MSA) of the FFT block is considered. MSA is the main block of butterfly module in the FFT. In this case study, 10 distinct individuals are created at design-time using a set of 10 or more faulty variations of the fundamental sub-circuit. In this experiment, GA based recovery operators are applied to regenerate the functionality in the affected individuals. Crossover rate of 97 % and mutation rate of 3 % is selected for the experiments. In order to simulate a hardware fault in the FPGA, a single stuck-at fault is inserted at a randomly chosen LUT input pin or its content.

5.1 Refurbishment of Case Study Circuits

The MSA circuit used here requires five 3-input LUTs, 7 inputs and two outputs. Of the seven inputs, two are control signals used to select addition, subtraction or multiplication. The remaining two inputs are from other preceding MSA units and three inputs are direct data inputs. The sum(or product) and the carry signals form the output. The 3-bit MSA has 9 inputs, 5 outputs and 12 LUTs. With a tournament size of 10 and a crossover rate of 97 %, most of the runs performed provide fully fit output for the applied test patterns. The maximum fitness of MSA module is 64. The number of generations required for healing various types of injected random faults obtained for 8-point FFT are plotted in Fig. 4. The number of experiments are taken along the x-axis and number of generations along the y-axis. The stuck-at-0 and stuck-at-1 faults are injected in a number of places randomly. It is clear from the results that, without pruning, the genetic algorithm achieves 100 % recovery using more number of generations for almost all cases of injected random faults. The implementation of the 8-point FFT required a total of 192 LUTs, it's composition being 12 butterfly units, each consisting of 4 MSA units. Each butterfly

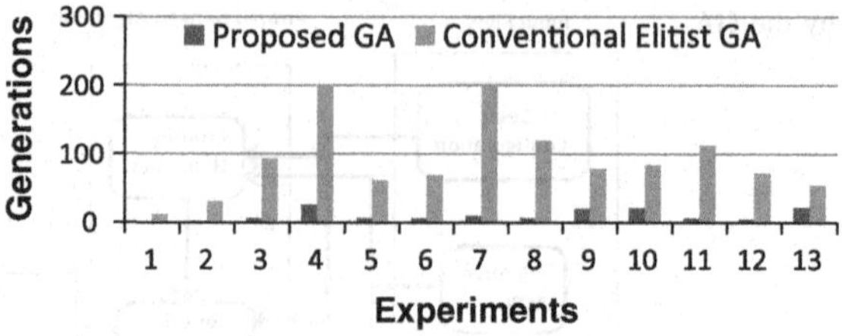

Fig. 4 8-Pont FFT fault generations

will take two inputs, each containing a real and imaginary portion, while providing two outputs in a similar format. The healing of the 8-point FFT module is done using the conventional GA and the modified GA. It is observed that, with pruning, the GA attains 100 % fitness within thirty generations. The conventional GA requires more than 100 generations to heal the FFT. It is observed that besides the merits provided by the GA, the proposed system is also capable of providing much faster healing of the individual, since only the faulty portions are modified to be corrected or are removed altogether. Another merit is that, the algorithm provides almost uniformly increasing fitness every generation. With pruning, since we are actively trying to reduce the number of marked LUTs in the new individual, full healing is achieved within a few generations. In the future, the proposed fault tolerant framework will be implemented in Xilinx Virtex-E FPGA using Verilog HDL. The reduction in test patterns for fault detection and fitness evaluation in the present approach will be analyzed in the future.

The simulations for the discussed architectures are performed using ModelSim. The simulation results of the preprocessing stage of the 8-point FFT are displayed in Fig. 5. The information of the fitness and the marked LUTs are passed on to further stages. The fitness of the elite individuals from the seed population is portrayed in Fig. 6. The fitness and M of the resulting individual from the mutation and crossover operations are obtained, to further decide this individual's use. The algorithm was so designed, that all the individuals can be produced in parallel to

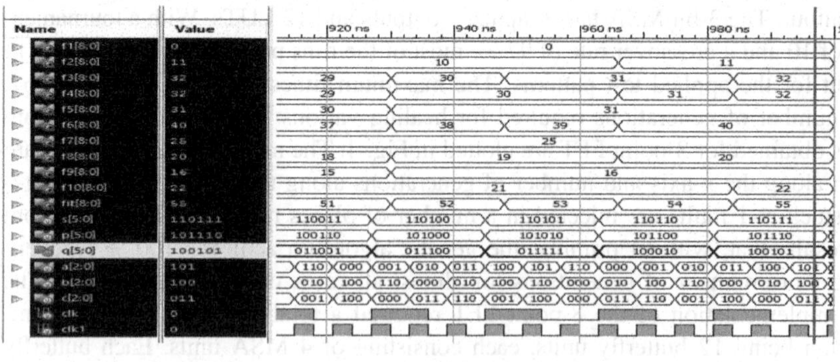

Fig. 5 Fitness and parameter estimation for the preprocessing stage

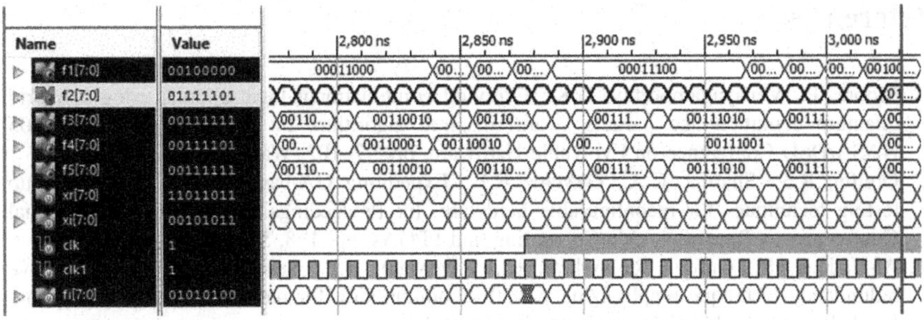

Fig. 6 Healed result of 8-point FFT using GA

Table 1 Hardware implementation results of the architecture on Virtex 6E

Implementation results of entire architecture for 8-point FFT	Hardware implementation
Number of slices	580
Number of 4 input LUTs	704
Number of bonded IOBs	54
Number of slice FFs	560
Delay overhead (ns)	5.330
Total power (static and dynamic) (W)	2.965

reduce the time period. The calculation of the fitness of each generation requires 1024 ns. It can be seen that, a satisfactory fitness level has been obtained in the third generation itself. Table 1 displays the hardware utilization details for the complete hardware GA implementation of the GA processor architecture on Virtex®-6 XC6VLX240T-1FFG1156 FPGA.

6 Conclusions

To circumvent the scalability hurdle of evolvable hardware, the design is partitioned into multiple groups of identical logic resources. In the future, the PDR implementations of the entire framework will be implemented in Virtex-6E FPGA and the power, area and speed of reconfiguration will be analyzed. In the proposed framework, upon detection of a fault, the affected resource alone can be evolved separately by means of dynamic partial reconfiguration. This generic framework using GA engine can be applied to heal larger signal/image processors.

References

1. Moore, P., Venayagamoorthy, G.K.: Evolving combinational logic circuits using a hybrid quantum evolution and particle swarm inspired algorithm. In: Proceedings of the 2005 NASA/DoD Conference on Evolvable Hardware, pp. 97–102, Washington (2005)
2. Oreifej, R.S., Sharma, C.A., DeMara, R.F.: Expediting GA-based evolution using group testing techniques for reconfigurable hardware. In: Proceedings of the IEEE International Conference on Reconfigurable Computing and FPGAs, pp. 1–8, San Louis Potosi (2006)
3. Al-Haddad, R., Oreifej, R., Ashraf, R.A., DeMara, R.F.: Sustainable modular adaptive redundancy technique emphasizing partial reconfiguration for reduced power consumption. Int. J. Reconfigurable Comput. **2011**, 1–12 (2011)
4. Li, Y., Mitra, S., Gardner, D., Kim, Y., Mintarno, E.: Overcoming early-life failure and aging challenges for robust system design. IEEE Des. Test Comput. **26**, 28–39 (2009)
5. Srinivasan, S., Krishnan, R., Mangalagiri, P., Xie, Y., Narayanan, V., Irwin, M.J., Sarpatwari, K.: Toward increasing FPGA lifetime. IEEE Trans. Dependable Secure Comput. **5**, 115–127 (2008)
6. Mintarno, E., Skaf, J., Zheng, R., Velamala, J.B., Cao, Y., Boyd, S., Dutton, R.W., Mitra, S.: Self-tuning for maximized lifetime energy-efficiency in the presence of circuit aging. IEEE Trans. Comput. Aided Des. Integr. Circuits Syst. **30**, 760–773 (2011)
7. Keymeulen, D., Stoica, A., Zebulum, R.: Fault-tolerant evolvable hardware using field programmable transistor arrays. IEEE Trans. Reliab. **48**, 305–316 (2000)
8. DeMara, R.F., Zhang, K.: Autonomous FPGA fault handling through competitive runtime reconfiguration. In: Proceedings of the 2005 NASA/DoD Conference on Evolvable Hardware, pp. 109–116, Washington (2005)
9. Zhang, K., DeMara, R.F., Sharma, C.A.: Consensus-based evaluation for fault isolation and on-line evolutionary regeneration. In: Proceedings of the International Conference in Evolvable Systems, pp. 12–24, Spain (2005)
10. Greenwood, G.W.: On the practicality of using intrinsic reconfiguration for fault recovery. IEEE Trans. Evol. Comput. **9**, 398–405 (2005)
11. Haddow, P.C., Tyrrell, A.M.: Challenges of evolvable hardware past, present and the path to a promising future. Genet. Program Evolvable Mach. **12**, 183–215 (2011)
12. Larchev, G.V., Lohn, J.D.: Evolutionary based techniques for fault tolerant field programmable gate arrays. In: Proceedings of the 2nd IEEE Conference Space Mission Challenges for Information Technology, pp. 321–329, Pasadena (2006)
13. Abramovici, M., Emmert, J.M., Stroud, C.E.: Roving STARs: an integrated approach to on-line testing, diagnosis, and fault tolerance for FPGAs in adaptive computing systems. In: Proceedings of the NASA/DoD Workshop on Evolvable Hardware, Long Beach (2001)
14. DeMara, R.F., Zhang, K., Sharma, C.A.: Autonomic fault-handling and refurbishment using throughput-driven assessment. Appl. Soft Comput. **11**, 1588–1599 (2011)
15. Oreifej, R.S., Al-Haddad, R.N., Tan, H., DeMara, R.F.: Layered approach to intrinsic evolvable hardware using direct bitstream manipulation of Virtex II Pro devices. In: Proceedings of the International Conference on Field Programmable Logic and Applications, pp. 299–304, Amsterdam (2007)
16. Partial Reconfiguration User Guide. Technical report, Xilinx. http://www.xilinx.com/support/documentation/sw_manuals/xilinx12_4/ug702.pdf (2010)
17. Ashraf, R.A., DeMara, R.F.: Scalable FPGA refurbishment using netlist-driven evolutionary algorithms. IEEE Trans. Comput. **62**, 1526–1541 (2013)

A Two Phase Approach for Efficient Clustering of Web Services

I.R. Praveen Joe and P. Varalakshmi

Abstract Sorting out a desired web service is a demanding concern in service oriented computing as the default keyword search options provided by UDDI registries are not so promising. This paper deals with a novel approach of employing an unsupervised neural network based clustering algorithm namely ART (Adaptive Resonance Theory) for service clustering. The input to the algorithm includes both functional characteristics which are quantified using the basic user requirements in phase 1 and non functional characteristics which are derived by means of swarm based techniques through appropriate mapping of metadata to swarm factors and thereafter updating the input in phase 2. Taking the advantages in being an unsupervised clustering algorithm, ART in a more potential way groups services eliminating a number of irrelevant services returned over a normal search and facilitates to rearrange the registry. Clustering depends on a threshold value namely vigilance parameter which is set between 0 and 1. Flocking of birds is the swarm behaviour considered.

Keywords Unsupervised clustering · ART network · Swarm based approach · Vigilance parameter

1 Introduction

This paper mainly focuses on web service clustering for rearranging services in the registry of the provider in order to facilitate web service discovery in a more desired way. 'Desired way' refers to considering the needs of the clients. An unsupervised clustering algorithm is expected to cater the requirement than a supervised one

I.R.P. Joe (✉) · P. Varalakshmi
Department of Computer Technology, MIT, Anna University, Chennai, India
e-mail: praveenjoeir@yahoo.com

P. Varalakshmi
e-mail: varanip@gmail.com

M. Senthilkumar et al. (eds.), *Computational Intelligence, Cyber Security and Computational Models*, Advances in Intelligent Systems and Computing 412, DOI 10.1007/978-981-10-0251-9_17

because it ends up in an unanticipated grouping where as the later demands fixed class labels in advance before grouping. Additionally metadata documents are parsed for data elements that would be mapped to swarm characteristics (flocking of birds) for further enhancement of input services feature in the already filtered set of services based on the basic functional requirements. Thus, input feature set for the unsupervised algorithm is prepared incorporating both functional (user's demand) and non functional (metadata) characteristics.

2 Related Work

The existing and already proposed articles about web service clustering have used the concept of ranking web services based on semantic relationships among web services. It evaluates the parameters—Relevance (Rel.), Specificity (Sp.), Span(S) for ranking web services [1]. For a given web service, that includes $\{t_1, t_2, \ldots, t_n\}$ concepts describing the service, the overall rank is expressed as $R\{t_i,t_j\}$, $R\{t_i, t_j\} = k_1 * Rel(t_i,t_j) + k_2 * Sp(t_i,t_j) + k_3 * S(t_i,t_j)$ where $0 < k_1 < k_2 < k_3$ and $k_1 + k_2 + k_3 = 1$. Hyper clique patterns are used for service discovery. Hyper clique patterns are based on the concepts of frequent items sets [2].

Taxonomy based clustering of web services have used the concept of match-making to a query description and a service description so that the middleware can decide whether a service is a candidate solution for the given query. This approach has also used various Taxonomy based algorithm for clustering [3].

Hierarchical clustering algorithms were also used for co-clustering with musical data after applying a supervised algorithm for eliminating noisy data further [4]. Temporal data or spatial data are added to the clustered data for further scrutiny and therefore fine clusters may be derived [5].

2.1 ART Algorithm and Swarm Algorithm

The merits of ART algorithm, working of ART and swarm algorithms are discussed in this section.

2.1.1 ART Algorithm—Merits

Unsupervised clustering
Data clustering is an unsupervised process of classifying patterns into groups (clusters), aiming at discovering structures hidden in a data set [6].

Fig. 1 Simple ART structure

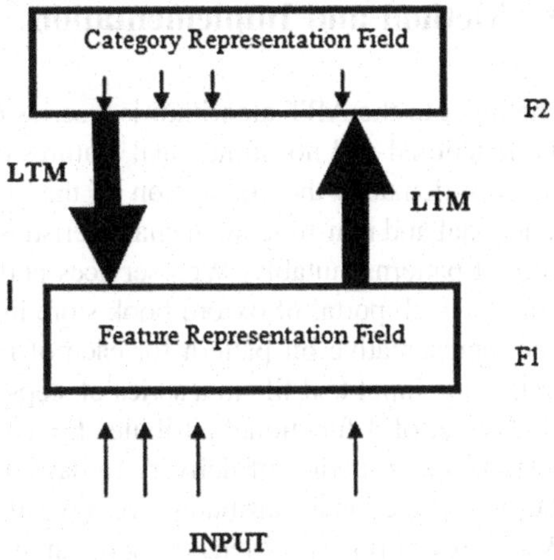

Non linearity
ART supports non linearity as most of the real world problems are non linear in nature [6].

Capability to handle huge and high dimensional data sets
It supports automatic discovery of clusters in different subspaces of a higher dimensional space using a density-based approach [6].

Vigilance parameter
It is a functional parameter to find out the degree of mismatch to be tolerated. ART network includes a choice process and a match process as its key parts (Fig. 1).

2.1.2 Swarm Algorithm

Swarm algorithms are biologically inspired algorithms for clustering data. Swarm intelligence is collective behaviour of decentralized, self-organized agents in a population or swarm. Some of the examples are ant colonies, flocking of birds, schooling of fishes and animal herding. The motivation often is derived from nature, particularly biological systems. The agents follow very simple rules, and even though there is no centralized control structure dictating how individual agents should behave, local, and to a certain degree random, interactions between such agents lead to the emergence of "intelligent" global behaviour, unknown to the individual agents [7]. For the experiment flocking of birds is considered.

3 Method and Implementation

The input to the ART algorithm is a series of bit patterns in a text file representing the functional and non-functional features of every web service. Input to the ART network demands the conversion of the data to zeros and ones and therefore the functional and non-functional characteristics of the web services are to be converted into bit patterns suitably. Web services pertaining to the online purchase of books from the web portal of oxford book store in Chennai is considered for the analysis. The representative bit pattern for each of the services has to be prepared for generating the input text file in a series of steps. The first four bits represent the service id. The set of 4 functional attributes F = {F1, F2, F3, F4} refer to author ranking, cost in rupees, period of delivery in days and year of publishing respectively. For the non functional attributes we consider the meta data documents namely WS-Policy (m1), WS-Metadata Exchange (m2), WSDL (m3) and XML Schema (m4) and the elements are mapped and quantified using swarm characteristics or factors here, flocking of birds. There are 9 factors that favour the flocking of birds namely 1. Velocity 2. Acceleration 3. Location 4. Species 5. Distance 6. Propulsive force 7. Speed 8. Position 9. Repulsive force. Each of the four metadata document is parsed for elements that would reflect the above said 9 characteristics, then quantified and then normalized either to 0 or 1 (Fig. 2).

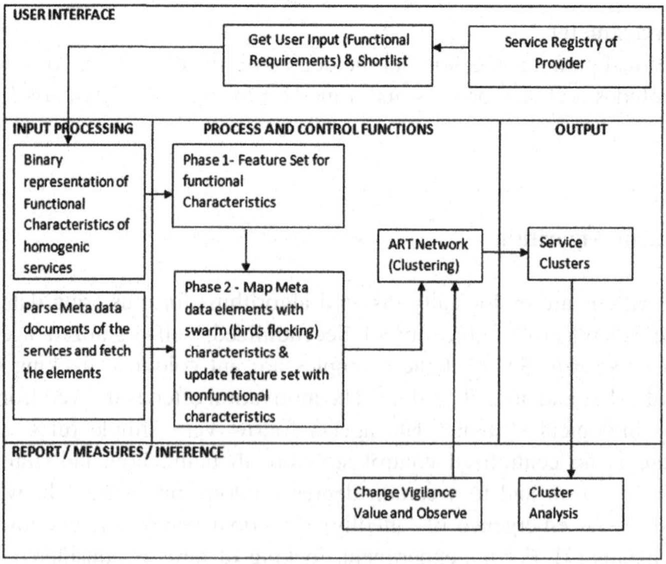

Fig. 2 Architecture of the proposed system

	vp = 0.6		*vp* = 0.7		*vp* = 0.8	
Service ID	**Cluster 0**	**Cluster 1**	**Cluster 0**	**Cluster 1**	**Cluster 0**	**Cluster 1**
s0012	**X**	-	**X**	-	**X**	-
s0002	**X**	-	-	**X**	-	**X**
s0015	-	**X**	**X**	-	**X**	-
s0014	-	**X**	**X**	-	**X**	-
s0025	-	**X**	**X**	-	-	**X**
s0006	-	**X**	**X**	-	-	**X**
s0017	-	**X**	-	**X**	**X**	-
s0008	**X**	-	-	**X**	**X**	-
s0011	-	**X**	-	**X**	**X**	-
s0010	-	**X**	-	**X**	-	**X**

Fig. 3 Cluster of services for varied vigilance parameters

4 Results and Interpretation

The results of how web services are clustered for varied vigilance values are tabulated below in Fig. 3. It is evident from the results that when the number of input increases the number of clusters formed also increases with increase in the vigilance value and especially when the non functional attributes are updated with the input there is a steady increase in the number of clusters formed as shown in Fig. 3. This means for the same vigilance it is possible to form more unique clusters similar to the example given in Fig. 3 when both functional and non-functional parameters are considered. The percentage of relevance indicated that a particular service fits into the same group irrespective of the change in vigilance value. On an average it is observed to be 74.7 %.

5 Conclusion and Future Enhancement

The clustering approach implemented would result in clusters of homogenic services in order to rearrange the registry. There are no parametric analysis of how to fix the vigilance parameter which can be considered as a problem for auto-tuning of the threshold.

References

1. Paliwal, A.V., Shafiq, B., Vaidya, J., Xiong, H., Adam, N.: Semantics-based automated service discovery. IEEE Trans. Services Comput. **5**(2), 260–275 (2012)
2. Mohana, R., Dahiya, D.: Optimized service discovery using qos based ranking: a fuzzy clustering and particle swarm optimization approach. In: 35th IEEE Annual Computer Software and Applications Conference Workshops (2011)
3. Dasgupta, S., Bhat, S.: Taxonomic clustering and query matching for efficient service discovery. In: International Conference on Web Services by IEEE Computer Society (2011)
4. Li, J., Shao, B., Li, T., Ogihara, M.: Hierarchical co-clustering: a new way to organize the music data. IEEE Trans. Multimedia **14**(2), 471–481 (2012)
5. Abu Sharkh, H., Fung, B.C.M.: Service-oriented architecture for sharing private spatial-temporal data. In: IEEE Conference on Cloud and Service Computing, Sept. 2011
6. Joe, I.R.P., Washington, G.: Art network—a solution for effective warranty management. CURIE J. **1**(2), 67–75 (2008)
7. Nagy, A., Oprisa, C., Salomie, I., Pop C.B.: Particle swarm optimization for clustering semantic web services. In: 10th International Symposium on Parallel and Distributed Computing (ISPDC), pp. 170–177, 6–8 July 2011

Elimination of Redundant Association Rules—An Efficient Linear Approach

Akilandeswari Jeyapal and Jothi Ganesan

Abstract Association rule mining plays an important role in data mining and knowledge discovery. Market basket analysis, medical diagnosis, protein sequence analysis, social media analysis etc., are some prospective research areas of association rule mining. These types of datasets contain huge numbers of features/item sets. Traditional association rule mining algorithms generate lots of rules based on the support and confidence values, many such rules thus generated are redundant. The eminence of the information is affected by the redundant association rules. Therefore, it is essential to eliminate the redundant rules to improve the quality of the results. The proposed algorithm removes redundant association rules to improve the quality of the rules and decreases the size of the rule list. It also reduces memory consumption for further processing of association rules. The experimental results show that, our proposed method effectively removes the redundancy.

Keywords Data mining · Association rule mining · Non-redundant rules · Market basket analysis · Frequent itemsets

1 Introduction

Agrawal et al. [1] has first introduced one of the most significant and glowing researched techniques in data mining called Association rule mining. It is a process to finding frequent patterns, associations, correlations, or causal structures among sets of items or features in transaction databases, relational databases, and other information repositories. Association rule mining technique [2], find out all frequent patterns among all transactions of data attributes. The entire frequent patterns are

A. Jeyapal (✉) · J. Ganesan
Department of IT, Sona College of Technology, Salem, Tamilnadu, India
e-mail: akilandeswari@sonatech.ac.in

J. Ganesan
e-mail: jothiys@gmail.com

M. Senthilkumar et al. (eds.), *Computational Intelligence, Cyber Security and Computational Models*, Advances in Intelligent Systems and Computing 412, DOI 10.1007/978-981-10-0251-9_18

presented in the form of rules. This powerful exploratory technique has been broadly used in many application domains such as market basket analysis, medical diagnosis, protein sequence analysis, social media analysis, census data processing, and fraud detection. The aim of association rule mining technique is to detect relationships or associations among specific values of categorical variables in large data sets. This is a frequent task in many data mining and knowledge discovery algorithms [3].

The frequent closed item sets based association rule mining algorithm generates huge number of rules, as it considers all subsets of frequent item sets as antecedent of a rule. If the number of frequent item sets increases then the total number of rules grows up. However, many of these rules have the same meaning or are redundant. In most of the cases, number of redundant rules is significantly larger than that of essential rules [4]. In many such cases enormous redundant rules often fades away the intention of association rule mining in the first place. Some of the disadvantages of association rule mining are generating non interesting rules, low algorithm performance, discovering many rules [5]. To overcome these limitations, our proposed framework generates the valid rules without loss of information or knowledge and eliminates redundant rules. Experimental results show the effectiveness of the proposed algorithm in terms of reduction in the number of rules presented to the user, and decrease in the time complexity.

The rest of this paper is organized as follows. Related work is discussed in Sect. 2. Section 3 provides an overview of the association rule mining preliminaries and the existing algorithm. Section 4 delineates our proposed approach with an example. Section 5 describes performance analysis. Section 6 presents concluding remarks.

2 Related Work

There are numerous algorithms for eliminating the redundancy of association rules in the literature. In [6], the redundancy of association rules is reduced using upper level closed frequent item set with the help of generator. The aim of this algorithm is to remove hierarchical duplicacy in multi-level datasets. It is also noted that, this algorithm improves the quality and efficiency of mining without any loss of information. HUCI-Miner algorithm was proposed by Sahoo et al. [7]. It has two steps: first step is to finding all high utility itemsets and then extracting all valid rules. This algorithm extracts the non-redundant rule set, the high utility closed itemset and associates the high utility generators to the corresponding closed itemset. Using this framework it mines all non-redundant rules efficiently. In [8], the non-redundant rule set is generated based on completeness and tightness properties of rule set. In this approach, the redundant association rules are removed based on frequent closed itemset mining (FCI) and lift value as the interesting measure for entering the interesting rule. Frequent closed itemsets and their generators are used to extract the non-redundant association rules with minimal antecedents and maximal consequents. This algorithm is carried out on real-life databases and the experimental result shows that relevant association rules efficiently generate [9]. Ashrafi et al. [10] initially eliminated the rules which have similar meanings. They categorize all the rules into two groups based on the same antecedent or consequent itemset present in that rule. From the groups each rule is analysed with the support and confidence and thus the redundant rules eliminated. This method eliminates the redundant rules without losing any higher confidence rules that cannot be substitute by other rules. The result demonstrates that, this method removed a significant number of redundant rules. shaw et al. [11] eliminate the hierarchical redundancy in multi-level datasets, thus reducing the size of the rule set to increase the quality and usefulness, without loss of any information. In this approach, frequent closed itemsets and generators are used to eliminate hierarchical redundancy. In [12], the proposed method removed the redundant rules and produces small number of rules from any given or frequent or frequent closed itemset generated. Zaki et al. [13] describes a new framework for association rule mining based on the novel concept of closed frequent itemsets. The traditional framework produces too many association rules. This new framework produces exponentially (in the length of the longest frequent itemset) fewer rules than the traditional approach, again without loss of information.

3 Preliminaries

Definition 1 (*Association Rule*) Let $I = \{i_1, i_2, \ldots, i_n\}$ be a set of n attributes called items. Let $D = \{t_1, t_2, \ldots, t_m\}$ be a set of transactions called the database. Each transaction in D has a unique transaction ID and contains a subset of the items in I. A rule is defined as an implication of the form $A \rightarrow B$ where $A, B\ I$ and $A \cap B = \emptyset$ [1].

To demonstrate the association rule concepts, we use a sample of five transactions from the groceries supermarket dataset which is shown in Table 1. The set of items I = {Whole milk, bread, yogurt, chocolate, coffee}. An example rule for the

Table 1 Sample dataset

Transaction ID	Items
1	Whole milk, bread
2	yogurt
3	chocolate, coffee
4	Whole milk, bread, yogurt
5	bread

supermarket could be {bread, yogurt} $\rightarrow$ {*whole milk*} meaning that if a customer buys bread and yogurt he/she may also buy whole milk.

Definition 2 (*Support*) Support of a rule $A \Rightarrow B$ is the probability of the itemset $\{A, B\}$. This gives an idea of how often the rule is relevant.

$$support\,(A \Rightarrow B) = P(\{A, B\}) \tag{1}$$

For example, the support of a rule {*whole milk*} $\rightarrow$ {bread} is 2/5.

Definition 3 (*Confidence*) Confidence of a rule $A \Rightarrow B$ is the conditional probability of B given A. This gives a measure of how accurate the rule is.

$$confidence\,(A \Rightarrow B) = P(B|A) = support\,(\{A, B\})/support\,(A) \tag{2}$$

For example, the confidence of a rule {*whole milk*} $\rightarrow$ {bread} is 2/2.

Definition 4 (*Redundant Rules*) Let $A \rightarrow B$ and $A' \rightarrow B'$ be two association rules with confidences con and con′ respectively, then $A \rightarrow B$ is said to be a redundant rule to $A' \rightarrow B'$ if $A' \subseteq A$, $B' \subseteq B$ and $con \leq con'$ [8]. For example, consider a rule set R having three rules such as,

$$\{whole\,milk,\, bread\} \rightarrow \{coffee\}, \{coffee,\, bread\} \rightarrow \{whole\,milk\}$$
$$and\{whole\,milk,\, coffee\} \rightarrow \{bread\}.$$

The last two rules do not convey any extra knowledge or information. These two rules are redundant rules.

3.1 Exiting Algorithm

In [8], present the rule pruning algorithm for removing redundancy. In this approach, the rule set is generated based on the frequent closed item set and then removing the redundancy using interesting measure called lift. Ashwini Batbarai et al. has discussed this algorithm in detail.

4 Proposed Algorithm

All the subsets of frequent itemsets are considered in most of the traditional association rule mining algorithms. Therefore, these algorithms generate thousand or even millions of rules. However, many of these rules have the same meaning or are redundant rules. In majority of the cases number of redundant rules is

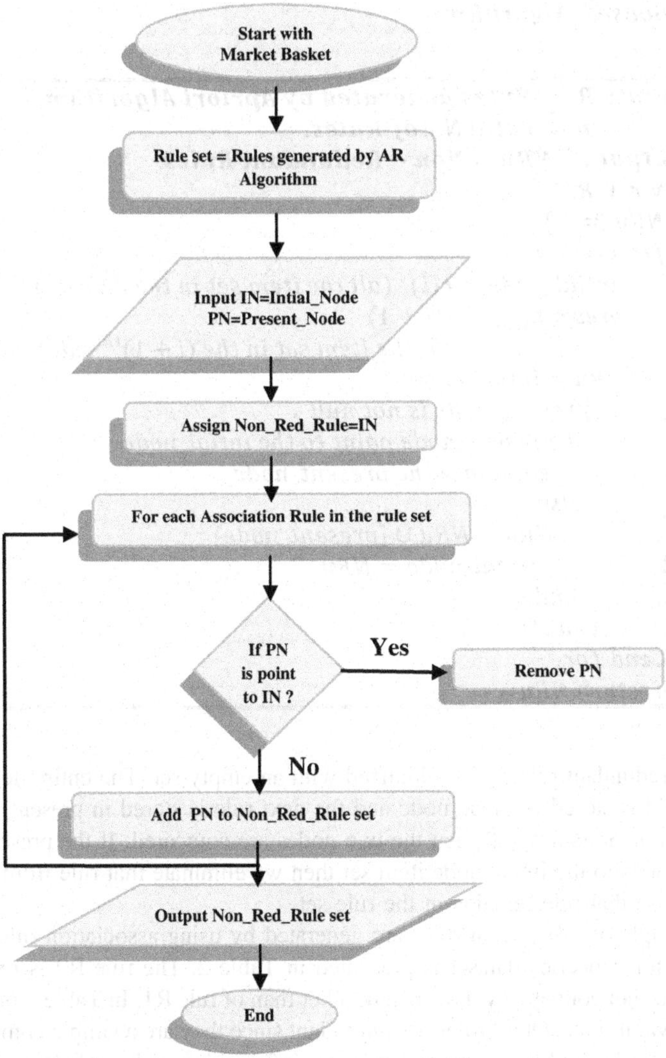

Fig. 1 Flowchart of the proposed methodology

significantly larger than that of essential rules. The redundant rules affect the importance of information. Thus, it is necessary to eliminate the redundant rules to improve the superiority of the information. The proposed algorithm removes the redundancy of association rules and improves the quality of rules and decrease the size of the rule list. The proposed methodology is presented in Fig. 1.

4.1 Proposed Algorithm

```
Input: R = Rules generated by Apriori Algorithm;
       n = Total No.of Rules.
Output : NRu = Non − Redundant Rules.
1. ∀ r ∈ R
2. NRu = {}
3. for i = 1: n
4.     intial_node = r(i)  (all the item set in the i^th rule)
5.     present_node = r(i + 1)
                  (all the item set in the (i + 1)^th rule)
6.     NRu = intial_node
7.     if present_node is not null
8.        if present_node point to the intial_node
9.           eliminate the present_node
10.       else
11.          NRu = NRu ∪ {present_node}
12.           intial_node = NRu
13.       end if
14.     end if
15. end for
16. return NRu
```

A non redundant rule set in initialized with an empty set. The entire item set of the first rule is stored in intial_node and the next rule is stored in present_node. If the present_node is not null then the two nodes are compared. If the present_node item set points to the intial_node item set then we eliminate that rule from the rule set otherwise that rule remains in the rule set.

The sample of association rules are generated by using association rule mining algorithm for geoceries dataset is presented in Table 2. The rule R2 is redundant since it does not convey any new insight other than of rule R1. In Table 2 rule 1 and rule 7 are valid rules. Other rules are redundant since they are a simple combination of the valid rules and they convey any extra information or knowledge.

Table 3 shows that rules generated by our proposed algorithm after efficiently eliminating the redundant rules. The rules generated by association rule mining algorithm are then fed as input to the proposed algorithm. The proposed algorithm selects the valid rules i.e. rule R1 and R7.

Table 2 Rules generated by association rule mining algorithm

Rules generated by association rule mining algorithm	
R1	Rolls/buns → sausage, organic sausage
R2	Sausage → rolls/buns, organic sausage
R3	Organic sausage → rolls/buns, sausage
R4	Rolls/buns, sausage → organic sausage
R5	Rolls/buns, organic sausage → sausage
R6	Sausage, organic sausage → rolls/buns
R7	Whole milk → chocolate marshmallow, specialty chocolate
R8	Chocolate marshmallow → whole milk, specialty chocolate
R9	Specialty chocolate → whole milk, chocolate marshmallow
R10	Whole milk, chocolate marshmallow → specialty chocolate
R11	Whole milk, specialty chocolate → chocolate marshmallow
R12	Chocolate marshmallow, specialty chocolate → whole milk

Table 3 Rules generated by the proposed algorithm

Rules generated by proposed algorithm	
R1	Rolls/buns → sausage, organic sausage
R7	Whole milk → chocolate marshmallow, specialty chocolate

5 Experimental Analysis and Discussion

5.1 Environment Used

All the experiments were performed on Intel Core3 2 GHz with 2 GB main memory running Windows XP. Algorithms were implemented in MATLAB.

5.2 Dataset Used

We used four different datasets Chess, Connect, Mushrooms and groceries. It is worth to state that a lot of association rule mining algorithms had used all of these datasets for testing their implementation. All datasets except the Groceries dataset are taken from the UCI Machine Learning Database Repository [14]. The characteristics of each of them are described in Table 4. Groceries dataset contains 1 month of real-world point-of-sale transaction data from a typical local grocery outlet [15]. The mushroom database contains characteristics of various species of mushrooms. Finally, connect and chess datasets are derived from their respective game steps.

Table 4 Datasets description

Dataset	#Items	Avg. length	#Transactions
Chess	75	37	3,156
Connect4	150	43	67,557
Mushroom	120	23	8,124
Groceries	169	48	9,835

5.3 *Performance Analysis*

For each dataset, association rule mining algorithm generate rules with different threshold i.e., support and confidence values. These rules are then fed as input to the proposed algorithm implementation. The efficiency of the proposed algorithm is compared to existing rule pruning algorithm and association rule mining algorithm from an application view point.

Figure 2 shows the total number of rules generated by the association rule mining algorithm and existing algorithm with our proposed methodology. From the graph it clear that the proposed algorithm significantly eliminates the total number of rules. It is noted that, the number of rules decreases linearly with the increase in the confidence threshold. The existing algorithm, removes the redundant rules, having the highest lift value. Again, non-redundant rules have same high lift value.

The proposed algorithm verifies each rule with a set of rules in order to find redundant rule. Therefore it produces only few rules from each frequent itemset. It is also noted that, in the case of mushroom dataset, all the rule pruning algorithms,

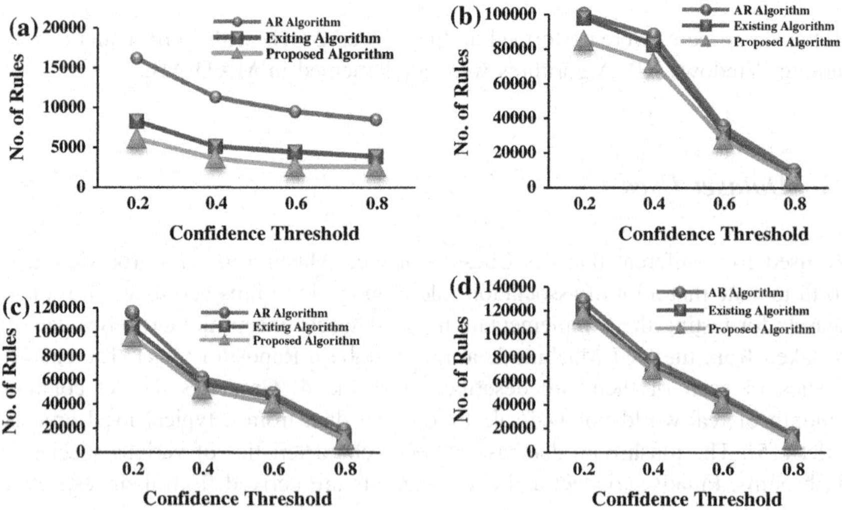

Fig. 2 Number of rules generated for various datasets. **a** Groceries dataset. **b** Chess dataset. **c** Connect4 dataset. **d** Mushroom dataset

including the proposed method have an almost identical number of rules. It is believed that, the proposed method reduces memory consumption for further processing of association rules.

6 Conclusion

There are many drawbacks found in association rule mining (1) producing non-fascinating rules, (2) reduce the algorithm performance and (3) discovering many rules. To overcome these limitations we proposed a novel non-redundant association algorithm to eliminate the redundant rules. Repetition in association rules mining decline the speed for rules generation. It causes a lot of rules to be produced for same set of attributes. The aim of our proposed algorithms is to remove redundancy in the frequent item set and decrease the time utilization of algorithm. The experimental results illustrate that, the proposed algorithm remove the redundant rules efficiently and improves the quality of mining without any loss of information or knowledge.

References

1. Agrawal, R., Imieliński, T., Swami, A.: Mining association rules between sets of items in large databases. ACM SIGMOD Record **22**(2), 207–216 (1993)
2. Ceglar, A., Roddick, J.F.: Association mining. ACM Comput. Surv. **38**(2), (2006)
3. http://www.statsoft.com/
4. Kannika Nirai Vaani, M., Ramaraj, E.: Effectiveness of ERules in generating non redundant rule sets in pharmacy database. Int. J. Sci. Res. (IJSR) **2**(9), 356–360 (2013). India Online ISSN:2319–7064
5. Kotsiantis, S., Kanellopoulos, D.: Association rules mining: a recent overview. GESTS Int. Trans. Comput. Sci. Eng. **32**(1), 71–82 (2006)
6. Chandanan, A.K., Shukla, M.K.: Removal of duplicate rules for association rule mining from multilevel dataset. Procedia Comput. Sci. **45**, 143–149 (2015)
7. Sahoo, J., Das, A.K., Goswami, A.: An efficient approach for mining association rules from high utility itemsets. Expert Syst. Appl. **42**(13), 5754–5778 (2015)
8. Batbarai, A., Naidu, D.: Approach for rule pruning in association rule mining for removing redundancy. Int. J. Innov. Res. Comput. Commun. Eng. **2**(5), 4207–4213 (2014)
9. Bastide, Y., et al.: Mining minimal non-redundant association rules using frequent closed itemsets. In: Computational Logic—CL 2000, pp. 972–986. Springer, Heidelberg (2000)
10. Ashrafi, M.Z., Taniar, D., Smith, K.: Redundant association rules reduction techniques. AI 2005. In: Advances in Artificial Intelligence, pp. 254–263. Springer, Heidelberg (2000)
11. Shaw, G., Xu, Y., Geva, S.: Eliminating redundant association rules in multi-level datasets, pp. 313–319. (2009)
12. Ashrafi, M.Z., Taniar, D., Smith, K.: A new approach of eliminating redundant association rules. Database and Expert Systems Applications, pp. 465–474. Springer, Heidelberg (2004)
13. Zaki, M.J.: Generating non-redundant association rules. In: Proceedings of the Sixth ACM SIGKDD International Conference on Knowledge Discovery and Data Mining, pp. 34–43 (2000)

14. Blake, C.L., Merz, C.J.: UCI Repository of machine learning databases. http://www.ics.uci.edu/~mlearn/
15. Hahsler, M., Hornik, K., Reutterer, T.: Implications of probabilistic data modeling for mining association rules. In: Spiliopoulou, M., Kruse, R., Borgelt, C., Nuernberger, A., Gaul, W. (eds.) From Data and Information Analysis to Knowledge Engineering, Studies in Classification, Data Analysis, and Knowledge Organization, pp. 598–605 (2006)

Clustering Techniques from Significance Analysis of Microarrays

K. Nirmalakumari, R. Harikumar and P. Rajkumar

Abstract Microarray technology is a prominent tool that analyzes many thousands of gene expressions in a single experiment as well as to realize the primary genetic causes of various human diseases. There are abundant applications of this technology and its dataset is of high dimension and it is difficult to analyze the whole gene sets. In this paper, the SAM technique is used in a Golub microarray dataset which helps in identifying significant genes. Then the identified genes are clustered using three clustering techniques, namely, Hierarchical, k-means and Fuzzy C-means clustering algorithms. It helps in forming groups or clusters that share similar characteristics, which are useful when unknown dataset is used for analysis. From the results, it is shown that the hierarchical clustering performs well in exactly forming 27 samples in first cluster (ALL) and 11 samples in the second cluster (AML). They will provide an idea regarding the characteristics of the dataset.

Keywords Microarray · Golub dataset · Significance analysis of microarrays · Significant genes · Hierarchical clustering · k-means clustering · Fuzzy C-means clustering

1 Introduction

Microarray is a glass surface consisting of small amounts of oligonucleotides attached to it [1]. Numerous copies of a single gene are represented in each spot and each microarray contains more than 20,000 of such gene spots [2]. The microarray

K. Nirmalakumari (✉) · R. Harikumar
Department of ECE, Bannari Amman Institute of Technology, Sathyamangalam, India
e-mail: nirmalakumarik@bitsathy.ac.in

R. Harikumar
e-mail: harikumarr@bitsathy.ac.in

P. Rajkumar
Robert Bosch Engineering and Business Solutions Limited, Coimbatore, India
e-mail: rajkumar.ppt@gmail.com

M. Senthilkumar et al. (eds.), *Computational Intelligence, Cyber Security and Computational Models*, Advances in Intelligent Systems and Computing 412, DOI 10.1007/978-981-10-0251-9_19

dataset is of high dimension, so gene selection plays an important role in selecting informative genes rather than using the whole dataset [3, 4, 5]. Significance Analysis of Microarrays (SAM) is a statistical technique [6] for identifying whether the modifications in gene expression are important or not. SAM was established in 2001 for identifying the statistically significant genes. The SAM's input is a set of expressions of genes from microarray experiments set, and also a response variable from every experiment. The response variable may be a mixture of like treated, untreated, a multiclass grouping, a variable which is quantitative or a possibly suppressed survival time. SAM is distributed in an R-software package [7] developed by Standford University. SAM performs well when compared with Bonferroni, step-down correction and Benjami and Hochberg methods [6]. SAM recognizes statistically significant genes by taking the T-tests on specific genes and computes a statistic d_i for each gene i, that measures the power of relationship between response variable and gene expression. While the data might not follow a normal distribution, the non-parametric statistics are used in the analysis. The response variable groups and describes the data based on experimental conditions. It uses continual permutations of the data to find if the gene expression is significant related to the response [8]. The utilization of permutation based analysis accounts for genes correlations and circumvents parametric assumptions regarding the individual gene distribution. Hence it is advantageous over other technique which assumes same variance or gene independence. The significance cutoff is determined by a tuning parameter delta, chosen based upon the false positive rate. Also, a fold change parameter can be chosen to ensure that called genes change, at least in a pre-specified time.

2 SAM Algorithm

SAM recognizes genes with statistically significant changes by carrying out a set of gene-specific tests. Each gene is specified a score based on the modifications in gene expression relative to the standard deviation of frequent measurements for that gene. A threshold is assumed and the genes with greater score than threshold are measured as potentially important. The percentage of those genes which are identified as fake is called False Discovery Rate (FDR) [8]. To estimate the FDR, fake genes are recognized by analyzing permutations of the measurements. The threshold is varied for identifying smaller or larger sets of genes and FDR are computed for every set. FDR is defined as the predictable percentage of false positive among all the declared positives. In practice, with the real FDR unknown, a predictable FDR can serve as a measure to evaluate the performance of a variety of statistical methods under the conditions in which the predictable FDR approximates the real FDR well, or at least, it does not improperly supports or not support any particular method. Permutations are the well definite methods for estimating FDR in genomic studies. The permutations are done at frequent times to determine the significance of gene expression related to the response. The permutation based method of analysis finds correlations in genes and avoids parametric assumptions of the individual genes distribution. FDR is the ratio of median or 90th percentile of the number of falsely called genes to the number of genes called significant. The SAM has the following advantages:

- Any type of data from cDNA arrays, oligo, SNP array, protein arrays etc., can be used in SAM.
- Expression data can be correlated to clinical parameters and time.
- Data permutation is used to estimate the false discovery rate for multiple testing.
- The local false discovery rate and miss rates can be reported.
- Can be worked with a blocked design for different treatments in batches of arrays.
- The significant genes can be determined by adjusting threshold.

3 Clustering

Three types of clustering namely Hierarchical, k-means and Fuzzy C-means clustering are chosen as they are simple and efficient than conventional methods and is well suitable for microarray datasets. These algorithms show the difference between Acute Lymphoblastic Leukemia (ALL), Acute Myeloid Leukemia (AML) samples. Clustering is defined as grouping objects of similar kind into respective categories. It is an unsupervised technique, in which data can be discovered without providing an explanation or interpretation. Three types of clustering using are Hierarchical, k-means and Fuzzy C-means clustering algorithms.

3.1 Hierarchical Clustering

In Hierarchical clustering, the data are partitioned into particular cluster, but not in a single step. A series of partitions takes place, in which it runs from a single cluster containing all objects to n clusters each containing a single object, it may be represented by a two dimensional diagram known as clustering heat map which explains the fusions or divisions made at each successive stage of analysis. Hierarchical clustering algorithm is of two types, namely, Agglomerative Hierarchical clustering algorithm or AGNES (agglomerative nesting) and Divisive Hierarchical clustering algorithm or DIANA (divisive analysis). The Divisive Hierarchical clustering algorithm is a top-down approach and is less commonly used, whereas the Agglomerative Hierarchical clustering is a bottom-up approach used for grouping the data one by one on the basis of the nearest distance measure of all the pairwise distance between the data point. The advantages are it can produce objects ordering that is informative for displaying data and aids in discovering using generated small clusters. So in this paper, Agglomerative Hierarchical clustering is used for clustering of microarray significance gene list.

3.2 k-Means Clustering

One of the unsupervised learning algorithms that determine the well known clustering problem is k means algorithm. It follows an easy way to categorize a given dataset through a specific number of clusters (assume k clusters) fixed previously. The solution is to describe k centers, each cluster contains one center. The centers should be placed in a dissimilar way since dissimilar location leads to different results. So, the enhanced choice is to put them far distance from each other. After that, take each point belonging to a specified dataset and relate it to the nearest center. When all points are completed, the first step is finished and an early grouping is done. Lastly, this algorithm intends at minimizing a squared error function known as objective function given in Eq. (1).

$$J(V) = \sum_{i=1}^{c} \sum_{j=1}^{c_i} \left(\left\| x_i - v_j \right\| \right)^2 \tag{1}$$

where, $\left\| x_i - v_j \right\|$ is the Euclidean distance between x_i and v_j. 'c_i' is the number of data points in ith cluster. 'c' is the number of cluster centers [9].

3.3 Fuzzy C-Means Clustering

Based on the distance between the data point and the cluster center, the membership for each data point is assigned related to each cluster center. If the data is very close to the cluster center the membership towards that specific cluster center will be high. Obviously, summation of membership of every data point must be equal to one. The cluster centers and membership are updated for iterations, according to the formulas [9] given in the Eqs. (2) and (3) respectively.

$$\mu_{ij} = \frac{1}{\sum_{k=1}^{C} \left(\frac{d_{ij}}{d_{ik}} \right)^{\left(\frac{2}{m} - 1 \right)}} \tag{2}$$

$$v_j = \frac{\left(\sum_{i=1}^{n} \left(\mu_{ij} \right)^m x_i \right)}{\left(\sum_{i=1}^{n} \left(\mu_{ij} \right)^m \right)}, \quad \text{for } j = 1, 2, \ldots.C \tag{3}$$

where, the number of data points is 'n', the jth cluster center is represented by 'v_j, the fuzziness index m € [1,∞], the number of cluster center is represented by

'C', the membership of ith data to jth cluster center is represented by 'μ_{ij}', the Euclidean distance between ith data and jth cluster center is represented by 'd_{ij}'. The foremost objective of fuzzy C-means algorithm is to reduce the function given in the Eq. (4).

$$J(U,V) = \sum_{i=1}^{n}\sum_{j=1}^{C}(\mu_{ij})^m(||x_i - v_j||)^2 \tag{4}$$

where, $||x_i - v_j||$ is the Euclidean distance between jth cluster center and ith data.

4 Implementation and Results

The dataset used in this work for testing the three clustering algorithms is the Golub et al. dataset [10] consists of gene expression levels of 3,051 genes and 38 samples. The first 27 samples belong to ALL (Acute Lymphoblastic Leukemia) and the next 11 samples belong to AML (Acute Myeloid Leukemia). SAM test is performed on the Golub dataset. This Golub dataset contains 3,051 genes and 38 samples among which the first 27 samples are in the control group and the rest are in the treatment group. The 'gene name' file for every one of 3051 genes is available in the first column and the respective log intensity values for 38 samples are available in columns 2 to 39. So the size of the dataset is 3051 × 38. Since microarray dataset contains row and column values, it is easy to use MATLAB than other software tools. The MATLAB implementation consists of defining the global variables and providing the values to those variables. Next, reading of samples and observation data from Golub dataset which contains two classes is done. Then SAM is performed on that dataset and the output contains the identified genes with the parameters and their corresponding values. Finally, the SAM results are plotted with observed expression scores against the expected expression scores and the obtained results are saved in a file. The result file after SAM test consists of various parameters.

4.1 SAM Results

The parameter D-value is the observed expression score, Std-dev is the standard deviation of repeated expression measurements. The Raw_P-value represents the unadjusted probability values. The Q-value is the lowest FDR at which a gene is called significant and Fold-Change is the antilog value of differences between average values of two classes. The parameter s_0 is called fugde factor, Δ is the threshold value, FDR is the False Discovery Rate and the lower and upper cutoff values are used for finding the significant repressed and induced genes respectively.

Table 1 SAM output for $\Delta = 0.4$

S. no.	Parameters	Value
1	Delta	0.4
2	Small positive constant s_0	0.0652801
3	Number of permutations	1000
4	Prior probability	5.169378e-002
5	Lower cutoff	−0.75582
6	Upper cutoff	0.867765
7	Significant genes	1810
8	Falsely called genes	0
9	FDR	0

The parameter values after running SAM algorithm are shown in Table 1. As discussed in the technical document, the number of permutations considered for this dataset is 1000 [11]. The threshold Δ is used for finding the number of significant genes. In the first case, the Δ value is chosen as 0.4. By varying the delta values in the range from 0.4 to 2.4, the optimum number of significant genes can be obtained.

Figure 1 shows the scatter plot of the expected relative difference $d_E(i)$ versus the observed relative difference d(i). The bold line indicates that $d(i) = d_E(i)$, but some genes (red colour square) are represented by points displaced from the $d(i) = d_E(i)$ line by a distance greater than a threshold Δ. The number of falsely significant genes can be determined by using horizontal cutoffs. For finding significantly induced genes, the horizontal cutoff value is 0.867765, defined as the smallest d(i), and the least negative d(i) is the lower cutoff value which is −0.75582, used for finding the significantly repressed genes.

Similarly for $\Delta = 1$, the parameter values obtained after SAM test is shown in Fig. 2. It is clear that as the Δ value is increased, the lower and upper cutoff values

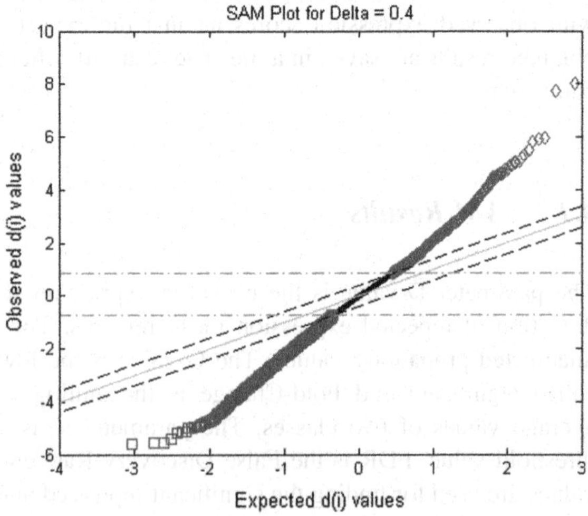

Fig. 1 Scatter plot of d(i) versus $d_E(i)$ for $\Delta = 0.4$

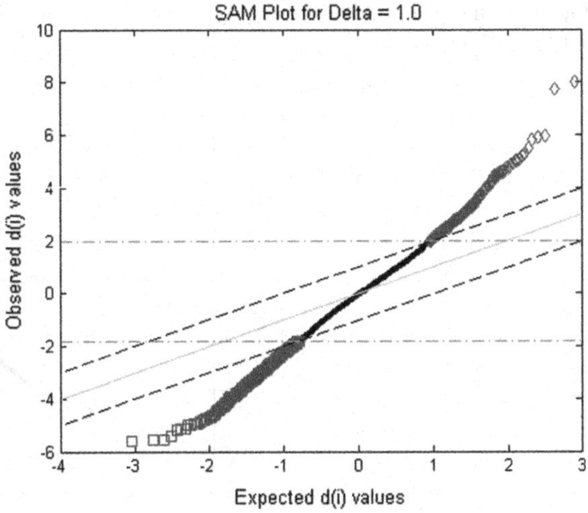

Fig. 2 Scatter plot of d(i) versus $d_E(i)$ for $\Delta = 1.0$

get increased leading to significant reduction in the gene numbers. It can be seen that the change in upper cutoff value of 1.96386 and lower cutoff value of -1.79455 reduces the significant gene numbers to 777 genes when compared with 1,810 genes for a Δ value of 0.4.

Figure 3 shows the scatter plot of d(i) versus $d_E(i)$ for $\Delta = 1.6$. The number of significant genes obtained is 267. Since the falsely called genes and FDR values are zero, the threshold value can be increased further to find a smaller number of significant genes. For another $\Delta = 2.2$, the number of significant genes obtained is 99, with falsely called genes and FDR value equals to zero.

Figure 4 shows that for $\Delta = 2.2$, the horizontal cutoffs are widened. Further increase in Δ value may lead to falsely called genes. For $\Delta = 2.5$. The significant

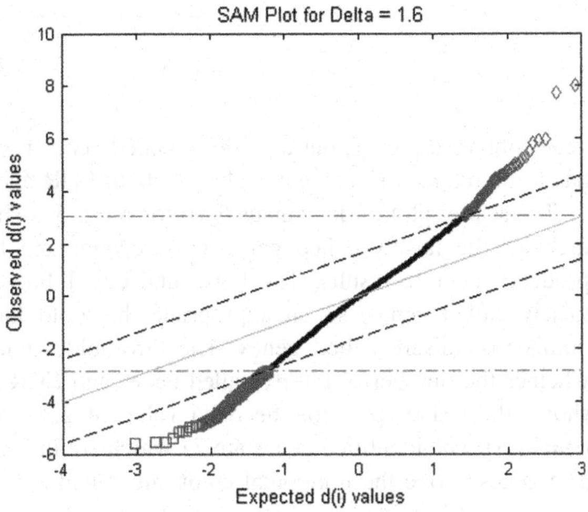

Fig. 3 Scatter plot of d(i) versus $d_E(i)$ for $\Delta = 1.6$

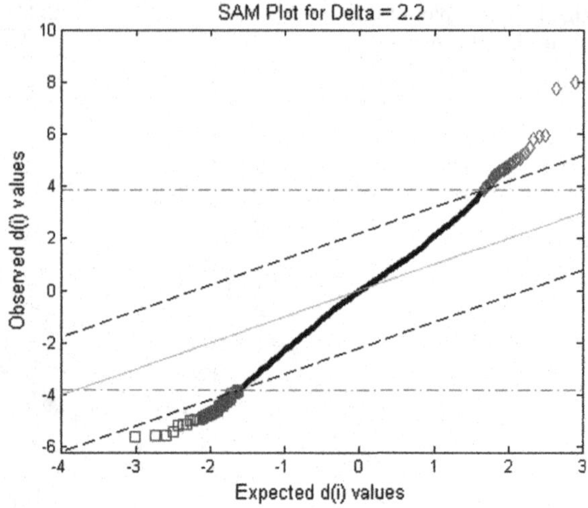

Fig. 4 Scatter plot of d(i) versus d_E(i) for Δ = 2.2

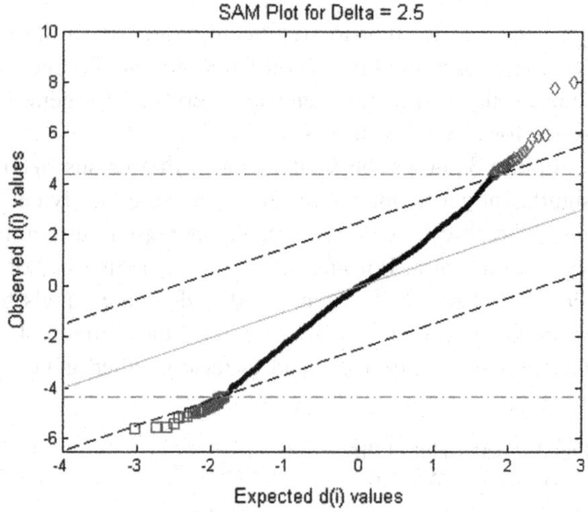

Fig. 5 Scatter plot of d(i) versus d_E(i) for Δ = 2.5

genes obtained are 61, but the falsely called genes become more. So this threshold has to be reduced to get zero value for both FDR and false called genes.

The scatter plot of d(i) versus d_E(i) for Δ = 2.5 is shown in Fig. 5. The threshold includes the falsely called genes into account and hence this threshold doesn't produce accurate results. The horizontal cutoff lines are widened, which covers falsely called genes, so an appropriate threshold has to be chosen to get zero number of falsely called genes. The threshold value is reduced to 2.3, to find whether the number of falsely called genes and FDR are becoming zero. Figure 6 shows the scatter plot for the delta value of 2.3. For most of the genes, the d (i) = d_E(i). The identified genes are 77 which are indicated by the red color squares. The genes above the horizontal cutoff are called as up regulated genes and genes

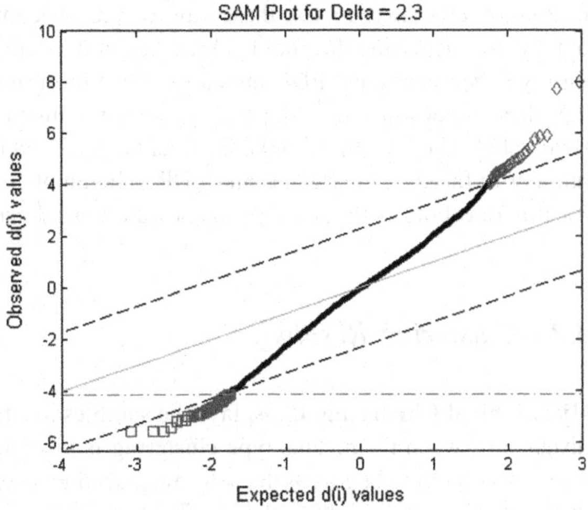

Fig. 6 Scatter plot of d(i) versus $d_E(i)$ for $\Delta = 2.3$

Table 2 SAM results for different delta values

S. no.	Delta value	Lower cutoff	Upper cutoff	Identified genes	Falsely called genes	FDR
1	0.4	−0.756	0.868	1810	0	0
2	0.6	−1.086	1.277	1374	0	0
3	0.8	−1.408	1.653	1048	0	0
4	1.0	−1.795	1.964	777	0	0
5	1.2	−2.132	2.310	565	108.104	0.0101038
6	1.4	−2.473	2.637	399	0	0
7	1.6	−2.832	3.032	267	0	0
8	1.8	−3.134	3.341	196	0	0
9	2.0	−3.526	3.633	132	0	0
10	2.1	−3.729	3.828	111	0.43	0
11	2.2	−3.824	3.882	99	0	0
12	2.3	−4.113	4.055	77	1137.14	0.779857
13	2.4	−4.146	4.166	73	1477.34	1.08278
14	2.5	−4.350	4.355	61	1570.04	1.39295

below lower cutoff value are called down regulated genes. In this case, there are 1137.14 falsely called genes and 0.779857 FDR value. So this threshold is also not chosen. The threshold which produces zero number of falsely called genes and FDR is 2.2 and hence the 99 significant genes obtained using this threshold is considered for clustering.

From Table 2, it is clear that when delta value is increased from 0.4 to 1.0, the false called genes and FDR values are zero. When delta value is increased to 1.2, the false called gene value and FDR starts to increase, but by further increasing the threshold value delta to 1.4, these values again becomes zero. So there are further options for increasing threshold values. When the delta value is increased to 2.3, the falsely called genes and FDR increases. Then by further increasing delta to 2.4 and 2.5, these values increases drastically. So the adjustment of threshold value can be terminated. The final threshold adjustment considered is $\Delta = 2.2$, as it produces zero number of falsely called genes and FDR. The number of significant genes obtained for this threshold is 99 and these genes are considered for clustering.

4.2 Clustering Results

Hierarchical Clustering. It displays the samples in a treelike structure based on the distance between them. This type clustering is visualized using the heat map [12]. The row clustering begins with each row placed in a separate cluster. The Euclidean distance measure is used to calculate the distance between all possible combinations of two rows. The two most similar clusters are grouped to form new clusters. In the following steps, the distance between new cluster and remaining clusters is again calculated and the number of cluster gets reduced by one in iterative steps [13]. At last, all rows are arranged to form one cluster. A similar procedure is used for column clustering.

The SAM test provides an output of 99 significant genes with 27 ALL and 11 AML samples. The samples are clustered using a hierarchical clustering algorithm. Figure 7 shows the clustering heat map of samples into two clusters ALL and AML. From the figure, it is clear than hierarchical clustering performs well by accurately

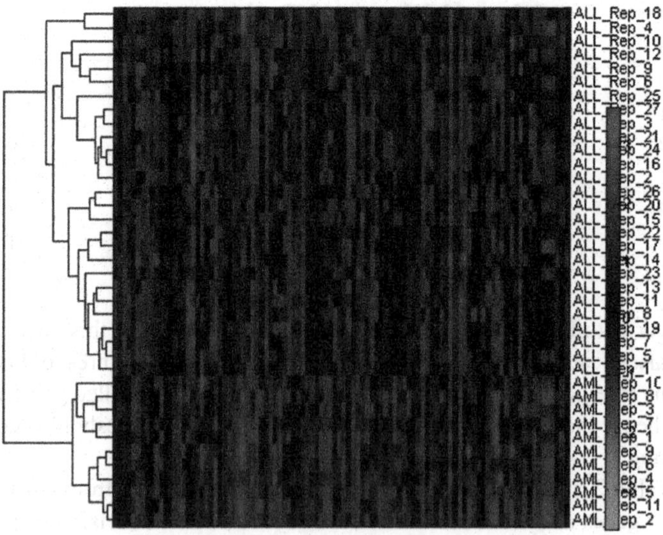

Fig. 7 Hierarchical clustering for significant gene set

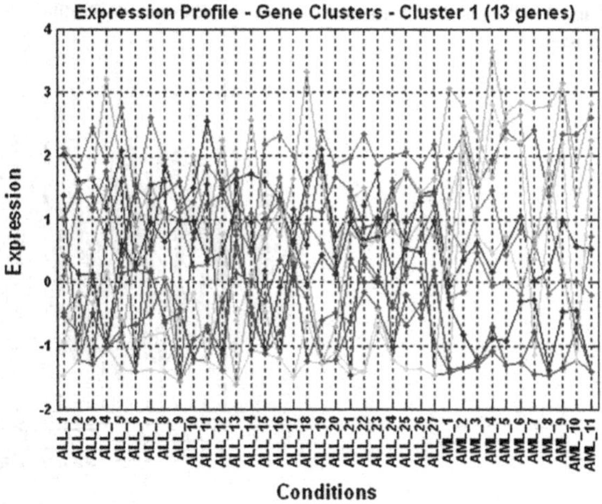

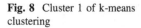

Fig. 8 Cluster 1 of k-means clustering

clustering 27 samples in ALL and 11 samples in AML. It is indicated by the tree like structure shown in blue color lines on the left part of the figure. So hierarchical clustering is better for this dataset.

k-means Clustering. The number of clusters, k, needs to be determined at the beginning itself. The aim is to divide the objects into k clusters such that the metric related to cluster centers is minimized. The next step is to assign each point in the dataset to the nearest centroid and grouping is formed. Then again recalculation of centroids is performed and grouped. This procedure is repeated until the centroids do not change positions. The k value is chosen as 2 and the cityblock manhatan metric is used.

Since there are only two groups in the Golub dataset with 27 ALL and 11 AML, the k value is chosen as 2. Figure 8 shows the formation of cluster 1 using k-means clustering algorithm. For 11 AML samples, 13 samples are grouped in first cluster (AML) adding two extra samples. Figure 9 shows the formation of cluster 2 using k-means clustering algorithm. For 27 ALL samples, 25 samples are grouped in the second cluster (ALL), so two samples of ALL group are left during the clustering procedure. This k-means left, 2 samples in ALL cluster and adds 2 samples in AML cluster. Thus the performance of hierarchical clustering is better when compared with k-means clustering algorithm.

Fuzzy C-means Clustering. The k-means clustering assigns each gene to a single cluster and they do not provide the information about the control of a given gene for the overall shape of clusters. This problem is solved using fuzzy C-means clustering, in which the genes are assigned cluster membership values. The gene, which has highest membership value in the cluster will belong to that cluster. The membership values for specific genes in other clusters are also checked [14] and they are clustered. Figure 10 shows the formation of cluster 1 using fuzzy C-means clustering algorithm. The fuzzy parameter and cluster values are set to 2. For 27 ALL samples, 26 samples are grouped in first cluster (ALL) leaving one sample.

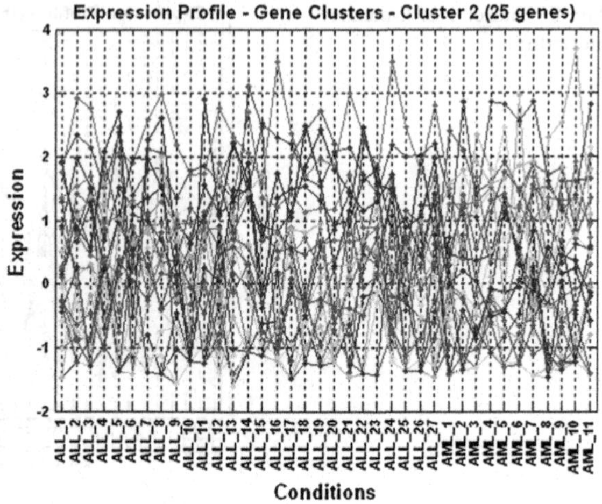

Fig. 9 Cluster 2 of k-means clustering

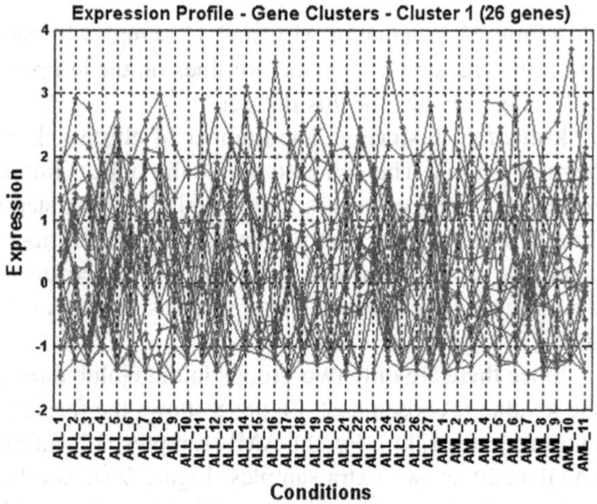

Fig. 10 Cluster 1 of fuzzy C-means clustering

Figure 11 shows the formation of cluster 2 using fuzzy C-means clustering algorithm. For 11 AML samples, 12 samples are grouped in the second cluster (AML), with one extra sample. Thus, fuzzy C-means results are good when compared with k-means clustering algorithm. But overall, the hierarchical clustering performs well in exactly forming 27 samples in first cluster (ALL) and 11 samples in the second cluster (AML). These three types of clustering algorithm are helpful when analyzing the unknown dataset. They will provide an idea regarding the characteristics of the dataset.

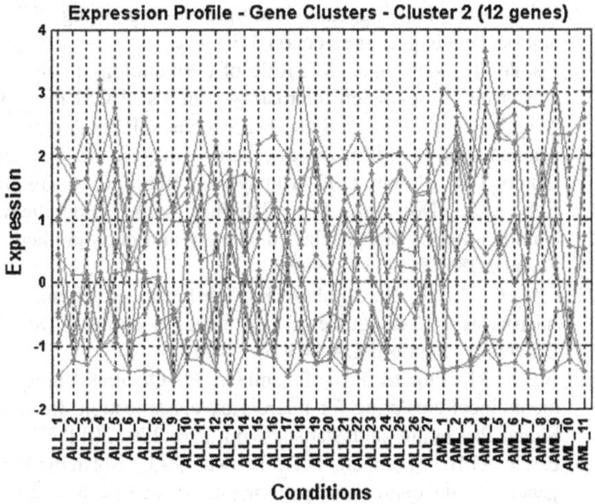

Fig. 11 Cluster 2 of fuzzy C-means clustering

5 Conclusion

In this paper, a statistical technique called SAM (Significance Analysis of Microarrays) is used for identifying the significant genes in Golub dataset, which contains 27 ALL samples and 11 AML samples. The SAM identifies 99 genes from 3,051 genes and these significant gene sets are clustered using three clustering techniques, namely, Hierarchical, k-means and Fuzzy C-means clustering algorithms for easy visualization of gene clusters, which aids in analyzing the unknown dataset. Out of these techniques, hierarchical clustering exactly clusters the ALL and AML samples.

References

1. Schena, M., Shalon, D., Davis, R.W., Brown, P.O.: Quantitative monitoring of gene expression patterns with a complemenary DNA microarray. Science **270**, 467–470 (1995)
2. Southern, E.M.: DNA microarray history and overview. Methods Mol. Biol. **170**, 1–15 (2001)
3. Chen, T.C., Hsieh, Y.C., You, P.S., Lee, Y.C.: Feature selection and classification by using grid computing based evolutionary approach for the microarray data. In: Third IEEE International Conference on Computer Science and Information Technology, pp. 85–89 (2010)
4. Hsu, H.H., Lu, M.D.: Feature selection for cancer classification on microarray expression data. In: Eighth International Conference on Intelligent Systems Design and Applications, pp. 153–158 (2008)
5. Soliman, T.H.A., Sewissy, A.A., AbdelLatif, H.: A gene selection approach for classifying diseases based on microarray datasets. In: Second International Conference on Computer Technology and Development, pp. 626–631(2010)
6. Tusher, V.G., Tibshirani, R., Chu, G.: Significance analysis of microarrays applied to the ionizing radiation response. PNAS **98**(9), 5116–5121 (2001)

7. Schwender, H.: Siggenes: multiple testing using SAM and Efron's empirical Bayes approaches. R package version 1.32.0
8. Damle, T., Kshirsagar, M.: Role of permutations in significance analysis of microarray and clustering of significant microarray gene list. IJCSI Int. J. Comput. Sci. **9**, 342–344 (2012)
9. Maguluri, L.P., Rajapanthula, K., Srinivasu, P.N.: A comparative analysis of clustering based segmentation algorithms in microarray images. Int. J. Emerg. Sci. Eng. **1**, 27–32 (2013)
10. Golub, T.R., Slonim, D.K., Tamayo, P., Huard, C., Gaasenbeek, M., Mesirov, J.P., Coller, H., Loh, M.L., Downing, J.R., Caligiuri, M.A., Bloomfield, C.D., Lander, E.S.: Molecular classification of cancer: class discovery and class prediction by gene expression monitoring. Science **286**, 531–537 (1999)
11. Chu, G., Li, J., Narasimhan, B., Tibshirani, R., Tusher, V.: SAM significance analysis of microarrays. Users guide and technical document, Department of Biochemistry, Stanford University, Stanford CA 94305 (2002)
12. Bosio, M., Pujalte, P.B., Salembier, P., Oliveras-Verges, A.: Feature set enhancement via hierarchical clustering for microarray classification. In: International Workshop on Genomic Signal Processing and Statistics, pp. 226–229 (2011)
13. Eisen, M.B., Spellman, P.T., Brown, P.O., Botstein, D.: Cluster analysis and display of genome-wide expression patterns. In: Proceeding of the National Academy of Sciences, pp. 14863–14868 (1998)
14. Maji, P.: Mutual information-based supervised attribute clustering for microarray sample classification. IEEE Trans. Knowl. Data Eng. **24**, 127–140 (2012)

Breast Region Extraction and Pectoral Removal by Pixel Constancy Constraint Approach in Mammograms

S. Vidivelli and S. Sathiya Devi

Abstract Nowadays, the breast cancer can be detected early with automated Computer Aided Diagnosis (CAD) system the best available technique to assist radiologist. For developing such an efficient computer-aided diagnosis system it is necessary to pre-process the mammogram images. Hence, this paper proposes a method for effective pre-processing of mammogram images. This method consists of two phases such as (i) Breast Region Extraction and (ii) Pectoral removal. In first phase, Adaptive Local Thresholding is used to binaries the image followed by morphological operations for removing labels and artifacts. Then the breast region is extracted by identifying and retaining the largest connected component of mammogram. The pectoral muscle which is the predominant density region of mammogram that should not carry any useful information and also affects the diagnosis is to be removed in phase two. A new method called Pixel Constancy Constraint at multi-resolution approach is introduced for pectoral removal. The proposed method is experimented with Mini-MIAS database (Mammographic Image Analysis Society, London, U.K.) and yields a promising result when compared with existing approaches.

Keywords Mammogram · Breast region extraction · Pectoral muscle · Segmentation · Pixel constancy constraint · Accuracy

---

S. Vidivelli (✉)
Department of Information and Communication Engineering,
Anna University, Chennai, India
e-mail: vidieng@gmail.com

S. Sathiya Devi
Department of Information Technology, University College of Engineering,
Anna University (BIT Campus), Trichy, India
e-mail: Sathyadevi.2008@gmail.com

M. Senthilkumar et al. (eds.), *Computational Intelligence, Cyber Security and Computational Models*, Advances in Intelligent Systems and Computing 412, DOI 10.1007/978-981-10-0251-9_20

1 Introduction

Globally, breast cancer is the leading cause of death in female population and statistical report shows that 1 out of 10 women are affected by breast cancer [1]. Screening Program based on mammography is one of the best and popular method for detection of breast cancer and early detection of breast cancer will reduce mortality. It also increases the chance of successful treatment and recovery of patient from breast cancer. To minimise the workload of radiologist and assist them, a computerised mammographic analysis system is having greater impact. Pre-processing is the initial step of automatic analysis of digitized mammogram and it involves in the segmentation of breast region and removal of pectoral muscles that can minimise the search area for abnormalities and make it limited to the relevant region of breast without excessive influence from the background of mammogram. Pectoral muscle appears as a triangular opacity across the upper posterior margin of the image. This pectoral muscle can bias and affect the result of any mammogram processing system, so it is necessary to automatically identify and segment the pectoral muscle. This paper proposes a method to extract the breast region and remove pectoral muscle. It is organized as follows: Sect. 2 discusses the related work, the proposed pre-processing method is presented in Sect. 3, Data set is explained in Sects. 4, and 5 discusses the Experimental result and the conclusion is described in Sect. 6.

2 Related Works

The state of art literature and the various methods related to pre-processing are reviewed in this section. The pre-processing can be carried out in two steps (i) Noise Removal and Breast Region Extraction, (ii) Pectoral Removal.

2.1 Noise Removal and Breast Region Extraction

Initially the breast border extraction and segmentation of mammogram images where performed by manual or semiautomatically. But it violates the requirement of CAD system and increases many problems. Mendez et al. [2] developed a fully automated technique to identify the breast border and nipple region. Gradient based on the grey level is computed for breast border identification and three algorithms are used (maximum height of the breast border, maximum gradient, and maximum second derivative of grey levels across the median-top section of the breast) for nipple region extraction. This method is evaluated by radiologist and they claim that 89 % of result is achieved in breast region extraction. Sreedevi et al. [3] proposed an algorithm to identify noisy pixels in the image by introducing statistical measure and a nonlocal means (NL-means) noise-filtering framework based on Discrete Cosine Transform (DCT).

2.2 *Pectoral Removal*

Detection and elimination of pectoral muscle is another challenging task in pre-processing. Ferrari et al. [4] proposed a method for the identification of the pectoral muscle in MLO mammograms using Gabor wavelets. This method overcomes the limitation of the straight-line representation considered in our initial investigation using the Hough transform. The method was applied to 84 MLO mammograms from the Mini-MIAS. Maitra et al. [5] proposed Contrast Limited Adaptive Histogram Equalisation (CLAHE) technique for image enhancement, which is a special case of the histogram equalization technique that functions adaptively on the image to be enhanced. And a rectangle method is used to isolate pectoral removal from region of interest using seeded region growing algorithm. In this algorithm the approximation of rectangle region was failed to generate satisfactory result for smaller and larger volume pectoral muscle. Liu et al. [6] proposed a combined approach of Otsu thresholding and morphological processing for pectoral border identification. But this method failed to produce good result in case of low contrast images. Vaidehi et al. [7] proposed a straight line method for removal of pectoral. This method approximates the pectoral muscle by a line. The method does not give desired result if the contour of the pectoral muscles is a curve. The algorithm tested on 120 mammogram images from Mini-MIAS database.

3 Proposed Breast Region Extraction and Pectoral Removal Method

Having discussed the related work of mammogram pre-processing in the previous section, this section proposes an efficient method for breast region extraction. This method consists of two phases (i) Breast Region Extraction and (ii) Pectoral muscle removal. The steps involved in these two phases are shown in Fig. 1a, b.

3.1 *Breast Region Extraction*

In the proposed method Breast Region can be extracted in four steps: (i) Local Adaptive Thresholding, (ii) Morphological operations, (iii) Connected Component identification and (iv) Retain the largest connected component. The detailed explanations of these steps are presented below.

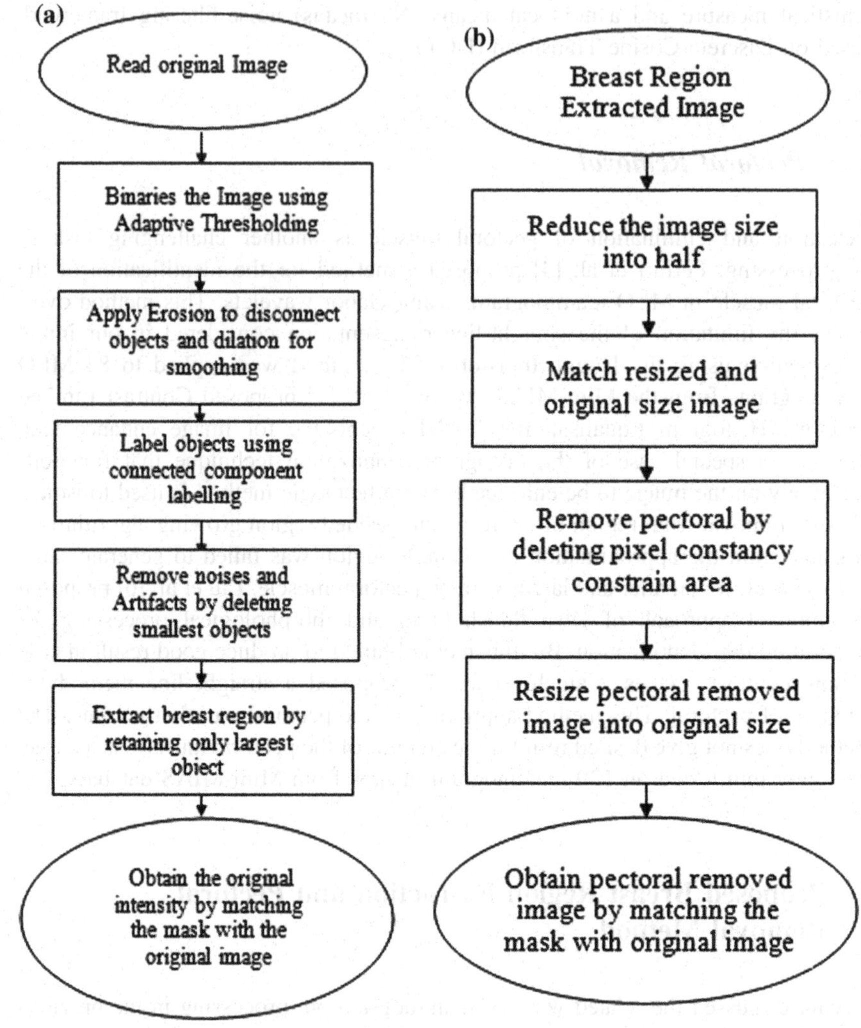

Fig. 1 **a** Breast region extraction. **b** Pectoral removal

3.1.1 Local Adaptive Thresholding

Binary image obtained by global thresholding can be very fast but produces poor result like noisy boundary and shifted edges in case of low contrast images. Hence, only this method implements local adaptive thresholding technique which produces good result with speed closer to global method. For binarising the image the pixel values are set to 0 or 1 as per Eq. (1).

$$b(i,j) = \begin{cases} 0 & \text{if } I(i,j) \leq T \\ 1 & \text{otherwise} \end{cases} \tag{1}$$

where *b(i,j)* is binarised image and *I(i,j)* is the intensity value of original image at the location of *(i,j)* and *T* is the threshold value. The binary segmentation technique computes the local threshold T of a pixel centred in a$(w \times w)$ window using the local mean m and the standard deviation *s* as per the Eq. (2).

$$T = m\left[1 + k\left(\frac{s}{R} - 1\right)\right] \tag{2}$$

where *R* is the maximum value of the standard deviation and k is a parameter which takes positive values in the range [0.2, 0.5].

3.1.2 Morphological Operations

The binary image obtained in the previous section contain objects such as (i) Artifacts, (ii) Labels, (iii) Breast region and (iv) Vertical and horizontal strips or noises. The morphological erosion followed by dilation is applied to disconnect objects from each other, remove noises and also eliminate the vertical and horizontal noisy strips. The dilation operation retains the useful information. The formula for erosion and dilation is given in Eqs. (3) and (4)

$$\mathrm{A} \oplus \mathrm{B} = \cup_{b \in \mathrm{B}} \mathrm{A}_b \tag{3}$$

$$\mathrm{A} \Theta \mathrm{B} = \cap_{b \in B} \mathrm{A}_{-b} \tag{4}$$

In these equations the A_b indicates the shift operation on *A* in direction of *b* and A_{-b} indicate shift in reverse direction.

3.1.3 Connected Component Identification

The output of morphological operation at previous step yields many number of disconnected objects. In our proposed work this connected component labelling algorithm [8] is used for identification of breast region, which is the largest connected component of the mammogram images.

3.1.4 Retain the Largest Connected Component

In previous step the individual objects are identified using connected component labelling algorithm, this step retains only the object with maximum number of label. The breast region is the biggest blob in the mammogram images and having maximum number of labels. Now, match the mask with largest connected component labelled object with original image so that the breast region can be extracted successfully.

3.2 Pectoral Removal

The breast region extracted in previous step contains a predominant density region known as pectoral muscle. This region affects the result of CAD system and it should be removed from the mammogram images. The proposed pectoral removal method comprised of the following steps: (i) Resize the original image, (ii) Pectoral positioning, (iii) Pixel Constancy Constraint Approach.

3.2.1 Resize the Original Image

The output image x obtained from the previous step of size (w × h) is taken as an input for this phase. The obtained image is divided into non-overlapping windows of size 2 × 2. In order to obtain the reduced image r of size (w/2 × h/2) the intensity values are computed from the window using the Eq. (5) as below.

$$r(i,j) = \frac{[x(2i,2j) + x(2i,2j + 1) + x(2i + 1,2j) + x(2i+1, 2j+1)]}{4} \tag{5}$$

The above algorithm is explained using a sample calculation in Fig. 2a, b. Using the Eq. (5) the values of R11, R12, R21, R22 in reduced image is calculated and explained through the Eqs. (6), (7), (8) and (9).

$$R11 = 1/4(A11 + A12 + A21 + A22) \tag{6}$$

$$R12 = 1/4(A13 + A14 + A23 + A24) \tag{7}$$

$$R21 = 1/4(A31 + A32 + A41 + A42) \tag{8}$$

$$R22 = 1/4(A33 + A34 + A43 + A44) \tag{9}$$

(a)

A11	A12	A13	A14
A21	A22	A23	A24
A31	A32	A33	A34
A41	A42	A43	A44

⇨

(b)

R11	R12
R21	R22

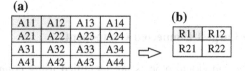

Fig. 2 **a** Original image. **b** Reduced image

3.2.2 Pectoral Positioning

In this step, the breast region extracted image and its reduced sized image is converted into binary format using the adaptive thresholding technique used in phase one. Then edges are identified in both the images using the Eqs. (10) and (11). To identify the pectoral position, summate the value of each column and store individually. Then find the column with highest accumulator value, definitely the column where the starting point of pectoral region resides will have highest value than others. Now, the starting position of the pectoral region is identified and hence matches that pectoral position of two images.

$$E_{r(i,j)} = |2 \times r(i,j) - r(i+1,j) - r(i,j+1)| \forall i,j \in 1, N-1 \quad (10)$$

where N = (w × h)

$$E_{x(i,j)} = |2 \times x(i,j) - x(i+1,j) - x(i,j+1)| \forall i,j \in 1, N-1 \quad (11)$$

where N = (w/2 × h/2)

3.2.3 Pixel Constancy Constraint Approach

For pectoral muscle removal from the breast region extracted image, this method uses the approach of pixel constancy constraint. Processing of any two images having same pixel values are said to be pixel Constancy constrain approach. In this method, the two image (i) breast region extracted image and (ii) resized image having same pixel values and produce zero when subtracted with each other. After the pectoral region positioning in previous step, the resized image is subtracted from breast region extracted image using Eq. (12) and hence the pectoral region of resized image got nullified. Then the resized image is doubled, so that a mask for pectoral removal is obtained. Match the mask with breast region extracted image and obtain the pectoral removed image as output.

$$pr(i,j) = x(i,j) - r(i,j) \quad (12)$$

where pr(i, j) is the pectoral removed image, x(i, j) is breast region extracted image and r(i, j) is the resized image.

4 Data Set

For the experiments, the mini-MIAS database was used (MIAS stands for Mammographic Image Analysis Society). It contains 322 craniocaudal mammographic images from left and right breast of 161 patients, some of them healthy and others containing lesions such as benign or malign tumours and calcifications. This is a reduced version of the original MIAS database reduced to 200 μ pixel edge and clipped or padded so that every image is 1024 × 1024 pixels.

5 Experimental Result

The Proposed algorithm is experimented on 322 images of mini-MIAS database to suppress pectoral muscle and remove the artifacts. The evaluation was done subjectively through inspection and comparison with expert report contained in the readme file of the MIAS database. The subjective evaluation is performed using the measure such as (i) Good, (ii) Acceptable, (iii) Unacceptable and (iv) Discarded.

5.1 Phase One: Breast Region Extraction

In this phase, the binarisation is performed by Local Adaptive Thresholding technique using the Eqs. (1) and (2) and the window size is taken as (3 × 3) in our proposed work. In the Eq. (2), the value for k is taken as 0.3 which produce good result as shown in Fig. 3b. Binary erosion and Dilation is applied as per the Eqs. (3)

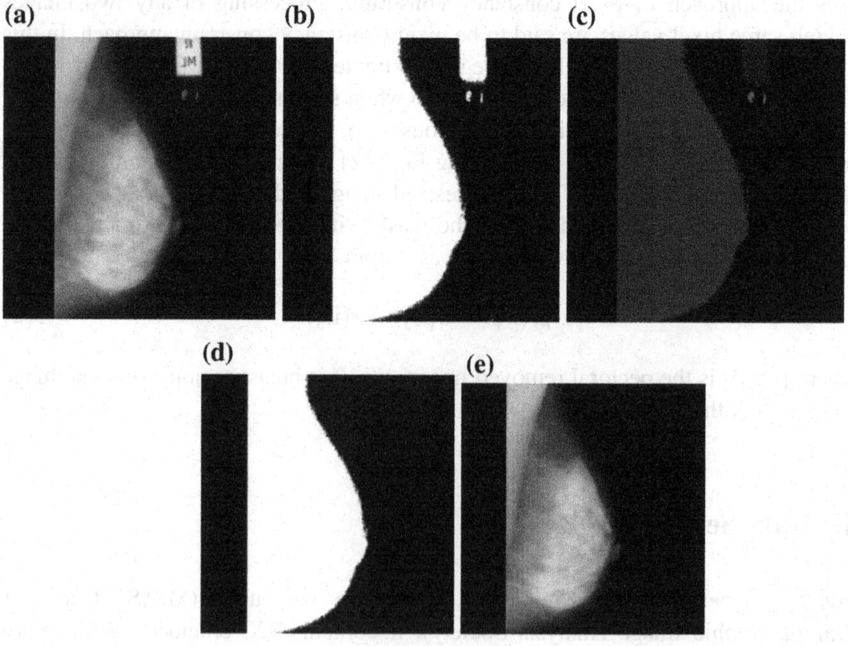

Fig. 3 **a** Original image. **b** Binary image. **c** Labelled image. **d** Breast region mask. **e** Breast region extracted

and (4), so that individual objects can be identified using connected components labelling algorithm as shown in Fig. 3c. Since the breast region is the biggest connected component of the image, delete all other objects and retain only the biggest connected object as in Fig. 3c. Match this mask with the original image (Fig. 3a), so that the breast region is extracted effectively as shown in Fig. 3e.

5.2 Phase Two: Pectoral Removal

In this phase, the resultant image of phase one (Fig. 3e) is resized using the Eq. (5) so that image height and width is reduced into (512 × 512) as shown in Fig. 4b.

Pectoral region in both the images are identified using edge detection technique by the Eq. (10) and (11) and matched by finding column with highest accumulator

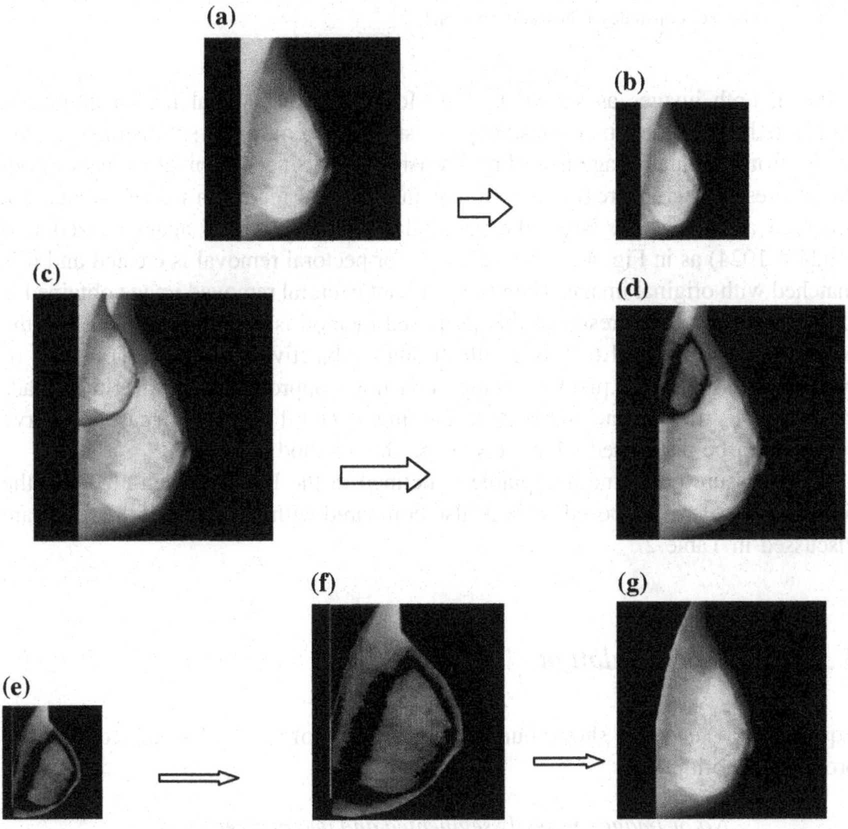

Fig. 4 **a** Breast region segmented image. **b** Resized image. **c** Pectoral region positioning. **d** Subtract segmented image from resized. **e** Pectoral region removed in resized image. **f** Double the size. **g** Pectoral removed image

Table 1 Result table of proposed method

Subjective evaluation	Result	Percentage (%)
Good	270	83.85
Acceptable	25	7.76
Unacceptable	23	7.14
Discarded	4	0.12

Table 2 comparative result analysis

S. no.	Method	Accuracy (%)	References	Year
1	AD method	81	Liu et al. [9]	2011
2	Straight line	85	Vaidehi et al. [7]	2013
3	Combination of global thresholding, canny edge detection and connected component labelling	90.06	Sreedevi et al. [3]	2014
4	Combination of adaptive thresholding and pixel constancy constraint approach	91.6	Proposed	2015

value in both images as shown in Fig. 4c. Then the pectoral muscle region is eliminated using the pixel constancy constraint approach by subtracting resized image from original image as in Fig. 4d using (12). Since the pixel values of both the images in pectoral region are similar the pectoral region in resized image got removed as in Fig. 4e. Now, the pectoral removed image is again resized into (1024 × 1024) as in Fig. 4f. A mask image for pectoral removal is created and it is matched with original image. Then the resultant pectoral removed image obtained is shown in Fig. 4g. The result of this proposed method is shown in Table 1 and the proposed work yield 91.6 % of result through subjective evaluation. This kind of pectoral removal using pixel constancy constraint approach overcomes the disadvantages of straight line method. Removing pectoral muscle in case of curve contour can be performed effectively using this method.

The measure good and acceptable is defined in the Eqs. (13) and (15) and the performance of the proposed work is also compared with [3, 7, 9] and the result are discussed in Table 2.

5.3 Sample Calculation

Equations (14) and (16) shows numerical evaluation of the result obtained from the proposed algorithm

$$\text{Good} = \frac{\textit{No of Images perfectly segmented and pectoral removed}}{\textit{Total no of Images}} \times 100 \qquad (13)$$

$$\text{Good} = \frac{270}{322} \times 100 = 83.85\,\% \tag{14}$$

$$\text{Acceptable} = \frac{\textit{No of Images Accepted on Subjective evaluation}}{\textit{Total no of Images}} \times 100 \tag{15}$$

$$\begin{aligned} \text{Acceptable} &= \frac{25}{322} \times 100 = 7.76\,\% \\ \text{Result achieved} &= (\text{Percentage of Good} + \text{Percentage of Acceptable}) \\ &= 83.85\,\% + 7.76\,\% \\ &= 91.6\,\% \end{aligned} \tag{16}$$

The proposed method is compared with other existing techniques like AD method, Straight Line method and a combinational method comprised of Thresholding, Canny Edge Detection and connected component labelling in Table 2. Compared with other existing techniques this proposed method yield good result.

6 Conclusion

In this paper, the proposed algorithm of pre-processing has been presented with noise removal, breast region extraction and pectoral removal. This extraction of breast contour is useful, because it limits the search-zone for abnormalities. The predominant pectoral muscle has to be identified and removed since it can affect the classification process of automatic breast cancer detection system. Our proposed work of adaptive thresholding and morphological operation for breast region extraction and pixel constancy constraint for pectoral removal achieve best result in pre-processing. Since a novel method is proposed for pectoral removal in our work, it achieves good result even for pectoral with curve contours. So in future we can implement the identification and classification of malicious tissues as the extension of automatic mammogram image analysis and classification system in an effective manner.

References

1. World Cancer Report: World Health Organization 2014. pp. Chapter 1.1. ISBN 92-832-0429-8
2. Mendez, A.J., Tahoces, P.G., Lado, M.J., Souto, M., Correa, J.L., Vidal, J.J.: Automatic detection of breast border and nipple in digital mammograms. Comput. Methods Programs Biomed. **49**, 253–262 (1996)
3. Sreedevi, S., Sherly, E.: A novel approach for removal of pectoral muscles in digital mammogram. In: International Conference on Information and Communication Technologies (ICICT 2014), Elsieve B.V (2015)

4. Ferrari, R.J., Rangayyan, R.M., Desautels, J.E.L., Borges, R.A., Frere, A.F.: Automatic identification of the pectoral muscle in mammograms. IEEE Tran. Med. Imaging **23**(2), 232–245 (2004)
5. Maitra, I.K., Nag, S., Bandyopadhyay, S.K.: Computer methods and programs in biomedicine. Elsevier Ireland **107**, 175–188 (2012)
6. Liu, Chen-Chung, Tsai, Chung-Yen, Liu, Jui, Chun-Yuan, Yu., Shyr-Shen, Yu.: A pectoral muscle segmentation algorithm for digital mammograms using Otsu thresholding and multiple regression analysis. Comput. Math Appl. **64**(5), 1100–1107 (2012)
7. Vaidehi, K., Subashini, T.S.: Automatic identification and elimination of pectoral muscle in digital mammograms. Int. J. Comput. Appl. **75**(14) (2013). 0975-8887
8. Appiah, K., Hunter, A., Dickinson, P., Meng, H.: Accelerated hardware video object segmentation: from foreground detection to connected components labelling. Comput. Vis. Image Understand. **114**, 1282–1291(2010)
9. Liu, L., Wang, J., Wang, T.: Breast and pectoral muscle contours detection based on goodness of fit measure. In: (iCBBE) 2011 5th International Conference on Bioinformatics and Biomedical Engineering, pp. 1–4, IEEE (2011)

Bridging the Semantic Gap in Image Search via Visual Semantic Descriptors by Integrating Text and Visual Features

V.L. Lekshmi and Ansamma John

Abstract To facilitate access to the enormous and ever–growing amount of images on the web, existing Image Search engines use different image re-ranking methods to improve the quality of image search. Existing search engines retrieve results based on the keyword provided by the user. A major challenge is that, only using the query keyword one cannot correlate the similarities of low level visual features with image's high-level semantic meanings which induce a semantic gap. The proposed image re-ranking method identifies the visual semantic descriptors associated with different images and then images are re-ranked by comparing their semantic descriptors. Another limitation of the current systems is that sometimes duplicate images show up as similar images which reduce the search diversity. The proposed work overcomes this limitation through the usage of perceptual hashing. Better results have been obtained for image re-ranking on a real-world image dataset collected from a commercial search engine.

Keywords Image re-ranking · Visual semantic descriptor · Semantic space · Perceptual hashing · Image search

1 Introduction

Web image search and retrieval has become an increasingly important research topic due to abundance in multimedia data on internet. Most of the web scale image search engines mainly use two schemes for searching for images on the web. In the first, keyword based scheme, images are searched for by a query in the form of

V.L. Lekshmi (✉) · A. John
Department of Computer Science and Engineering, TKM College of Engineering, Kollam, Kerala, India
e-mail: lekshmivl20@gmail.com

A. John
e-mail: ansamma.john@gmail.com

M. Senthilkumar et al. (eds.), *Computational Intelligence, Cyber Security and Computational Models*, Advances in Intelligent Systems and Computing 412, DOI 10.1007/978-981-10-0251-9_21

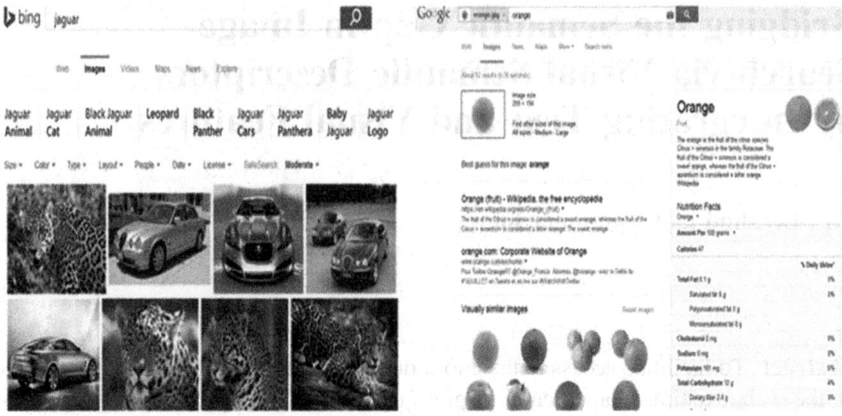

Fig. 1 Illustration of the keyword based and example based image search scheme

textual keyword provided by the user. The second, example based scheme allows the users to search for similar images by providing an example serving as query image both schemes are illustrated in Fig. 1.

Independent of which search scheme is deployed, an image search engine generally operates in two main steps: the offline and the online step. For many query keywords the image retrieval performance is good, but the precision of the returned results is still relatively low. They suffer from the ambiguity of query keywords, because it is difficult for users to accurately describe the visual content of target images only using query keywords. One of the major challenges is the conflict between the content of the image and the webpage textual information. This paper attempts to resolve this bottleneck by depending on both the textual information and visual information. Another major challenge in the existing systems is that its similarities of low level visual features may not correlate with image's high level semantic meanings. To reduce this semantic gap, visual features are mapped to a predefined attributes known as visual semantic descriptors.

In this paper we propose a web image search approach, which requires both query keyword and query image. First a text based search is performed by using a query keyword. From the pool of images retrieved, the user is asked to select the query image. Images in the pool are re-ranked based on the visual semantic descriptors of the query image. This query-specific visual semantic descriptor effectively reduces the gap between low-level visual features and semantic categories, and makes image matching more consistent with visual perception.

Another major issue in web image search is removing near-duplicate images. We address the diversity problem to make the search result more diverse and to improve efficiency of the search engine. Identifying distinct images can prevent duplicate images from being used once they are uploaded. In this paper we use Perceptual hash method to remove duplicate images and improve search diversity. Image features are used to accomplish this and experimental results show the better improvement of diversity in the search result. Encouraging results are obtained for proposed image re-ranking method on a real world image dataset collected from commercial search engines.

The rest of this paper is organized as follows. We discuss related works in Sect. 2 and follow up with the details of our method in Sect. 3 and present evaluation results in Sect. 4 and finally conclusions and future works are offered in Sect. 5.

2 Related Work

Most of the internet scale image search engines like Google Image Search and Bing primarily depend on the textual information. The annotation error due to the subjectivity of human perception is one of the major disadvantage of this approach. Text based image search suffers from the ambiguity of query keywords [1]. Content based image retrieval (CBIR) was introduced in early 1980s, to overcome the limitations of text based image search. Visual features of images are used in CBIR to evaluate image similarity.

Generally there are three categories of visual re-ranking methods classification based, clustering based and graph based. Classification-based methods first select some pseudo-relevant samples from the initial search result. Deng [2] learned to produce a similarity score for retrieval by using a predefined comparison function based on a known hierarchical structure. In clustering based methods the pool of images in the initial search result are first grouped into different clusters. According to the cluster conditional probability, order the clusters and re-ranked result list is created. By using the cluster membership value order the samples within each cluster. Graph based methods are more recently proposed and have received increased attention. In this method first a graph is built with the images in the initial search result serving as nodes. If two images are visual neighbors of each other, an edge is defined between those images and the edges are weighted by the visual similarities between the images. For instance, either a random walk over the graph or an energy minimization problem, re-ranking can be formulated.

Cui [1, 3] classified query images into eight predefined adaptive weight categories, inside each category a specified pre-trained weight schema is used to combine visual features. For reducing the semantic gap, query specific semantic signatures was first proposed in [4]. The proposed visual semantic descriptor is effective in reducing the semantic gap when computing the similarities of images. One shortcoming of the existing system [1, 3, 5] is that sometimes duplicate images show up as similar images to the query. The proposed system overcomes this problem by adding a perceptual hash method to detect duplicate images. Our work incorporates both textual and visual information for re-ranking, moreover incorporation of textual information and duplicate detection significantly improves the search result and was not considered in previous work.

3 Method

We use both query keyword and query image for searching the images. The images for re-ranking are collected from different search engines like Google, Bing etc. User first submits a query keyword, from the pool of images returned after text based search, user is asked to select a query image which reflects the user's search intention. Based on the visual similarity with the query image, images in the pool are re-ranked. The semantic classes associated with query keywords are discovered by expanding the query keywords from the result obtained from the text based image search by utilizing both text and visual features. This expanded keyword defines the semantic classes for the query keywords. For each semantic class the training images are collected. A multilayer perceptron on the text and visual features is trained from the training sets of the semantic classes and its output is stored as visual semantic descriptor which describe the visual content from different aspects.. We combine the described features together to train a single classifier, which extracts a single semantic descriptor for an image and the extracted descriptors are stored. These semantic descriptors are used to compute the similarities for re-ranking. It provides much better re-ranking accuracy, since training the classifiers of reference classes captures the mapping between the visual features and semantic meanings [5]. The Flowchart of our approach is shown in Fig. 2.

3.1 Feature Design

We adopt a set of features which describe the visual content of an image in different aspects. For reducing the semantic gap both high level features and low level features are used to capture the user's search intention. Here we briefly explain the features used in this work.

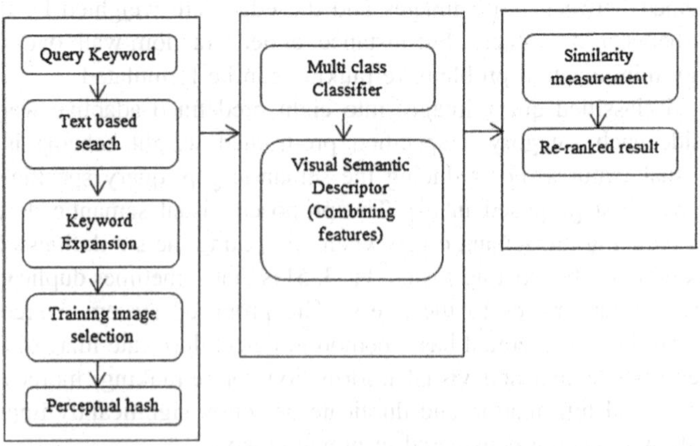

Fig. 2 Overall block diagram

- *Color*: The most widely used color feature used in image retrieval is color. One color description that is good to dispense efficiency and effectiveness in illustrating the distribution of colors of an image is color moment [6]—Mean, standard deviation, skewness. Mathematically those moments are defined as follows [7].
- *Texture*: Texture is one of the major feature used in recognizing objects [8]. Here we use co-occurrence matrix for image texture analysis.
- *Attention Guided Color Signature*: The color composition of an image describes color signature [9]. Clusters centers and their relative proportions are taken as the signature.
- *Histogram of Gradient (HoG)*: HoG reveals distributions of edges over different parts of an image, and is especially effective for images with strong long edges [10].
- *Wavelet*: We use the 2nd order moments of wavelet coefficients in various frequency bands to characterize the texture properties in the image [11].

3.2 Expanding the Keywords and Collection of Training Images

The keyword expansions most relevant to the query keyword are identified to define the semantic classes associated with each query keyword (For example, for a query keyword '*diamond*' some of the semantic classes identified are *black diamond, diamond jewelers, red diamond* etc.). To find the keyword expansions first a text based search is performed and the top ranked images are retrieved from the text based search. Keyword expansions are found from words extracted from the top retrieved images. The T most frequent words $W_1 = \{w_I^1, w_I^2, \ldots, w_I^T\}$ among top D re-ranked images are identified and these words are used for expanding the query keyword. The extracted words are sorted according to the frequency of words become available among the D images from higher to lower. If a word w is among the top ranked image, it has a ranking score $r_i(w)$ according to its ranking order. Otherwise ranking score must be zero [5].

$$r_I(w) = \begin{cases} T - j & w = w_I^j \\ 0 & w \notin W_I \end{cases} \tag{1}$$

To obtain the training images of semantic classes, we combine the words obtained from the keyword expansion with original query keyword and it is used as the query to the search engine. From the search result top K images are taken as the training images for the semantic classes. To reduce the computational cost remove the similar semantic classes this is both semantically and visually similar.

3.3 Perceptual Hash Method

To identify the duplicate images here we use perceptual hash method. Duplicate images significantly reduce the storage space and decrease the search diversity. Perceptual image hash functions create hash values based on the image's visual appearance. This function computes similar hash values for similar images, whereas for different images dissimilar hash values are calculated. Finally, using a similarity function to compare two hash values, it decides whether two images are different or not. The different steps involved in the method are as follows:

- Reduce size: Method starts with a small image. 32 × 32 is a good size; this step is done to simplify the DCT (Discrete Cosine Transform) computation.
- Reduce color: The image is reduced to grayscale just to further simplify the number of computations.
- Compute the DCT: DCT disparate the image into a collection of frequencies and scalars, here uses a 32 × 32 DCT.
- Reduce the DCT: While the DCT is 32 × 32, to represent lowest frequencies in the image keep the top-left 8 × 8.
- Calculate the average value: Mean DCT value is computed.
- Further reduce the DCT: Depending on whether each of the 64 DCT values is above or below the average value set the 64 hash bits to 0 or 1.
- Construct the hash: Set 64 bits into a 64-bit integer. To understand what this fingerprint looks, set the values (based on whether the bits are 1 or 0 this uses +255 and −255) and convert from the 32 × 32 DCT (with zeros for high frequencies) back into the 32 × 32 image. Construct the hash from each image to compare two images, and count the number of bit positions that are different. (Hamming distance) A distance of zero indicates that it is very similar pictures thus identifies the duplicate images and eliminate it thereby improving the search results.

3.4 Combining Features and Visual Semantic Descriptor Extraction

The above described features characterize the images from different perspective of color, shape and texture. The objective of using a visual semantic descriptor is to capture the visual content of an image. Here we combine all the visual features to train a single multilayer perceptron better distinguishing reference classes. The output of the classifier is taken as the visual semantic descriptor. If S semantic classes for a query keyword, classifier on the visual features is trained and it outputs an S dimensional vector v, which indicates the probability of the image belonging to different semantic classes and it is stored as the visual semantic descriptor of the image. The distance between two images I^a and I^b are measured as the $L1$ distance between their semantic descriptors v^a and v_b,

$$d(I^a, I^b) = \left\| v^a - v^b \right\|_1 \quad (2)$$

4 Experimental Results

From different search engines, the images are collected for testing the performance of re-ranking. Averaged top m precision is used as the evaluation criterion. Top m precision is defined as the proportion of relevant mages among top m re-ranked images. Averaged top m precision is obtained by averaging over all the query images. The averaged top m precisions on dataset are shown in Fig. 3. The improvements of averaged top 10 precisions on the 10 query keywords on dataset by comparing text and visual based methods are shown in Fig. 4. Here we choose 10 semantic classes for a single query keyword and only one classifier is trained combining all types of features, so the semantic descriptors are of 10 dimensions on average. For a particular semantic class fifty images are selected, so for a single

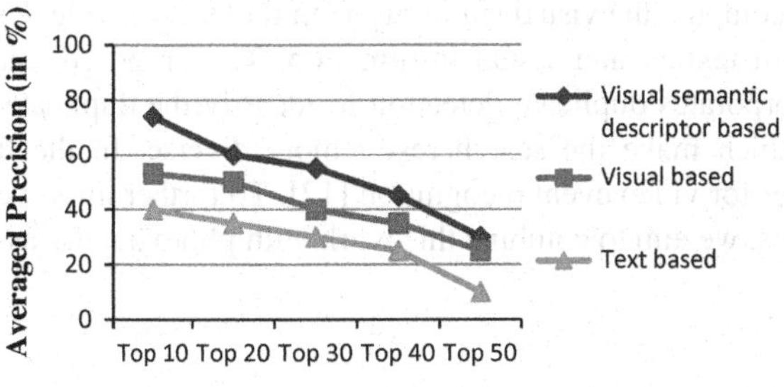

Fig. 3 Averaged top m precisions on different methods

Fig. 4 Improvements of averaged top 10 precisions on the 10 query keywords

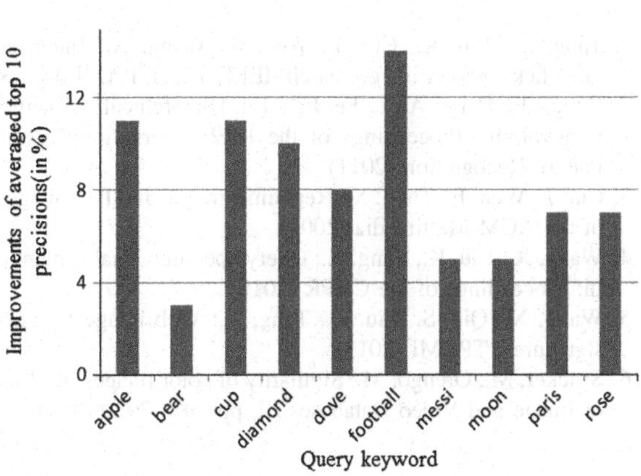

query keyword 500 images are used. For training, 75 % of the images were used and remaining 25 % were utilized for testing.

In this work we combine all the extracted features to train a single classifier to reduce the computational overhead caused when training separate classifiers for each feature. The dimensionality of extracted features is very large, so in the existing systems more computational cost is for comparing the visual features of images when re-ranking. Here we overcome this computational overhead by the use of visual semantic descriptors which have very less dimension compared to the visual features. The use of perceptual hashing method increases the diversity in search result by removing the duplicate images. The proposed work significantly outperforms the text based and visual based methods which directly compare visual features. The averaged top 10 precision is enhanced from 53 % (visual based) to 74 %. 21 % relative improvement is achieved.

5 Conclusion and Future Work

We propose a Web image search approach which requires both query keyword and query image. Previous methods only use either textual information or example input by the user. Visual semantic descriptors are proposed to combine the visual features and to compute the visual similarity with the query image. User intention is captured by both textual and visual information. The proposed image re-ranking framework incorporates duplicate detection to identify the duplicate images to the query image which make the search result more diverse. In the future we plan extend this paper for video event recognition [12]. To further improve the quality of re-ranked images, we aim to combine this work with photo quality assessment work in [13].

References

1. Tang, X., Liu, K., Cui, J., Wen, F., Wang, X.: Intentsearch: capturing user intention for one-click internet image search. IEEE Trans. PAMI **34**, 1342–1353 (2012)
2. Deng, J., Berg, A.C., Fei-Fei, L.: Hierarchical semantic indexing for large scale image retrieval. In: Proceedings of the IEEE International Conference on Computer Vision and Pattern Recognition (2011)
3. Cui, J., Wen, F., Tang, X.: Real time google and live image search re-ranking. In: Proceedings of the ACM Multimedia (2008)
4. Wang, X., Liu, K., Tang, X.: Query-specific visual semantic spaces for web image re-ranking. In: Proceedings of the CVPR (2010)
5. Wang, X., Qiu, S., Liu, K., Tang, X.: Web image re-ranking using query-specific semantic signatures. TPAMI (2013)
6. Stricker, M., Orengo, M.: Similarity of color images. In: IS&T and SPIE Storage and Retrieval of Image and Video Databases III, pp. 381–392 (1995)

7. Maheshwary, P., Sricastava, N.: Prototype system for retrieval of remote sensing images based on color moment and gray level co-occurrence matrix. IJCSI Int. J. Comput. Sci. Issues **3**, 20–23 (2009)
8. Haralick, R.M., Shanmugam, K., Its'Hak, D.: Textural features for image classification. IEEE Trans. Syst. Man Cybernetics **3**(6), 610–621 (1973)
9. Rubner, Y., Guibas, L., Tomasi, C.: The earth movers distance, multi-dimensional scaling, and color-based image retrieval. In: Proceedings of the ARPA Image Understanding Workshop (1997)
10. Dalal, N., Triggs, B.: Histograms of oriented gradients for human detection. In: Proceedings of the IEEE International Conference on Computer Vision and Pattern Recognition (2005)
11. Unser, M.: Texture classification and segmentation using wavelet frames. IEEE Trans. Image Process. **4**(11), 1549–1560 (1995)
12. Duan, L., Xu, D., Tsang, I.W., Luo, J.: Visual event recognition in videos by learning from web data. In: Proceedings of the IEEE Computer Society Conference on Computer Vision and Pattern Recognition, pp. 1959–1966 (2010)
13. Ke, Y., Tang, X., Jing, F.: The design of high-level features for photo quality assessment. In: Proceedings of the IEEE International Conference on Computer Vision and Pattern Recognition (2006)

Adaptive Equalization Algorithm for Electrocardiogram Signal Transmission

L. Priya, A. Kandaswamy, R.P. Ajeesh and V. Vignesh

Abstract The proposed work is based on the mathematical analysis of steepest descent stochastic gradient weight updating method for the adaptive cancellation of intersymbol interference in Electrocardiogram (ECG) signal transmission over wireless networks. The major challenge associated with the physiological signal transmission over high data rate band limited digital communication systems is prone to Intersymbol Interference (ISI). In those cases, adjacent symbols on the output of the signal overlap each other which results in irreducible errors as a result of ISI. This work investigates the performance of proposed weight updating method based adaptive equalization technique for estimating the original transmitted ECG signal from the noise corrupted channel output signal. Adaptive filters are designed and implemented to minimize the error at the receiver side, thus making data to be of error free. The results are investigated to validate the operational parameters such as Mean Square Error (MSE), computational complexity, correlation coefficient and convergence rate of the proposed method and their comparative performances over other equalization methods. Simulation results indicated that, the proposed adaptive linear equalization method has good extraction performance than other nonlinear and blind equalization methods.

Keywords Electrocardiogram signal · Intersymbol interference · Steepest descent · Mean square error · Correlation · Convergence rate · And computational complexity

L. Priya (✉) · A. Kandaswamy · R.P. Ajeesh · V. Vignesh
PSG College of Technology, Coimbatore, Tamil Nadu, India
e-mail: lpm@bme.psgtech.ac.in

A. Kandaswamy
e-mail: akswamy@bme.psgtech.ac.in

R.P. Ajeesh
e-mail: ajeeshrp7@gmail.com

V. Vignesh
e-mail: vignesh7117@gmail.com

M. Senthilkumar et al. (eds.), *Computational Intelligence, Cyber Security and Computational Models*, Advances in Intelligent Systems and Computing 412, DOI 10.1007/978-981-10-0251-9_22

1 Introduction

Advancements in the field of information technology and telecommunication, wireless communication finds significant potential in supporting diverse intricate and advanced services for health care [1]. In the Wireless Body Area Networks (WBAN), human body acts as a communication channel for electrical signals which offers efficient data communication in biomedical monitoring systems. The vital signals from body sensors are transmitted to a nearby body node coordinator (BNC) via Ultra-Low-Power Short-Haul radios. These signals are then transmitted to a remote terminal like a hospital via the Internet [2]. To detect atrial fibrillation using mobile phones, Beijin described the design and implementation of wearable and wireless ECG intelligent system [3]. Motoi design, a portable device for ambulatory Electrocardiogram recording is now routinely used to detect infrequent, asymptomatic arrhythmias and to monitor the effects of cardiac drugs or surgical procedures [4]. It has been shown that the development of real time patient monitoring using wireless technologies is entirely possible [5–8].

The motivation of this work is to make the received ECG data at the receiver end to be error free for cardiovascular disease diagnosis and to relieve patients from the need of visiting hospitals frequently and also allows the continuous and ubiquitous monitoring of their ECG. All physical channels tend to exhibit Intersymbol Interference. Multipath delay spread occurs due to reflections of radio signals from concrete structures that results in multiple copies of the same signal to be received by the receiver at different delays. Delay spread along with fading cause ISI thereby limiting the maximum symbol rate of the digital multipath channel. Estimation and equalization is necessary when data is travelled across band limited channel in presence of noise. The equalization filter in the receiver side is the inversion of the channel filter. Lucky processed a zero forcing algorithm for automatically adjusting tap weights of a transversal equalizer [9]. In this algorithm, the performance index called peak distortion, which is directly proportional to the maximum value of ISI can occur. The tap weights are adjusted to minimize the peak distortion. This effectively forces the ISI, due to those adjacent pulses contained in the equalizer to become zero. The proposed method finds the solutions for the above constraints in linear channel equalizers by updating adaptive filter coefficients using steepest descent stochastic gradient algorithm. This method estimate the ECG signal amplitude and phase values from the noisy band limited channel output. In this work, we obtained the fast convergence rate and high correlation output which are the keen essentials for the telemedicine applications.

2 Methodology

The proposed system is shown in Fig. 1. In this communication system, we will analyze the amplitude and phase fluctuations of the transmitted ECG signal through the propagation channel and also estimate the transmitted signal with minimum error using steepest gradient estimation method.

2.1 Normal ECG Signal

Electrocardiogram (ECG) is the electrical manifestation of the contractile activity of the heart. ECG analysis helps in the diagnosis of numerous cardiac related disorders. These signals if transmitted wirelessly can pave way for the early diagnosis and hence for the treatment of cardiovascular diseases. The signal used in this work is the ECG test data from record number 14046 of the MIT-BIH Long Term ECG database. This Normal Discrete ECG (Lead-I) signal is downloaded from the Physiobank ATM of 4 s duration. The sampling frequency is 128 Hz. 1200 samples are used for processing in this work. Thus the sampling interval is 0.0078125 s. The maximum amplitude of an ECG signal ranges from −3.5 to 1.5 mV [10].

2.2 Mathematical Model of Wireless Channel

In the communication system, channel is a unknown linear time invariant channel (LTI) and h(n) which represents all the interconnections between the transmitter and

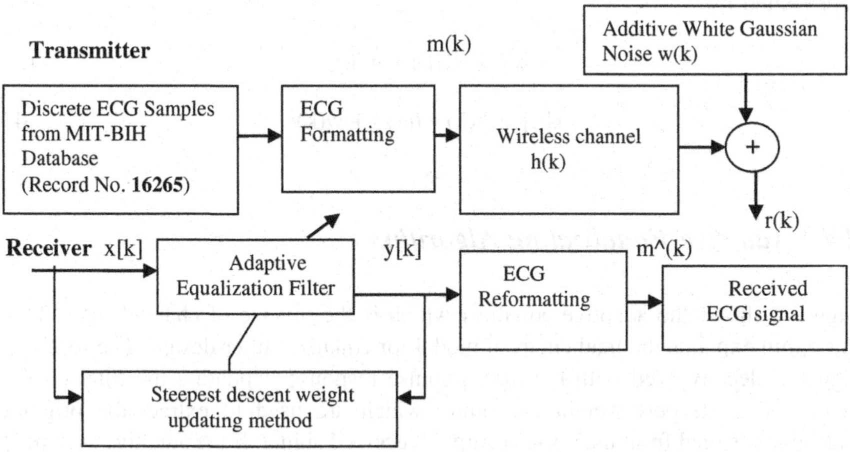

Fig. 1 Base band communication system

receiver baseband. This LTI channel can be modeled using Finite Impulse Response (FIR) filter whose channel impulse response as h(n).

The LTI Discrete channel in z transform notation as

$$H(z) = \sum_{k=0}^{\infty} h(k)z^{-k} \tag{1}$$

In this design, the signal bandwidth (Bs) of 60 Hz is greater than the channel filter bandwidth (Bc) of 40 Hz. For the given design considerations of FIR filter, the intersymbol interference is introduced in the ECG signal that travels through the wireless channel.

Design Specifications

Type: Low pass FIR filter, Passband frequency = 40 Hz, Order = 8,

Number of coefficients = 9.

The channel output signal is obtained by taking convolution between the source ECG signal and channel impulse response.

$$r[k] = h(k) * m(k) \tag{2}$$

White Gaussian Noise, (w(k)) is the white noise which has the Gaussian distribution of amplitudes that is widely present in the digital communication signals. Additive refers to the linear addition of white Gaussian noise with the transmitted signal.

2.3 *Received Signal*

The receiver receives the White Gaussian noise corrupted ECG signal. It can be represented by

$$x[k] = r(k) + w(k) \tag{3}$$

$$x[k] = h(k) * m(k) + w(k) \tag{4}$$

2.4 *Adaptive Equalization Algorithm*

Figure 4 shows the adaptive equalizer which is the inverse of channel filter. This algorithm explains the mathematical model for equalizer filter design. The received signal is deconvolved with the filter impulse response. Updating the filter coefficients using steepest weight algorithm, which are used to extract the original transmitted signal from the noise corrupted received signal. If the adaptive system is an adaptive linear combiner, and if the input vector x[k] and the desired response d[k] are available at each iteration, the Steepest descent is generally the best choice for many different applications of adaptive signal processing. Adaptive linear combiner was applied in two basic ways, depending on whether the input is available in parallel or serial form. In both the cases we have the combiner output as the deconvolved signal y[k] as a linear combination of input samples. Then we calculate the error signal as e[k] = d[k] − y[k] [11]. If the error is maximum, the steepest descent algorithm uses a special estimate of the gradient that is valid for the adaptive linear combiner method. It does not require any offline repetitions of the data. The gain constant (μ) regulates the speed and stability of adaptation.

2.4.1 Steepest Gradient Search Method

It is a method of gradient search that also causes all components of the weight vector to be changed at each step or iteration cycle in the direction of negative gradient of the performance surface. This surface can be represented as

$$\xi = \xi_{min} + \lambda(w - w^*) \tag{5}$$

where λ is eigen value, w^* is optimal weight value in which the mean square error is minimized.

With only a single weight, the repetitive or iteration gradient search procedure described as

$$w_{k+1} = w_k - 2\mu\lambda(w_k - w^*) \tag{6}$$

2.4.2 Algorithm Steps

Step1 Get x[k] and d[k]

Step2 Find y[k] = deconvolve x[k] with h(k) (Or)
Using vector notation:

$$y_k = x_k^T h_k \tag{7}$$

Step3 Calculate the error

$$e[k] = d[k] - y[k] \tag{8}$$

(Or)

$$e_k = d_k - y_k \tag{9}$$

Step4 Find new coefficients of the filter using steepest descent stochastic gradient estimation

$$h[k + 1] = h[k] + 2\mu e[k]y[k] \tag{10}$$

(Or)

$$h_{k+1} = h_k + 2\mu e_k x_k \tag{11}$$

μ is the step size, $1/\lambda_{max} > \mu > 0$
λ_{max} is the largest eigen value

Step5 Find the filter output using updated weights
deconvolve x[k] with h(k + 1) (Or)

$$y_k = x_k^T h_{k+1} \tag{12}$$

3 Results and Discussion

Figure 2 shows the source ECG signal which is downloaded from the Physionet Longterm ECG database (rec no. 14046). This Lead 1 ECG signal is recorded for 4 s duration and its amplitude ranges from −3.5 to 1.5 mV. The sampling rate is 128 Hz and sampling interval is 0.0078125 s. The gain of the amplifier is 200. After amplification, the ECG signal amplitude ranges from −350 to 150 mV. The simulation is averaged over 1200 samples [10].

Figure 3 shows the channel output signal which has the intersymbol interference and Additive White Gaussian noise. The channel output signal shows the resultant of multiple delayed versions of the source ECG signal which has random amplitude and phase.

Figure 4 shows the adaptive equalizer output in time domain. This figure shows that the equalizer output ECG signal amplitude with respective samples is identical with that of original ECG signal. The proposed method overcomes the fluctuations

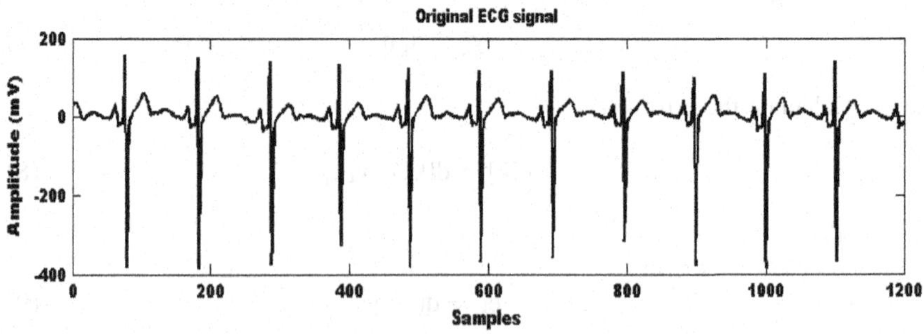

Fig. 2 Original ECG signal

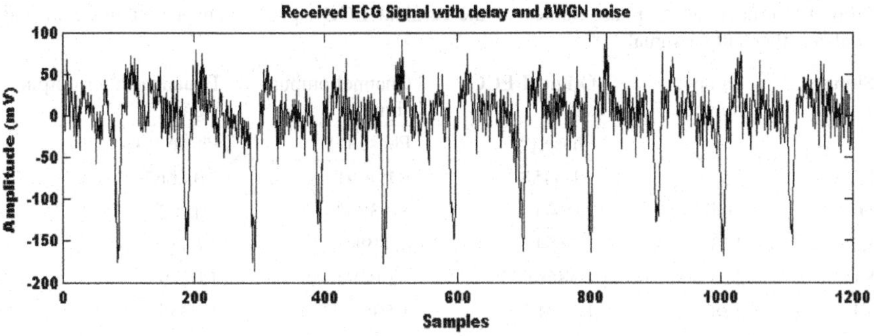

Fig. 3 Wireless channel output signal

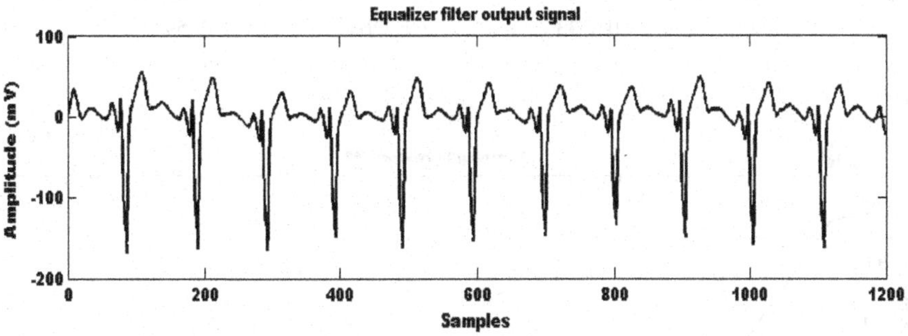

Fig. 4 Adaptive equalizer filter output signal

in the transmitted signal amplitude better than that of non linear equalization methods.

From Table 1, the two sample sets of ECG peaks such as P1, Q1, R1, S1, T1 and P2, Q2, R2, S2, T2 phase values with respective samples for the original signal, channel output signal and equalizer filter output signal are tabulated. ECG signal was transmitted through the band restricted channel which introduces the phase fluctuations in the received signal. The received signal is processed by adaptive equalizer by updating 9 coefficients of the filter. The equalizer output signal phase values are similar that of the original ECG signal.

Figure 5 depicts the learning curve that determines the mean square error over iteration number. The learning behaviors of proposed algorithm are perfectly robust. We observed that steepest algorithm outperforms blind equalization algorithm in terms of convergence rate, because proposed method mean square error is reduced to zero at 4th iteration but blind method mean square error is reduced to zero after 3×10^5 iterations [12].

Table 2 shows the comparative performance of proposed method with conventional zero forcing equalization method under various signal parameters. From

Table 1 Comparison of phase values of the original ECG signal, channel output signal and equalizer filter output signal

Signal events	Sample number	Original ECG signal	Channel output signal	Equalizer filter output signal
		Phase (°)	Phase (°)	Phase (°)
P1	75	−1.5458	−1.4991	−0.8667
Q1	104	2.0145	−0.3863	1.1802
R1	113	1.2154	0.4398	1.648
S1	123	1.9866	−1.0694	1.0279
T1	188	1.7671	1.5963	1.4258
P2	229	−2.7563	−1.855	−1.2317
Q2	258	2.1357	−2.6646	2.1181
R2	267	−2.6415	2.855	−2.139
S2	277	2.0274	2.1308	2.2103
T2	342	0.0733	−1.2265	0.2659

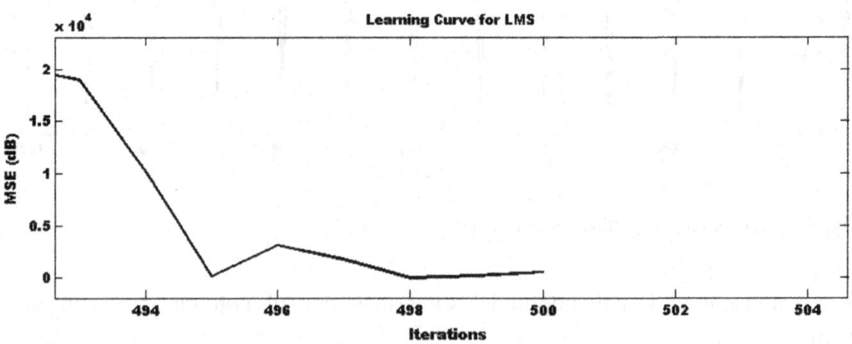

Fig. 5 Averaged mean square error trajectory for steepest descent algorithm

the table, autocorrelation and cross correlation coefficients confirms that the received signal is highly correlated with the original transmitted signal. It shows that the mean, standard deviation and signal to noise ratio of the filter output signal is less deviated from the desired signal. It can be seen that equalizer with proposed algorithm achieves high signal to noise ratio than zero forcing equalizer in removing the effect of ISI.

Table 2 Various parameter of the signal under consideration

Signal specification	Parameters	Proposed method	Conventional zero forcing equalization method
ECG signal (MIT-BIH data base)	Auto correlation coefficient	1	1
	Cross correlation coefficient	0.489	0.39
Duration = 4 s	Mean (mV)	Original signal: 11.83	Original signal: 11.83
Samples = 1200		Filter output signal: 14.75	Filter output signal: 6.8
Sampling frequency = 128 Hz	Standard deviation (mV)	Original signal: 82.86	Original signal: 82.86
		Filter output signal: 59.08	Filter Output signal: 54.6
	Signal to noise ratio	Original signal: 0.14	Original signal: 0.14
		Filter output signal: 0.25	Filter output signal: 0.12

4 Conclusion

In this work, we had investigated the performance of adaptive channel equalization for wireless Electrocardiogram signal transmission that helps in the effective functioning of Wireless Body Area Network. Correlation results showed that, the transmitted signal is effectively recovered from the channel output signal. With respect to convergence rate and signal to noise ratio, the steepest descent method has superior performance than zero forcing technique. This scheme is used to transmit the real time error free patient's ECG to the hospital server.

References

1. Kim, K.: A study on a health care system using smart clothes. In: Journal of Electrical Engineering and Technology, Jan (2014)
2. Aris, S., Lalosa, C., Alonso, L., Verikoukisb, C.: Model based compressed sensing reconstruction algorithms for ECG telemonitoring in WBANs. Digital Signal Process. **35**, 105–116, (2014)
3. Beijin, B., Daja, N., Reljin, I.: Telemonitoring in cardiology-ECG: transmission by mobile phone. Ann. Acad. Studennica **4**, 63–66 (2001)
4. Motoi, K., Nogawa, M., Tanaka, S., Yamakoshi, K.: A new portable device for ambulatory monitoring of human posture and walking velocity using miniature accelerometers and gyroscope, In: Proceedings of the 26th Annual International Conference of the IEEE EMBS, San Francisco, pp. 1–5. CA, USA (2004)
5. Lin, C.-T.: An intelligent telecardiology system using a wearable and wireless ECG to detect atrial fibrillation. IEEE Trans. Inf. Technol. Biomed. **14**(3), (2010)

6. Jasemian, Y.: Design and implementation of a wireless telemedicine system applying Bluetooth technology and cellular communication network: new approach for real time remote patient monitoring, Ph.D. Thesis, Aalborg University, Denmark, (2005)
7. Alm, A., Gao, T., Greenspan, D., Juang, R.R., Welsh, M.: Vital signs monitoring and patient tracking over a wireless network. In: 27th Annual International Conference of the IEEE Engineering in Medicine and Biology Society, (2005)
8. Arredondo, M.T., Fernández, C., Guillén, S., García, J.M., Traver, V.: Multimedia tele homecare system using standard TV set. IEEE Trans. Biomed. Eng. **49**(1), (2002)
9. Lucky, R.W., Rudin, H.R.: An automatic equalizer for general purpose communication channels. Bell Syst. Tech. J. **46**, 2179–2208 (1967)
10. Physiobank, P.: Physiologic signal archives for biomedical research, http://www.physionet.org/physiobank/, viewed Jul (2013)
11. Widrow, B.: Adaptive signal processing. Pearson Education, Englewood Cliffs (2006)
12. Rupp, M.: Convergence properties of adaptive equalizer algorithms. IEEE Trans. Signal Process. **59**(6), (2011)

An Efficient Approach for MapReduce Result Verification

K. Jiji and M. Abdul Nizar

Abstract Hadoop follows a master-slave architecture and can process massive amount of data by using the MapReduce paradigm. The major problem associated with MapReduce is correctness of the results generated. Results can be altered and become wrong by the collaboration of malicious slave nodes. Credibility-based result verification is one of the effective methods to determine such malicious nodes and wrong results. The major limitation of the approach is that, it depends on the complete results of long-running jobs to identify malicious nodes and hence holds valuable resources. In this paper, we propose a new protocol called *Intermediate Result Collection and Verification (IRCV) Protocol* that prunes out unnecessary computations by collecting results for verification earlier in the execution line. In addition, unlike the previous approach, IRCV uses only a subset of nodes for the purpose. Our simulation experiments suggest that the new approach has improved performance and will lead to better utilization of resources.

Keywords Hadoop · MapReduce · Collusion · Result verification

1 Introduction

Big Data analytics is the iterative process of analyzing *vast* and complex data from multiple data sources. Hadoop [1], which is an open source, distributed, master-slave architecture running on commodity hardware, has been effectively used along with the MapReduce paradigm [1] to process large volume of data efficiently. In order to achieve reliability, the framework replicates the same job on multiple slave nodes, which operate in parallel.

K. Jiji (✉) · M. Abdul Nizar
College of Engineering Trivandrum, Thiruvananthapuram 695016, India
e-mail: jijikarikkad@gmail.com

M. Abdul Nizar
e-mail: nizar@cet.ac.in

M. Senthilkumar et al. (eds.), *Computational Intelligence, Cyber Security and Computational Models*, Advances in Intelligent Systems and Computing 412, DOI 10.1007/978-981-10-0251-9_23

One of the main problems associated with MapReduce is result alteration, due to the presence of malicious nodes. This becomes more dangerous when malicious nodes *collude* [2] and return wrong results. One of the recent approaches to address collusion is *credibility-based result verification* [3]. This method proceeds by running 'quiz' jobs on the slave nodes. Based on the results of quizzes and *regular* MapReduce jobs, the master node assigns credibility values to slave nodes, which are later used to verify the correctness of results produced by the nodes. The limitation of the approach is that, the result verification has to wait until the regular Mapreduce jobs, which are long-running, complete their execution. This adversely affects the turn-around time, resource utilization and throughput of the system. In this paper, we propose a new approach called *Intermediate Result Collection & Verification (IRCV)* protocol that prunes out unnecessary computations by collecting results for verification early in the execution line. In addition, unlike the previous approach, IRCV uses only a subset of nodes for the purpose. Our experiments show that the new approach reduces turn-around time and resource usage considerably without compromising on accuracy. We make the following contributions:

1. We propose a new protocol to collect the intermediate results from running jobs and use it for result verification and to prune out unnecessary computations as early as possible.
2. We implement grouping of slave nodes to reduce the resource requirement for quiz jobs, thereby improving resource utilization and throughput.
3. We compare, through simulation experiments, the performance of the proposed approach against existing approaches.

The rest of the paper is organized as follows. Section 2 gives the background, presents the related work and motivates the proposal. The proposed approach is discussed in Sect. 3. Experiments and results are presented in Sect. 4. Finally, Sect. 5 concludes the paper.

2 Background, Related Work and Motivation

Hadoop is composed of two main components [1, 4] namely, Hadoop Distributed File System (HDFS) and MapReduce Paradigm. HDFS is built on the top of the master-slave architecture, which consists of two types of nodes: *NameNode* (master) and *DataNodes*[1] (slaves). Figure 1a shows the architecture of Hadoop. HDFS distributes the submitted data across various DataNodes by breaking down each input file into chunks of equal size, known as *blocks*. The blocks are stored in a replicated manner on different DataNodes. The NameNode has a daemon called *Jobtracker* intended to perform job scheduling, resource

[1]DataNodes are also termed as WorkerNodes.

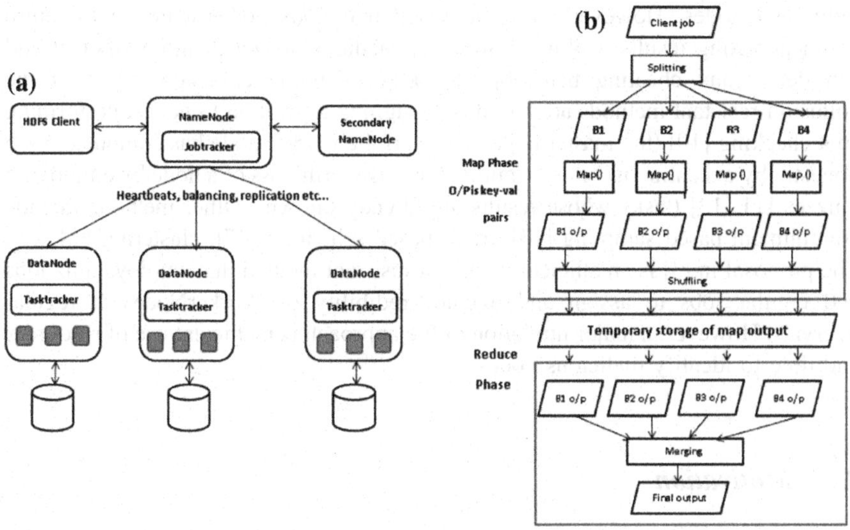

Fig. 1 Hadoop and Map-Reduce. **a** Hadoop architecture. **b** MapReduce principle

allocation, job execution handling etc. Another daemon called *Tasktracker*, residing in each DataNode, looks into processing the tasks assigned by Jobtracker. Execution of of job is initiated by the job tracker, which splits the data as said earlier and deploys it on the free WorkerNodes. It also replicates the job on all such worker nodes. Each replication is known as a task, which can be one of two types: *map* and *reduce* (see Fig. 1b). In map phase, each WorkerNode applies the map function to the data block assigned to it. The *map()* function consists of any user-defined function suitable for that particular client job. This flexibility of Hadoop allows the user to process any job as MapReduce job, just by converting the input into key-value pairs. The default mapping functions can be overridden as required. Output of each *map()* function is also key-value pairs, which subsequently undergoes *shuffling*, where they are sorted on the basis of keys. The shuffled output is fed to the reducers. Normally *reduce()* functions are intended to merge the different map outputs. The results from different reducers are combined to produce the final result.

2.1 Related Work

Result verification schemes can be categorized as general schemes and result-based schemes. General methods are based on either probabilistic/statistical methods [5] or involves water marking [6] and do not rely on the results generated.

Result-based schemes are collusion susceptible or collusion-resistant. Collusion-susceptible methods [2, 7–9] can detect malicious results returned by individual nodes. However, they fail when malicious nodes team up to return identical wrong results. Collusion resistant methods are much more efficient and can detect any cheating behavior of nodes including collusion. Most of the collusion-resistant methods are based on assigning reputation to nodes, performing spot checking [10] (the master node verifies if the worker node has computed a job honestly by executing it) or determining the trustworthiness of a node by employing quizzes [11, 12] (tasks whose results are already known). Other methods include commitment-based sampling [13], trust- based scheduling [7], clustering [14] etc. The proposal in [3] is an efficient collusion-resistant method that employs quiz jobs and regular jobs to assign and update credibility of WorkerNodes at regular intervals. However, a major limitation of the approach is inefficient use of resources and time to identify malicious nodes.

2.2 *Motivation*

As mentioned above, the credibility-based result verification scheme in [3] is very good in accuracy, compared to other collusion-resistant methods. However it shows poor turn-around time and resource utilization. The identification of malicious results and collusion are possible only after completing all the replicated jobs. Normally MapReduce jobs are long-running jobs, which take hours or days. Result verification has to wait until the jobs are completed. In some cases, the master node has to recompute the results before arriving at a decision on the validity of results. These limitations of [3] motivate us to develop a new protocol that identifies malicious nodes early in the execution line. The protocol collects intermediate output for result verification and prunes out erroneous computation. In addition, unlike previous approach, which uses the entire set of nodes to run quizzes and normal jobs, for the new protocol partitions WorkerNodes into groups, thereby improving throughput.

3 Proposed System

As in [3] we employ quiz jobs and actual MapReduce jobs for credibility updation and result verification. Quiz jobs are scaled down MR jobs, similar in structure and nature to the MR Jobs. This is to prevent WorkerNodes from distinguishing between quizzes and MR jobs. In addition, a job when replicated on many nodes will have different job ids, so that the nodes cannot identify similar replicas.

We compute credibility using the same approach as in [3] as follows: Each of set of quiz jobs is replicated on the WorkerNodes and the results returned by the nodes are compared against actual known results. After all jobs complete computation, the credibility value of each node is computed as the ratio of the number of jobs for which the node returned correct results to the total number of jobs submitted to the node. However, we use only a subset of WorkerNodes to run quizzes so that the remaining nodes can run normal jobs based on previous credibility of those nodes (details below).

In order to effectively and efficiently perform the credibility updation of nodes based on results of MR jobs and subsequent result verification, we introduce the concept of 'grouping'. Recall from Sect. 2 that, all the tasks corresponding a job are executed in parallel on all the available WorkerNodes. For each job, we create multiple execution groups which are replicas of one another. Each execution group consists of a subset of WorkerNodes that perform all the map tasks for the job. The number of replicas is user-defined. Number of replicated groups increases accuracy of result verification. Grouping also has the advantage that more than one MR job can run in parallel leading to efficient utilization of resources and increased throughput.

3.1 Intermediate Result Collection and Verification

Result collection and verification are performed by IRCV protocol (Algorithm 1). It collects the intermediate result from each group. Note that, for any job that runs on Hadoop, the map phase contains most of the job-specific execution. Therefore the processing time and resource utilization are more in this phase. However, the protocol can be easily extended to the reduce phase with limited modifications.

Algorithm 1: Intermediate Results Collection

Input: G: Groups which run the MR Job, n: size of a group
Output: H :Vector with hash result and group credibility pairs

```
Begin
H ← 0
for group g ∈ G do
    c_g ← 0
    c_t ← 0
    g_r ← 0
    for worker node w ∈ g do
        interResult ← IRCV protocol collects intermediate result from w
        add interResult to the array g_r
        c_t ← c_t + c_w
    end
    c_g ← c_t/n
    add < hashvalue(g_r), c_g > to array H
end
```

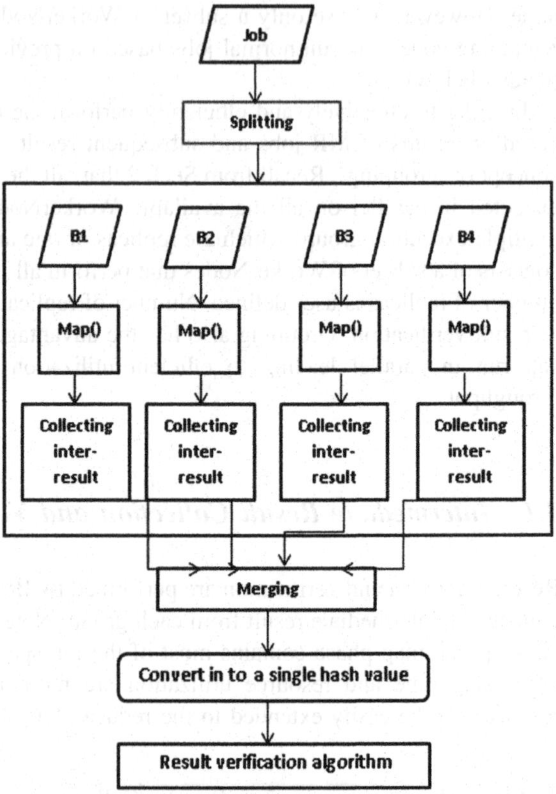

Fig. 2 Intermediate result collection for a group

At random intervals, the protocol collects intermediate results from all the WorkerNodes in each group and send to the result verification algorithm. The intention is to decide the cheating behavior of WorkerNodes as early as possible and prune out the execution thereby save resources. Figure 2 shows the intermediate results collection for a group. The protocol monitors the mappers of a group and collects a fraction of key-value pairs from each mapper as intermediate results. These results are merged and converted into a hash value. Similarly the hash values from the other replicated groups are computed and outlined in Algorithm 1 (line 3–10). This algorithm also computes the credibility of the group as the average of the credibility of the WorkerNodes in the group (line 11). The pair <*hashvalue*, *groupcredibility*>, are cached to a vector (line 12) for further processing by the result verification algorithm (Algorithm 2).

The result verification algorithm proceeds as follows: For each pair $<h, c_g>$ from the vector H returned by Algorithm 1, where h represents hash value of the group result and c_g represents the group credibility, we compute the following:

(i) *frequency* (f): the number of groups that returned the result (line 6). (ii) *result credibility* (r_g): average of the credibility of the groups which returned the result (line7). (iii) *threshold count* (t): number of groups which returned the same result as *h*, but have credibility value greater than credibility of *h*, i.e., r_g.

The final result chosen by our algorithm as follows: (i) Choose the result with maximum frequency (f) and maximum result credibility (r_g) as the final result (line 10). (ii) If there is not any result with maximum frequency and maximum credibility, suspend the job execution in all group replicas and recompute the results (lines 11–12). (iii) If there is more than one such result, choose the one with highest threshold count (lines 17–20). (iv) If there is a tie on the threshold count also, suspend the job execution in all group replicas and recompute the results (lines 17–21).

Note that, if the result is accepted and has the highest result credibility, it is rational to give an incentive to the WorkerNodes of the group which have lower value of credibility. Though such nodes have lower value of credibility due to their previous performance, they collaborated well in returning the correct result; hence their credibility has to be updated. We do this by setting the credibility of WorkerNodes with credibility value <0.5 to the credibility of result.

4 Simulation Experiments and Results

IRCV protocol was implemented in an *Intel Core i3* machine with 4 GB memory using Java on the top of a simulated environment that imitates Hadoop framework. Nodes were simulated as operating system processes. The system was modelled by running any type of job as MapReduce job, just by converting its input into key-value pairs. In the experiments, we used a Word-Count job[2] as MR job for input files of 10,000 lines. Quizzes were similar in structure and nature as MR jobs, and their results were computed before-hand. Number of quiz jobs, replication factor and number of MR jobs were controlled through suitable user input. We used maximum of 100 MR jobs. Based on the status of the result, the credibility of the nodes were also updated. We compared the proposed approach against the credibility-based result verification scheme [3] as well as majority voting scheme [6].

[2]https://hadoop.apache.org/docs/stable/hadoop-mapreduce-client/hadoop-mapreduce-client-core/MapReduceTutorial.html #Example:_WordCount_v1.0.

Algorithm 2: Credibility updation in MR jobs

Input: M: MR Job, G: set of groups of worker nodes for executing M
begin
  Execute M on each node in G Generate hash H of results H = Intermediate Result Collection() **for** *distinct* $h \in H$ **do**
    Let $X \subseteq G$ be the set of groups which returned h Frequency of $h \leftarrow |X|$ Credibility of $h \leftarrow$ *average*$\{x \in X$: credibility of $x\}$ Threshold Count of $h \leftarrow$ *count*$\{x \in X$: credibility of $x >$ credibility of $h\}$
  end
  Let $H_m = \{h | h \in H$ and h has max frequency and max credibility$\}$ **if** $H_m = \emptyset$ **then**
    Recompute M
  else if $|H_m| = 1$ **then**
    Return h
    TASK_STATUS $\leftarrow$ *'accepted'*
  end
  else
    Let $H_T = \{h | h \in H_m$ and h has max threshold count$\}$ **if** $|H_T| = 1$ **then**
      Return h
      TASK_STATUS $\leftarrow$ *'accepted'*
    else
      Recompute M
    end
  end
  if *TASK_STATUS* = *'accepted'* **then**
    Let $X \subseteq G$ be the set of groups which returned the accepted result **for** *group* $g \in X$ **do**
      for *worker node* $w \in g$ **do**
        if *credibility of* $w < 0.5$ **then**
          credibility of $w \leftarrow$ credibility of accepted result
      end
    end
    Let $\bar{X} \subseteq G$ be the set of remaining group which returned incorrect result
    for *group* $g \in \bar{X}$ **do**
      suspend the exectution of M on g
    end
end

4.1 Accuracy

Accuracy of the IRCV protocol in result verification and malicious node detection was assessed by running 100 jobs. Experiments were carried out for replications factors 10 and 15 and for cases where 10, 20 or 40 % of the jobs are quizzes. Figure 3 shows the results. It was observed that accuracy of the credibility based

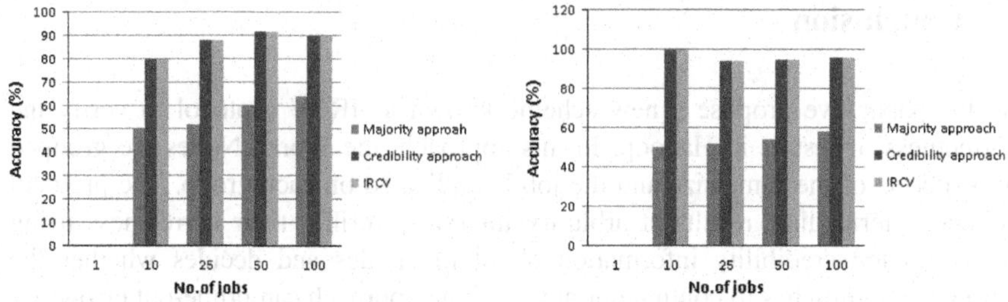

Replication factor =10 Replication factor =15

Fig. 3 Accuracy

scheme and IRCV scales up equally, with increasing fractional percentage of quizzes compared to majority approach. The reason for improved accuracy is that credibility of each WorkerNode would be updated and hence more accurate idea of its trustworthiness is available.

4.2 Turn-Around Time

IRCV prunes out unnecessary executions at the earliest possible to bringing drastic performance improvement. This is corroborated by our experiments which compute the average turn around time. We used 300 jobs for the experiments. The results are shown as graphs in Fig. 4. Note that our approach scales far better than the rival systems. The reason is two-fold: (i) We collect intermediate results for verification, wherever possible, thereby pruning out unnecessary computations (ii) We group WorkerNodes thereby bring concurrency in the execution of quiz jobs and MR jobs.

Fig. 4 Average turn-around time

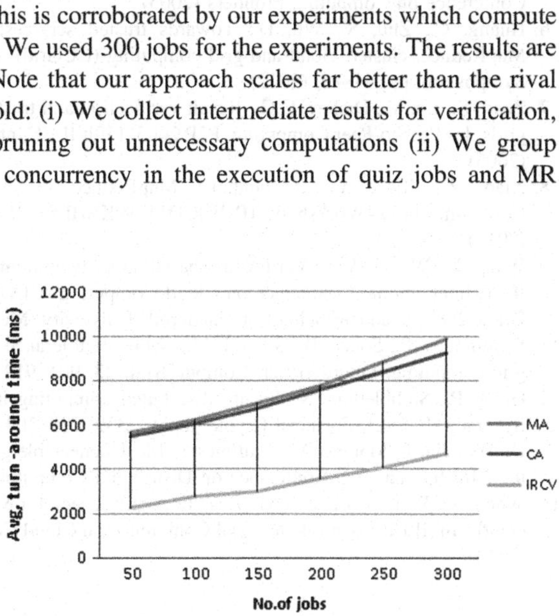

5 Conclusion

In this paper, we propose a new scheme known as IRCV protocol to verify the correctness of results in Hadoop. In this approach, the WorkerNodes are grouped into subsets of the same size and the job is replicated on each group. The protocol collects intermediate results at arbitrary intervals, verifies their correctness using pre-computed credibility information about the nodes and decides whether the current execution has to continue or not. Such an approach can prune-out erroneous computation, and thus save valuable resources. Our experiments show that the proposed approach outperforms credibility based scheme by wide margins while at the same time maintains the accuracy of the latter.

References

1. White, T.: Hadoop: the definitive guide. O'Reilly Media Inc. **1** (2004)
2. Wei, W., Du, J., Yu, T., Gu, X.: SecureMR: a service integrity assurance framework for MapReduce. In: Computer Security Applications Conference. ACSAC '09. Annual, pp. 73–82 (2009)
3. Samuel, T.A.; Nizar, M.A.: Credibility-based result verification for map-reduce. In: India Conference (INDICON), Annual IEEE, pp. 1–6, (2014)
4. HDFS Architecture Guide. http://hadoop.apache.org/docs/r1.2.1/hdfs_design.html
5. Germain-Renaud, C., Monnier-Ragaigne, D.: Grid result checking. In: Proceedings of the 2nd Conference on Computing Frontiers (2005)
6. Huang, C., Zhu, S., Wu, D.: Towards trusted services: result verification schemes for MapReduce. cluster, cloud and grid computing (CCGrid). In: 12th IEEE/ACM International Symposium, pp. 41–48, (2012)
7. Zhao, S., Lo, V., Dickey, C.G.: Result verification and trust-based scheduling in peer-to-peer grids. In: Peer-to-Peer Computing, P2P 2005. Fifth IEEE International Conference, pp. 31–38, (2005)
8. Xiao, Z., Xiao, Y.: Accountable MapReduce in cloud computing. In: Computer Communications Workshops (INFOCOM WKSHPS), IEEE Conference, pp. 1082–1087, (2011)
9. Wang, Y., Wei, J.: Viaf: Verification-based integrity assurance framework for MapReduce. In: IEEE International Conference on Cloud Computing (CLOUD), pp. 300–307, (2011)
10. Grant, P.C.: Graduate School of Vanderbilt University, Masters thesis (2006)
11. Domingues, P., Sousa, B., Silva, L.M.: Sabotage tolerance and trust management in desktop grid computing. Future Gener. Comput. Syst. **23**, 904–912 (2007)
12. Golle, P., Stubblebine, S.: Secure distributed computing in a commercial environment. In: Syverson, P. (ed.), Financial Cryptography **2339**, 289–304 (2002)
13. Du, W., Jia, J., Mangal, M., Murugesan, M.: Uncheatable grid computing. In: Proceedings of the 24th International Conference on Distributed Computing Systems, pp. 4–11 (2004)
14. Wang, Y., Wei, J., Srivatsa, M.: Result integrity check for MapReduce computation on hybrid clouds. In: IEEE Sixth International Conference on Cloud Computing (CLOUD), pp. 847–854 (2013)

Improving Lifetime of Memory Devices Using Evolutionary Computing Based Error Correction Coding

A. Ahilan and P. Deepa

Abstract Error correction coding (ECC) plays an important role in the reliability improvement of circuits having application in space and mission critical computing-, low-power CMOS design-, microprocessor based computing-, and nanotechnology-based systems. Conventional ECC are not suitable for multiple bit detection and correction. A memory circuit holds both instruction and data of the given system and it is susceptible to multiple bit soft error problems. To mitigate such kind of problems in memory circuit, an evolutionary computing based new ECC called reconfigurable matrix code (RMC) is suggested in this paper. The proposed RMC are evaluated in terms of error correction coverage. The results show that the proposed RMC technique can drastically increase the Mean-Error-To-Failure (METF) and Mean-Time-To-failure (MTTF) up to 50 % and hence the life time of the memory devices is more compared to conventional coding techniques based memories.

Keywords Computing · Error correction coding · Reliability · Memory · Soft error · Multiple bit upset · Fault coverage · Redundant bits

1 Introduction

A foremost risk of recent system on chips comes from discontinuous faults induced by ecological circumstances; for example particle strikes on the semiconductor circuit nodes, causing soft errors, and Soft errors have usually been a main concern for semiconductor memories. Soft errors (transient errors) caused by harsh radiation particles are becoming an ever more important issue for memory consistency. Due

A. Ahilan (✉) · P. Deepa
Government College of Technology, Coimbatore, India
e-mail: listentoahil@gmail.com

P. Deepa
e-mail: deepap05@gmail.com

M. Senthilkumar et al. (eds.), *Computational Intelligence, Cyber Security and Computational Models*, Advances in Intelligent Systems and Computing 412, DOI 10.1007/978-981-10-0251-9_24

to this soft error problem original data stored in the memory will be altered through flipping [1]. To avoid this altered form data issue in memories, one option is to use Error Correction Codes. In 1950, the Hamming codes were invented and they have been widely used to protect the soft errors formed by a radiation particle hitting memory array system and flipping the value stored in them [2]. Recent important issue in memory system design is that as technology scales (down scaling of transistors), it is more expected for a particle strike to produce multiple soft errors in close proximity memory cells, a occurrence known as Multiple Bit Upset (MBU) [3–5]. To mitigate this problem various techniques presented in literature [6–12].

In 2014, Jing Guo et al. proposed decimal matrix code (DMC) technique to correct MBU, but it uses more number of parity bits and it can correct maximum of 5 bits. The same row decimal computation of DMC code words will fail to detect the soft error for same index bits when the memory is affected by radiation effect [12]. To overcome these problems, the paper proposes a novel evolutionary Computing technique called Reconfigurable Matrix Code (RMC). RMC technique uses less redundant bits compared to existing DMC and it can able to detect and correct more than 16-bit error in a memory word. Most soft error mitigation techniques are essential but recent technique decimal matrix code h as higher reliability compared to other techniques. The following problem often occurs and makes DMC less reliable towards error detection and correction. The problems occurred are

1. Best case error detection capability is limited to 16 bits, and detection capability is degraded to zero in worst case due to complete same index errors. In best case error correction capability is limited to 5 bits for a given 32 bit configuration word.
2. More number of redundant bits required.
3. Possibility of decoding errors for inconsecutive error from inconsecutive symbols.
4. Decimal addition limits the performance.

The work on RMC will pay an attention for high reliability point of view with minimum redundant bits. The problems discussed above can be analyzed in two different cases. Even though the DMC technique requires high redundant bits, its error detection capability is limited. The main reason for this is that its error detection mechanism is based on same row decimal computation. Another draw-back related to same row computation with DMC is illustrated in Fig. 1. The redundant bits were calculated using Eqs. (1)–(5)

$$B_{11}B_{10}B_9B_8 \quad \text{xor} \quad B_3B_2B_1B_0 = h_1[H_4H_3H_2H_1H_0] \tag{1}$$

$$B_{15}B_{14}B_{13}B_{12} \quad \text{xor} \quad B_7B_6B_5B_4 = h_2[H_9H_8H_7H_6H_5] \tag{2}$$

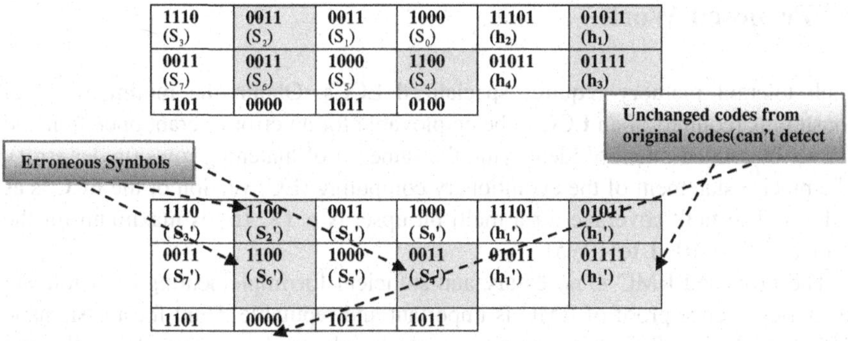

Fig. 1 Limitation of error detection in DMC **a** original bits and redundant bits **b** Erroneous bits and erroneous redundant bits due to radiation

$$B_{27}B_{26}B_{25}B_{24} \quad \text{xor} \quad B_{19}B_{18}B_{17}B_{16} = h_3[H_{14}H_{13}H_{12}H_{11}H_{10}] \tag{3}$$

$$B_{31}B_{30}B_{29}B_{28} \quad \text{xor} \quad B_{23}B_{22}B_{21}B_{20} = h_4[H_{19}H_{18}H_{17}H_{16}H_{15}] \tag{4}$$

$$Vn = Bn \quad \text{xor} \quad Bn + 16 \tag{5}$$

where $B_0 \ldots B_{31}$ are original bits, H_0 to H_{19} are horizontal redundant bits, h_1 to h_4 are symbol representation of horizontal redundant bits, Vn is vertical redundant bits. The horizontal and vertical syndrome are computed using Eqs. (6) and (7)

$$\Delta hn = h_n \quad \text{xor} \quad h_n' \tag{6}$$

$$\Delta vn = v_n \quad \text{xor} \quad v_n' \tag{7}$$

Δh_n is the horizontal syndrome for nth redundant symbol where n ranges from 1 to 4. Δv_n is the vertical syndrome for nth bit. For S_4, S_6 and S_4', S_6' though there is an error, the Δh_n and Δv_n are equal to zero and hence these errors cannot be detected by DMC.

Hence a suitable reliable correction code is needed to overcome the above problems. In this paper an evolutionary computing based RMC has been proposed to overcome such failures. The work on RMC will pay an attention for high reliability in safety critical applications. More specifically, the characteristics of evolutionary computation are incorporated in designing the proposed RMC architecture.

The rest of this paper is organized as follows. The proposed evolutionary computing mechanism based RMC and memory organization is explained in Sect. 2. Implementation and detail analysis of proposed work is given in Sect. 3. Finally, concluding notes are made in Sect. 4.

2 Proposed Work

Fault tolerant memory requires specialized ECCs. Obtain the minimum set of constraints required of an ECC to be employable for an error tolerant operation and deliver our stated intent of identifying the same set of matching rows (under error). The precise statement of the evolutionary computing (EC) version of the ECC is as follows; The fault coverage for a multi bit upset N > {8, 16} is maximum for the error rate (λ = 10−1 to 10−5).

The Proposed RMC is necessary and sufficient for implementing error tolerant memories. Hence proof of RMC is important and reputable. Hunt for a systematic ECC which is efficient to encode and decode and satisfies the following requirements.

- X_1 (order preservation without errors): Let B be a completely specified data word (DW), B′ an arbitrary DW and B ≡ B′. Then, the code words (CW's) should be in the same order, i.e., (B, H) ≡ (B′, H′).
- X_2 (order preservation under errors): Let B be a completely specified data word (DW), B′ an arbitrary DW such that B ≢ B′, the original code word co = (B, c) and the arbitrary code word c1 = (B′, c′) where the original and arbitrary redundant bits c = (H, V), c′ = (H′, V′) respectively. Let c1 and co differ in at most 2t + 3 bit locations, i.e., the distance between co and c1 is ≤ 2t + 3. Then co ≢ c_1.
- X_3 (No order preservation under errors): Let B be a completely specified data word (DW), B′ an arbitrary DW such that B ≢ B′, c_o = (B, c) and c_{11} = (B′, c′) be their respective CW's where c = (H, V) and c″ = (H′, V′). Let c_{11} be an altered CW and differ from co in at most 2t + 3 bit locations and same index errors, i.e., the distance between c_o and c_{11} is ≤ 2t + 3. c_o ≢ c_{11}.
- X_4 (More than 16 bit errors may contain same index errors): Let B be a completely specified data word (DW), B′ an arbitrary DW such that B ≢ B′ and c_o = (B, c) and c111 = (B′, c′) be their respective CW's where c = (H, V) and c″ ′ = (H′, V′). Let c_{111} be an altered CW and differ from co in at most 2t + 16 bit locations, may contain same index errors, i.e., the distance between c_o and c_{111} is ≤ 2t + 16. Then c_o ≢ c_{111}.

Requirement X_1 says that the CWs for the relevant DWs be analogous. X_2 is a requirement on the CWs for the relevant DWs are not corresponding. X_3 is a requirement on the CWs for the relevant DWs are not consequent (faulty match) with same index errors. X_3 is a requirement on the CWs for the relevant DWs are not consequent with more than 8 bit errors (t ≥ 3) may contain same index errors. X_4 is a requirement on the CWs for the relevant DWs are not consequent with more than 16 bit errors (t ≥ 16) may contain same index errors.

The summary of evolutionary computing based RMC is given as follows;

EC1 Let Đ be a systematic ECC capable of correcting t bit errors, t ≥ 3. Đ is an error tolerant SCM if and only if conditions X_1 and X_2 are satisfied i.e., Đ is a Multi Bit ECC (MBECC) and Single Bit ECC (SBECC).

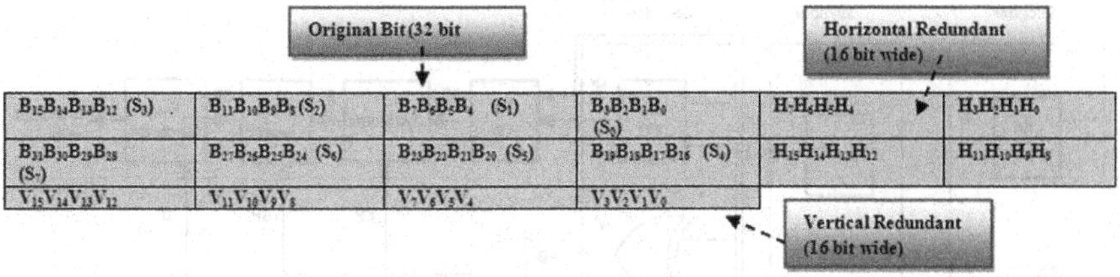

Fig. 2 32-bit memory word ($k = 2 \times 4$ and $m = 4$)

EC2 Let Đ be a systematic ECC capable of correcting t bit errors, $t \geq 8$ with same index errors. Đ is an error tolerant SCM if and only if conditions X_1, X_2 and X_3 are satisfied i.e., Đ is a Multi Bit ECC (MBECC) and Single Bit ECC (SBECC).
EC3 Let Đ be a systematic ECC capable of correcting t bit errors, $t \geq 16$ with same index errors. Đ is an error tolerant SCM if and only if conditions X_1, X_2, X_3 and X_4 are satisfied i.e., Đ is a Multi Bit ECC (MBECC) and Single Bit ECC (SBECC).

2.1 Memory Organization

In this proposed work the Memory word is arranged in the matrix format with divide by Symbol approach. N-bit code word is divided into K symbols and m bits/symbol and the K symbols are represented by Eq. (8).

$$K = R * C \tag{8}$$

R, C represents the Number of Rows and Columns which makes the matrix format. Finally the complete N-bit word is organized as given in Eq. (9) (Fig. 2).

$$N = R * C * m \tag{9}$$

2.2 Proposed Memory Protection Model

Fault tolerant encoder converts the input data into appropriate symbols. An appropriate symbol pair has been selected for redundant bit computation based on the symbol nature. Proposed RMC uses three different methods based on three different symbol pair to compute the write and read redundant bits for soft error detection (Fig. 3).

The number of horizontal redundant bits has been reduced to 16 compared to DMC and it can be computed using Eqs. (10)–(13) to increase the performance.

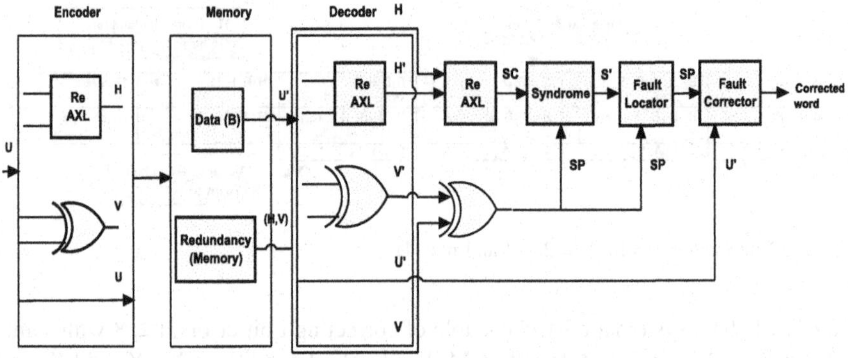

Fig. 3 Proposed memory protection model using RMC

$$B_{11}B_{10}B_9B_8 \quad \text{xor} \quad B_3B_2B_1B_0 = H_3H_2H_1H_0 \tag{10}$$

$$B_{15}B_{14}B_{13}B_{12} \quad \text{xor} \quad B_7B_6B_5B_4 = H_7H_6H_5H_4 \tag{11}$$

$$B_{27}B_{26}B_{25}B_{24} \quad \text{xor} \quad B_{19}B_{18}B_{17}B_{16} = H_{11}H_{10}H_9H_8 \tag{12}$$

$$B_{31}B_{30}B_{29}B_{28} \quad \text{xor} \quad B_{23}B_{22}B_{21}B_{20} = H_{15}H_{14}H_{13}H_{12} \tag{13}$$

Encoder uses ReAXL and XOR gate to compute the horizontal (H) and vertical (V) write redundant bits, each of 16 bit wide. These horizontal and vertical redundant bits called as Write Redundancy (WR) are stored along with the data into the memory for protection of LUT's. If any heavy radiation particle strikes the memory, soft errors occur immediately [13]. SRAM memory has chosen as a realistic test case to validate the RMC, where both original bits (U) and redundant codes (H, V) are stored. On the decoder, the reconfigurable encoder recalculate the redundant codes (H′, V′) for the bits read from the memory using ReAXL and XOR gates. Then syndrome calculator uses horizontal syndrome also called as syndrome check (SC) and vertical syndrome also called as syndrome parity (SP) to detect and locate the error. The difference between read and write redundancy gives the fault alarm and it is achieved by ReAXL. If there is any non zero bits in final syndrome, it will confirmed as the Soft error. Also the precise location of multiple bit upsets will be given by vertical fault detector. Finally fault corrector corrects the erroneous word (U′) based on vertical syndrome.

3 Implementation Results and Analysis

In this section the proposed RMC has been simulated and verified by injecting thousands of random radiation errors. Based on the simulation results Maximum error correction coverage has been found. This Maximum error correction coverage improves the life time of memory devices. The correction capability of soft error for hamming, Matrix, DMC and RMC are shown in Fig. 4. The hamming can correct only one bit at a time [2]. Matrix codes can correct up to two bits [11]. The nature of these codes limits the correction rate. Recently DMC can able to correct up to 8 bits [12]. The evolutionary computing based RMC corrects a maximum of up to 32 bits.

For reliability testing random errors are injected into the SCM and the METF for various ECCs are calculated. The METF of Hamming, Matrix codes, DMC and proposed RMC are calculated using 15000 trials for different memory sizes shown in Fig. 5, for more details, refer to [14].

$$MTTF = \frac{\text{METF}}{\lambda \times MemorySize} \quad (14)$$

The enhancement for 32-bit codeword in proposed RMC is more than 30 and 50 % compared to DMC and matrix codes respectively. Finally the MTTF values has been calculated for Hamming codes, Matrix codes, DMC and proposed RMC using Eq. 14 and shown in Fig. 6. The MTTF of the "RMC" for the memory ranges

Fig. 4 Error correction percentage comparison graph for various ECC's

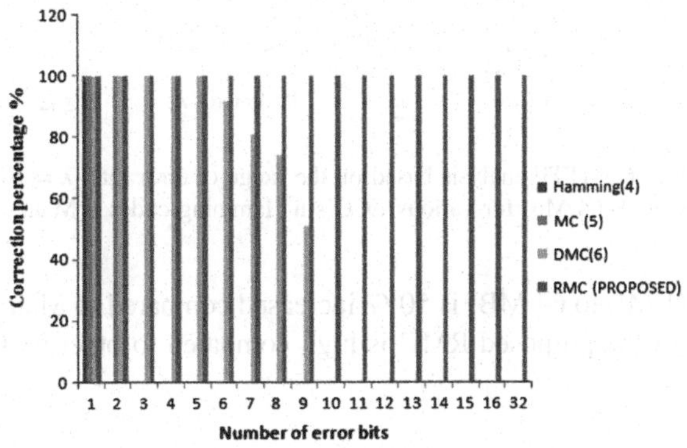

Fig. 5 METF for different memory sizes

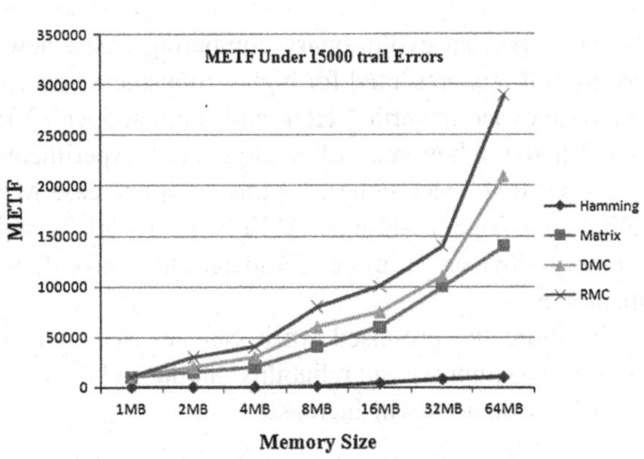

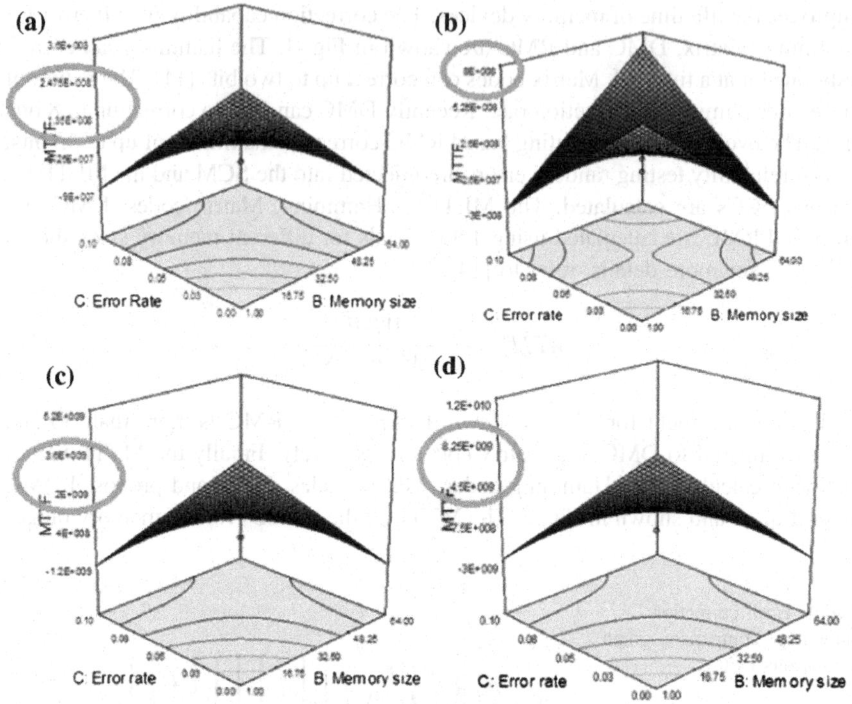

Fig. 6 MTTF analysis based on the range of error rate ($\lambda = 10^{-1}$ to 10^{-5}) and different memory size (1–64 Mb) for various ECC's **a** Hamming codes **b** Matrix codes **c** DMC **d** RMC

(1 MB to 64 MB) is 50 % increased compared to Matrix Codes. Thus the reliability for the proposed RMC is high compared to other ECCs.

4 Conclusion and Future Work

In this paper, an evolutionary computing based new ECC, reconfigurable matrix code (RMC) is presented for higher robustness against MBUs. The proposed RMC has been coded in verilog HDL and simulated with XILINX-ISIM and the behavior is verified for hundreds of random fault experiments. The results show that the proposed RMC technique can drastically increase Mean-Error-To-Failure (METF) and Mean-Time-To-failure (MTTF) up to 50 % and hence the life time of the memory devices is more compared to conventional coding techniques based memories.

In future the proposed work can be used in field programmable gate arrays (FPGAs) to improve the reliability similar to [15, 16] works. Hence the life time of the FPGA devices will increase.

Acknowledgements The authors are grateful for the financial support provided by the University Grant Commission (UGC), under National fellowship Scheme for PhD students, India, for this research.

References

1. Baumann, R.C.: Radiation-induced soft errors in advanced semiconductor technologies. IEEE Trans. DeviceMater. Reliab. **5**(3), 301–316 (2005)
2. Hamming, R.W.: Error detecting and error correcting codes. Bell Syst. Tech. J. **26**(2), 147–160 (1950)
3. Ibe, E., Taniguchi, H., Yahagi, Y., Shimbo, K., Toba, T.: Impact of scaling on neutron-induced soft error in SRAMs from a 250 nm to a 22 nm design rule. IEEE Trans. Electron. Devices **57**(7), 1527–1538 (2010)
4. Hands, A., Morris, P., Dyer, C., Ryden, K., Truscott, P.: Single event effects in power MOSFETs and SRAMs due to 3 MeV, 14 MeV and fission neutrons. IEEE Trans. Nucl. Sci. **58**(3), 952–959 (2011)
5. Ebrahimi, M., Asadi, H., Tahoori, M.B.: A layout-based approach for multiple event transient analysis. In: Proceedings of 50th Annual Design Automation Conference (DAC), pp. 1–6 (2013)
6. Baeg, S., Wen, S., Wong, R.: SRAM interleaving distance selection with a soft error failure model. IEEE Trans. Nucl. Sci. **56(4)**, pt. 2, pp. 2111–2118 (2009)
7. Dutta, A., Touba, N.A.: Multiple bit upset tolerant memory using a selective cycle avoidance based SEC-DED-DAEC code. IEEE VLSI Test Symp. **25**, 349–354 (2007)
8. Liu, S.-F., Reviriego, P., Maestro, J.A.: Efficient majority logic fault detection with difference-set codes for memory applications. IEEE Trans. Very Large Scale Integr. (VLSI) Syst. **20**(1), 148–156 (2012)
9. Reviriego, P., Member, IEEE, Flanagan, M., Senior Member, IEEE, Maestro, J.A.: A (64,45) triple error correction code for memory applications. IEEE Trans. Device Mater. Reliab. **12**(1), 101–106 (2012)
10. Neuberger, G., Kastensmidt, D.L., Reis, R.: An automatic technique for optimizing reed-solomon codes to improve fault tolerance in memories. IEEE Design Test Comput. **22** (1), 50–58 (2005)
11. Argyrides, C., Pradhan, D.K., Kocak, T.: Matrix codes for reliable and cost efficient memory chips. IEEE Trans. Very Large Scale Integr. (VLSI) Syst. **19**(3), 420–428 (2011)
12. Guo, J., Xiao, L., Mao, Z., Zhao, Q.: Enhanced memory reliability again st multiple cell upsets using decimal matrix code. IEEE Trans. Very Large Scale Integr. (VLSI) Syst. **22**(1), 127–135 (2014)
13. Legat, U.: SEU recovery mechanism for SRAM-based FPGAs. IEEE Trans. Nucl. Sci. **59**(5), 2562–2571 (2012)
14. Maestro, J.A., Reviriego, P.: Study of the effects of MBUs on the reliability of a 150 nm SRAM Device. In: Proceedings of 45th Annual Design Automation Conference (DAC), pp. 930–935 (2008)
15. Ahilan, A., Deepa, P.: Design for built-in FPGA reliability via fine-grained 2-D error correction codes. Microelectronics reliability. http://dx.doi.org/10.1016/j.microrel.2015.06.075 (2015)
16. Ahilan, A., Deepa, P.: A reconfigurable virtual architecture for memory scrubbers (VAMS) for SRAM based FPGA's. Int. J. Appl. Eng. Res. **10** (2015) ISSN 0973-4562

Comparison of Machine Learning Techniques for the Identification of the Stages of Parkinson's Disease

P.F. Deena and Kumudha Raimond

Abstract Parkinson's Disease (PD) is a degenerative disease of the central nervous system. This work performs a four-class classification using the motor assessments of subjects obtained from the Parkinson's Progressive Markers Initiative (PPMI) database and a variety of techniques like Deep Neural Network (DNN), Support Vector Machine (SVM), Deep Belief Network (DBN) etc. The effect of using feature selection was also studied. Due to the skewness of the data, while evaluating the performance of the classifier, along with accuracy other metrics like precision, recall and F1-score were also used. The best classification performance was obtained when a feature selection technique based on Joint Mutual Information (JMI) was used for selecting the features that were then used as input to the classification algorithm like SVM. Such a combination of SVM and feature selection algorithm based on JMI yielded an average classification accuracy of 87.34 % and an F1-score of 0.84.

Keywords PPMI · Motor assessments · Multi-class classification · Feature selection · SVM · DNN · Dropout · DBN

1 Introduction

PD is a degenerative disease of the central nervous system. It is manifested in the form of movement-related disorders like tremor, rigidity, postural instability, etc. It also results in thinking and behavioural problems like dementia, depression, etc. PPMI [1] is a collaboration between funders, researchers and study participants that aims to identify various progressive markers that can be associated with the changes

P.F. Deena (✉) · K. Raimond
Department of Computer Science and Engineering, Karunya University, Coimbatore, India
e-mail: deena.francis@gmail.com

K. Raimond
e-mail: kraimond@karunya.edu

M. Senthilkumar et al. (eds.), *Computational Intelligence, Cyber Security and Computational Models*, Advances in Intelligent Systems and Computing 412, DOI 10.1007/978-981-10-0251-9_25

in the stage of PD. Some of the kinds of data provided by PPMI include Magnetic Resonance Imaging (MRI) data, motor assessments, non motor assessments, Single Photon Emission Computed Tomography (SPECT) DaTSCAN images as well as Striatal Binding Ratio (SBR), etc.

The paper is organized as follows. The related previous works are described in Sect. 2. The description about the database and the data set are given in Sect. 3. The basic block diagram of the proposed work is described in Sect. 4. The classifiers used are described in Sect. 5. The results obtained and comparison of the different algorithms are given in Sects. 6 and 7 gives a discussion of the results obtained, and Sect. 8 concludes the paper.

2 Literature Survey

In some earlier works, voice recordings were used to diagnose PD [2–7]. [5] employed Sparse Multinomial Logistic Regression (SMLR) and obtained a classification accuracy of 100 % on the speech data. [8] used tremor time series of the subjects to detect PD. Some works such as [2, 9] have used gait signal measurements for PD detection. [2] also used gene expression data for PD detection.

Most of the previous works [2–7, 9–13] were aimed at classifying the subjects as PD or non-PD. Although binary classification of the subjects is important, accurately identifying the various stages of PD is much more important. One of the reasons is that the correct diagnosis of the stages of the disease may help to prevent unnecessary therapies, medication, side-effects and expenses. Table 1 shows the comparison of related works and this work.

The first attempt made to analyse the relationship between motor and non-motor symptoms of the disease, and to predict the severity stage of patients based on non-motor symptoms was done by [14]. Later works on PD identification such as [10, 11, 13] used DaTSCAN SPECT images. Such images are available from PPMI [1]. [12] used SBR values obtained from PPMI to classify subjects as early PD or healthy individuals using SVM. [15] used MR images to classify subjects as early PD or healthy individuals using PBL-McRBFN with RFE.

It has been reported that motor symptoms of the subjects appear long before non-motor symptoms start appearing. Besides, the progression of the disease is associated with varying degrees of motor decline. No previous work has been attempted the classification of the subjects into severity stages based on the motor assessments of subjects. So, in this proposed work, multi-class classification is performed using the motor assessments of subjects obtained from PPMI.

Table 1 Comparison with related works

Work	Algorithm used	Performance	Type of classification performed
[14]	Naïve bayes, KNN, linear discriminant analysis, decision trees and NN using a wrapper feature selection scheme	Accuracy 72–92 %	Multi-class
[15]	Projection-based learning for meta-cognitive radial basis function network (PBL-McRBRFN) with recursive feature elimination, SVM, independent component analysis (ICA)	Classification performance was 82.32 % on voxel based morphometric (VBM) features	Binary
[2]	PBL-McRBFN, ICA, extreme learning machine (ELM), self adaptive resource allocation network, SVM	PBL-McRBFN gave the best result in all cases, on gene expression data: average per-class accuracy of 97.17 % obtained with statistically selected genes, on gait dataset: average per-class accuracy of 84.36 %, on speech dataset: average per-class accuracy of 99.35 %	Binary
[8]	Multi state Markov model	Maximum error of 1.33 %	Multi-class
[3]	DMNeural, NN, regression and decision tree	Accuracy of 92.9 % using NN	Binary
[5]	NNs, SVM, LogitBoost, AdaBoost M1, Furia, ensemble selection, Pegasos, rotation forest, SMLR	SMLR with Laplacian-direct kernel gave 100 % test accuracy	Binary
[16]	ANN, SVM	Accuracy ~90 %	Binary
Proposed work	SVM, feature selection, ANN, DBN, DNN, stacked autoencoder	SVM with JMI based feature selection: accuracy of 87.34 %, F1-score of 0.84	Multi-class

3 Data Set

Data used in the preparation of this article were obtained from the Parkinson's Progression Markers Initiative (PPMI) database (www.ppmi-info.org/data). For up-to-date information on the study, visit www.ppmi-info.org. The data set consists of the motor assessments of 750 subjects obtained from PPMI [1]. The data downloaded from PPMI consists of motor assessments available in csv format that contains 4118 rows and 55 columns. Each column represents an attribute (columns 1 to 54) or motor assessment, and each row holds one subject's record. So there are

Table 2 Number of instances of each class

Class	# Instances	% of the total no. of instances (%)
0	608	14.76
1	1269	30.81
2	2167	52.62
3	74	1.79

54 attributes in the data. The severity stage of the patient is indicated by the column named NHY which stands for Hoehn and Yahr (HY) scale [17]. HY scale is used to indicate the severity of PD, and it is based on the motor symptoms of the subject. In general, PD subjects can be grouped into five stages: stages 1, 2, 3, 4, and 5. In addition to the stages, another class '0' is also included in the study. Class'0' indicates the case of 'no PD/Healthy'. Further, classes 1 and 2 indicate the early stages of PD, and 3, 4 and 5 refers to the advanced stages. Due to the presence of imbalance in the number of instances of each class, classes 3, 4 and 5 were combined together and treated as one class (class '3'). The number of instances of each class is shown in Table 2. So, the dataset used in this work consists of 4 classes (Classes 0, 1, 2, and 3).

Due to this skewed nature of the data, accuracy alone may not be sufficient to estimate the performance of the classifier used. The performance metric must also take into account of how many instances are being correctly classified for each class.

4 Proposed Approach

The dataset is first subjected to a pre-processing stage before it is used in the classifier, by removing the rows that contained null values. Further, the dataset is normalized using the following formula:

$$X = \frac{X - mean(x)}{std.dev(x)} \tag{1}$$

The basic block diagram of the proposed work is shown in Fig. 1. After the pre-processing stage, feature selection may be used. If feature selection is used, then those selected attributes alone are used by the classifier. If no feature selection is used then the classifier takes as input all the attributes of the input data. After classification is performed, the performance is evaluated using various metrics.

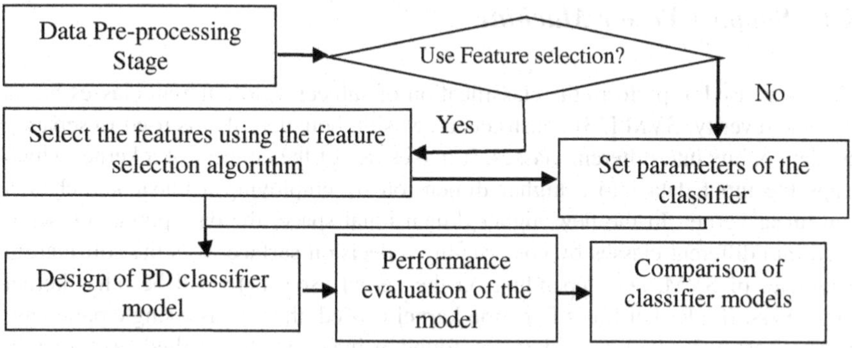

Fig. 1 Basic block diagram of the proposed system

5 Techniques Used

5.1 Feature Selection Techniques

A variety of feature selection techniques are used in this work. They are based on criteria such as Conditional Mutual Information Maximisation, Joint Mutual Information etc. In this work, the following criteria are used:

- CMIM: Conditional Mutual Information Maximisation is introduced by [18]. The conditional mutual information is the criterion used to decide which attribute is good. A feature x' is said to be good only if it carries information about the label vector Y, and if this information has not been captured by any other attributes already picked.
- JMI: Joint Mutual Information criterion introduced by [19]. The relevance of a set of attributes is determined by their joint mutual information.
- DISR: Double Input Symmetrical Relevance criterion is introduced by [20]. The idea behind it is that a set of variables can give information about the output class better than one variable taken alone. DISR criterion can be used to select a set of variables that provide maximum information about the output class.
- MIM: The concept of Mutual Information was introduced by [21]. Mutual Information Maximisation criterion is used to pick relevant features. The mutual information score of the attributes are calculated, and the attributes having the highest scores are chosen.

The MATLAB toolbox by [22] which incorporated the above mentioned criteria was used for performing feature selection.

5.2 Support Vector Machine

SVM was used to perform the classification of subjects into different classes based on their severity. SVM [23] constructs a decision boundary that is used to separate the data points into different classes. It makes use of the concept of a kernel which maps the input data into a higher dimension by employing a linear kernel or a non-linear kernel. In the new higher dimensional space, the data points are separated into different classes by constructing a decision surface. C is the soft margin parameter of SVM, and depending on the kernel used there will be other kernel parameters. If a Radial Basis Function kernel is used, then it has a single parameter γ. The best values of C and γ are found using a method called grid search. MATLAB implementation of LIBSVM [24] was used for performing SVM.

5.3 Stacked Autoencoder

A stacked autoencoder (SAE) [25] is a neural network consisting of multiple layers of sparse autoencoders in which the outputs of each layer are connected to the inputs of the successive layer. An autoencoder is a neural network which is used to learn efficient codings. An autoencoder is trained to reconstruct its own inputs x by minimizing the reconstruction error, generally computed as the cross entropy between x and a reconstructed version z (of x). However, it was found that reconstruction error alone was not enough to guarantee the extraction of useful features. Partially corrupting the input with noise and then performing the mapping to a representation y and then to z gives a better result. So, the data is first corrupted with noise using a stochastic mapping. This corrupted input is then mapped to a hidden representation y, and it is then mapped to z. The denoising autoencoder can also be used in a stacked manner. The advantage of using a stacked autoencoder is that it learns useful higher level representations from the input data. A deep learning MATLAB toolbox implemented by [26] was used.

5.4 Deep Neural Network with Dropout

A DNN is neural network with multiple layers of hidden nodes. Dropout [27] is a technique that is used to prevent overfitting in deep neural networks. It provides a way of approximately combining exponentially many different neural network architectures efficiently [27]. In this technique visible as well as hidden units are dropped or temporarily removed along with their incoming as well as outgoing connections. The deep learning toolbox by [26] was used in this work.

5.5 *Deep Belief Network*

A DBN [28] consists of networks such as Restricted Boltzmann Machines (RBM). They are generative, probabilistic, graphical models that can learn deep hierarchical representations from the training data. They model the joint distribution between observed vector x and the l hidden layers h^k A DBN can learn to probabilistically reconstruct its inputs when trained in an unsupervised manner. After this learning step, a DBN can be further trained in a supervised way to perform classification. DNN classification was done by means of the MATLAB deep learning toolbox by [26].

6 Results

The performance metrics used to evaluate the method are precision, recall and accuracy. Due to the skewed nature of data, accuracy alone cannot be used as a measure of the performance of the classifier, instead the F1-score can be used to indicate how well the classifier classifies the instances of each class. All average classification accuracies have been reported after performing 10-fold cross validation. The average precision, recall, and F1-scores were computed by averaging the values obtained for all the four classes.

6.1 *Classification Using SVM*

Multi-class classification was done using SVM as well as many feature selection algorithms with SVM. The value of the parameters used are $C = 2^1$, and $\gamma = 2^{-5}$. The accuracy of SVM has been verified using 10-fold cross-validation. The average accuracy obtained is 87.27 %. The result of applying SVM is shown in Table 3.

Feature Selection and Classification Using SVM

Different feature selection algorithms based on criteria such as CMIM, JMI, MIM, DISR were used. The result of applying feature selection followed by SVM

Table 3 Performance evaluation using SVM

Class	Precision	Recall	F1-score
0	0.95	0.92	0.94
1	0.83	0.82	0.82
2	0.88	0.90	0.89
3	0.83	0.20	0.32
Average accuracy	Average precision	Average recall	F1-score
87.27 %	0.87	0.71	0.78

Table 4 Results of performing feature selection followed by SVM classification

Method	#Features	Avg accuracy (%)	Avg precision	Avg recall	F1-score
CMIM	15	85.40	0.85	0.81	0.83
CMIM	20	85.98	0.85	0.81	0.83
CMIM	25	86.37	0.86	0.81	0.83
CMIM	40	86.86	0.86	0.76	0.79
JMI	20	85.52	0.81	0.66	0.66
JMI	25	87.34	0.87	0.82	0.84
JMI	40	87.22	0.85	0.77	0.80
DISR	15	85.42	0.85	0.81	0.83
DISR	20	86.35	0.86	0.82	0.83
DISR	25	86.73	0.86	0.81	0.83
DISR	40	87.83	0.87	0.75	0.78
MIM	20	85.79	0.89	0.66	0.66
MIM	40	87.42	0.86	0.77	0.80
MIM	50	87.44	0.85	0.72	0.76

classification are shown in Table 4. Here SVM with RBF kernel was used, and the values of its parameters C and γ are $2^1$and 2^{-5} respectively.

The best result obtained after the feature selection is the average accuracy of 87.34 % and an F1-score of 0.84 for the combination of JMI based feature selection and SVM.

6.2 *Classification Using Neural Network*

NN was also used for multi-class classification of the motor assessments. The result of applying NN is shown in Table 5. The average classification accuracy obtained in this case is 82.21 % for 27 hidden layer neurons, 200 epochs, learning rate = 0.5 and momentum = 0.5.

Table 5 Identification of PD stages using NN

Class	Precision	Recall	F1-score
0	0.88	0.89	0.89
1	0.75	0.74	0.74
2	0.84	0.86	0.85
3	0.61	0.36	0.45
Avg. accuracy (%)	Avg. precision	Avg. recall	Avg. F1-score
82.21	0.77	0.71	0.73

Table 6 Identification of PD stages using stacked autoencoder followed by NN classification

Class	Precision	Recall	F1-score
0	0.92	0.89	0.90
1	0.76	0.78	0.76
2	0.86	0.87	0.86
3	0.79	0.47	0.58
Avg. accuracy (%)	Avg. precision	Avg. recall	Avg. F1-score
84.06	0.83	0.75	0.77

Stacked Autoencoder Pre-trainer and Neural Network

The result of using stacked autoencoder pre-trainer followed by neural network classification is shown in Table 6. The average classification accuracy obtained in this case is 84.06 %, using 45 hidden units in the stacked autoencoder.

6.3 Deep Neural Network with Dropout

A DNN with different values of dropout fractions and architectures was used for the multi class classification. The parameters of the DNN like learning rate, momentum and number of epochs are 0.5, 0.8, and 300 respectively. Different values of dropout were tried, and the value of 0.3 gave the best result. The DNN architecture of [54-100-60-50-4] gave an average F1-score of 0.82. Note: In the architecture, the first and last numbers indicate respectively the number of input and output units, the numbers in between indicate the number of hidden units. Table 7 shows the various deep neural network architectures and the dropout fractions used

6.4 DBN

M_DBN denotes the momentum of DBN, LR_DBN denotes the learning rate of the DBN, LR_NN, M_NN denote the learning rate and momentum of the neural network respectively. Many different configurations of DBN and NN layers, as well as different values of momentum, learning rate, number of epochs were tried, and the

Table 7 Result of applying NN with various dropout fractions

Architecture	Dropout	Avg. accuracy (%)	Avg. precision	Avg. recall	F1-Score
54-100-60-50-4	0.2	86.55	0.83	0.80	0.82
54-100-60-50-4	0.3	86.50	0.85	0.80	0.82
54-100-60-4	0.3	86.57	0.83	0.79	0.81

Table 8 DBN + ANN versus ANN alone for [15 15] architecture of DBN

Method	Avg. accuracy (%)	Avg. precision	Avg. recall	F1-score
DBN + ANN	86.91	0.84	0.81	0.83
ANN	86.56	0.84	0.81	0.82

configuration that gave the best result had DBN parameters: [15 15] architecture, 300 epochs, learning rate of 0.7, momentum of 0.2, and NN parameters: 200 epochs, learning rate of 0.5, and momentum of 0.3. The average classification accuracy obtained in this case is 86.91 %.

The average accuracy obtained for DBN followed by ANN classification was 86.91 % with an F1-score of 0.83. ANN classification without the DBN was also attempted. The details are shown in Table 8. It can be observed that, the contribution of DBN toward the subsequent neural network classification is very small.

7 Discussion

Table 9 shows the average classification accuracy, precision and recall obtained by all the classifiers used in this work. Classification accuracy is a commonly used performance metric for describing the performance of a binary classifier. But due to the imbalance in the number of instances in the data set, other metrics like F1-score is used. A good classifier should produce results with good precision and recall, and F1-score being the harmonic mean of theses two values, captures this information. From Table 9 it can be observed that SVM with JMI based feature selection algorithm gave a good improvement in the F1-score. See the first two rows of the table, and observe that there is almost a 10 % improvement in the F1-score. Although DBN followed by ANN classifiaction gave a comparable result, the combined training times of the DBN and ANN was higher than the total runnning time of the other techniques.

Table 9 Comparison of performance metrics obtained from different methods

Method	Avg. accuracy (%)	Avg. precision	Avg. recall	F1-score
SVM	87.27	0.87	0.71	0.78
JMI criterion based feature selection + SVM	87.34	0.87	0.82	0.84
NN	82.21	0.77	0.71	0.73
SAE + NN	84.06	0.83	0.75	0.77
DNN with dropout	86.50	0.85	0.80	0.82
DBN + ANN	86.91	0.84	0.81	0.83

Table 10 Paired t-test on classifier results

Output of classifier	Classifiers compared	p-value	t-value	Null hypothesis
F1-score	SVM + Feature selection, SVM	0.015	2.565	Rejected
F1-score	SVM + Feature selection, DBN + NN	0.14281	1.17198	Rejected
F1-score	SVM + Feature selection, SAE + NN	0.00036	5.29159	Rejected

Statistical Significance Test

A paired t-test was also done in order to statistically compare the results of the classifers. The input to the test is the per fold F1-score obtained after doing 10-fold crossvalidation. The results are shown in Table 10.

Average F1-score obtained during each fold after performing 10-fold crossvalidation with the classifier was used to perform the paired t-test at the significance level of 0.05. SVM with feature selection using JMI criterion was found to be more statistically significant than SVM since the computed p value in this case is smaller than the t-value. At a significant level of 0.05, the null hypothesis was rejected in this case, which shows that using feature selction along with SVM improved its performance. The performance of feature selection with SVM was statistically compared against other classifers like DBN with NN, and SAE with NN. In all the comparisons the former classifer output was found to be statistically significant than the others.

8 Conclusions

Previous works mainly focused on the binary classification of subjects as either PD or non-PD. This work classifies subjects into different classes based on the severity of the disease. Generally, there are five stages of PD which are defined using a scale called HY scale. Identifying the stages of the disease can help clinicians prescribe appropriate medications to the patients and thus avoid unnecessary treatments. This work deals with the multi-class classification of subjects using 54 of their motor assessments. Due to the skewed nature of data, accuracy alone cannot be used to evaluate the performance of the classification algorithm. F1-score that combines the precision and recall of the classifier for each class was also used to evaluate the performance of the classifier. The effect of using feature selection along with SVM classification was studied and it was found that a feature selection algorithm based on JMI criterion gave the best result. The effect of using dropout in a deep neural network was also studied. It was found that a DNN with dropout performed better than a shallow one without dropout by giving an F1-score of 0.82 and accuracy of 86.50 %.

Acknowledgments PPMI—a public-private partnership—is funded by the Michael J. Fox Foundation for Parkinson's Research and funding partners, including [list the full names of all of the PPMI funding partners found at http://www.ppmi-info.org/fundingpartners].

References

1. Marek, K., Jennings, D., Lasch, S., Siderowf, A., Tanner, C., Simuni, T., et al.: The parkinson progression marker initiative (PPMI). Prog. Neurobiol. **95**, 629–635 (2011)
2. Babu, G.S., Suresh, S.: Parkinson's disease prediction using gene expression-A projection based learning meta-cognitive neural classifier approach. Expert Syst. Appl. **40**, 1519–1529. Elsevier (2013)
3. Das, R.: A comparison of multiple classification methods for diagnosis of Parkinson disease. Expert Syst. Appl. **37**, 1568–1572. Elsevier (2010)
4. Hazan, H., Hilu, D., Manevitz, L., Ramig, L.O., Sapir, S.: Early diagnosis of parkinson's disease via machine learning on speech data. In: 2012 IEEE 27th Convention of Electrical and Electronics Engineers in Israel (IEEEI), pp. 1–4. IEEE (2012)
5. Mandal, I., Sairam, N.: Accurate telemonitoring of Parkinson's disease diagnosis using robust inference system. Int. J. Med. Informatics **82**, 359–377 (2013)
6. Rustempasic, I., Can, M.: Diagnosis of Parkinson's disease using fuzzy C-means clustering and pattern recognition. SouthEast Europe J. Soft Comput. **2**, 42–49 (2013)
7. Tsanas, A., Little, M.A., McSharry, P.E., Spielman, J., Ramig L.O.: Novel speech signal processing algorithms for high-accuracy classification of Parkinson's disease. IEEE Trans. Biomed. Eng. **59**, 1264–1271 (2012)
8. Costin, H., Geman, O.: Parkinson's disease prediction based on multistate markov models. Int. J. Comput. Commun. Control **8**, 525–537 (2013)
9. Daliri, M.R.: Chi-square distance kernel of the gaits for the diagnosis of Parkinson's disease. Biomed. Signal Process. Control **8**, 66–70. Elsevier (2013)
10. Haller, S., Badoud, S., Nguyen, D., Garibotto, V., Lovblad, K.O., Burkhard, P.R.: Individual detection of patients with Parkinson disease using support vector machine analysis of diffusion tensor imaging data: initial results. AJNR Am. J. Neuroradiol. **33**, 2123–2128 (2012)
11. Martínez-Murcia, F.J., Górriz, J.M., Ramírez, J., Illán, I.A., Ortiz, A.: The Parkinson's progression markers initiative.: automatic detection of parkinsonism using significance measures and component analysis in DaTSCAN imaging. Neurocomputing **126**, 58–70. Elsevier (2014)
12. Prasanth, R., Roy, S.D., Mandal, P.K., Ghosh, S.: Automatic classification and prediction models for early Parkinson's disease diagnosis from SPECT imaging. Expert Syst. Appl. **41**, 3333–3342. Elsevier (2014)
13. Segovia, F., Górriz, J.M., Ramírez, J., Álvarez, I., Jiménez-Hoyuela, J.M., Ortega, S.J.: Improved Parkinsonism diagnosis using a partial least squares based approach. Med. Phys. **39**, 4395–4403 (2012)
14. Armañanzas, R., Bielza, C., Chaudhuri, K.R., Martinez-Martin, P., Larrañaga, P.: Unveiling relevant non-motor Parkinson's disease severity symptoms using a machine learning approach. Artif. Intell. Med. **58**, 195–202. Elsevier (2013)
15. Babu, G.S., Suresh, S., Mahanand, B.S.: A novel PBL-McRBFN-RFE approach for identification of critical brain regions responsible for Parkinson's disease. Expert Syst. Appl. **41**, 478–488. Elsevier (2014)
16. Gil, David, Johnson, Magnus: Diagnosing parkinson by using artificial neural networks and support vector machines. Glob. J. Comput. Sci. Technol. **9**(4), 63–71 (2009)
17. Hoehn, M.M., Yahr, M.D.: Parkinsonism: onset, progression and mortality. Neurology **17**, 427–442 (1967)

18. Fleuret, F.: Fast binary feature selection with conditional mutual information. J. Mach. Learn. Res. **5**, 1531–1555 (2004)
19. Yang, H.H., Moody, J.E.: Data visualization and feature selection: new algorithms for nongaussian data. In: NIPS, pp. 687–702 (1999)
20. Meyer, P.E., Bontempi, G.: On the use of variable complementarity for feature selection in cancer classification. In: EvoWorkshops 2006: EvoBIO, LNCS, pp. 91–102. Springer, Heidelberg (2006)
21. Shannon, C.E.: A mathematical theory of communication. Bell Syst. Tech. J. **27**, 379–423 (1948)
22. Brown, G., Pocock, A., Zhao, M.J., Luján, M.: Conditional likelihood maximisation: a unifying framework for information theoretic feature selection. J. Mach. Learn. Res. **13**, 27–66 (2012)
23. Cortes, C., Vapnik, V.: Support-vector networks. Mach. Learn. **20**, 273–297. Springer (1995)
24. Chang, C.C., Lin, C.J.: LIBSVM: a library for support vector machines. ACM Trans. Intell. Syst. Technol. **2**(27), 1–27 (2011)
25. Vincent, P., Larochelle, H., Bengio, Y., Manzagol, P.A.: Extracting and composing robust features with denoising autoencoders. In: ICML '08 Proceedings of the 25th International Conference on Machine Learning, pp. 1096–1103. ACM, New York (2008)
26. Palm, R.B.: Prediction as a candidate for learning deep hierarchical models of data. Master's Thesis. Technical University of Denmark. IMM-M.Sc.-2012-31 (2012)
27. Srivastava, N., Hinton, G., Krizhevsky, A., Sutskever, I., Salakhutdinov, R.: Dropout: a simple way to prevent neural networks from overfitting. J. Mach. Learn. Res. **15**, 1929–1958 (2014)
28. Hinton, G.E., Osindero, S., Teh, Y.: A fast learning algorithm for deep belief nets. Neural Comput. **18**, 1527–1554 (2006)

Security Constrained Unit Commitment Problem Employing Artificial Computational Intelligence for Wind-Thermal Power System

K. Banumalar, B.V. Manikandan and K. Chandrasekaran

Abstract In this article, an effective hybrid nodal ant colony optimization (NACO) and real coded clustered gravitational search algorithm (CGSA) is involved in producing a corrective/preventive contingency dispatch over a specified given period for wind integrated thermal power system. High wind penetration will affect the power system reliability. Hence, the reliability based security-constrained unit commitment (RSCUC) problem is proposed and solved using bi-level NACO-CGSA hybrid approach. The RSCUC problem comprises of reliability constrained unit commitment (RCUC) as the master problem and the sub problem as a security constrained economic dispatch (SCED). NACO solves master problem and the sub problem is solved by real coded CGSA. The objective of RSCUC problem model is to obtain the economical operating cost, while maintaining the system security. The proposed solution for the hourly scheduling of generating units is based on hybrid NACO-CGSA. Case studies with IEEE 118-bus test system are presented in detail.

Keywords Bi-level optimization techniques · Nodal ant colony optimization · Reliability and security constrained unit commitment · Clustered gravitational search algorithm · Security constrained economic dispatch · Transmission constraints

K. Banumalar (✉) · B.V. Manikandan
Department of Electrical and Electronics Engineering,
Mepco Schlenk Engineering College, Sivakasi, Tamil Nadu, India
e-mail: banumalar.234@gmail.com

B.V. Manikandan
e-mail: bvmani73@yahoo.com

K. Chandrasekaran
Department of Electrical and Electronics Engineering,
National Institute of Technology, Karaikal, Puducherry, India
e-mail: chansekaran23@gmail.com

M. Senthilkumar et al. (eds.), *Computational Intelligence, Cyber Security and Computational Models*, Advances in Intelligent Systems and Computing 412, DOI 10.1007/978-981-10-0251-9_26

List of Symbols

$\mathrm{AvgC}, \mathrm{Max}_{\mathrm{it}}$	Average minimum cost($) obtained in 10 simulations and Maximum number of iterations
$\mathrm{BF}_{i,t}, \mathrm{BF}_{\mathrm{i}}^{\max}$	Flow through branch i at time t (MVA) and Maximum flow limits for branch I (MVA)
$\mathrm{C}_i(\mathrm{P}_{(\mathrm{i,t})})$	Production cost ($) $C_i(P_{(i,t)}) = a + b^{*}P_{(i,t)} + c * P_{(i,t)}^2$
$\mathrm{P}_{(\mathrm{i,t})}, \mathrm{P}_{(\mathrm{i\,min})}\mathrm{P}_{\mathrm{imax}}$	Power level Minimum and Maximum power output of ith generator unit (MW)
a, b, c	Cost co-efficient of ith generator unit in ($/hr), ($/MWhr) and ($/MW^{2}hr)
$\mathrm{D}_{\mathrm{t}}, \mathrm{Fit}_{\mathrm{p}}$	Total system demand at time t *and* Fitness value of the solution p
DR(i), UR(i)	Ramp-down rate limit of ith generator unit and Ramp-up rate limit of ith generator unit
$\mathrm{G}_{\mathrm{ij}}\mathrm{B}_{\mathrm{ij}}$	Conductance and susceptance between bus i and bus j
$\mathrm{I}_{(i,t)}\mathrm{L}_{\mathrm{gb}}$	Commitment state of ith unit at tth hour and Maximum total profit incurred till the current tour
$\mathrm{N}, \mathrm{N}_{\mathrm{ants},}\mathrm{N}_{\mathrm{B}}$	Total number of generating units, Total number of ants and Number of busses
$\mathrm{N}_{\mathrm{B-1}}, \mathrm{N}_{\mathrm{PQ}}$	Number of buses excluding slack bus, Number of **PQ** bus,
$\mathrm{P}_{\mathrm{Gi,t,}}\mathrm{Q}_{\mathrm{Gi,t}}$	Active and reactive power generation at bus i at time t, *r*espectively
$\mathrm{P}_{\mathrm{Di,t,}}\mathrm{Q}_{\mathrm{Di,t}}$	Minimum and maximum active power generation limit tor unit i
$\mathrm{Pr}_{\mathrm{rs}}^{\mathrm{k}}(\mathrm{st})$	Transition probability of kth ant from stage r to s
$\mathrm{Q}_{\mathrm{Gi}}^{\min}, \mathrm{Q}_{\mathrm{Gi}}^{\max}$	Minimum and maximum reactive power generation limit for unit i
$\mathrm{R}_{(\mathrm{i,t})}, \mathrm{SR}_{\mathrm{t}}$	System spinning reserve at tth hour (MW/hour) and Total system spinning reserve at time t
$\mathrm{SU}_{\mathrm{i}}^{\mathrm{t}}, \mathrm{SD}_{\mathrm{i}}^{\mathrm{t}}$	Start up cost and shut down cost of unit i at time t,
TS, T, TC	Total number of stages, Dispatch period in hours and Total cost ($)
$\mathrm{T}^{\mathrm{on}}(\mathrm{i}), \mathrm{T}^{\mathrm{off}}, \mathrm{TC}$	Minimum up-time of ith generator unit and Minimum down-time of ith generator unit
$\mathrm{V}_{\mathrm{i,t}}, \mathrm{V}_{\mathrm{pq}}$	Voltage magnitude of bus i at time t (pu) and Modified position of employed or onlooker bees
$\mathrm{V}_{i}^{\min}, \mathrm{V}_{i}^{\max}$	Minimum and maximum voltage magnitude limit at bus i (pu)
$\mathrm{X}^{\mathrm{on}}(\mathrm{i,t}), \mathrm{X}^{\mathrm{off}}(\mathrm{I,t})$	"ON" duration of ith generator unit till time t *and* "OFF" duration of ith generator unit till time t
A, B, P	Relative importance pheromone trail intensity and Relative importance of heuristic function
P, i, t	Evaporation factor, Index for generator unit and Index for time
$\tau_{\mathrm{rs}}(\mathrm{st}), \Delta\tau_{\mathrm{rs}}$	Heuristic function of stage (st) r to s and The updating co-efficient

1 Introduction

The main goal of unit commitment problem (UCP) is to determine on when to switch on and switch off the generator units and the sub problem is to determine the dispatch of the committed units to meet the electricity demands. A literary review of UCP and the solution techniques are given in Refs. [1, 2]. Techniques like dynamic programming [3], Lagrangian relaxation [4], and mixed integer programming [5] are the widely used conventional techniques. The main disadvantage of dynamic programming method is that the computational time increases rapidly with the system size which is termed as curse of dimensionality. Though LR method provides a fast solution, it may encounter feasibility problem due to the dual nature of the method. MIP guarantees convergence to the optimal solution in a finite number of steps while providing flexible and accurate framework. The major drawback of using MIP formulation is that the execution time needed to find a good solution.

Artificial intelligence techniques like genetic algorithm and integer coded genetic algorithm [6], simulated annealing [7], fuzzy logic [8], particle swarm optimization [9], and ant colony optimization [10, 11] are used. However, the UCP solution is not accurate since security constraints are not considered. SCUC problem is a generalization of UCP, which considers transmission network security limits in addition to UCP constraints [12]. In a deregulated power market, the day-ahead schedule are planned by independent system operator (ISO using SCUC problem. However, SCUC problem is a nonlinear, non-convex, non-smooth, and mixed-integer optimization problem with discontinuous solution space [13].

In [14], UCP is solved by a GA and then the optimal power flow (OPF) including power flow constraints and line flow limits is solved by another GA. However, when using these existing methods, the SCUC problems encounter some inherent limitations, that of the unreasonable relaxations for the discrete variables, unstable computing efficiency, or excess decomposition for the problem model.

Proposed Work: The recent trend adopted worldwide in the new structured power system is the integration of renewable energy resources with fossil fuel plants. The reliability of the generation system is influenced greatly by high penetration of wind energy due to the random nature of the wind availability. The introduction of renewable energy resources into conventional utilities creates new concerns for power engineers. Hence, this paper solves the RSCUC problem through effective application of hybrid NACO-CGSA algorithm. As hybridization, NACO solves the master UC problem and real coded CGSA solves the security constrained economic dispatch sub-problem. The performance of the hybrid NACO-CGSA in terms of solution quality is compared with that of other algorithms reported in literature.

2 Over View of Proposed Methodology

2.1 Theory of Nodal Ant Colony Optimization (NACO)

ACO is an intelligent optimization algorithm that searches the optimal solution, mimicking real ants. Ants choose their own path to arrive at a destination. Initially the movement of ants is random i.e., once an ant comes across any barrier it chooses its path randomly. While moving from source to destination and vice versa, ants deposit a chemical substance called pheromone on the path, forming a pheromone trail. Ants which move in the shortest path, reach the destination faster than the ants which move in other possible paths. More pheromone is deposited in the shortest path. By smelling the pheromone trail, ants find their way to the source by their nest mates. The ant's search is based on pseudo-random probability function based on problem-dependent heuristic and the amount of pheromone previously deposited in this trail. The pheromone updation is based on evaporation rate and quality of the current solution which is given detailed in Ref. [10, 11]. The existing ACO approach for solving UCP mainly consists of two phases; namely Ant Search Space (ASS) and Ant's Search (AS), respectively [15]. If N generating units are available for commitment for a particular hour, then the possible combinations of states at that hour is $(2^N - 1)$. For a 24 h time horizon, the ASS matrix space can be formed as $(2^N - 1) \times 24$. During the tour of an ant, it chooses a state from the large number of available states in the matrix column.

2.2 Gravitational Search Algorithm

The algorithm was proposed by Rashedi et al. [16, 17] that uses the Newton's Gravitational Principle to search the optimum solution. In this algorithm, the coordinates or the agents in the search space are considered as masses. All these masses attract each other according to laws of Gravity and form a direct means of communication through it. Recently, in many article GSA is successfully implemented for solving discrete and continuous optimization problem. In the proposed work new methodology is implemented to improve the convergence and solution quality of GSA which is discussed in next section.

2.2.1 Clustered Gravitational Search Algorithm

To keep the exploration in GSA alive without killing the exploitation, in this paper, a group method is proposed. Here, the whole population is divided into three basic groups: namely Leader, Follower and Freelancer. The Leaders are the best particles obtained at the end of the first iteration. Each leader particle shall lead a group of optimizers. The Leader and the optimizer group together shall work like a simple

GSA population thereafter. In this way there would be some independent GSA populations led by their leader that will search for the optimum solution. The last group, the freelancers shall be randomly initiated every iteration and in this way they shall keep the search alive. Each group those led by a leader and the freelancers shall have a best particle. The best out of these bests shall be the final best particle of the iteration. Depending on the requirements of the function, the ratio of the population of Leader, follower and the freelancer can be adjusted.

2.2.2 Pseudo-code of the Real-Coded CGSA Algorithm

```
Step 1. Initialize the random population of agents.
Step 2. Evaluate the population on the given function.
Step 3. Sort the population on the fitness values.
Step 4. The first 10% of the population(size can vary according to func-
tion) is called the Leader.
Step 5. The next 80% of the population(size can vary according to func-
tion) is called the Follower.
Step 6. The last 10% of the population(size can vary according to func-
tion) is called the Freelancer.
Step 7. To each of the agent in the leader group a set of agents from
follower group is allotted and they together make a single sub popula-
tion.
 For i=1 to max no of iterations.
 {
    For j=1 to max of sub population
      {
        Run GSA for each of the subgroup.

      }

Evaluate each particle in the freelancer group. Find the minimum of the
best fitness values amoung all sub groups and the freelancer group.
Again randomly initialize the entire freelancer population.
}
Step 8. The best population is thus the minimum of all the best fitness
values at the end of the iteration
```

3 Problem Formulation

The objective of RSCUC is to minimize the fuel cost, simultaneously satisfying equality and inequality constraints with security and transmission network constraints. The objective function of the RSCUC can be formulated as follows;

Minimize

$$\sum_{t=1}^{T}\sum_{i=1}^{N}[F_i(P_{(i,t)}) * I_{(i,t)} + SU_i^t * I_{(i,t)} * (1 - I_{(i,t-1)}) + SD_i^t * (1 - I_{(i,t)})] \quad (1)$$

$$F_i(P_{(i,t)}) = a + b * P_{(i,t)} + c * P_{(i,t)}^2 \tag{2}$$

Subject to the following constraints.

3.1 System Constraints

3.1.1 Power Balance Constraints

$$P_{Gi,t}PWG_{i,t} - P_{Di,t} = V_{i,t}\sum_{j=1}^{N} V_{j,t}(G_{ij}\cos\theta_{ij} + B_{ij}\sin\theta_{ij}) \quad i \in N_{B-1} \tag{3}$$

$$Q_{Gi,t} - Q_{Di,t} = V_{i,t}\sum_{j=1}^{N} V_{j,t}(G_{ij}\sin\theta_{ij} - B_{ij}\cos\theta_{ij}) \quad i \in N_{PQ} \tag{4}$$

3.1.2 Spinning Reserve Constraints

$$\sum_{i=1}^{N}(P_{i,\max} * I_{(i,t)}) + PWG_{i,t} \geq Load_t, \ k \in [1, T] \tag{5}$$

3.2 Unit Constraints

3.2.1 Minimum Up and Down Time Constraints

$$[X^{on}(i,t-1) - T^{on}(i)] * [I_{(i,t-1)} - I_{(i,t)}] \geq 0 \tag{6}$$

$$[X^{off}(i,t-1) - T^{off}(i)] * [I_{(i,t)} - I_{(i,t-1)}] \geq 0 \tag{7}$$

3.2.2 Unit Ramp Constraints

$$P_{(i,t)} - P_{(i,t-1)} \leq UR(i) \tag{8}$$

$$P_{(i,t-1)} - P_{(i,t)} \leq DR(i) \tag{9}$$

3.3 Security Constraints

$$V_i^{\min} \leq V_{i,t} \leq V_i^{\max} \quad i \in N_{B-1} \tag{10}$$

$$\left|BF_{i,t}\right| \leq BF_i^{\max} \quad i \in N_B \tag{11}$$

These security constraints are related to steady state conditions and it should be considered in the post-contingency states.

3.3.1 Reactive Power Generation Limits

$$Q_{Gi}^{\min} \leq QGi, t \leq Q_{Gi}^{\max} \tag{12}$$

3.4 Reliability Constraints

In wind integrated thermal power system, high wind penetration can lead to high-risk level in power system reliability. The wind energy dispatch and reliability level of the power system are considered as in Ref. [18] and is given in Eq. (13).

$$\text{Min}_{EENS} = \sum_{j=LC} PR_j L_j \quad (\text{Kwh}) \tag{13}$$

4 Implementation of Hybrid NACO-CGSA for RSCUC Problem

Repair mechanism is utilized to overcome the violations with the minimum up/down time constraints, reserve constraint violation and SCED problem violation. Repair mechanism implemented to overcome the above violation is shown in Fig. 1. Also the step by step procedure for implementation of NACO-CGSA for RSCUC problem is given in flowchart shown in Fig. 2.

5 Case Study

The matlab coding is developed for SCUC problem using hybrid NACO-CGSA in Pentium-IV, 3 GHz, 1 GB RAM processor. To validate the proposed algorithm, three test cases (UCP, SCUC and RSCUC problem) are carried out on the modified

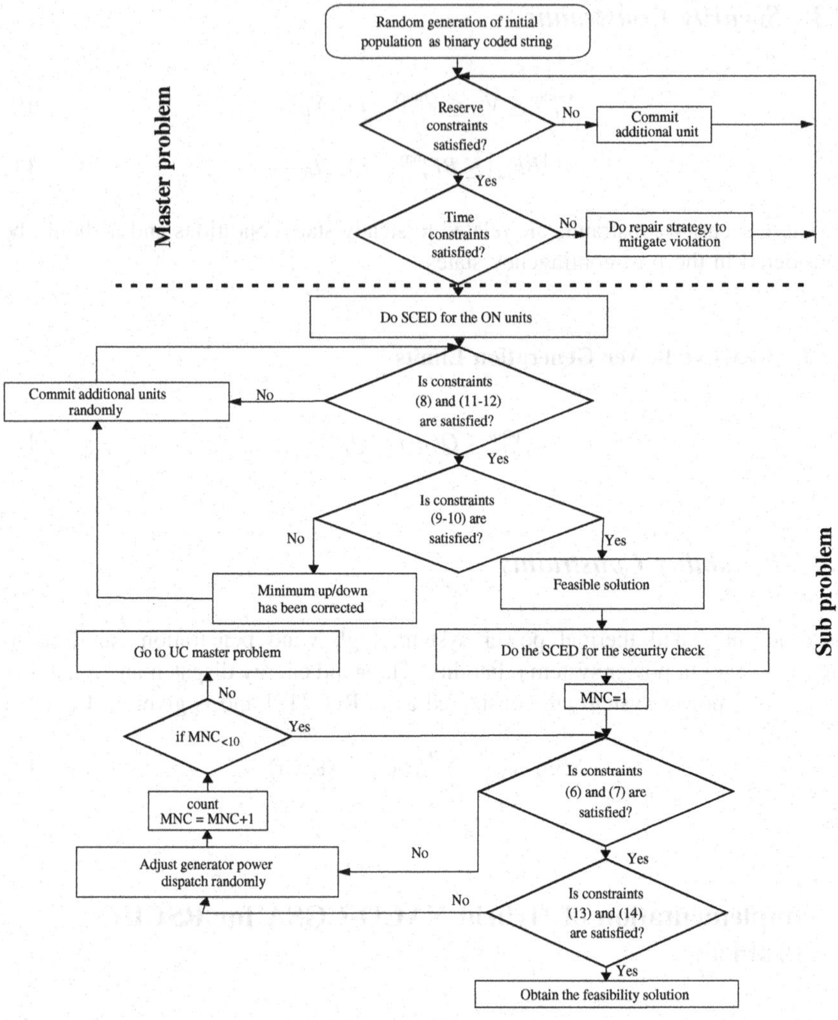

Fig. 1 Flowchart of constraint handling mechanism

IEEE 118 standard test systems with 54 generators [19, 20] for a scheduling time horizon of 24 h. It is the largest SCUC problem test system, which is available in the existing literature. The test systems are subject to (n − 1) security criterion [12, 21], characterized by the loss of a generation unit.

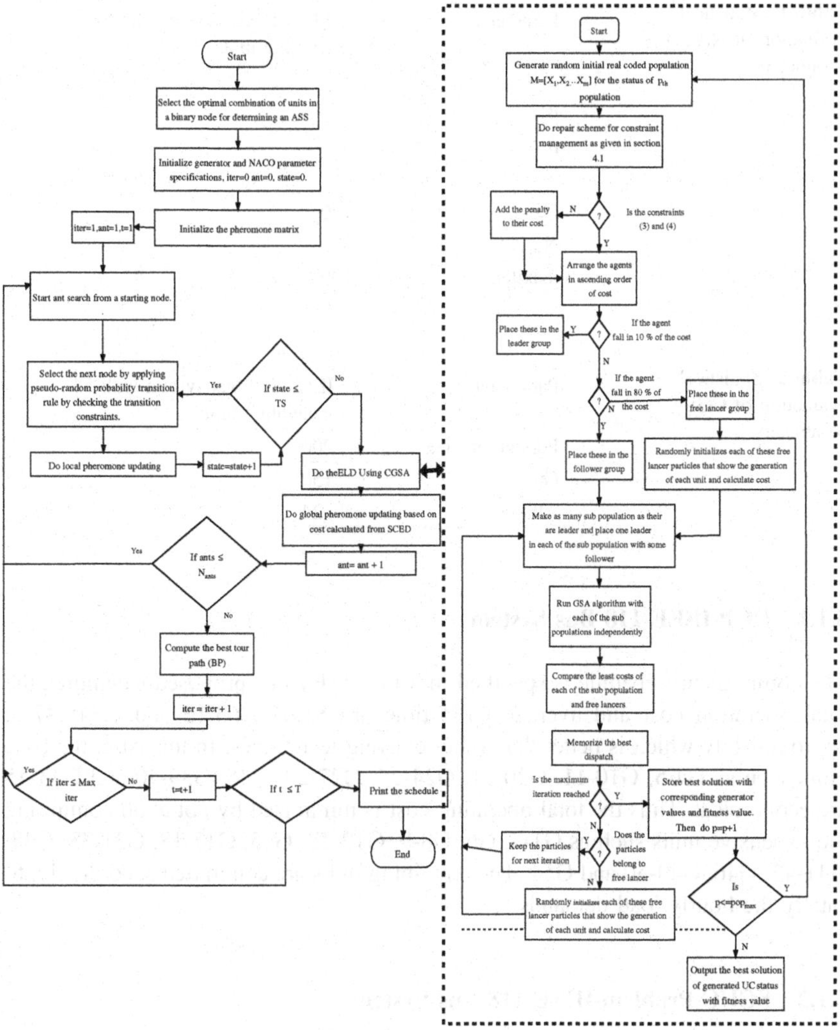

Fig. 2 Flow chart of Hybrid NACO-CGSA for SCUC problem

5.1 IEEE-118 Bus System

To validate the efficiency of hybrid NACO-CGSA in a large scale system, a modified IEEE-118 bus test system is used to solve UC and SCUC problems. The NACO and CGSA parameters are tabulated in Tables 1 and 2. The test data and the load data for 24 h duration along with reserves for IEEE 118-bus system are given in http://motor.ece.iit.edu/data/IEEE118data.xls and http://motor.ece.iit.edu/data/JEAS/ IEEE. The detailed data of the system is found in [19, 21].

Table 1 Statistical evaluation of NACO parameters

Parameters	IEEE 118-bus system with 54 generating units
α	2
β	18
ρ	0.9
n	3
C	100
N_{ants}	50
maxiter	100

Table 2 Statistical evaluation of CGSA parameters

Parameters	IEEE 118-bus system with 54 generating units
Population size	200
G0	150
Alfa	24

5.1.1 UCP-IEEE 118 Bus System

The commitment schedule is specified in Table 3. By this proposed technique, the total operation cost and average CPU time are $1,644,263.13 and 1,348.47 s, approximately which is better than other existing techniques. In this case, the base units such as G4-5, G10-11, G20-21, G24-25, G27-29, G36, G39-40 and G43-45 are economical. Thus the total operating cost is minimized by not at all commiting the expensive units such as G1-3, G6, G8-9, G12-13, G15, G17-18, G31-33, G38, G41-42, G46, G49-50 and G54. The remaining units are committed accordingly, to satisfy the hourly load demands.

5.1.2 SCUC Problem-IEEE 118 Bus System

The SCUC problem is solved by NACO as a master problem and SCED is solved as a sub-problem using real coded CGSA. In this case the economical units such as G4-5, G10-11, G20-21, G24-25, G27-29, G36, G39-40 and G43-45 are used as base units. Even though the generating units G1-3, G12, and G13 are expensive, they are committed to satisfy the transmission constraints. However the other expensive units G8-9, G15, G17-18, G31-33, G38, G41-42, G49-50 and G54 are not committed at all. The remaining units are committed accordingly to satisfy the hourly load demands. The commitment schedule and the total operating cost consumed with the available techniques in the literature are given in Tables 4 and 5,

Table 3 Commitment schedule of UCP IEEE 118-bus

Time (h)	Total operating cost = $1644263.13
	Unit ON/OFF status
1	0001100001100000000110011011100000010011001110000000000
2	0001100001100000000110011011100000010011001110000000000
3	0001100001100000000110011011100000010011001110000000000
4	0001100001100000000110011011100000010011001110000000000
5	0001100001100000000110011011100000010011001110000000000
6	0001100001100000000110011011100000010011001110000000000
7	0001100001100000000110011011100000010011001110000000000
8	0001101001100001000110111011100000010011001110000000000
9	0001101001100101000111111011100000011011001110000000000
10	0001101001100101000111111011100000111011001110000001110
11	0001101001100101000111111011100000111011001110000001110
12	0001101001100101000111111011100000111011001110000001110
13	0001101001100101000111111011100000111011001110000001110
14	0001101001100101000111111011100000110011001110000001110
15	0001101001100101000111111011110001110011001110110011110
16	0001101001100101000111111111110001110011001110110011110
17	0001101001100101000111111111110001110011001110110011110
18	0001101001100101001111111111110001110011001110110011110
19	0001101001100101001111111111110001111011001110110011110
20	0001101001100101001111111111110001111011001110110011110
21	0001101001100101001111111111110001111011001110110011110
22	0001101001100100001111111111110000111011001110100000100
23	0001101001100100001111110111000001110110011101000000100
24	0001101001100100000111011011100000010011001110100000000

respectively. The total operation cost and average CPU time are $1,713,321.62 and 12,986.48 s, approximately. Figures 3 and 4 exposed the Good convergence behaviour and rapid decrease of the objective function (cost function) for UC and SCUC problems. The results in Table 5 show the supremacy of the proposed hybrid NACO-CGSA method with the other methods which are presented in the literature.

5.1.3 RSCUC Problem-Wind Integrated Thermal Power System

In case 3, reliability RSCUC problem is solved using NACO-CGSA for 118 bus system. The reliability level of system is maintained at each hour based on the case 2. The reliability data for the conventional generating unit is adapted from the ref. http://motor.ece.iit.edu/data/IEEE118data.xls. Here it is assumed that the wind farm with 200 identical 1.75 MW WTG units with FOR of 0.04. The wind penetration in this case is 5 % of the system load at each hour of the dispatch period. The total

Table 4 Commitment schedule of SCUCP IEEE 118-bus

Hr	Total operating cost = $1713321.62
	Unit ON/OFF status
1	000110000110000000011001101110000001001100111000000000
2	000110000110000000011001101110000001001100111000000000
3	000110000110000000011001101110000001001100111000000000
4	000110000110000000011001101110000001001100111000000000
5	000110000110000000011001101110000001001100111000000000
6	000110100110000000011001101110000001001100111000000000
7	000110100110000000011001101110000001001100111000000000
8	000110100110000100011011101110000001001100111000000000
9	000110100110010100011111101110000011101100111000001110
10	000110100110010100011111101110000011101100111000001110
11	000110100110010100011111101110000011101100111000001110
12	000110100110010100011111101110000011101100111000001110
13	000110100110010100011111101110000011101100111000001110
14	000110100110010100011111101111000111001100111011001110
15	000110100110010100011111111111000111001100111011001110
16	000110100110010100011111111111000111001100111011001110
17	000110100110010100111111111111000111001100111011001110
18	000110100110010100111111111111000111101100111011001110
19	000110100110010100111111111111000111101100111011001110
20	111110100111110100111111111111000111101100111011001110
21	111110100111110100111111111111000111101100111011001110
22	000110100110010000111111111111000011101100111010000100
23	000110100110010000111111011100000111011001110100000100
24	000110100110010000111011011100000110011001110100000000

Table 5 Comparison of results for IEEE 118-bus system

Solution method	Total operating cost ($)	
	UCP	SCUCP
SDP [21]	1,645,444.98	1,714,255.63
CGSA-LR [22]	1,644,269.00	–
Hybrid NACO-CGSA	1,644,263.13	1,713,321.62

number of dispatching unit in the system is fifty five, including one wind energy resources. It is noted that, a cost saving of 1707602.62 $ is obtained per day for the proposed hybrid generation system and therefore a significant amount of saving will be reflected per annum.

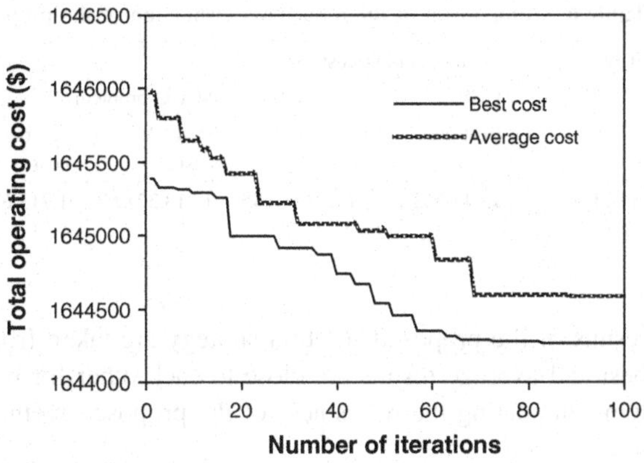

Fig. 3 Convergence graph for UCP

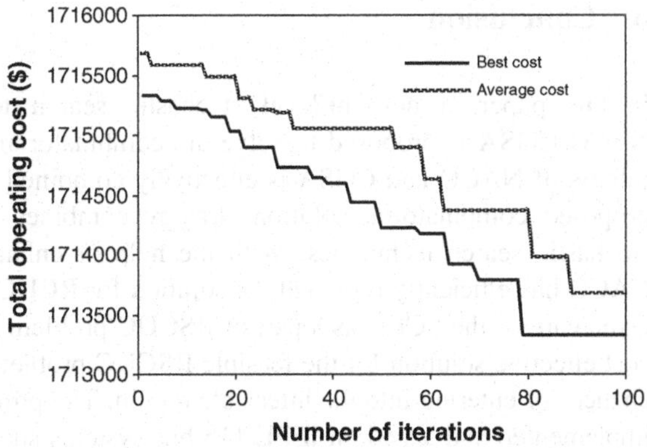

Fig. 4 Convergence graph for SCUCP

5.2 Performance Analysis

The results achieved with the proposed approach is shown in Table 5. They are the best among the 30 trial runs. The proposed approach reveals its supremacy for both UC and SCUC problems when compared with the other methods available in the literature. Also, in the SCUC problem, cost of each method is more than its UCP cost due to the addition of more practical constraints in the SCUC model. With the same reliability level, greater cost saving is achieved by integrating thermal and wind power. This proposed work presented, i.e., UCP, SCUC and RSCUC problem satisfy all the constraints based on the suggested hybrid NACO-CGSA method. The convergence graph of Figs. 3 and 4 show the clearness for the best cost and average cost of the test systems and rapid decrease of the objective function (cost function) for both UC and SCUC problems of the test systems. Also for search efficiency, robustness is another important factor for a stochastic search technique, since these methods start from random initial points. To estimate the robustness of the proposed method, the best cost and average results are tabulated in Table 6. The tabulated

Table 6 Comparison of the robustness evaluation of the proposed solution strategy

Test systems	Total operating cost ($)					
	UCP		SCUC problem		RSCUC problem	
	Best cost	Average cost	Best cost	Average cost	Best cost	Average cost
118 bus system	1,644,263.13	1,644,580.45	1,713,321.62	1,713,676.9	1,707,602.62	1,711,073.07

results of the proposed solution strategy are taken from 30 trial runs. As seen, the best and average results are close to each other for both the UC and SCUC problems, indicating the robustness of the proposed method.

6 Conclusion

In this paper, a new bi-level stochastic search technique named as hybrid NACO-CGSA is proposed to solve the combinatorial RSCUC problem. The usefulness of NACO and CGSA is effectively combined to address the problem. The proposed combinatorial solution strategy combines the advantages of both the stochastic search techniques. With the help of enhanced evolutionary operators, NACO has efficiently provided the solution for RCUC problem and the CGSA has taken care of the SCED as a part of RSCUC problem. This method has provided a cost effective solution for the feasible RSCUC problem solution space without any chances of entering into the infeasible region. The proposed methodology has been implemented and tested on IEEE 118 bus systems successfully. With the proposed method being executed for UC problem, operating cost is $1,644,263.13 and the cost saving is $1181.95 compared with the SDP method. Moreover, with the hybrid NACO-CGSA method applied to SCUC problem, saving in cost is $934.01. Also, it is observed that for thermal integrated wind power system, cost saving due to the wind power is exorbitantly higher than the cost incurred due to additional reserves and increase in losses. It is easily visible from the numerical results obtained with the application of NACO-CGSA approach. Therefore, the proposed hybrid NACO-CGSA approach can be readily implemented in the load dispatch centers for greater saving in cost.

References

1. Padhy, N.P.: Unit commitment—A bibliographical survey. IEEE Trans. Power Syst. **19**(2), 1196–1205 (2004)
2. Saravanan, B., Das, S., Sikri, S., Kothari, D.P.: A solution to the unit commitment problem—a review. In: Weng, S., Ni, W., Peng, S. (eds.) Frontiers in Energy, vol. 7, Issue 2, pp. 223–236. Springer, Heidelberg (2013)

3. Singhal, P.K., Sharma, R.N.: Dynamic programming approach for large scale unit commitment problem. In: Proceedings of International Conference on Communication Systems and Network Technologies, pp. 714–717. Katra, Jammu (2011)
4. Chuang, C.S., Chang, G.W.: Lagrangian relaxation based unit commitment considering fast response reserve constraints. Energy Power Eng. **5**, 970–974 (2013)
5. Chang, G.W., Tsai, Y.D., Lai, C.Y., Chung, J.S.: A practical mixed integer linear programming based approach for unit commitment. In: Proceedings of IEEE Power Engineering Society General Meeting, pp. 221–225. Piscataway, USA (2004)
6. Pan, Q., He, X., Cai, Y.Z., Wang, Z.H., Su, F.: Improved real-coded genetic algorithm solution for unit commitment problem considering energy saving and emission reduction demands. In: Zheng, H. (eds) Journal of Shanghai Jiaotong University (Science), vol. 20, Issue 2, pp. 218–223, Springer, Heidelberg (2015)
7. Purushothama, G.K., Jenkins, L.: Simulated annealing with local search—A hybrid algorithm for unit commitment. IEEE Trans. Power Syst. **18**(1), 273–278 (2003)
8. Saneifard, S., Prasad, N.R., Smolleck, H.A.: A fuzzy logic approach to unit commitment. IEEE Trans. Power Syst. **12**, 988–995 (1997)
9. Gaddam, R.R., Jain, A., Belede, L.: A PSO based smart unit commitment strategy for power systems including solar energy. In: Bansal, J.C., Singh, P.K., Deep, K., Pant, M., Nagar, A. (eds.) (BIC-TA 2012) Volume 201 of the series Advances in Intelligent Systems and Computing, pp. 531–542 (2012)
10. Stutze, T., Lopez-Ibanez, M., Pellegrini, P., Maur, M., Montes de Oca, M.A., Birattari, M., Dorigo, M.: Parameter adaptation in ant colony optimization. In: Hamadi, Y., Momfroy, E., Saubion, F. (eds.) Autonomous Search. Springer, Berlin (2012)
11. Hao, Z.-F., Huang, H., Qin, Y., Cai, R.: An ACO algorithm with adaptive volatility rate of pheromone trail. In: Shi, Y., van Albada, G.D., Dongarra, J., Sloot, P.M.A. (eds.) ICCS 2007, Part IV. LNCS, vol. 4490, pp. 1167–1170. Springer, Heidelberg(2007)
12. Amjady, N., Nasiri-Rad, H.: Security constrained unit commitment by a new adaptive hybrid stochastic search technique. Energy Conv. Manage. **52**, 1097–1106 (2009)
13. Li, Z., Shahidehpour, M.: Security-constrained unit commitment for simultaneous clearing of energy and ancillary services markets. IEEE Trans. Power Syst. **20**, 1079–1088 (2005)
14. Senthil Kumar, V., Mohan, M.R.: Solution to security constrained unit commitment problem using genetic algorithm. Electr. Power Energy Syst. **32**, 117–125 (2010)
15. Christopher Columbus, C., Chandrasekaran, K., Simon, S.P.: Nodal ant colony optimization for solving profit based unit commitment problem for GENCOs. Appl. Soft Comput. **12**, 145–160 (2012)
16. Rashedi, E., Nezamabadi, H., Saryazdi, S.: GSA: a gravitational search algorithm. Inf. Sci. **178**, 2232–2248 (2009)
17. Halliday, D., Resnick, R., Walker, J.: Fundamentals of Physics. John Wiley and Sons, New York (1993)
18. Hu, P., Karki, R., Billinton, R.: Reliability evaluation of generating systems containing wind power and energy storage. IET Gener. Transm. Distri. **3**, 783–791 (2009)
19. Fu, Y., Shahidehpour, M., Li, Z.: Security-constrained unit commitment with AC constraints. IEEE Trans. Power Syst. **20**, 1538–1550 (2005)
20. Grey, A., Sekar, A.: Unified solution of security-constrained unit commitment problem using a linear programming methodology. IET Gener. Transm. Distrib. **2**, 856–867 (2008)
21. Bai, X., Wei, H.: Semi-definite programming-based method for security-constrained unit commitment with operational and optimal power flow constraints. IET Gener. Transm. Distrib. **3**, 182–197 (2009)
22. Chandrasekaran, K., Hemamalini, S., Simon, S.P., Padhy, N.P.: Thermal unit commitment using binary/real coded artificial bee colony algorithm. Elect. Power Syst. Res. **84**, 109–119 (2009)

Human Gait Recognition Using Fuzzy Logic

Parul Arora, Smriti Srivastava, Abhishek Chawla and Shubhkaran Singh

Abstract In this paper, an efficient technique has been implemented for gait based human identification. This paper proposes a human identification system based on human gait signatures extracted using topological analysis and properties of body segments. The gait features extracted are height, hip, neck and knee of the human silhouette and a model-based feature i.e. area under hermite curve of hip and knees. The experimental phase has been conducted on the SOTON covariate database, which comprises of eleven subjects. The database also takes into account different factors that vary in terms of apparel, carrying objects etc. Subject classification is performed using fuzzy logic and compared against the nearest neighbor method. From the conducted experimental results, it can be accomplished that the stated approach is successful in human identification while some analysis prove that specific number of input variables and membership functions help to elevate the accuracy level.

Keywords Gait · Hermite curve · Fuzzy logic · Nearest neighbor

P. Arora (✉) · S. Srivastava · A. Chawla · S. Singh
Netaji Subhas Institute of Technology, New Delhi, India
e-mail: parul.narula@gmail.com

S. Srivastava
e-mail: smriti.nsit@gmail.com

A. Chawla
e-mail: chawla.abhishek96@gmail.com

S. Singh
e-mail: shubhksingh1994@gmail.com

M. Senthilkumar et al. (eds.), *Computational Intelligence, Cyber Security and Computational Models*, Advances in Intelligent Systems and Computing 412, DOI 10.1007/978-981-10-0251-9_27

1 Introduction

Human Gait is an image based human identification technique in which the pattern of synchronized and cyclic movement of the legs is used to recognize the person. It can be used to identify persons by using the video recording of the way a person walks. The gait analysis includes the extraction of some specific unique features of the persons under study to draw conclusions for the identification [1]. Human gait recognition can be performed without the subject being aware of the surveillance. However, the gait unique features may be affected by unwanted footwear, fatigue and injury as well. The continuous increase in the number of people has led to a need for security and surveillance. This has made gait analysis a recent area of research.

Gait recognition has been classified into two types—model based and model free. Model based methods capture the human body configuration in motion. The features that are extracted during the motion are modeled and matched to the designed model features. It amalgamates the knowledge of the body shape and the kinematic gait features extracted from the model parameters. The main advantages of the model-based approach is that it can reliably handle occlusion (especially self-occlusion), noise, scale and rotation well, as opposed to silhouette-based approaches [2]. However, it creates many parameters from extracted gait features and hence resulting in a complex model. Due to the complexity involved in analysis of the model based approach, model free approaches are more widely used. Model free gait recognition methods or appearance based methods work directly on the gait sequences [3, 4]. They don't consider a model for the human body to rebuild human walking steps. They have the advantage of low computational cost in comparison with model-based approaches and they also have the disadvantage of sensitivity to cloth and appearance changing. Normally, its parameters are obtained from the static gait features like centroid, width and height of the silhouette.

Because of the better results and high accuracy, a combination of model based and model free features are extracted and used as input for the fuzzy system for identification of the person. The model free features that are extracted are height, neck, hip and knee. The model based feature is the area under the hermite curve formed from hip and knee joints during the whole motion. Fuzzy inference system is used to classify the subjects based on these evaluated features. For the learning part, rules are made for each of the features extracted and the testing is performed with data available in the knowledge base.

2 Data Preprocessing

In this paper, the SOTON small dataset is used. It consists of gait silhouette sequences of 11 subjects with shoulder bag (accessory) and clothing variations from the side view. Every frame is selectively used such that the start of the gait cycle for each individual is made similar. The complete binary image shown in Fig. 1 is used without the need for defining any region of interest or image cropping [4].

Fig. 1 Sample frame of SOTON data

3 Gait Feature Extraction

3.1 Model Free Features

Human gait is the process in which the center of mass of the whole body switches alternately on the right hand side and left hand side during the whole cycle. Gait recognition can either use the individual frames or several frames over a gait cycle. We have aimed at the individual frame approach, even though increases complexity, greatly improves accuracy rates by computing the variation in heights for the person with each step taken during the entire gait cycle.

A normal human body is a structure where the set ratios with respect to height elucidates the approximate position of various other body joints like neck, shoulder, waist, elbows, pelvis, knees, ankles, etc. [5] We have considered the 2D stick model to determine the heights of the body joints. Since our derived heights are flexible, i.e. continuously changes for each person from frame to frame as he walks, so are our various other positions of the body joints. For horizontal mapping, a bidirectional scanning consisting of calculating the white pixel intensity is used. This gives the location of the desired point. Thus the first pixel (x_s) and last pixel (x_l) on the scan line gives the initial and final point. The total number of pixels is used to determine the required thickness. The exact point location [5] is then obtained by successively dividing with the integer 2 or 4 as shown in Eq. 1.

$$x_{center} = x_s + (x_l - x_s)/2 \tag{1}$$

3.2 Model Based Features

Hermite Curve. We have used a cubic Hermite curve model [6] described by four control points to give a vivid picture of motion statistics in a gait cycle. For a cubic curve, the control points are supposed to be the hip point (starting point), tangent joining hip and left knee (initial tangent), hip point (ending point) and the tangent

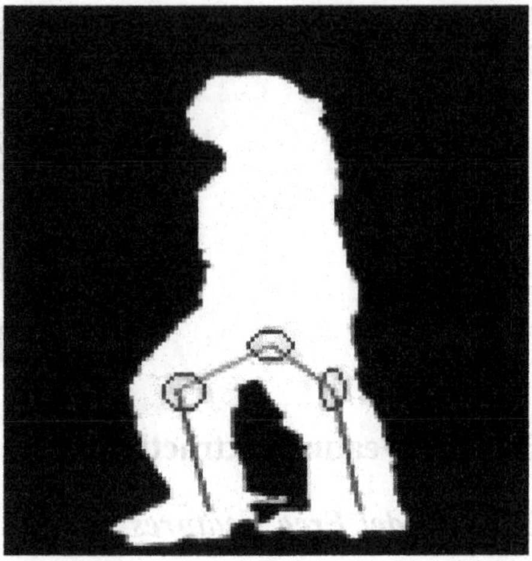

Fig. 2 Hermite curve control points

joining hip and right knee (final tangent) as shown in Fig. 2. The equation defining a parametric cubic curve is shown in Eq. 2:

$$\begin{cases} x(t) = a_3t^3 + a_2t^2 + a_1t + a_0 \\ y(t) = b_3t^3 + b_2t^2 + b_1t + b_0 \\ z(t) = c_3t^3 + c_2t^2 + c_1t + c_0 \end{cases} \quad (2)$$

Here the parametric equations of the curve are given separately as x(t), y(t), z(t) and 200 points are taken to form the curve.

Cubic Hermite splines are typically used for numeric data interpolation specified at given argument values, to obtain a smooth curve as shown in Fig. 3.

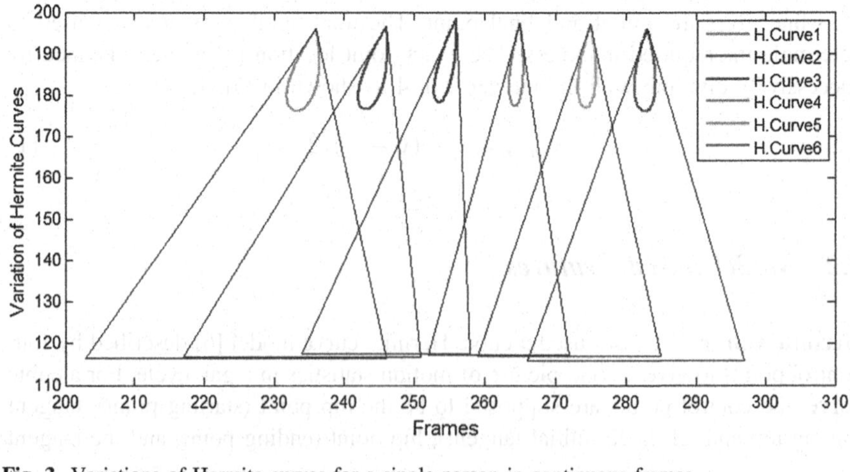

Fig. 3 Variations of Hermite curves for a single person in continuous frames

The resultant equation of Hermite curve is given below in Eq. 3.

$$x(t) = (2t^3 - 3t^2 + 1) * xhip - (2t^3 - 3t^2) * xhip + (t^3 - 2t^2 + t) * (xleftknee - xhip) + (t^3 - t^2) * (xhip - xrightknee) \tag{3}$$

4 Gait Recognition Using Fuzzy Approach

Fuzzy logic is an approach of determining the degree of truthfulness rather than the Boolean logic of 0 or 1. It is a form of knowledge representation that is applicable to notions which are not precisely defined. The fuzzy input variables can take any real value between 0 and 1. This approach is used for identifying persons based on the extracted features [7].

4.1 Membership Functions

The membership function $\mu_A(x)$ in fuzzy logic represents the quantification of grade of membership of the elements x in fuzzy set A. The extracted features used as inputs are given overlapped Gaussian membership function and non-overlapped triangular membership function are used for the persons to be identified. The Gaussian membership function is specified by two parameters [m, σ] and is defined as follows:

$$function(x : m, \sigma) = e^{\frac{-(x-m)^2}{\sigma^2}} \tag{4}$$

where m and σ denote the mean and standard deviation of the function, respectively. As the Gaussian functions facilitate smooth, continuously differentiable hypersurfaces of a fuzzy mode. It provides less degree of freedom. Moreover, it is a general representation of membership functions, where the same can be converted into other shapes i.e. trapezoidal, triangular etc. by changing the term α in $e^{-\left(\frac{x-c}{\sigma}\right)^{\alpha}}$.

4.2 Mamdani Fuzzy Inference

Mamdani fuzzy module is the most commonly used fuzzy interference system (FIS). The purpose of fuzzification is to map the inputs between 0 and 1 using a set of input membership functions. The input membership functions are used to demonstrate fuzzy concepts such as very small, small, average, large and very large etc. Maximum aggregation method is used by which the fuzzy sets that represent the outputs of each rule are combined into a single fuzzy set. The defuzzification is done by the Mean of maximum method, in other words, the fuzzy controller uses the typical value of the consequent term of the most valid rule as the crisp output value.

4.3 Methodology

Our proposed methodology focuses on the feature: Area under the curve formed by Hermite parametric equations for various frames. The control points used here are the hip, right knee and left knee. This feature is used in combination with many other extracted features such as the height, neck height and hip to make a fuzzy inference system. The whole range for each feature is determined and used as fuzzy inputs for the identification of the persons. These inputs are then equally divided into 5 fuzzy sets as very small, small, average, large and very large to make the rules for identification of persons. We have considered the normal gait sequences, which consists of 3 samples. We used 2 samples for training and 1 sample for testing. The results are based on the rule structure shown below.

Rule Structure:

If (neck is very small) and (knee is very small) and (hip is very small) and (area is very small) and (height is small) then (output1 is person1).

4.4 Algorithm

The main steps of our methodology is framed into an algorithm, explained below:

1. The features like height, neck, hip, knee and hermite curve for 25 frames are extracted from the SOTON database.
2. The area under hermite curve is determined by using the trapezoidal function in MATLAB.
3. The extracted features have used Gaussian membership functions in input range.
4. The input range is determined from mean of the first two samples of each person.
5. Non-overlapped triangular membership functions are used in output range for all eleven persons.
6. The aforesaid rule structure is followed to make rules for five inputs to identify each person.
7. The feature inputs of third sample are checked using the fuzzy inference system and results are recorded accordingly.

5 Experimental Results

To show the efficacy of our work, the whole work is analysed in three experiments. First experiment is done to show the performance of the proposed features. Second experiment is done to reflect the effectiveness of our proposed classifier and the third experiment is done to show the importance of number of fuzzy sets.

5.1 Experiment 1

To show the performance of particular feature, we have used different combinations of input features at 2 levels. At first level, we have combined five inputs in the group of three at a time, which results in 10 groups. Table 1 shows the identification of subjects (P1–P11) with different feature combinations of 3 at a time using fuzzy logic. The groups (1–10) refer to the various combinations mentioned below:

1. Neck, knee and hip 2. Neck, knee and area 3. Neck, knee and height 4. Knee, area and height 5. Knee, hip and area 6. Knee, hip and height 7. Hip, area and height 8. Neck, hip and area 9. Neck, hip and height 10. Neck, area and height.

The recognition of a person is identified by symbols ✓ and × represents that person is identified or not.

Correct classification rate (CCR) can be expressed as follows:

$$R.R. = \frac{N_c}{N} \times 100\,(\%) \tag{5}$$

Where N_c is the total number of subjects which are correctly recognized. N represents the total number of gait subjects.

Figure 4 shows the recognition rate of each combination. The combination of neck, knee, area under Hermite (group-2) and knee, area under Hermite, height

Table 1 Recognition results from proposed feature combinations

Groups	P1	P2	P3	P4	P5	P6	P7	P8	P9	P10	P11	R. R (%)
1	×	×	×	✓	✓	×	✓	×	×	×	✓	36.3
2	✓	×	✓	✓	✓	×	✓	×	✓	✓	✓	**72.7**
3	✓	×	✓	×	✓	✓	✓	×	✓	×	×	54.5
4	✓	✓	✓	✓	×	×	✓	×	✓	✓	✓	**72.7**
5	✓	×	×	✓	×	×	✓	×	✓	✓	✓	54.5
6	✓	×	×	✓	✓	×	✓	×	✓	×	×	45.4
7	✓	✓	×	✓	×	×	✓	×	✓	×	×	45.4
8	✓	×	×	✓	✓	×	✓	×	✓	✓	✓	**63.6**
9	✓	×	×	✓	✓	×	✓	×	×	✓	×	45.4
10	✓	×	✓	✓	×	×	✓	✓	✓	×	✓	**63.6**

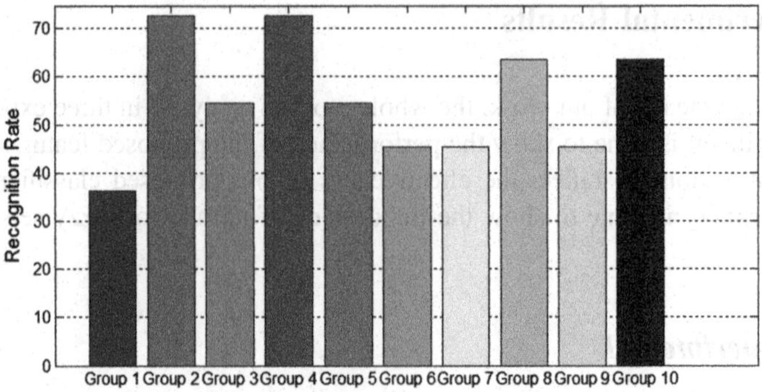

Fig. 4 Correct recognition rate of each combination of 3 inputs

(group-4) provides similar and most accurate results when applied as inputs than the other combinations. Hence, the enhanced accuracy of area under Hermite and knee inputs can be well noted. Second highest recognition accuracy is achieved in group 8 and group 10, which also shows the involvement of neck and area under Hermite, so overall Hermite area is the best feature so far.

Similarly, for four inputs, an analysis can be drawn at second level. There are 5 unique combinations available shown below.

1. Neck, knee, area and height 2. Neck, knee, area and hip 3. Neck, area, height and hip 4. Neck, knee, hip and height 5. Knee, area, height and hip.

Table 2 gives the details of subjects identified in each group. Figure 5 suggests that the combination of neck, knee, area, hip i.e. group 2 gives best recognition rates among the 5 combinations mentioned and group 4 (combination of Neck, knee, hip and height) reflects the absence of area in his accuracy value. This area feature captures the dynamics of the structural characteristics over the gait cycles of different individuals thus possessing the discrimination power.

Table 2 Recognition results for 4 input combinations

Groups	P1	P2	P3	P4	P5	P6	P7	P8	P9	P10	P11	R. R (%)
1	✓	×	✓	✓	✓	×	✓	×	✓	×	×	54.5
2	✓	×	✓	✓	✓	×	✓	×	✓	✓	✓	**72.7**
3	✓	×	×	✓	×	×	✓	×	✓	×	✓	45.4
4	✓	×	×	✓	✓	×	✓	×	×	×	×	36.3
5	✓	✓	✓	✓	×	×	✓	×	✓	×	✓	63.6

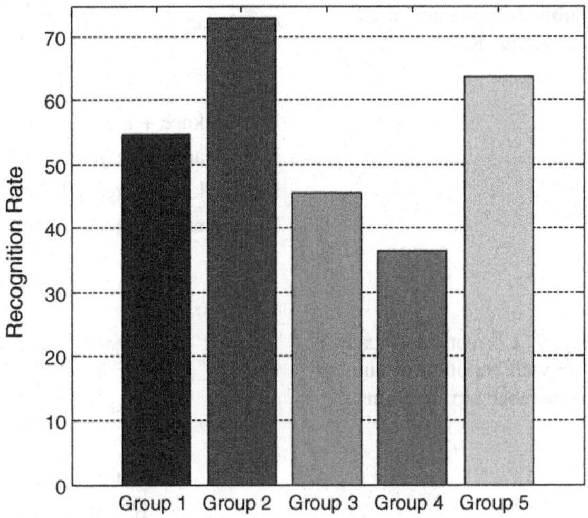

Fig. 5 Correct recognition rate of each combination of 4 inputs

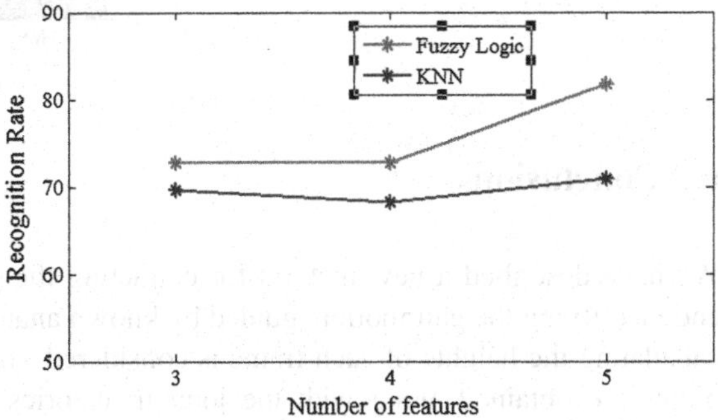

Fig. 6 Comparison between KNN and fuzzy logic

5.2 *Experiment 2*

Here, we have compared the combination which gave the best result for 3 input and 4 input combinations found using fuzzy logic with the nearest neighbor (NN) method and graphically depicted in Fig. 6 (Table 3).

5.3 *Experiment 3*

We have divided input range into three, four and five fuzzy sets to show the importance of them. From Fig. 7, it is observed that five fuzzy sets give better accuracy.

Table 3 Comparison of fuzzy with KNN

Combination	Classification rate (%)	
	KNN	Fuzzy
Neck + knee + area	69.69	72.73
Knee + area + height	69.67	72.73
Neck + knee + area + height	68.18	72.73
Knee + neck + area + height + hip	70.91	81.82

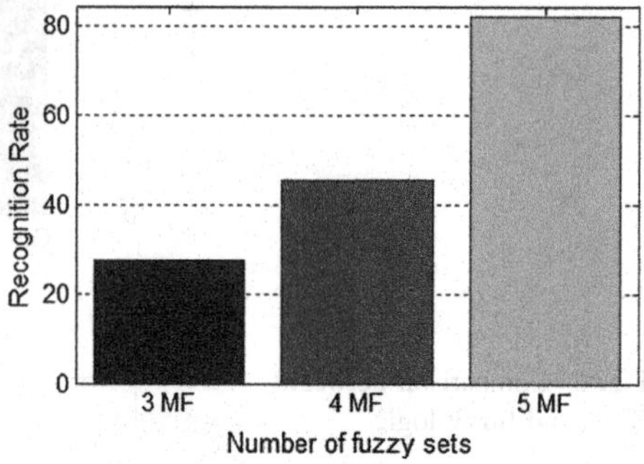

Fig. 7 Correct classification rate with variation in number of membership functions

6 Conclusions

We have described a new method for extracting the gait signatures for analyzing and identifying the gait motion, guided by known anatomy. An adaptive method for calculating the heights of each frame is considered. Subsequently, all other control points are obtained along with the joint trajectories. The resultant body control points are used to create the Hermite model for gait analysis. We have done an exhaustive study to analyze the importance of our proposed feature and classifier and it is concluded that the area under hermite curve is one of the features developed. The fuzzy approach gives better result than the nearest neighbor approach.

References

1. Dawson, M.R.: Gait Recognition. Final Thesis Report, Department of Computing, Imperial College of Science, Technology & Medicine, London (2002)
2. Attwells, R.L., Birrell, S.A., Hooper, R.H., Mansfield, N.J.: Influence of carrying heavy loads on soldiers' posture. Movements Gait. Ergonomics **49**(14), 1527–1537 (2006)
3. Arora, P., Srivastava, S.: Gait Recognition Using Gait Gaussian Image. In: 2nd International Conference on Signal Processing and Integrated Networks, pp. 915–918. IEEE, India (2015)
4. Arora, P., Hanmandlu, M., Srivastava, S.: Gait based authentication using gait information image features. Pattern Recogn. Lett. (2015)

5. Yoo, J.H., Nixon, M.S., Harris, C.J.: Extracting human gait signatures by body segment properties. In: 5th IEEE Southwest Symposium on Image Analysis and Interpretation, pp. 35–39. IEEE Press, Sante Fe, New Mexico (2002)
6. Hearn, D., Baker, M.P.: Computer graphics. C version, second edition thirteenth impression Pearson education India
7. Maduko, E.I.: Pattern recognition of human gait signatures. Thesis, university of Texas, El Paso

Detection and Diagnosis of Dilated and Hypertrophic Cardiomyopathy by Echocardiogram Sequences Analysis

G.N. Balaji, T.S. Subashini, N. Chidambaram and E. Balasubramaiyan

Abstract Automating the detection and diagnosis of cardiovascular diseases using echocardiogram sequences is a challenging task because of the presence of speckle noise, less information and movement of chambers. In this paper an attempt has been made to classify the normal hearts, and hearts affected by dilated cardiomyopathy (DCM) and hypertrophic cardiomyopathy (HCM) by automating the segmentation of left ventricle (LV). The segmented LV from the diastolic frames of echocardiogram sequences alone is used for extracting features. The statistical features and Zernike moment features are obtained from extracted diastolic LV and classified using the classifiers namely support vector machine (SVM), back propagation neural network (BPNN) and probabilistic neural network (PNN). An intensive examination over 60 echocardiogram sequences reveals that the proposed method performs well in classifying normal hearts and hearts affected by DCM and HCM. Among the classifiers used the BPNN classifier with the combination of Zernike moment features gave an highest accuracy of 92.08 %.

Keywords Echocardiogram · Left ventricle (LV) · Dilated cardiomyopathy (DCM) · Hypertrophic cardiomyopathy (HCM) · Support vector machine (SVM) · Back propagation neural network (BPNN) · Probabilistic neural network (PNN)

G.N. Balaji (✉) · T.S. Subashini
Department of CSE, Annamalai University, Annamalai Nagar, India
e-mail: balaji.gnb@gmail.com

T.S. Subashini
e-mail: rtramsuba@gmail.com

N. Chidambaram · E. Balasubramaiyan
Department of Cardiology, Annamalai University, Annamalai Nagar, India

M. Senthilkumar et al. (eds.), *Computational Intelligence, Cyber Security and Computational Models*, Advances in Intelligent Systems and Computing 412, DOI 10.1007/978-981-10-0251-9_28

1 Introduction

The Dilated Cardiomyopathy (DCM) and Hypertrophic Cardiomyopathy (HCM) are principal types of cardiomyopathy with predominant myocardial involvement [1]. Dilated cardiomyopathy is a muscular heart disease, usually starting in the heart's main pumping chamber (left ventricle). The left ventricle pumps oxygenated blood to the entire body through the aorta. The amount of blood which is pumped into aorta during each cardiac cycle, referred to as Stroke Volume (SV), depends on the change produced in LV Volume. Thus the left ventricle shape plays an important role in diagnosis and in arriving at the therapeutic index for evaluation of cardiac diseases like cardiomyopathy.

The cardiac cycle can be divided into a contracting phase (systole) and a relaxing phase (diastole)

(1) In early systole as the myocardium starts to compress, the tricuspid and mitral valves close.
(2) At the end of systole the pulmonary and aortic valves open and blood flows out through the aorta and the pulmonary artery.
(3) In the beginning of diastole the ventricular muscles relaxes and the atrio-ventricular-plane springs upwards.
(4) At the end of diastole, the pressure is higher in the atria than in the ventricles. The tricuspid and mitral valves open up and blood flows into the ventricles.

In the case of dilated cardiomyopathy the heart can't pump blood as a healthy heart could, since the ventricle stretches and thins (dilates), where in the case of hypertrophic cardiomyopathy the ventricular septum gets thicker. Figure 1 shows the (a) normal heart, (b) Dilated Cardiomyopathy and (c) Hypertrophic Cardiomyopathy. To visualize and asses the heart anatomy and function, echocardiography is used. Echocardiography is of low cost, portable, real time, flexible modality to image the heart and has no known negative effects for patients. The clinical parameters such as Ejection Fraction (EF), Myocardial Mass (MM) are calculated by the cardiologist from this image which helps in improving diagnostic

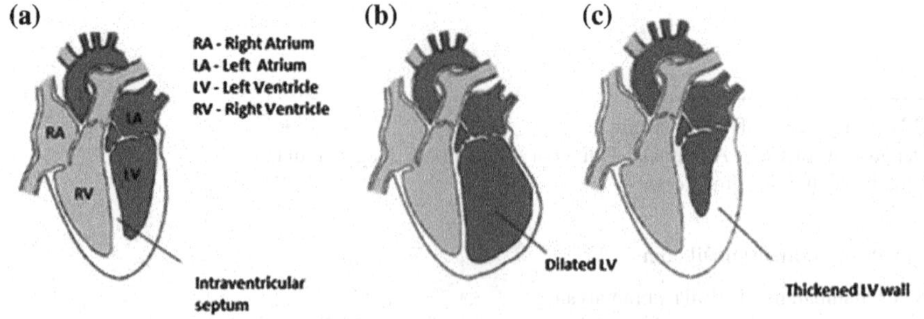

Fig. 1 **a** Normal heart. **b** Dilated cardiomyopathy and **c** Hypertrophic cardiomyopathy

accuracy. Though several views of the heart could be imaged the main echocardiogram views are parasternal, apical and subcostal view [2].

In this paper features extracted from end-diastolic frame i.e., segmented relaxed LV is used to classify normal, dilated cardiomyopathy and hypertrophic cardiomyopathy. The changes in LV parameters of patients with cardiac abnormality with that of normal subjects can be used in determining the type of abnormality [3].

The paper is organized as follows: In Sect. 2 previous work done in this area is described, Sect. 3 explains the various stages involved in the development of the proposed system, the experimental results are discussed in Sects. 4, and 5 concludes the paper.

2 Previous Work

Measurements of cardiac chamber size, ventricular mass and LV function are the most clinically important and most commonly requested tasks of echocardiography and are important clinical variables with respect to diagnosis, management, and to evaluating prognosis in patients with cardiac disease [4]. Heart chamber volumes provide information which will aid the cardiologist in disease diagnosis [5, 6]. A neural network based border detection method was suggested by authors in [7]. The work in [8] reports an interesting approach to detect LV boundary of short axis echocardiographic sequences using a multiple active contour model which is an extension to the original model proposed in [9]. The application of neural network for echocardiography image segmentation is desirable because neural network is insensitive to noise, which is a major problem of echocardiography images. Turning angle functions [10] have been used to describe the boundary of a shape by measuring the angle of the counter-clockwise tangent as a function of the arc-length. The centroid-radii model [11] employs an ordered sequence of radii lengths drawn from a shape's centroid to its periphery at regular intervals. To perform CBIR, some low level feature descriptors which include color, texture and shape features are used. Some commonly used low level feature description methods are described in [12]. Characterization of spatio-temporal motion patterns of heart regions from echocardiography sequences is used for disease discrimination in [13]. An effort has been made by the authors in [14] to automate the detection and diagnosis of myocardial ischemia using composite wall motion analysis, which succeeds in classifying normal and abnormal (Myocardial Ischemia) hearts.

3 Methodology

In this paper the left ventricular segmentation is automated in parasternal short axis echocardiogram sequences using which the end diastolic left ventricular (EDLV) frame is extracted and the features (statistical and Zernike moments) are obtained

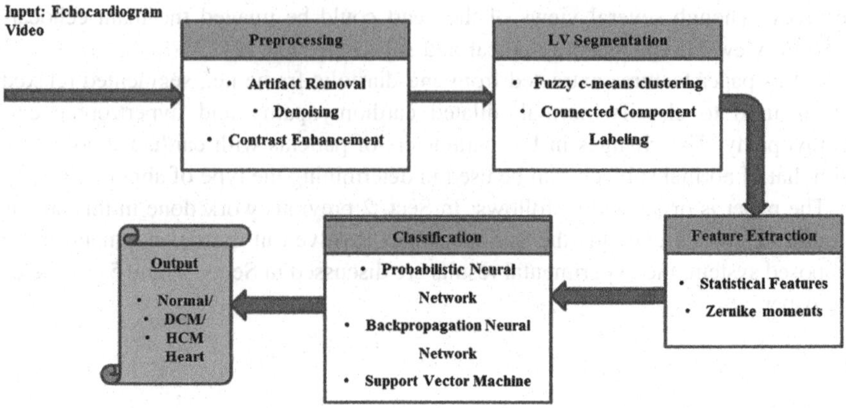

Fig. 2 Block diagram of the proposed system

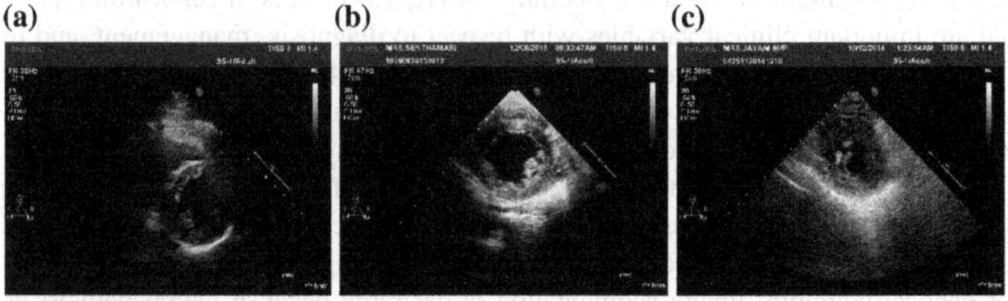

Fig. 3 Sample Echocardiogram frames. **a** Normal heart. **b** Heart with DCM and **c** Heart with HCM

from the EDLV. The extracted features are fed into the BPNN, SVM and PNN classifiers to classify the normal hearts and hearts affected by DCM and HCM.

The block diagram of the proposed system is shown in Fig. 2. The proposed method involves four steps namely preprocessing, LV segmentation feature extraction and classification. Figure 3 shows a sample of the normal, DCM and HCM frame extracted from the echocardiogram video. It can be seen that the size of the left ventricle in DCM and HCM varies when compared to the left ventricle of the normal heart.

3.1 Preprocessing

When the echocardiogram sequences are given as an input to the proposed system, all the frames are extracted and the artifact is removed in order to make the segmentation easier. The artifacts are undesirable labels and wedges which hinders the

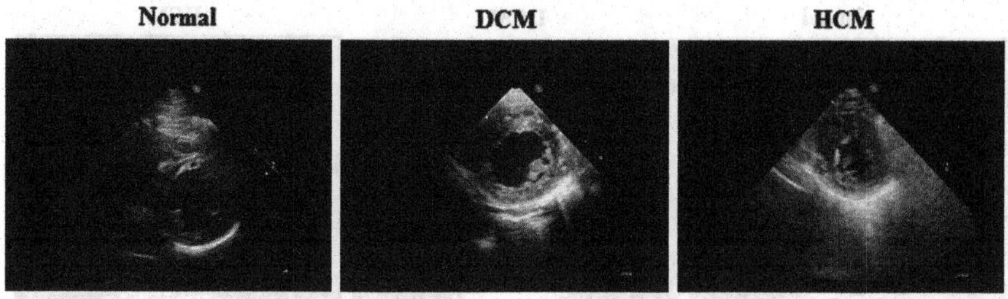

Fig. 4 Sample frames after artifact removal and denoising

segmentation process. After empirically analyzing numerous echocardiogram frames, the pixels lying on the boundaries at a width less than or equal to 80 pixels are set to black. In the next step, denoising is carried out using Speckle Reducing Anisotropic Diffusion (SRAD) filter. The SRAD filter removes the speckle noise present in the frames which makes the segmentation easier. The sample frames after preprocessing is shown in Fig. 4.

3.2 Segmentation

Following the preprocessing step segmentation of left ventricle is carried out and the process involved in segmenting the left ventricle is shown in Fig. 5.

FCM clustering is used to partition N objects into C classes. In our method, N is equal to the number of pixels in the image i.e. $N = X \times Y$ and $C = 3$ for 3-class FCM clustering. The FCM algorithm uses iterative optimization of an objective function based on a weighted similarity measure between the pixels in the image and each of the C-cluster centers. After performing FCM clustering, finally each

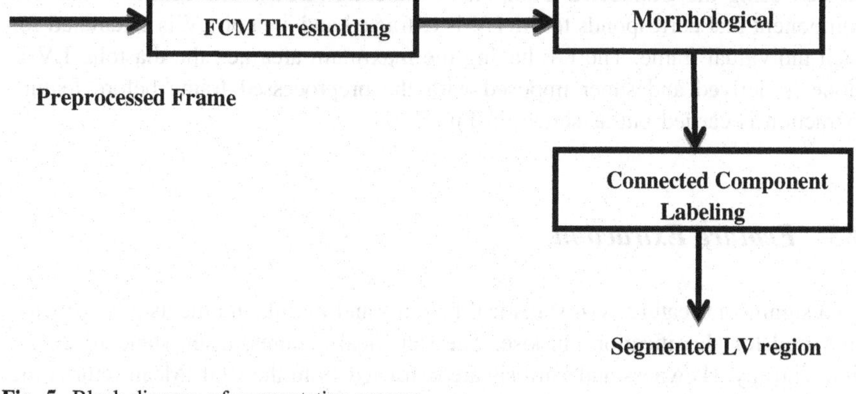

Fig. 5 Block diagram of segmentation process

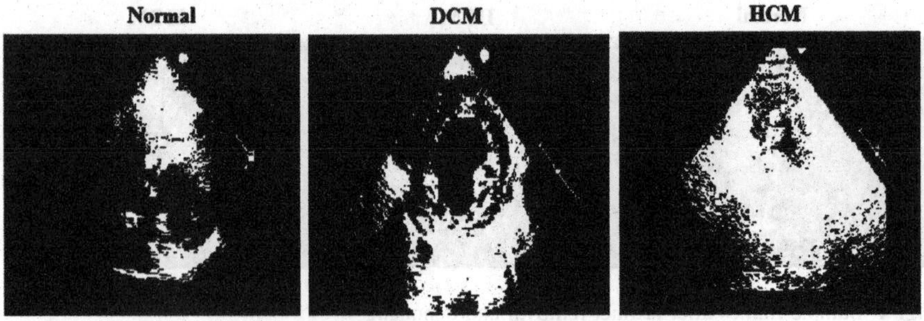

Fig. 6 Result of FCM thresholding

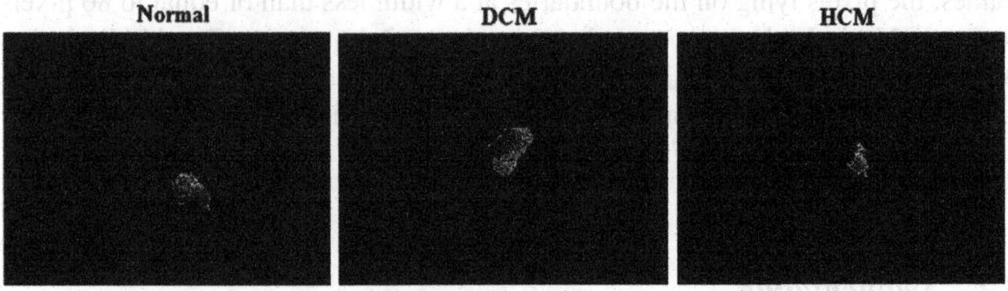

Fig. 7 Left ventricle extracted and super imposed

pixel is assigned to the cluster for which its membership value is maximum. Based on the intensity distribution obtained using histogram of the image, the threshold value is calculated by taking mean of maximum of cluster 1 and minimum of cluster 2 or maximum of cluster 2 and minimum of cluster 3. This method of threshold selection takes into account the intensity distribution in the image. This choice helps in obtaining optimum threshold values for different images obtained under different conditions. The result of FCM thresholding is shown in Fig. 6. To obtain the cardiac cavity (left ventricle) the connected components in the resultant image is labeled using the Connected Component Labeling (CCL). The central connected component that corresponds to the LV is extracted and area of LV is calculated for each individual frame. The LV having the maximum area i.e., the diastolic LV is alone is derived and super imposed with the preprocessed frame before feature extraction is carried out as shown in Fig. 7.

3.3 Feature Extraction

Two significant features viz, statistical features and Zernike moments are extracted and used for classification purpose. The statistical features mean, standard deviation, entropy, skewness and kurtosis are extracted from the ROI. Mean returns the average intensity value of the extracted region of interest. The standard deviation gives the information regarding how the data is dispersed from the mean. Entropy is a statistical measure of randomless that can be used to characterize the texture of the input image. Kurtosis gives an idea about the shape of the probability distribution. Skewness is a measure which tells how the data are symmetrically arranged about its mean. Zernike moments have been suggested to be a good descriptor for shape feature extraction [15–17]. Zernike moments are computed by projecting the image onto a set of complex Zernike polynomials which satisfy the orthogonal property. The Zernike moments computation is described in these steps: computation of radial polynomials, computation of Zernike basis functions and computation of Zernike moments by projecting the image onto the Zernike basis functions. The discrete formulation to obtain Zernike moment of a N × N image is as follows

$$\begin{aligned} Z_{n,m} &= \frac{n+1}{\lambda_N}\sum_{c=0}^{N-1}\sum_{r=0}^{N-1} f(x,y)V_{n,m}^{*}(x,y) \\ &= \frac{n+1}{\lambda_N}\sum_{c=0}^{N-1}\sum_{r=0}^{N-1} f(x,y)R_{n,m}\left(\rho_{xy}\right)e^{-jm\theta_{cr}} \end{aligned} \tag{1}$$

where $0 \le \rho_{xy} \le 1$, and λ_N is a normalization factor. The order of the radial polynomial is represented by a non-negative integer n, while the integer m satisfies the constraints $n - |m| = even$ and $m \le n$ representing the repetition of the azimuthal angle. $R_{n,\ m}$ is radial polynomial and $V_{n,\ m}$ is 2-D Zernike basis function.

The list of Zernike polynomials up to 4th order implemented in this work is given in Table 1.

Table 1 List of Zernike polynomial up to 4th order

Index	N	M	Zernike Polynomials
0	0	0	1
1	1	−1	$2\rho sin\theta$
2	1	1	$2\rho cos\theta$
3	2	−2	$\sqrt{6}\rho^2 sin2\theta$
4	2	0	$\sqrt{3}(2\rho^2 - 1)$
5	2	2	$\sqrt{6}\rho^2 cos2\theta$
6	3	−3	$\sqrt{8}\rho^3 sin3\theta$
7	3	−1	$\sqrt{8}(3\rho^3 - 2\rho)sin\theta$
8	3	1	$\sqrt{8}(3\rho^3 - 2\rho)cos\theta$
9	3	3	$\sqrt{8}\rho^3 cos3\theta$
10	4	−4	$\sqrt{10}\rho^4 sin4\theta$
11	4	−2	$\sqrt{10}(4\rho^4 - 3\rho^2)sin2\theta$
12	4	0	$\sqrt{5}(6\rho^4 - 6\rho^2 + 1)$
13	4	2	$\sqrt{10}(4\rho^4 - 3\rho^2)cos2\theta$
14	4	4	$\sqrt{10}\rho^4 cos4\theta$

3.4 Classification

The literature strongly made with regard to the classification of medical images reveal that supervised learning methods such as neural networks and SVM gives promising classification accuracy compared to other non-supervised techniques used for classification. So in this work, SVM and two neural network architectures PNN and BPNN were employed to determine the efficacy of the statistical features and zernike moment in correctly classifying hearts. Also Experiments were done to see how well SVM, BPNN and PNN were able to model the extracted features to classify accurately the normal hearts and hearts affected by DCM and HCM.

3.4.1 Back Propagation Neural Network (BPNN)

The backpropagation networks are typically multi-layer ANNs, usually with an input layer, one or more hidden layers and an output layer. For the hidden layer neurons to serve any useful purpose, they must have non-linear activation (or transfer) function. The most common non-linear activation functions include the: log-sigmoid, tan-sigmoid, Gaussian and softmax transfer functions. In a back propagation neural network (BPNN), learning is formulated as follows.

Firstly, a training pattern is presented to the input layer of the BPNN. The network propagates the input pattern from layer to layer until the output pattern is generated by the neurons in the output layer. If the output pattern is different from the desired output, an error is calculated. This error is then propagated backwards through the network to the input layer. As the error is propagated backwards, the weights connecting the neurons are adjusted by the back propagation algorithm. Figure 8 shows a typical architecture of a BPNN [18].

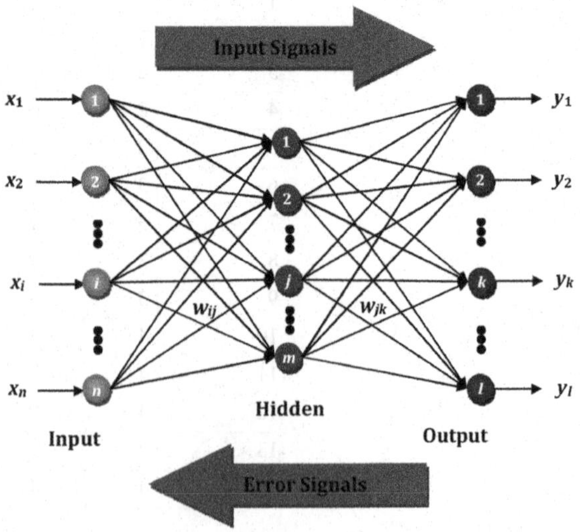

Fig. 8 General architecture of BPNN

3.4.2 Support Vector Machine (SVM)

SVM is a supervised learning algorithm, for binary classification is the simplest form. From the training samples SVM tries to models the two different classes. So that it generalize well to test data. The principle of SVM is to find the hyperplane which maximize the distance between the two classes. The hyperplane generated depends on the samples which are a subset of the two classes. The samples which lie near the hyperplane are called support vectors [19]. Support vectors are the training samples that define the optimal separating hyperplane and are difficult patterns to classify. When the training samples are linearly separable it is easy to classify them, however when the samples are linearly inseparable then the kernel function is used to separate the two classes. Some common kernels are Gaussian radial basis function, polynomial (inhomogeneous), polynomial (homogeneous) and sigmoid kernel. In this proposed work polynomial and radial basis function kernel of SVM is used.

3.4.3 Probabilistic Neural Network (PNN)

PNNs can be used for solving classification problems. In this setting, a training set consisting of known input variables and corresponding outputs, is used to estimate a probability density function (PDF). Each output in the training set belongs to some class. When evaluated on data outside the training set, the PNN then classifies the input variables using the estimated PDF. A class is assigned, corresponding to that with the highest probability of occurrence.

4 Results and Discussion

A database of 70 echocardiogram videos (20 normal, 30 DCM and 20 HCM) videos acquired using Philips iE33 xMATRIX echo system from the Department of Cardiology, Raja Muthaiah Medical College Hospital, Annamalai University is used for this study. Each video are up to 2 s having 46 frames per second. Table 2 compares the performance of Speckle Reducing Anisotropic Diffusion (SRAD) denoising filter proposed in this work with other filters seen in the literature. The SRAD filter removes the speckle noise and at the same time preserves the edge which improves the quality of the echocardiogram frames and this phenomenon leads the SRAD filter to give better PSNR value, as could be seen in Table 2.

The rand index (RI), boundary displacement error (BDE), variations of information (VOI) and global consistency error (GCE) were calculated to evaluate the performance of the proposed segmentation method. If the RI value is higher and GCE, BDE and VOI are lower, then the accuracy of the segmentation method is better [23]. From each of the three class, The measures RI, GCE, BDE and VOI were calculated using the manually segmented LV and the LV segmented using the

Table 2 Performance of denoising filters

Denoising filter	PSNR value	References
Wavelet	29.14	[20]
Wiener	32.14	[21]
High boost filter	33.16	[22]
Speckle reducing anisotropic diffusion	**44.21**	**Proposed method**

Table 3 Segmentation performance measures

Methods	References	RI	GCE	BDE	VOI
Morphological operations + CCL	[22]	0.9728	0.0178	0.1469	10.013
K-means + active contour	[23]	0.9507	0.0239	0.2171	16.6471
FCM + Morphological operations	**Proposed**	**0.9737**	**0.0171**	**0.1400**	**9.3908**

Table 4 Performance measures of BPNN, SVM and PNN with statistical and Zernike moment features

Feature	Classifier	Accuracy	Sensitivity	Specificity
Statistical	BPNN	88.94	84.29	91.47
Zernike moments	BPNN	**92.08**	**88.51**	**93.94**
Statistical	SVM	90.00	85.71	92.31
Zernike moments	SVM	86.87	81.43	89.84
Statistical	PNN	82.29	75.36	86.18
Zernike moments	PNN	80.10	72.86	84.30

proposed method. The average values are depicted in Table 3. Table 3 also compares the performance of the proposed method with existing works and from the table it could be concluded that the proposed segmentation method which employs FCM and CCL is able to detect LV more accurately compared to the other methods.

The extracted statistical and Zernike moment features are fed to the different classifiers namely back propagation neural network, support vector machine and probabilistic neural network. The performance measures accuracy, sensitivity and specificity are used to evaluate the performance of the classifiers and are tabulated in Table 4.

From Table 4 it can be seen that the combination of Zernike moment feature with BPNN classifier gives the best accuracy of 92.08 % with 88.51 % sensitivity and 93.94 % specificity, since the shape of LV varies between the normal heart affected by DCM and HCM, the Zernike moment explains the shape very well and it is modeled effectively by BPNN compared to statistical features. The combination of statistical features with the SVM classifier gives the second highest values while the PNN performs poorly when compared to SVM and BPNN.

5 Conclusion

A new method to automatically detect and diagnose dilated cardiomyopathy and hypertrophic cardiomyopathy is proposed in this paper. The input echocardiogram sequences are preprocessed, segmented and the diastolic frame alone is extracted from the sequence. The statistical and Zernike moments feature are extracted from the diastolic frames and used for classifying the heart as normal or affected by DCM or HCM with the help of the classifiers BPNN, SVM and PNN. Among the classifiers used the BPNN classifier with Zernike moment features performs best and the proposed method can be used, as a diagnostic tool for detecting and diagnosing DCM and HCM. In future, the calculation of global LV parameters can be automated for normal, DCM and HCM which will help in improving the diagnostic accuracy. Content based image retrieval (CBIR) systems could be developed which will help the cardiologist to refer and retrieve the required patient data.

References

1. Verhaert, D., Gabriel, R.S., Johnston, D., Lytle, B.W., Desai, M.Y., Klein, A.L.: Advances in cardiovascular imaging. J. Am. Heart Association, May 2010
2. Thomas, J.D., Popović, Z.B.: Assessment of left ventricular function by cardiac ultrasound. J. Am. Coll. Cardiol. **48**(10), 2012–2025 (2006)
3. SOLVD Investigators: Effect of enalapril on mortality and the development of heart failure in asymptomatic patients with reduced left ventricular ejection fractions. N. Engl. J. Med. **327**, 685–691 (1992)
4. Sun, W., Cetin, M., Chan, R., Reddy, V., Holmvang, G., Chandar, V., Willsky, A.: Segmenting and tracking the left ventricle by learning the dynamics in cardiac images. In: Information Processing in Medical Imaging, pp. 553–565. Springer Berlin Heidelberg (2005)
5. Meluzín, J., Spinarová, L., Bakala, J., Toman, J., Krejcí, J., Hude, M.: Pulsed Doppler tissue imaging of the velocity of tricuspid annular systolic motion; a new, rapid, and non-invasive method of evaluating right ventricular systolic function. Eur. Heart J. **22**, 340–348 (2001)
6. Pritchett, A.M., Jacobsen, S.J., Mahoney, D.W., Rodeheffer, R.J., Bailey, K.R., Redfield, M. M.: Left atrial volume as an index of left atrial size: a population-based study. J. Am. Coll. Cardiol. **41**, 1036–1043 (2003)
7. Wu, E.J.H., De Andrade, M.L., Nicolosi, D.E., Pontes, S.C. Jr.: Artificial neural network: border detection in echocardiography. Med. Biol. Eng. Comput. **46**, 841–848 (2008)
8. Chalana, V., Linker, D.T., Haynor, D.R., Kim, Y.: A multiple active contour model for cardiac boundary detection one echocardiographic sequences. IEEE Trans. Med. Imaging **15**, 290–298 (1996)
9. Kass, M., Witkin, A., Terzopoulos, D.: Snakes: active contour models. Int. J. Comput. Vis. **1**, 321–331 (1988)
10. Arkin, E.M., et al.: An efficiently computable metric for comparing polygonal shapes. IEEE Trans. PAMI **13**(3), 209–216 (1991)
11. Tan, K.L., Ooi, B.C., Thiang, L.F.: Retrieving similar shapes effectively and efficiently. Multimedia Tools Appl. **19**, 111–134 (2003)
12. Parekh, Ranjan: Principles of Multimedia. Tata McGraw Hill, New Delhi (2006)

13. Syeda-Mahmood, T, Wang, F., Beymer, D., London, M., Reddy, R.: Characterizing spatio-temporal patterns for disease discrimination in cardiac echo videos. In: Ayache, N., Ourselin, S., Maeder, A. (eds.) MICCAI 2007, Part I, LNCS 4791, pp. 261–269 (2007)
14. Balaji, G.N., Subashini, T.S., Chidambaram, N.: Detection of heart muscle damage from automated analysis of echocardiogram video. IETE J. Res. **61**(3), 236–243 (2015)
15. Wee, Chong-Yaw, Paramesran, Raveendran: On the computational aspects of Zernike moments. Image Vis. Comput. **25**(6), 967–980 (2007)
16. Tahmasbi, Amir, Saki, Fatemeh, Shokouhi, Shahriar B.: Classification of benign and malignant masses based on Zernike moments. Comput. Biol. Med. **41**(8), 726–735 (2011)
17. Wei, C.-H., Li, Y., Chau, W.Y., Li, C.-T.: Trademark image retrieval using synthetic features for describing global shape and interior structure. Pattern Recogn. **42**(3), 386–394 (2009)
18. Yegnanarayana, B.: Artificial Neural Networks. PHI Learning Pvt. Ltd., New Delhi (2004)
19. Boser, B.E., Guyon, I., Vapnik, V.: A training algorithm for optimal margin classifiers. In: Proceedings of the 5th ACM Workshop on Computational Learning Theory, pp.144–152. ACM Press, New York (1992)
20. Zhang, F., Yoo, Y.M., Mong, K.L., Kim, Y.: Nonlinear diffusion in Laplacian pyramid domain for ultrasonic speckle reduction. IEEE Trans. Med. Imaging **26**(2), 200–211 (2007)
21. Junior, S.A.M., Macchiavello, B., Andrade, M.M., Carvalho, J.L., Carvalho, H.S., Vasconcelos, D.F., Berger, P.A., Da Rocha, A.F., Nascimento, F.A.: Semi-automatic algorithm for construction of the left ventricular area variation curve over a complete cardiac cycle. Biomed. Eng. Online **9**(5), 1–17 (2010)
22. Balaji, G.N., Subashini, T.S.: Detection of cardiac abnormality from measures calculated from segmented left ventricle in ultrasound videos. In: Mining Intelligence and Knowledge Exploration, pp. 251–259, Springer (2013)
23. Mobahi, H., Rao, S.R., Yang, A.Y., Sastry, S.S., Ma, Y.: Segmentation of natural images by texture and boundary compression. Int. J. Comput. Vision **95**(1), 86–98 (2011)

An Elitist Genetic Algorithm Based Extreme Learning Machine

Vimala Alexander and Pethalakshmi Annamalai

Abstract Extreme Learning Machine (ELM) has been proved to be exceptionally fast and achieves more generalized performance for learning Single-hidden Layer Feedforward Neural networks (SLFN). In this paper, a Genetic Algorithm (GA) is proposed to choose the appropriate initial weights, biases and the number of hidden neurons which minimizes the classification error. The proposed GA incorporates a novel elitism approach to avoid local optimum and also speed up GA to satisfy the multi-modal function. The experimental results indicate the superior performance of the proposed algorithm with lower classification error.

Keywords Neural networks · Extreme learning machine · Genetic algorithm · Elitism

1 Introduction

Huang et al. [1] proposed a novel learning method called Extreme Learning Machine (ELM) to train Single-hidden Layer Feedforward Neural networks (SLFNs) faster than any other learning approaches. In case of classical learning algorithms, the learning process is done by iterative process which would be slower, however, the ELM simplifies the learning process in one step through a simple generalized inverse calculation [2]. Unlike other learning methods, ELM doesn't requires any parameters to be tuned except network architecture. However, ELM has its own limitation that the random selection of input weights and biases might

V. Alexander (✉)
Department of Computer Science, Fatima College, Madurai, Tamil Nadu, India
e-mail: cscfcvimala@gmail.com

P. Annamalai
Department of Computer Science, MVM Government Arts College (W), Dindigul, Tamil Nadu, India
e-mail: pethalakshmi@yahoo.com

M. Senthilkumar et al. (eds.), *Computational Intelligence, Cyber Security and Computational Models*, Advances in Intelligent Systems and Computing 412, DOI 10.1007/978-981-10-0251-9_29

reduce the classification performance [3, 4]. Therefore, it is important to modify the ELM to overcome this inadequacy [5, 6]. Global searching methods such as Evolutionary Algorithms (EA) were used in the past decades for optimizing the weights and biases of the neural networks. In [7–11], Differential Evolution (DE) is adopted as a learning algorithm for SLFNs, but there is a chance that the DE algorithms may result in premature or slow convergence. Lahoz et al. [12] and Suresh et al. [13] used a Genetic Algorithm (GA) for optimizing the weights and number of hidden nodes.

This paper proposes a novel GA based learning for ELM. Genetic Algorithms (GAs) have proven useful in solving a variety of search and optimization problems. A major problem might occur in GA is that it might provide local optimum solution. The elitist strategy is widely adopted in the GAs' search processes to recover the chance of finding the global optimal solution. Elitism is able to preserve promising individuals from one generation to the next and maintain the diversity of the population [14, 15]. A novel elitist strategy is proposed in this paper, which improves the performance of the GA towards improving the classification accuracy of ELM.

The rest of the paper is organized as follows: Sect. 2 describes the background terminologies about ELM and GA, and explains the proposed EGA-ELM subsequently. Section 3 discusses the databases and parameter settings used for the experiments. Section 4 brings out the results and analyzes the performance of the proposed system with the existing ELM algorithms. Section 5 concludes the paper.

2 Materials and Methods

2.1 Basic ELM

ELM reduces the learning time of SLFNs by finding the weights using a simple inverse generalization calculation. Consider N arbitrary training samples (x_i, t_i), where x_i are the decision attributes, $x_i = [x_{i1}, x_{i2}, \ldots, x_{in}]^T \in R^n$ and ti are the target values or class attributes $t_i = [t_{i1}, t_{i2}, \ldots, t_{im}]^T \in R^m$. A typical SLFNs with $\tilde{N}$ number of hidden nodes and activation function $f(x)$ are defined as

$$\sum_{i=1}^{\tilde{N}} \alpha_i f_i(x_j) = \sum_{i=1}^{\tilde{N}} \alpha_i f(w_i.x_j + b_i) = o_j, j = 1, \ldots, N \tag{1}$$

where $w_i = [w_{i1}, w_{12}, \ldots, w_{in}]^T$ is the weight matrix between *i*th hidden and input nodes, $\alpha_i = [\alpha_{i1}, \alpha_{i2}, \ldots, \alpha_{im}]^T$ is the weight matrix between *i*th hidden and output nodes, and b_i is the bias value of the *i*th hidden node. $w_i \cdot x_j$ denotes the inner product of w_i and x_j.

The classical learning algorithm repeats the learning process till the network error mean becomes zero, that is

$$\sum_{j=1}^{\tilde{N}} \left\| o_j - t_j \right\| \cong 0 \tag{2}$$

The target value is the mapping between the weights and the activation function as defined below

$$\sum_{i=1}^{\tilde{N}} \alpha_i f\left(w_i \cdot x_j + b_i\right) = t_j, j = 1, \ldots, N \tag{3}$$

The above N equation could be simplified as Hα = T, where

$$H\left(w_1, \ldots, w_{\tilde{N}}, b_1, \ldots, b_{\tilde{N}}, x_1, \ldots, x_N\right) = \begin{bmatrix} f(w_1 \cdot x_1 + b_1) & \cdots & f(w_{\tilde{N}} \cdot x_1 + b_{\tilde{N}}) \\ \vdots & \cdots & \vdots \\ f(w_1 \cdot x_N + b_1) & \cdots & f(w_{\tilde{N}} \cdot x_N + b_{\tilde{N}}) \end{bmatrix}_{N \times \tilde{N}} \tag{4}$$

where H is called the hidden layer output matrix of the neural network [16, 17]; *i*th column of H is the ith hidden node output with respect to inputs x_1, x_2, …, x_N. If the activation function f is infinitely differentiable we can prove that the required number of hidden nodes $\tilde{N} \leq N$. The main objective of any learning algorithms is to find the specific $\hat{w}_i, \hat{b}_i, \hat{\alpha}$ to reduce the network mean error, defined as

$$\left\| H\left(\hat{w}_1, \ldots, \hat{w}_{\tilde{N}}, \hat{b}_1, \ldots, \hat{b}_{\tilde{N}}\right)\hat{\alpha} - T \right\| = \min_{w_i, b_i, \alpha} \left\| H\left(\hat{w}_1, \ldots, \hat{w}_{\tilde{N}}, \hat{b}_1, \ldots, \hat{b}_{\tilde{N}}\right)\hat{\alpha} - T \right\| \tag{5}$$

This can be written as a cost function

$$E = \sum_{j=1}^{N} \left(\sum_{i=1}^{\tilde{N}} \alpha_i f\left(w_i \cdot x_j + b_i\right) - t_j \right)^2 \tag{6}$$

The gradient-descent learning algorithms generally search for the minimum of $H\beta - T$, for that the weights and biases are iteratively adjusted as follows:

$$W_k = W_{k-1} - \eta \frac{\partial E(W)}{\partial W} \tag{7}$$

Here η is a learning rate. BP is one of the popular learning algorithm for SLFNs, which has the following issues [18, 19]:

- The learning rate has to be carefully chosen, the minor η slows down the learning algorithm, and however, if it is too large then the learning becomes unstable.

- Training with more number of iteration leads to inferior performance.
- Gradient-descent based learning is very time-consuming in most applications.

According to Huang et al. [2] the w_i and b_i of SLFNs could be assigned with random values rather than iteratively adjusting them, and the output weights can be estimated as

$$\hat{\alpha} = H^{\dagger}T \tag{8}$$

where $H^{\dagger}$ is the Moore–Penrose (MP) generalized inverse [20] of the hidden layer output matrix H. There are several methods to calculate the MP generalized inverse of a matrix, among that the Singular Value Decomposition (SVD) is known for the best inverse and thus is used in most of the ELM implementations [18, 21]. The ELM can be summarized as follows:

Pseudo code for Basic ELM

Given training set N samples, output function $f(x)$ and $\tilde{N}$ hidden nodes.

1. Randomly assign input weights w_i and bias b_i, $I = 1, \ldots, \tilde{N}$.
2. Calculate the hidden layer output matrix H.
3. Calculate the output weight $\hat{\alpha} = H^{\dagger}T$

2.2 Genetic Algorithm

Holland [22] introduces the basic principle of Genetic Algorithm (GA). GA is one of the best known search and optimization methods widely used to solve complex problems [23, 24]. The detailed description of the real coded GA is given in this section.

Procedure Standard GA

```
t ← 1          // iteration number
P ← initial population with randomly generated real values
Calculate the fitness value of each parent in P, g(P)
while (not termination condition) do
    t ← t + 1
    C ← { }     // Initialize children population
    While |C| < |P| do
        Select a pair of parents for crossover
        Mate the parents to create children c1 and c2
        C ← C ∪ { c1, c2 } and perform mutation
    end
    P ← C, and evaluate P, g(P)
end        // Next generation
```

Initial Population. The initial population is a possible solution set P, which is a set of real values generated randomly, P = {p_1, p_2, ..., p_s}.

Evaluation. A fitness function should be defined to evaluate each chromosome in the population, can be written as fitness = g(P).

Selection. After the fitness value is calculated, the chromosomes are sorted based on their fitness values, then the tournament selection method is widely used for parent selection.

Genetic Operators. The genetic operators are used to construct some new chromosomes (offspring) from their parents after the selection process. They include the crossover and mutation [25]. The crossover operation is applied to exchange the information between two parents, which are selected earlier. There are number of crossover operators are available, here we use arithmetical crossover. The crossed offspring will then apply with mutation operation, to change the genes of the chromosomes. Here we have used uniform mutation operation.

After the operations of selection, crossover and mutation, a new population is generated. The next iteration will continue with this new population and repeat the process. Each time the best chromosomes from the current or new population is selected and compared with the best from the previous population to maintain the global best chromosome. This iterative process could be stopped either the result converged or the number of iterations exceeds the maximum limit.

Elitism—The genetic operations such as crossover and mutation may produces weaker children than the parents. But the Evolutionary Algorithms (EA) could recover from this problem after consequent iterations but there is no assurance. Elitism is the known solution to solve this problem. The basic idea is to retain a copy of best parents from the previous population to the subsequent populations. Those parents are copied without changing them, helps to recover EA from re-discovering weaker children. Those elitist parents are still eligible to act as parents for the crossover and mutation operations [26]. Elitism can be implemented by producing only (s—ep) children in each generation, where ‘s’ is the population size and ‘ep’ is the user-defined number of elite parents. In this case, at each iteration, after the evaluation step, the parents are sorted based on their fitness values, and ‘ep’ number of best parents is selected for next generation, and only (s—ep) number of children are reproduced with genetic operations. Then the next iteration is continued with the population contains union of elitist parents and the children. The following pseudo code illustrates the elitist strategy.

```
Procedure Elitist GA
// Initialization
while (not termination condition) do
    t ← t + 1
    Elites ← Best 'ep' parents
    C ← { }     // Initialize children population
    While |C| < |P| – ep  do
        Select a pair of parents for crossover
        Mate the parents to create children c1 and c2
        C ← C ∪ { c1, c2 } and perform mutation
    end
    P ← C ∪ Elites and evaluate P
end        // Next generation
```

2.3 The Proposed Elitism Strategy

In classical elitist strategy, the number of elites are fixed and continued till the genetic algorithm terminates. Our idea is to make 'ep' as a variable depends on the age of the population. And it is noted that, the elitist parents what we are assuming at the initial iterations may not the best parents globally, this point makes the proposed approach to keep increasing the number of elites while the iteration grows. Initially we started less number of elite parents (for example 10 % of the population size), over a period of time, the number of elites are gradually increased to reach up to 50 % of the population size. The increment is done with a constant value scheduled to happen at regular interval, say after some 'q' number of iterations.

2.4 Elitist Genetic Algorithm Based ELM

The proposed Elitist GA (EGA) strategy is applied to find the optimal weights between input and hidden layers and bias values. The weights has the bound of [−1, 1], so the lbound and ubound takes the values −1 and 1 respectively. The population dimension depends on the number of input variable to ELM. Initially a population is generated with a set of random numbers and these values are fed to ELM and find the classification accuracy as fitness for each chromosome. Then the genetic iteration is continued with the new population as discussed earlier with genetically optimized weights. After termination, the proposed algorithm will provide the optimum weights and bias values for the ELM.

3 Experimental Setup

The performance of the proposed EGA-ELM for SLFN is evaluated on the benchmark datasets described in Table 1, which includes five classification applications. The datasets are received from UCI machine learning repository. Table 2 summarizes the ELM and Elitist GA parameter values used for the experiments.

4 Results and Discussions

EGA-ELM performance is compared with the evolutionary based ELMs such as SaE-ELM [10], E-ELM [9], LM-SLFN [8], and compared with general Elitist GA based ELM (Elitist-ELM), also with traditional ELM [1], based on classification accuracy measure. Table 3 summarizes the classification accuracy received from each ELM algorithms on different datasets. The results clearly indicate that the proposed EGA-ELM outperforms other ELM approaches in terms of higher classification accuracy. Comparatively, only the EGA-ELM and the Elitist-ELM improves the classification accuracy consistently, the rest of the algorithms' performance oscillates between a ranges, this way the proposed algorithm proves its stable performance.

Table 1 The datasets used for performance evaluation

Datasets	#Attributes	#Classes	#Samples
Breast cancer Wisconsin	10	2	699
Ecoli	8	6	336
Iris	4	3	150
Parkinsons	23	3	197
Pima	8	2	768

Table 2 Summary of Ex-ELM and elitist GA parameters settings

Ex-ELM		Genetic algorithm	
Parameters	Values	Parameters	Values
#Input neurons	#Input attributes	Number of iterations	1000
#Hidden neurons	Ñ	Crossover probability	0.8
Bias values	0 to 1	Mutation probability	0.01
Input weights	−1 to 1	Population size	100
Target outputs range	−1 to 1	Initial #elite parents	10
#Output neurons	#Class values		
Activation function	Sigmoid		
#Hidden layers	1 to 20		
Hidden layer weights	−1 to 1		

Table 3 Summary of classification accuracy from ELM algorithms

Datasets	Testing accuracy					
	EGA-ELM	Elitist-ELM	SaE-ELM	E-ELM	LM-SLFN	ELM
WBC	89.54	89.01	87.24	87.20	87.15	87.39
Ecoli	96.73	96.13	94.96	94.92	95.07	94.86
Iris	98.91	98.75	98.31	98.35	98.40	98.34
Parkinsons	90.36	89.36	87.22	86.52	86.77	86.87
Pima	82.07	80.67	77.69	77.49	77.55	77.38

5 Conclusions

Extreme Learning Machine (ELM) is one of the fastest learning algorithms used for classification and regression. One of the limitations of ELM is the random choice of input weights, might be results in lower prediction accuracy. Evolutionary algorithms have been implemented to optimize those weights. This paper proposed a novel Elitism based Genetic Algorithm (GA) based weight optimization strategy for ELM. The performance of the proposed algorithms is tested with five datasets from UCI machine learning repository against four other different evolutionary and traditional ELM algorithms. The result shows that the proposed EGA-ELM algorithm surpasses the other algorithms with superior classification accuracy in stable.

References

1. Huang, G.B., Zhu, Q.Y., Siew, C.K.: Extreme learning machine: theory and applications. Neurocomputing **70**(1), 489–501 (2006)
2. Huang, G., Huang, G.B., Song, S., You, K.: Trends in extreme learning machines: a review. Neural Networks. **61**, 32–48 (2015)
3. Wang, Y., Cao, F., Yuan, Y.: A study on effectiveness of extreme learning machine. Neurocomputing **74**(16), 2483–2490 (2011)
4. Bartlett, P.L.: The sample complexity of pattern classification with neural networks: the size of the weights is more important than the size of the network. IEEE Trans. Inf. Theory **44**(2), 525–536 (1998)
5. Levenberg, K.: A method for the solution of certain nonlinear problems in least squares. Quart. Appl. Math. **2**, 164–168 (1994)
6. Hagan, M.T., Menhaj, M.B.: Training feedforward networks with the Marquardt algorithm. IEEE Trans. Neural Networks **5**(6), 989–993 (1994)
7. Ilonen, J., Kamarainen, J.K., Lampinen, J.: Differential evolution training algorithm for feed-forward neural networks. Neural Process. Lett. **17**(1), 93–105 (2003)
8. Subudhi, B., Jena, D.: Differential evolution and Levenberg Marquardt trained neural network scheme for nonlinear system identification. Neural Process. Lett. **27**(3), 285–296 (2008)
9. Zhu, Q.Y., Qin, A.K., Suganthan, P.N., Huang, G.B.: Evolutionary extreme learning machine. Pattern Recogn. **38**(10), 1759–1763 (2005)
10. Cao, J., Lin, Z., Huang, G.B.: Self-adaptive evolutionary extreme learning machine. Neural Process. Lett. **36**(3), 285–305 (2012)

11. Qin, A.K., Huang, V.L., Suganthan, P.N.: Differential evolution algorithm with strategy adaptation for global numerical optimization. IEEE Trans. Evol. Comput. **13**(2), 398–417 (2009)
12. Lahoz, D., Lacruz, B., Mateo, P.M.: A multi-objective micro genetic ELM algorithm. Neuro-computing **111**, 90–103 (2013)
13. Suresh, S., Babu, R.V., Kim, H.J.: No-reference image quality assessment using modified extreme learning machine classifier. Appl. Soft Comput. **9**(2), 541–552 (2009)
14. Li, J.P., Balazs, M.E., Parks, G.T., Clarkson, P.J.: A species conserving genetic algorithm for multimodal function optimization. Evol. Comput. **11**(1), 107–109 (2003)
15. Liang, Y., Leung, K.S.: Genetic Algorithm with adaptive elitist-population strategies for multimodal function optimization. Appl. Soft Comput. **11**(2), 2017–2034 (2011)
16. Huang, G.B., Babri, H.A.: Upper bounds on the number of hidden neurons in feedforward networks with arbitrary bounded nonlinear activation functions. IEEE Trans. Neural Networks **9**(1), 224–229 (1998)
17. Huang, G.B.: Learning capability and storage capacity of two-hidden-layer feedforward networks. IEEE Trans. Neural Networks **14**(2), 274–281 (2003)
18. Huang, G.B., Chen, L., Siew, C.K.: Universal approximation using incremental constructive feedforward networks with random hidden nodes. IEEE Trans. Neural Networks **17**(4), 879–892 (2006)
19. Hykin, S.: Neural Networks: A Comprehensive Foundation. Printice-Hall. Inc., NJ (1999)
20. Rao, C.R., Mitra, S.K.: Generalized Inverse of Matrices and Its Aplications, vol. 7. Wiley, New York (1971)
21. Corwin, E.M., Logar, A.M., Oldham, W.J.: An iterative method for training multilayer networks with threshold functions. IEEE Trans. Neural Networks **5**(3), 507–508 (1994)
22. Holland, J.H.: Adaptation in Natural and Artificial Systems. MIT Press (1992)
23. Mohamed, M.H.: Rules extraction from constructively trained neural networks based on genetic algorithms. Neurocomputing **74**(17), 3180–3192 (2011)
24. Bi, C.: Deterministic local alignment methods improved by a simple genetic algorithm. Neurocomputing **73**(13), 2394–2406 (2010)
25. Michalewicz, Z.: Genetic Algorithms + Data Structures = Evolution Programs, pp. 111–112. Springer Science & Business Media (1996)
26. Simon, D.: Evolutionary Optimization Algorithms, pp. 188–189. Wiley (2013)

Formulation and Enhancement of User Adaptive Access to the Learning Resources in E-Learning Using Fuzzy Inference Engine

V. Senthil Kumaran and RM. Periakaruppan

Abstract The learning resources used in teaching and learning process are highly important and so, are considered to be primary building blocks of the learning environment. This paper is devoted to propose a methodology to support and enhance adaptive access to the knowledge residing in a learning resource based on varying user contexts. In the proposed approach, user and learning resources are represented using ontology and fuzzy logic is employed to identify the knowledge level of the student to improve the adaptability of access to the learning resource. Experiment results show that this approach enhances adaptive access to the knowledge in learning resource. This proposed approach outperforms the baseline approach that uses ontology mapping.

Keywords Learning resource · Adaptive e-learning · Fuzzy inference engine · Student ontology · Learning resource ontology

1 Introduction

In recent days, the impact of advancements in information and communication technology (ICT) has also improved learners' activities in e-learning. E-learning systems are commonly used to assist the day-to-day teaching and learning activity. E-Learning is popular educational trend of this century. The wide variety of supporting tools developed for e-learning in recent years has enhanced the teaching and learning process, thereby increasing the knowledge and skill level of students, which in return gives a better experience and motivation [1]. These e-learning activities and resources involved in e-learning environments are required to be managed efficiently [2]. The most significant, but difficult task in e-learning is

V. Senthil Kumaran (✉) · RM. Periakaruppan
Department of Applied Mathematics and Computational Sciences,
PSG College of Technology, Coimbatore 641004, India
e-mail: vsk.mca@gapps.psgtech.ac.in

M. Senthilkumar et al. (eds.), *Computational Intelligence, Cyber Security and Computational Models*, Advances in Intelligent Systems and Computing 412, DOI 10.1007/978-981-10-0251-9_30

finding the appropriate learning resource. So, it is important to provide adaptive access to these learning resources [3].

An educational digital library consisting of diverse educational resources targeting wide range of audiences, right from young school children to researchers, is present in an e-learning system [4]. With the increase in the volume and the diversity of the digital library, it becomes highly challenging for the users to find resources that are relevant to their quest. Use of learning resources in e-learning environment, will enhance the teaching and learning process. Quantitative and qualitative analyses made by [5, 6] showed that students accessed various learning resources frequently in order to improve their learning. It is also suggested that the students should learn how to access these resource-rich environments efficiently, for an effective learning. Students have to find the most suitable learning resource for them. Nowadays, information systems at libraries or the Web efficiently manage the large volume of documents they hold, but a problem emerges when access has to be given to relevant resources that satisfy their information needs of the user [7].

Ontology is a conceptualization of a domain into a human understandable, machine-readable format consisting of entities, attributes, relationships, and axioms [8]. Ontology is being used in many applications for domain knowledge representation. In e-learning, ontology-based user profiles are being widely used for representing the student context information [7]. Ontology can also be used to represent the learning resources. Anyway, knowledge representation using ontologies is essentially a subjective process. So, ontology produced by different people for the same domain, will be different depending on their sensitivity, their background etc. However, the conceptual formalism supported by typical ontology, may not be sufficient to represent the uncertainty in information commonly found in many application domains, due to the lack of clear-cut boundaries between concepts of the domains [9]. Representing knowledge in the form of concepts (i.e. conceptualization) is crucial for the automatic processing of the information in the Web [10]. However, ontologies also enhance the management, distribution and retrieval of the learning material within a Learning Management System (LMS) [4], which plays a significant role in eLearning. Ontology exhibits a method of sharing and reusing knowledge in e-learning [3]. A fuzzy ontology is an extension of domain ontologies for solving the problems related to uncertainty [4, 11].

1.1 The Problem Area

Nowadays many e-learning systems provide access to the repository of learning resources and their digital resources. The learning resources are designed by the subject experts. Usually, students access these learning resources based on the learning path prescribed by the experts or they can browse for learning resources. Finding the most suitable learning material in the repository is a very tedious task for students. It has been already realised that digital libraries should incorporate personalization approaches to serve their users better. Hence, the access to the digital libraries should adapt itself to specific characteristics of each student like knowledge level, confidence, efficiency in learning and learning goals. Personalization is particularly important for accessing digital resources in e-learning. This paper is devoted to propose a methodology to support and enhance adaptive access to the knowledge residing in digital resources based on varying user contexts whose uncertainties are resolved by using a fuzzy system.

1.2 The Proposed Approach

To tackle the problem of adaptive access to learning resources in e-learning system, one possible solution is to use ontology for describing the user profile, the relationships between all learning objects in learning resources and to use ontological mapping for personalization. But there exists an uncertainty in inferring learning behaviour of students from ontology. So, we incorporate fuzzy logic as an inference engine into ontology to handle uncertainty data.

2 System Architecture

The proposed system is composed of several components which collectively provide adaptive access to learning resources. Figure 1 shows the architecture of the proposed system conceptually in which the core components are represented. The core components are designed to provide the following functionalities.

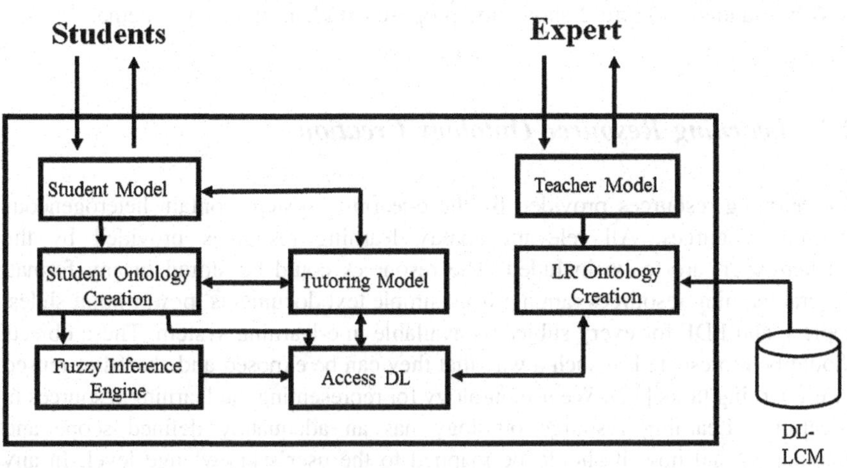

Fig. 1 Proposed system architecture

- *Student Model* understands the student's learning style, behaviour, performance, involvement, interactions, cognitive and adaptive information and sends to the Student Ontology creation component. It also provides the necessary user interface for students and invocates appropriate handlers.
- *Student Ontology Creation* component will generate/update the ontology that represents the students.
- *Fuzzy Inference Engine* deals with the uncertainty in student information and provides rule set to discover the knowledge level of the student.
- *AccessToDL* component will locate the respective learning resources that are suitable for the student and will be provided to the student.
- *LROntologyCreation* component will be created and updated by the teacher or subject expert through the teacher model component. This ontology will have the representation of learning resources with suitable classification.
- *Tutoring Model* is responsible for providing necessary information to the student through student model.

2.1 Student Ontology Creation

In this section we shall give a description of student ontology that has been developed in order to capture the main concepts of a student. Our focus is not only restricted to the static profile of the student, but also encompasses dynamic characteristics of the student. We encompassed the personal, cognitive and adaptive information about the students. First, we discuss the student's characteristics retrieved by the student model along with the possible values. Table 1 illustrates the personal, cognitive and adaptive information retrieved by the student model.

In order to build the ontology, we adopted web ontology language (OWL) which is W3C standard. Figure 2 shows the proposed student ontology schema.

2.2 Learning Resource Ontology Creation

The learning resources provided by the e-learning system contain heterogeneous learning resources. All relevant quality learning resources provided by the teacher/expert are to be included. The resources could be stored in any format. Several learning resources ranging from simple text documents, power point slides, applets, and PDF for every subject is available in e-learning system. These objects should be represented in such a way that they can be exposed and shared or reused by e-Learning tasks [12]. We use ontology for representing the learning resources in e-learning. Learning resource ontology has an adequately defined scope and granularity, and how it should be mapped to the user's knowledge level. In any e-Learning system, learners will search for learning resources. The knowledge

Table 1 Student's characteristics and their possible values

Student's characteristics	Values
Educational level	School/UG/PG/research
Course information	Course name
Performance	Excellent/outstanding/good/satisfactory/poor
Learning goal	High/moderate/low
Learning preference	Simulation/conceptual map/synthesis/explanation/example
Learning style	Active/reflective/sensing/intuitive/visual/verbal/sequential/global
Previous experience	Yes/no
Confidence factor in pre-test during course registration	High/low
Time for study	No time/little/much
Reason for study	Career development/career change/self improvement
Involvement	
Forum participation	No. of threads
Interaction with teacher	Not at all/rarely/frequently
Student current activity	
Current module	Module name
Performance in previous module	Excellent/outstanding/good/satisfactory/poor
Session goal	High/moderate/low
Student's feedback	Excellent/outstanding/good/satisfactory/poor

engineering methodology steps proposed by Noy and McGuinnessb [8] are used for constructing Ontologies. We have created two different ontologies. (i) Curriculum Ontology, for defining the goals and pedagogy for the curriculum of study, and (ii) Learning Resource Ontology, for representing the learning objects, learning path, learning materials, level.

2.3 *Fuzzy Inference System*

Inferring is a process of drawing conclusions from a set of inputs. A fuzzy inference system comprises of a knowledge-base which is nothing but a set of rules that maps fuzzy inputs to an appropriate output [13]. The input space is usually referred to as universe of discourse. The knowledge-base is a set of if-then conditions that facilitate the mapping process. There are two main fuzzy inference models namely Mamdani's fuzzy inference method and Takagi-Sugeno-Kang fuzzy inference method, out of which, we use the former method in this paper. Figure 3 depicts the steps in fuzzy inference engine.

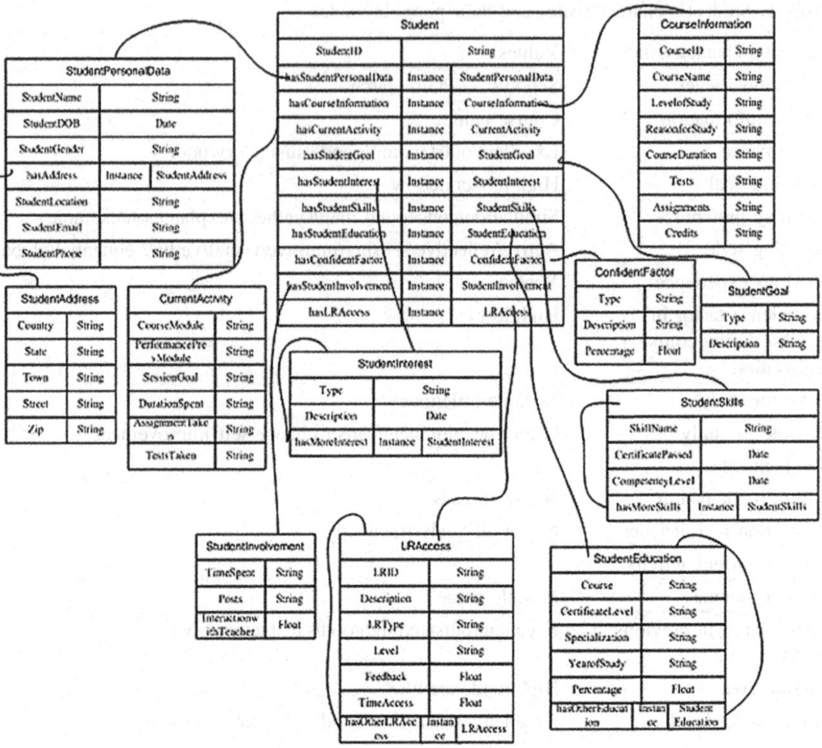

Fig. 2 Student ontology schema

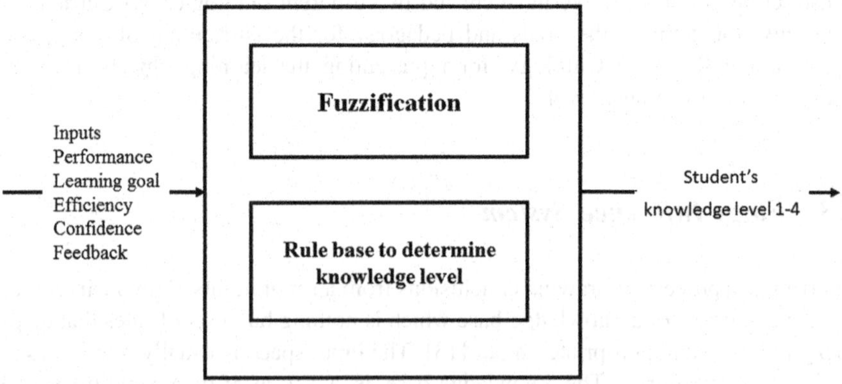

Fig. 3 Fuzzy inference engine

Algorithm

(i) A dataset containing the fuzzy parameters namely performance, feedback, goal, efficiency and confidence of each student required for inference are collected and passed to the inference engine.
(ii) This dataset is passed to a fuzzification module, where the inputs are fuzzified according to their membership functions chosen, such as triangular and trapezoidal membership functions.
(iii) The output of the above is passed to the rule engine, which is the heart of the inference process, containing all combinations of the input parameters and a knowledge level assigned to each of them.
(iv) The rule engine finds a match to the inputs in the dataset and determines the output level.
(v) Usually, the outputs from the rule engine are defuzzified. But in our system, we only need the output level and hence defuzzification is not required.
(vi) The output of the fuzzy inference engine is the output of the rule engine.

In this paper, fuzzy logic is used an extension to ontology in order to access learning resources. The learning resource is a collection of wide range of information collected from numerous sources. The students, who access them, vary by their knowledge and experience levels. Hence, it is important that we take preventive measures to not complicate the learning process for the students. The learning process should be kept as simple as possible, for the students to appreciate and understand the contents being taught. Therefore, we have to analyze the current status of every student seeking education via e-learning systems. The analysis will not only help the teacher understand the student's capability but also identify portions of the subject which can be understood by the student from his knowledge level. Therefore, we can avoid taxing the student with higher standard contents of the learning resource.

Various realistic parameters from goal to capacity of the student is thoroughly analyzed and graded. Though these grades are individual values, their real time meaning still has some fuzziness. For example, if we say the student has a capacity, the value though crisp can never be concluded to be exactly high or low. Thus, we induce partial membership in the sets high and low, denoting that 8 % can partially belong to each of these sets. To overcome this fuzziness in the analysis, we use fuzzy logic to conclude the status of the student. The linguistic variables and the fuzzy sets for each of these parameters used to analyze a student are given in Table 2.

A membership function is usually a graphical representation of fuzziness in linguistic variables. The graph looks like a geometric curve spun over various scales. The scale is decided based on the range of each inputs and may differ from each other. The triangular membership function is a triangular curve with three parameters a, b and c representing the two ends and the peak of a triangle. The trapezoidal membership function has four parameters a, b, c and d representing the two peaks at the top and the base line of a trapezoid.

Table 2 Linguistic variables and the fuzzy set

Inputs	Values
Performance	Excellent/outstanding/good/satisfactory/poor
Learning goal	High/moderate/low
Learning efficiency	High/moderate/low
Feedback	Excellent/ outstanding/ good/ satisfactory/ poor
Confidence factor in pre-test during course registration	High/ low

Table 3 Output of fuzzy inference engine

Level	Meaning
Level 0	students without special background
Level 1	students with some interest in learning
Level 2	students with special interest and good specialized knowledge
Level 3	students with high interest and expectations
Level 4	Researchers with high professional level of education.

A membership function is chosen to fuzzify the above inputs. The most common membership functions include triangular, trapezoidal, bell curve, etc. In this paper we use triangular and trapezoidal membership functions for the parameters listed in Table 3. The membership function graphs used for our fuzzy input parameters are shown in Fig. 4.

Based on the given input parameters, the knowledge of the student is categorized into five levels. The output fuzzy sets are listed in Table 3.

The fuzzy knowledge-base is a set of rules that determines the output for the given fuzzy inputs. The knowledge-base maps each possibility of the combinations with an appropriate output. The rules usually comprise of all possible combinations of the input fuzzy sets and their corresponding outputs. All the combinations of the input variables are stored in the rule base. The combinations for every fuzzy set for the performance variable except for excellent and poor, with all the other sets in the remaining variables are added to the rule base and an appropriate output has to be assigned to them. If the performance is Excellent or Poor then their other parameters are not considered, the output level is directly assigned to 4 or 0 respectively.

The combinations are represented as if-then rules in the fuzzy knowledge-base in order to obtain the knowledge level of the student having one of the stored input combinations. For our research, we primarily focus on categorizing each student into one of the five knowledge levels mentioned in Table 3. Hence, we need not defuzzify the output from the knowledge-base. The outputs obtained are further used in the later phases.

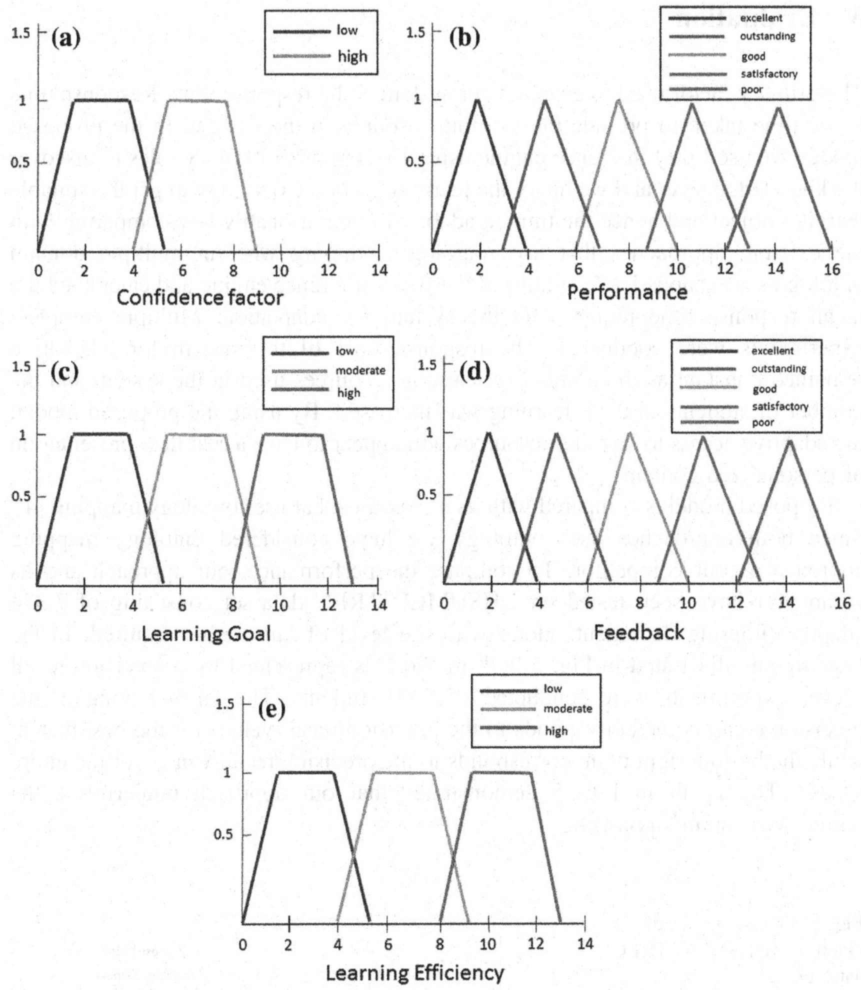

Fig. 4 Membership function for the linguistic variables. **a** Confidence factor, **b** performance, **c** learning goal, **d** feedback, **e** learning efficiency

2.4 Access to Learning Resources

This component collects the knowledge level from fuzzy inference engine and other context information from the student ontology and proceeds with searching for the learning resources that are suitable for the student in the learning resources ontologies. Later, it retrieves the learning objects from the learning resource store and returns it to the student. We use SPARQL query language to query the learning resources ontologies.

3 Evaluation

The primary factor used to evaluate our system is the response time. Response time is the time taken to provide the learning resource to the student. In the proposed model, we use fuzzy inference engine using fixed number of fuzzy rules to discover the knowledge level and we query the learning resource ontology to get the suitable learning object and hence the time to adapt will be reasonably low comparing with the existing approaches that uses ontology mapping wherein multiple domain ontologies are mapped. We simulated the fuzzy inference engine and calculated the mean response time required by the system for adaptation. Multiple complete experiments were conducted. The response time of the system for adaptation remained constant as the number of learning resources used in the system and the number of students in the e-learning was increased. By using the proposed model, the adaptive access to learning resources will appear to have a seamless presentation of personalized content.

Proposed model is compared with its competitor that uses ontology mapping [4]. Since both approaches uses ontology we have considered ontology mapping approach as our competitor. To compare the performance, our approach and its competitor have been tested on OHSUMED TREC data set consisting of 7,374 adaptive filtering documents along with the level of knowledge required. In the experiments illustrated in Fig. 5 both methods is represented by a precision/recall curve. Experiments were conducted with 50 students. The top-left point of the precision/recall curve corresponds to the precision/recall values for the best match, while the bottom right point corresponds to the precision/recall values for the entire access. The result in Fig. 5 demonstrates that our approach outperforms the ontology mapping approach.

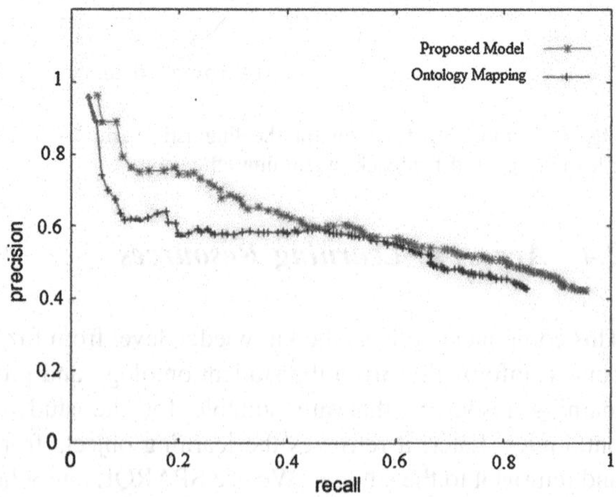

Fig. 5 Precision—recall graph for retrieval on TREC data set

4 Conclusion

We proposed, in this paper, a fuzzy based inference engine to discover the knowledge level of students from student ontology and the discovered context information is then used to provide adaptive access to the knowledge residing in learning resources provided by the e-learning system. The discovered knowledge level of the student could be used for personalization in any e-learning activity. The proposed approach collects all the characteristics of the student using an e-learning methodology, which are considered important for a learning resources to be fully adaptive to the needs of the students and creates student ontology. We use fuzzy inference engine to deal with the characteristics that are fuzzy in nature. One of the main advantages of the proposed model is the integration of ontology and fuzzy rules. The fuzzy inference engine for ontology, which is responsible for inferring the knowledge level of the students and other context information, are incorporated in the ontology in order to provide additional knowledge. The most challenging part in our work is the selection of characteristics to be included in the student ontology and the classification of learning resources in learning resources ontology. We plan to extend this work by using some machine learning approach to dynamically select the linguistic variables. This will enhance the results further.

References

1. Cheng, Y.H., Cheng, J.T., Chen, D.J.: The effect of multimedia computer assisted instruction and learning style on learning achievement. WSEAS Trans. Inf. Sci. Appl. **9**(1), 24–35 (2012)
2. Pramitasari, L., Hidayanto, N.A., Aminah, S., Krisnadhi, A.A., Ramadhanie, A.M.: Development of student model ontology for personalization in an E-learning system based on semantic web. In: International Conference on Advanced Computer Science and Information Systems (ICACSIS09), Indonesia, pp. 7–8 (2009)
3. Chung, H.S., Kim, J.M.: Learning ontology design for supporting adaptive learning in e-Learning environment. In: International Conference on Information and Computer Networks (ICICN 2012) IPCSIT 27, 2012
4. Gašević, D., Hatala, M.: Ontology mappings to improve learning resource search. Br. J. Educ. Technol. **37**(3), 375–389 (2006)
5. Kay, J., Lum, A.: Ontologies for scrutable student modelling in adaptive e-learning. In: Proceedings of the Adaptive Hypermedia and Adaptive Web-Based Systems Workshop on Semantic Web for E-Learning (2004)
6. Ferreira-Satler, M., Romero, F.P., Menendez-Dominguez, V.H., Zapata, A., Prieto, M.E.: Fuzzy ontologies-based user profiles applied to enhance e-learning activities. Soft Comput. **16**(7), 1129–1141 (2012)
7. Clemente, J., Ramírez, J., De Antonio, A.: A proposal for student modeling based on ontologies and diagnosis rules. Expert Syst. Appl. **38**(7), 8066–8078 (2011)
8. Noy, N.F., McGuinness, D.L.: Ontology development 101: A guide to creating your first ontology (2001)
9. Gaeta, M., Orciuoli, F., Paolozzi, S., Salerno, S.: Ontology extraction for knowledge reuse: The e-learning perspective. IEEE Trans. Syst. Man Cybern. Part A Syst. Hum. **41**(4), 798–809 (2011)

10. Gaeta, M., Orciuoli, F., Ritrovato, P.: Advanced ontology management system for personalised e-Learning. Knowl.-Based Syst. **22**(4), 292–301 (2009)
11. Yildirim, Y., Yazici, A.: Automatic semantic content extraction in videos using a fuzzy ontology and rule-based model. IEEE Trans. Knowl. Data Eng. **25**(1), 47–61 (2013)
12. Cakula, S., Salem, A.B.M.: E-learning developing using ontological engineering. WSEAS Trans. Inf. Sci. Appl. **10**(1), 14–25 (2013)
13. Mendes, M.E.S., Sacks, L.: Dynamic knowledge representation for e-learning applications. Enhancing the Power of the Internet, pp. 259–282. Springer, Berlin (2004)

Part III
Cyber Security

A Robust User Anonymity Preserving Biometric Based Multi-server Authenticated Key Agreement Scheme

Mrudula Sarvabhatla, M. Chandra Mouli Reddy, Kodavali Lakshmi Narayana and Chandra Sekhar Vorugunti

Abstract Due to the intense evolution of IoT (Internet of Things) technology, in which currency notes to bicycles will form a vital part of Internet to exchange their information among the surrounding environments. IoT will definitely consequences in an augmented level of users connecting to the remote servers (via insecure public internet). Connecting through less resource and portable devices requires a light weight, identity preserving and highly secured key agreement and authentication mechanism. In this framework, recently Mishra et al. proposed an anonymity preserving biometric based authentication scheme and claimed that their scheme resists all major cryptographic attacks. In this, manuscript we will demonstrate that, on thorough analysis, Mishra et al. scheme is liable to 'Forward Secrecy' attack and based on that, the attacker can realize all key attacks. As a part of our contribution, we will propose a light weight, strongly secure user authentication and key agreement scheme which is best outfits for IoT environment.

Keywords Biometric · Authentication · Smart card · Cryptography · Security

M. Sarvabhatla (✉)
N.B.K.R IST, Nellore 524413, A.P, India
e-mail: mrudula.s911@gmail.com

M.C.M. Reddy
Vel Tech Technical University, Avadi, Chennai 382007, India

K.L. Narayana
SITAMS, Chittoor, A.P, India
e-mail: kodavali.lakshmi@gmail.com

C.S. Vorugunti
Dhirubhai Ambani Institute of Information and Communication Technology, Gandhi Nagar, Gujarat 382007, India
e-mail: vorugunti_chandra_sekhar@daiict.ac.in

M. Senthilkumar et al. (eds.), *Computational Intelligence, Cyber Security and Computational Models*, Advances in Intelligent Systems and Computing 412, DOI 10.1007/978-981-10-0251-9_31

1 Introduction

With the advancement of mobility and internetworking technologies, there is an exponential increase in data request load on remote servers from devices which are mobile. To handle the extreme data load on servers, all major E-Commerce companies opted for a multi-server architecture. The rapid growth of Internet of Things technology further increased the data load on servers, in which an owen or car engine can connect to remote server via insecure Internet to exchange data with outside surroundings. Increased modes of connectivity (mobiles, sensors, IoT devices) resulted in authentication of users as an indispensable mechanism.

In 2013, Chaturvedi et al. [1] proposed an authentication scheme based on biometrics and claimed that their scheme proposed scheme achieves efficient login and password alter mechanism, in which incorrect password can be detected quickly and user is provided with a facility to alter the password freely. Very recently in 2014, Jiang et al. [2] proposed a two factor authentication scheme and clearly depicted the security requirements need to be achieved by the authentication schemes.

In 2014, Chuang et al. [6] proposed a user anonymity preserving multi-server key agreement scheme based on smart card, password and biometrics. Chuang et al. [6] scheme present a competent resolution for multi-server environment, where a user can interact with any server based on a single registration. In the year 2014, Mishra et al. [7] performed cryptanalysis on Chuang et al. scheme and demonstrated that Chuang et al. scheme is vulnerable to major attacks like user impersonation attack, stolen smart card attack and Denial-of-Service (DoS) attack.

In this paper, we pointed out that Mishra et al. [7] scheme still suffer mainly from smart card breach attack and forward secrecy attack. We proposed a new scheme which is more robust than Mishra et al. scheme. The remainder of this manuscript is structured as follows. In Sect. 2, we briefly clarify the Mishra et al. scheme. In Sect. 3, we evaluate the vulnerabilities of Mishra et al. scheme. In Sect. 4, we discuss our proposed scheme. In Sect. 5, we explain security strengths of our proposed scheme. In Sect. 6, we do a comparison study of our scheme with current schemes and conclusion is given in Sect. 7.

2 Assessment of Mishra et al. Scheme

In this segment, we examine the Mishra et al. [7] Biometric based multi user authenticated key agreement scheme using smart cards. Mishra et al. system is divided into the five stages, which are server registration, user registration, login, mutual authentication and password change stages. The notations used in Mishra et al. scheme are listed below:

U_i ith user
ID_i Unique identification of U_i

PW_i	Unique password of U_i
S_j	Server j
T_r	Registration Time
SC	Smart card
RC	Registration Center
SID_j	Server 'J' identity
PSK	Pre shared master Key by R.C to S_j
$PBIO_i$	Personal biometric of user U_i
x	Server secret key
h(.)	One way hash function
H(.)	Biohash function
$\oplus$	Exclusive-OR operation
$\|$	String concatenation operation

2.1 Server Registration Phase

In this phase server S_j sends a registration request to the registration center to get authorization permission. Then registration center authorizes the server by assigning the key PSK to the server using Key Exchange Protocol. Sever can use PSK to authorize a legitimate user.

2.2 User Registration Phase

This phase is raised whenever a new user U_i wishes to access server services. Initially new user must register with the registration center (RS) as follows.

(R1) U_i opts his/her password PW_i, identity ID_i and random number N_i.
(R2) U_i computes $V_1 = h(PW_i \| N_i)$ and $V_2 = h(ID_i \oplus N_i)$ and directs the registration request message $\{ID_i, V_1, V_2\}$ to the registration center RC via a secure channel.
(R3) RC computes $A_i = h(ID_i \| x \| T_{REG})$, $B_i = h(A_i) = h^2(ID_i \| x \| T_{REG})$, $X_i = B_i \oplus V_1$, $Y_i = h(PSK) \oplus V_2$ and $Z_i = PSK \oplus A_i$.
(R4) The RC supply a tamper-proof smart card SC with the parameters: $\{X_i, Y_i, Z_i, h(.))$ and send to U_i over a secure channel.
(R5) Upon accepting the SC from RC, U_i imprints his/her personal biometric $PBIO_i$ at the sensor, and computes $N = N_i \oplus H(PBIO_i)$ and $V = h(ID_i \| N_i \| PW_i)$.
(R6) U_i lastly stocks N and V into his/her smart card. Thus SC contains the information $\{X_i, Y_i, Z_i, h(.), N, V\}$.

2.3 Login Phase

Once the user desires to login into the server, the user U_i enclosures his/her SC into the smart card reader and pass in his ID_i, PW_i and also imprints his biometric $PBIO_i$ at the sensor. Then the smart card undertakes the following calculations:

(L1) SC computes $N_i^* = N \oplus H(PBIO_i)$, $V^* = h(ID_i||N_i||PW_i)$ and verifies whether V^* equals to the V stored in the U_i smart card. If the verification fails, the session is ended immediately.

(L2) SC additionally computes $V_1 = h(PW_i||N_i)$, $V_2 = h(ID_i \oplus N_i)$, $B_i = X_i \oplus V1$and $h(PSK) = Yi \oplus V_2$.

(L3) SC also produces an arbitrary number u_r, and computes $M_1 = h(PSK) \oplus u_r$, $M_2 = ID_i \oplus h(u_r||B_i)$ and $M_3 = h(ID_i||u_r||B_i)$.

(L4) Finally, SC directs the login request message $\{Z_i, M_1, M_2, M_3\}$ to server S_j via a public channel. (Z_i is stored in U_i S.C and passed as part of login request message of U_i. Hence, an attacker, if gets the smart card of U_i, and intercepts Z_i, can trace the communication messages of U_i based on Z_i. Hence, strong user anonymity is not achieved in Mishra et al. scheme).

2.4 Mutual Authentication Phase

Subsequently getting U_i's login request message, the server S_j concludes the succeeding steps:

(A1) S_j uses its PSK to retrieve $A_i = Z_i \oplus PSK$ and computes $r_u = M_1 \oplus h(PSK)$ and $ID_i = M_2 \oplus h(u_r||h(A_i))$, $B_i = h(A_i)$, $M_3^* = h(ID_i||u_r||h(A_i))$.

(A2) S_j verifies M_3^* equals the received M_3 in the login request. If this equality fails, the session is dismissed immediately. Otherwise, S_j proceeds to generate a random number s_r, and computes the session secret key $SK_{ji} = h(ID_i|| SID_j||B_i||u_r||s_r)$ to be shared with the user U_i.

(A3) S_j computes $M_4 = s_r \oplus h(ID_i||u_r)$ and $M_5 = h(SK_{ji}||u_r||s_r)$ and then directs login acknowledge message $\{SID_j, M_4, M_5\}$ to the smart card SC of the user U_i.

(A4) SC retrieves the value $s_r = M_4 \oplus h(ID_i||u_r)$, and then computes the session key $SK_{ij} = h(ID_i||SID_j||B_i||u_r||s_r)$ to be shared with the server S_j.

(A5) SC computes $M_5 = h(SK_{ij}||u_r||s_r)$ for server authentication. If this test does not hold, the session is terminated, Otherwise, U_i considers SK_{ij} as the session key.

(A6) SC computes $M_6 = h(SK_{ij}||s_r||u_r)$ and directs the authentication reply message $\{M_6\}$ to S_j via a public channel.

(A7) Server S_j verifies $M_6 = h(SK_{ji}||s_r||u_r)$. If this test does not hold, the session is terminated, Otherwise, S_j considers the SK_{ij} as valid key and also U_i as valid user.

3 Analysis of Security Weakness of Mishra et al. Scheme

Threat Model

In this manuscript, we will follow the similar threat model practiced by Mishra et al. and other users [1–7]:

1. An attacker or adversary or legal user can extract the information cached in the smart card by several techniques such as power consumption or leaked information etc.
2. An attacker can passive monitor or eavesdrop all the login request, login reply messages forwarded between user and server over a public channel which is Internet.
3. An attacker can alter, remove, replay or resend, forward the eavesdropped messages.

3.1 Failure to Counter Smart Card Breach Attack

As discussed, On completing the registration process, S_j responds to U_i a smart card S.C, which is cached with the information {X_i, Y_i, Z_i, h(•), N, V}, where Yi = h(PS K) ⊕ W_2, W_2 = h(IDi ⊕ Ni). A legal user U_i, can extract the stored or cached values in the S.C through various techniques like power analysis attack [8] etc. U_i knows his identity ID_i and random value N_i. U_i can compute $Y_i \oplus W_2$ = h(PS K). PSK is a secret key assigned to the server S_j, by registration center RC. As PSK is S_j secret key, h(PSK) is common for all users registered with S_j. The legal user U_i, on intercepting the common value h(PSK), can use it while attacking the other legal users as discussed below.

3.2 Susceptible to Forward Secrecy Attack

Forward secrecy is a security vulnerability in which, An opponent 'E' can try to frame the established session key SK_{ij} framed between U_i and S_j on getting the long-term secret key of U_i i.e. B_i.

As discussed in the threat model, the attacker 'E' can apprehend all the communication messages exchanged between the U_i and S_j. i.e. {Z_i, M_1, M_2, M_3}, {S ID_j, M_4, M_5}.

Step1 Get the h(PSK) from Y_i = h(PSK) ⊕ V_2, (U_i knows his V_2) which is a hash of server S_j secret key and common to all users.

Step 2 retrieve: u_r = M_1 ⊕ h(PS K).

Step 3 Compute ID_i = M_2 ⊕ h(r_u||B_i). 'E' obtains u_r from step 2, assuming that the U_i long term secret key B_i is leaked to 'E'.

Step 4 retrieve $s_r = M_4 \oplus h(ID_i||u_r)$. 'E' obtains M4 from the intercepted login reply message.

Step 5 As 'E' got all the required values to compute a S.K, 'E' can frame $S\,K_{ij} = h(ID_i||S\,ID_j||B_i||u_r||s_r)$. As illustrated above, in Mishra et al. scheme, the attacker can compute the session key framed between the user U_i and S_j. On computing the session key, the attacker can decrypt all the messages exchanged between U_i and S_j. Hence, Mishra et al. scheme failed to achieve the fundamental requirement of a key agreement scheme i.e. secure data transfer.

4 Our Enhanced Scheme

4.1 Registration Stage

When a new user U_i intends to access the remote server resources, from a set of multiple servers $\{S_1, S_2, \ldots, S_r\}$ registered with R.C.

U_i opts his identity ID_i and password PW_i of his/her choice. U_i registration process consists of following steps.

(R1) The user U_i produce an arbitrary number N_i, and calculate $V_1 = h(PW_i||N_i)$ and $V_2 = h(ID_i \oplus N_i)$. U_i then forwards the registration appeal message $\{ID_i, V_1, V_2\}$ to the registration center via a safe channel.

(R2) On receiving the registration request from U_i, RC proceeds to check whether the ID_i is in the registered list or not, if it exists, 'RC' discards the registration request else it continue as follows:
Compute: $A_i = h(ID_i||x||T_{REG})$, $B_i = h(A_i) = h_2(ID_i||x||T_r)$, $M_i = B_i \oplus V_1$, $N_i = h(PSK||T_{REG}) \oplus V_2$ and $P_i = h(T_{REG}||PSK) \oplus A_i$, where T_{REG} denotes the registration time of U_i.

(R3) RC–> U_i, a smart card containing $\{h(.), M_i, N_i, P_i, T_{REG}\}$ to the user U_i via a secured communication channel. On getting the smart card, U_i computes R, V, T values and inserts into it. i.e. S.C contains $\{h(.), M_i, N_i, P_i, R, S, T\}$ where $R = R_i \oplus h(PBIOM_i)$, $V = h(ID_i||N_i||PW_i)$, $T = T_{REG} \oplus h(ID_i||N_i)$ and removes T_{REG} from S.C.

4.2 Login Stage

Each time user U_i wishes to get the remote server S_j resources, he/she put his smart card into the card reader and put forward his ID_i, PW_i, $PBIOM_i$, later the smart card performs the subsequent tasks.

(L1) Compute: $N_i^* = N \oplus h(PBIOM_i)$, $V^* = h(ID_i||N_i^*||PW_i)$, check whether the computed V^* equals to V stored in the S.C. If both are identical, the legitimacy of the user is validated and smart card proceeds further to compute: $V_1 = h(PW_i||N_i)$ and $V_2 = h(ID_i \oplus N_i)$. $B_i = M_i \oplus V_1$ and $h(PS\ K|| T_{REG}) = N_i \oplus V_2$.

(L2) Choose a random number u_r Compute: $M_1 = h(PSK||T_{REG}) \oplus u_r$, $M_2 = ID_i \oplus h(u_r||B_i)$, $M_3 = h(ID_i||u_r||h(PSK||T_{REG})||B_i)$. $M_4 = P_i \oplus h(PSK|| T_{REG})$ and smart card sends to server 'S_j' the login request message $\{M_1, M_2, M_3, M_4, T_{REG}\}$.

4.3 Authentication Stage

In receipt of the login appeal message from U_i, S_j performs the successive tasks:

(A1) compute: $T_1 = h(PS\ K||T_{REG})$, $T_2 = h(T_{REG}||PSK)$, $P_i = M_4 \oplus T_1$, $u_r = M_1 \oplus T_1$, $A_i = P_i \oplus T_2$, $ID_i = M_2 \oplus h(u_r||h(A_i))$.

(A2) On getting these values, S_j computes: $M_3^* = h(IDi||u_r||h(PSK||T_{REG})||B_i)$ and check computed M_3^* equals the M_3 received through login request message. If both are equal, U_i is authenticated. Then S_j advance as follows:

(A3) Choose a random number s_r. Compute: session key $S.K = h(ID_i||SID_j||B_i||u_r|| s_r||h(PSK||T_{REG}))$, $M_5 = s_r \oplus h\ (ID_i||r_u||B_i)$, $M_6 = h(S.K_{ij}||u_r||s_r||T_{REG}^*)$, $M_7 = T_{REG}^* \oplus h(B_i||ID_i||u_r||s_r)$ and replies back with the message $\{SID_j, M_5, M_6, M_7\}$ to user U_i.

(A4) Following the login reply message, user U_i proceeds as follows: U_i retrieves s_r from M5 as $s_r = M_5 \oplus h(u_r||ID_i||T_2)$.

(A5) Computes: $S.K_{ij}^* = h(ID_i||SID_j||B_i||u_r||s_r||h(PSK||T_{REG}))$, $T_{REG}^* = M_7 \oplus h(B_i|| ID_i||u_r||s_r)$, $M_6^* = h(S.K_{ij}||u_r||s_r||T_{REG}^*)$ and compare the computed M_6^* with the received M_6^*. If both are equal then S_j is authenticated by U_i.

(A6) Further, U_i updates the T value in smart card with $T^* = T_{REG}^* \oplus h(ID_i||N_i)$.

5 Security Analysis of Improved Scheme

5.1 Counter Smart Card Breach Attack

On successful registration, a legal user 'E' will be assigned a smart card which is stored with the values $\{M_e, N_e, P_e, R, S, T\}$, where $A_e = h(ID_e||x||T_{REG})$, $B_e = h(A_e) = h_2(ID_e||x||Te)$, $M_e = B_e \oplus V_1$, $N_e = h(PSK||T_{REG}) \oplus V_2$, $P_e = h(T_{REG}|| PSK) \oplus A_e$, $R = R_i \oplus h(BIOM_i)$, $V = h(ID_i||N_i||PW_i)$, $T = T_{REG} \oplus h(ID_i||N_i)$.

Assume that 'E' extracted the secret data stored in his smart card by some means [1–9]. 'E' can get values mentioned above.

In Mishra et al. scheme, the legal user can get h(PS K), which is a secret key of server S_j and common to all users. In our scheme, we are concatenating PSK with T_{REG}, where T_{REG} is a registration time of individual user and not unique. Hence 'E' gets $h(PSK||T_{REG})$ which is specific to himself. It is not possible for 'E' to use h $(PSK||T_{REG})$ for attacking other users messages. Hence, our scheme counters smart card breach attack.

5.2 Resistance to Stolen Smart Card Attack

Assume that if a legal attacker 'E' acquires the smart card of any legal user U_i of the system for a while or snips the card, 'E' can excerpt the secret data kept in U_i's smart card by some resources [1–9]. 'E' can get {h(.), M_i, N_i, P_i, R, S, T}. 'E' can try to frame a valid login message as sent by U_i and login into the server 'S'. To frame a valid login message {M_1, M_2, M_3, M_4, T_{REG}} where $M_1 = h(PSK|| T_{REG}) \oplus u_r$, $M_2 = ID_i \oplus h(u_r||B_i)$, $M_3 = h(ID_i||u_r||h(PSK||T_{REG})||B_i)$ and $M_4 = P_i \oplus h(PSK||T_{REG})$. To compute M_3, 'E' must know ID_i, B_i. ID_i is protected by $h(u_r||B_i)$, where $u_r = M1 \oplus h(PS\ K||T_{REG})$. As discussed to get $h(PS\ K||T_{REG})$ or $h(T_{REG}||PS\ K)$, 'E' needs V_2 and B_i. The secret key of U_i i.e. B_i is protected with $B_i = M_i \oplus V1$ = where $V1 = h(PW_i||N_i)$ and $N_i = N \oplus h(PBIOM_i)$. Since the $PBIOM_i$ and PWi are known only to user U_i, 'E' doesn't have a chance to get these values. Therefore, we can conclude that, our proposed scheme counter the stolen smart card attack.

6 Cost and Security Analysis

As discussed in Table 1, it is impossible for an attacker to achieve any unknown value, even he extracted the information stored in his S.C, any other legal user (U_i) S.C and intercepted all the communication messages exchanged between U_i and server. As discussed above, Mishra et al. [7] scheme requires totally 24 Hash, where as our scheme requires 31 hash operations. Our scheme is lighter compared to [3–6], it is slightly costlier compared to Mishra et al. scheme by seven hash operations, which is negligible in current computer and network advancements,

Table 1 Types of users and the values they can intercept

Attacker (E)	Messages 'E' can intercept	Values 'E' can achieve
'E' S.C values	{M_e, N_e, P_e, R_e, S_e, T_e}	{ID_i, PW_i, B_i, PS K, h (PSK)} of U_i
U_i S.C values	{M_i, N_i, P_i, R_i, S_i, T_i}	
Communication messages	{M_1, M_2, M_3, M_4, T_{REG}}, {SID_j, M_5, M_6, M_7}	

complexity can be easily managed with improved and advanced technology from both hardware and software point of view. Our scheme provides robust security compared to all other relevant schemes as discussed above. So we can conclude that our improved scheme is more robust and best suits for complex multi server environment.

7 Conclusion

In this manuscript, we have briefly discussed the advantages and shortcomings of the current multi-server password and biometric oriented authentication schemes in the literature. The examination indicates that the existing schemes have failed to achieve security strengths which are majorly required for IoT environment. We have scrutinized the security pitfalls freshly proposed Mishra et al. scheme and demonstrated that an attacker, on getting the user long term secret key, can perform all major attacks. As a part of our contribution, we have proposed a strongly secure multi-server biometric authentication scheme, which resists all major cryptographic attacks.

References

1. Chaturvedi, A., Mishra, D., Mukhopadhyay, S.: Improved biometric-based three-factor remote user authentication scheme with key agreement using smart card. International Conference on Information Systems Security—ICISS, vol. 8303, pp 63–77, Springer—LNCS (2013)
2. Jiang, Q., Ma, J., Li, J., Yang, L.: Robust two-factor authentication and key agreement preserving user privacy. Int. J. Netw. Secur. **16**, 321–332 (2014)
3. Yang, D., Yang, B.: A biometric password-based multi-server authentication scheme with smart card. In: ICCDA 2010, vol. 5, pp. 554–559 (2010)
4. Yoon, E.J., Yoo, K.Y.: Robust biometrics-based multi-server authentication with key agreement scheme for smart cards on elliptic curve cryptosystem. J. Supercomputing, pp. 235–255 (2011) (Springer)
5. Kim, H., Jeon, W., Lee, K., Lee, Y., Won, D.: Cryptanalysis and improvement of a biometrics-based multi-server authentication with key agreement scheme. In: ICCSA 2012, Springer, pp. 391–406 (2012)
6. Chuang, M.C., Chen, M.C.: An anonymous multi-server authenticated key agreement scheme based on trust computing using smart cards and biometrics. Elsevier J. Expert Syst. Appl., pp. 1411–1418 (2014) (Impact factor:2.254)
7. Mishra, D., Das, A.K., Mukhopadhyay, S.: A secure user anonymity-preserving biometric-based multi-server authenticated key agreement scheme using smart cards. Elsevier J. Expert Syst. Appl. **41**(18), 8129–8143 (2014) (Impact factor:2.254)
8. Kocher, P., Jaffe, J., Jun, B.: Differential power analysis. In: Proceedings of Crypto 99, pp. 388–397, Springer (1999)
9. Messerges, T.S., Dabbish, E., Sloan, R.H.: Examining smart-card security under the threat of power analysis attacks, IEEE Trans. Comput. **51**(5), 541–552 (2002)

Extended Game Theoretic Dirichlet Based Collaborative Intrusion Detection Systems

Sayan Paul, Tushar Makkar and K. Chandrasekaran

Abstract Security has always been one of the key issues of any man-made system, this paved the way for a submodule or application or a device to monitor or system for malicious activities. This system or submodule or device is known as Intrusion Detection System (IDS). As technology evolves so does the associated threats and thus the intrusion detection system needs to evolve. Game theory throws in a different perspective which have not been looked upon much. Game theory provides a way of mathematically formalizing the decision making process of policy establishment and execution. Notion of game theory can be used in intrusion detection system in assisting in defining and reconfiguring security policies given the severity of attacks dynamically. We are trying to formulate a robust model for the theoretical limits of a game theoretic approach to IDS. The most important flaw of game theory is that it assumes the adversary's rationality and doesn't take into consideration multiple simultaneous attacks. Therefore, a collaborative trust and Dirichlet distribution based robust game theoretic approach is proposed which will try to resolve this issue. Reinforced learning approaches using Markov Decision Process will be utilized to make it robust to multiple simultaneous attacks.

Keywords Intrusion detection system · Dirichlet based trust management · Collaborative trust management · Game theory · Nash equilibrium

S. Paul (✉) · T. Makkar · K. Chandrasekaran
Department of Computer Science and Engineering, National Institute of Technology Karnataka, Surathkal, India
e-mail: sayan.paul6@gmail.com

T. Makkar
e-mail: tusharmakkar08@gmail.com

K. Chandrasekaran
e-mail: kchnitk@ieee.org

M. Senthilkumar et al. (eds.), *Computational Intelligence, Cyber Security and Computational Models*, Advances in Intelligent Systems and Computing 412, DOI 10.1007/978-981-10-0251-9_32

1 Introduction

Intrusion Detection Systems plays a key role in security of modern day software applications/systems. They compare observable behaviour in the system against suspicious patterns to identify any kind of intrusions. There are two variances of IDS: Network based (NIDS) or Host based (HIDS). Traditional IDSs have a problem that they work in isolation and therefore have higher chance of getting compromised by unknown or new threats. A Collaborative IDS solves this problem by having peer IDS help each other out and get aided by shared collective knowledge and experience from peers. This increases both the accuracy and the ability to detect new intrusion threats. Collaborative IDS assumes that all IDSs will honestly cooperate. The lack of trust management leaves the system vulnerable to malicious peers [1].

Few IDSs have been produced to cooperate honestly based on trust and/or distributed trust models but they have not incorporated any kind of incentives for IDS collaboration. Incentives are important criteria any collaborative system otherwise it will suffer from "free rider problem" in which certain group of IDSs only keep on asking for assistance but may not actually contribute to the system. Thus, this leads to degradation of performance of the system. So we need to take care of incentives while designing such a system [2]. The distributed collaborative is preferred over centralized as it need rely on a central server to gather and analyze alerts and thus avoiding any bottleneck problems [3]. We just need to take care of any malicious or malfunctioning IDS which can degrade the performance of the system (Fig. 1).

In this paper, we propose a trust-based IDS collaboration network system with incentive based scheme for resource allocation. In this the amount of resources

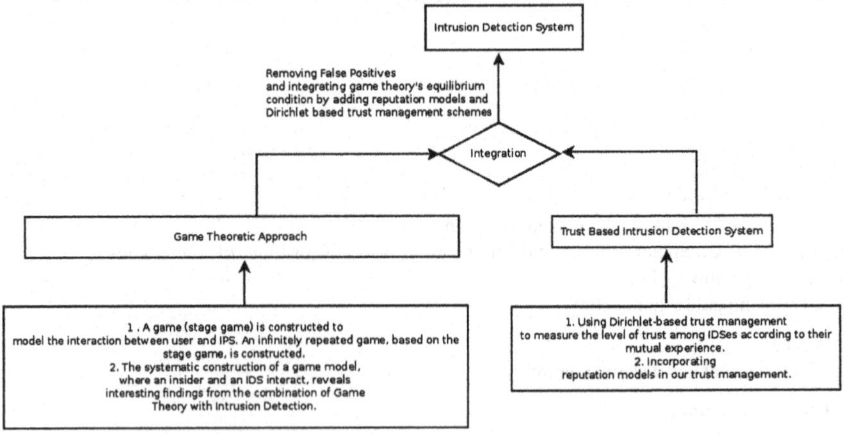

Fig. 1 Overview of system

allocated by each IDSs to their neighbors is dependent on the trustworthiness and resources already allocated by its neighbors to help it.

An optimization problem is constructed to aid the IDS to optimally allocate resource to maximize the satisfaction level of each of its peers. We show that under certain controllable system conditions, there exists a unique Nash Equilibrium. We use a Bayesian trust management model based on Dirichlet family of probability density functions to predict the likely future behaviour of an IDS based on its past records.

The uncertainty associated with IDS gets pacified due to this model. Acquaintance is managed between the IDSs using the estimated trust values to enhance the accuracy of the system. Dirichlet function with game theoretic approaches makes the system even more robust and efficient. We now try to introduce the concept of reinforcement learning to make the system more secure and robust. Reinforcement learning takes environment into account to maximise the overall satisfaction levels. Markov decision process (MDP) are generally used to formulate the environment. Since Nash Equilibrium technique is limited for a single attack scenario we will be considering each attack as a separate state and model it as a MDP for effective detection of Intrusion in our network.

Section 2 tells us about the related work which have been done in this field. Section 3 discusses about the motivation to carry out this project. Section 4 describes the Architectural Overlay of the system. Section 5 discusses our proposed work in detail. Section 6 gives us an analysis of why the proposed work is better than any other current methods. Section 7 gives us the conclusion and the plans for future work.

2 Related Work

In one of the work presented by Zhu and Tamer [4], they propose a trust management system which involves IDSs to build trust rapport by exchanging test messages. In this each IDS sends a trace of possible attacks from its own database (one with risk level of each attack), to its acquaintances to test their trustworthiness. Each of these peers then sends back the risk values of each of those attack in trace. The sender then cross-checks these values with its own knowledge database and generates a satisfaction level for each feedback using "a satisfaction mapping function". In [5], we see the use of a simple weighted average model to estimate the trust value while in [6] Bayesian statistics model is used to calculate the trust value.

In Bartos et al. [7], a distributed model is presented which helps to collaborate multiple ids sensors which are heterogeneous in nature. This model assumes to be able to monitor ids in different locations using multiple detection sensors. It uses game theoretic approaches in dynamic environments to optimize behavior among the sensors. They propose a hand between defenders and attackers as a model along with a trust based model e-fire to collaborate in such a highly dynamic environment while trying to prevent any kind of poisoning or manipulation by malicious attackers.

In Fung et al. [3], they propose a Trust management using Dirichlet distribution, to calculate trust based on mutual experience between ids. An acquaintance algorithm is proposes to manage its peers based on the trust value. Intrusion detection system is like a two player game between the attacker and the intrusion detection system as two opposing players. In Alpcan and Basar [8], a two person, finite game has been portrayed between each sensor of the ids based on cooperative game theory. In papers [9–11], we see the use of non-cooperative game frameworks in intrusion detection system. In Liu and Comaniciu [12], we can see the use of Bayesian game techniques in intrusion detection for ad hoc networks, using a two-person non-zero-sum incomplete information game as a framework for the system.

In Zhu et al. [2], they show that there exists a Nash equilibrium state and its unique, amongst the peers so they can communicate in cooperative manner. They were able to develop an algorithm which is iterative and converges geometrically to the equilibrium. In Allazawe et al. [13], we get an overview of traditional IDS and its challenges and how game theoretic approaches help and its limitations.

In Lye and Wing [14], they show the use of game theory in the field of security of computer networks. They construct a two-player stochastic game model to visualize the interactions between attacker and administrator. They use non-linear program to evaluate the Nash equilibrium state or the best possible strategies for both players taken into account. These strategies can then be used by administrators to improve the security of the system.

In Alpcan and Tamer [15], A game theoretic based model is made for the sensors observing and reporting attacks to the IDS as a finite Markov chain. Therefore a two player stochastic Markov game is observed depending on the information on the players. It captures various intricacies of the system. Both MDP and Q-learning methods are used to build the foundation for development of various strategies for the players. As we can see trust management and game theoretic approaches have not been used in together as a collaborative system to enhance the robustness of intrusion detection system as a whole.

3 Motivation

The most important flaw of Intrusion Detection Systems based out of game theory is that it assumes the adversary's rationality. It assumes that it would take steps to obtain a maximum gain (or near maximum) for itself or its interests. However, in reality, this may not always be the case as human behavior is still unpredictable. So it is hard to discern whether the users attack is rational or whether he/she is simply trying to confuse the IDS by employing another attack vector. Another area of weakness of the game theoretic approach to intrusion detection is the unsolved approach on how to detect and handle simultaneous attacks [14].

Most of the systems available are either optimized for Collaborative Intrusion Detection Networks [CIDN] (group of Intrusion Detection System (IDS)) or for a single IDS. There is no generalized system available for both CIDN and IDS.

4 Architecture

Our architecture can be quite similar to Fig. 2 and also to Fig. 3. The consolidated architecture can be described as network of different IDS. The system will have a group of IDS's who are connected to each other via Internet. Each IDS will in turn consist of group of computers which are mostly connected via LAN. So in short, there will be communication between clients in a network forming a IDS as well as multiple IDS communicating together.

There will be 2 types of messages which are being transferred through our system:

1. Local Messages
2. Global Messages

Local Messages are the messages which will be transferred in between a given IDS (LAN connected) and Global messages will be transferred in between different IDS. The messages which are transferred is explained in depth in the next section. Our system can be a mixed of Anomaly detection system and Misuse detection system.

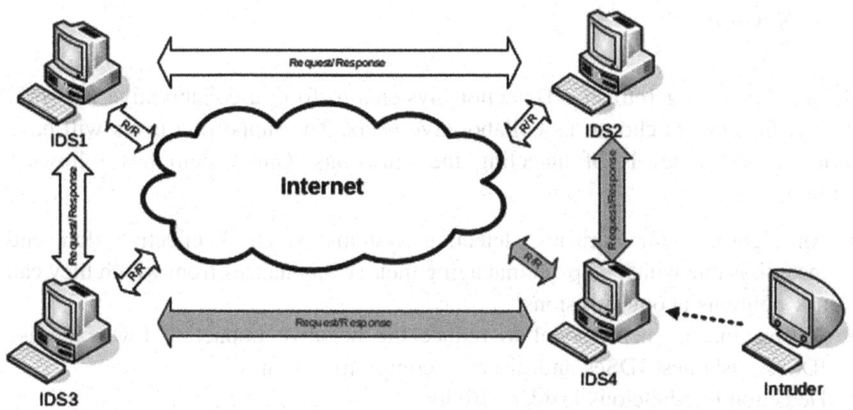

Fig. 2 Individual IDS

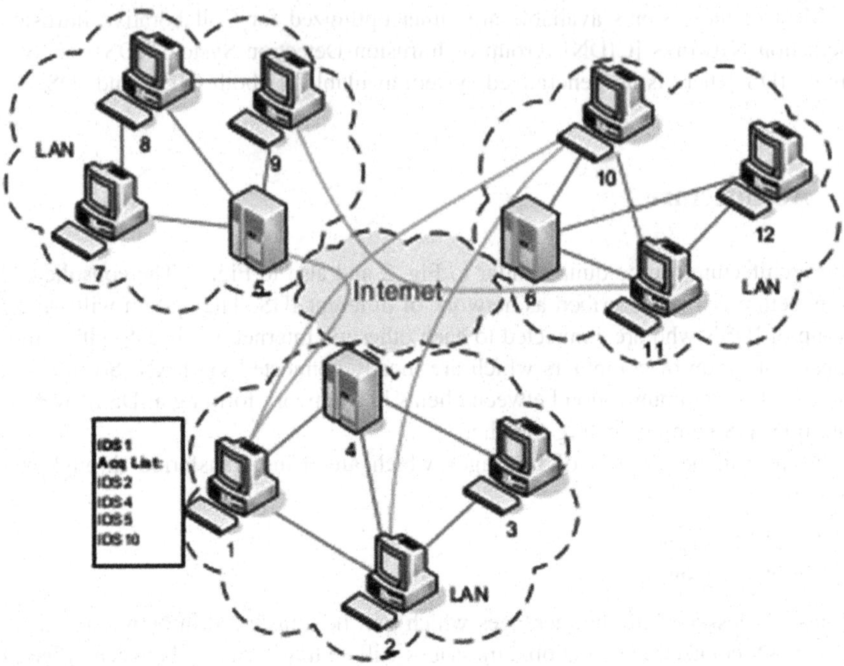

Fig. 3 Architecture overlay

5 Proposed Work

5.1 Dirichlet Based Collaborative Intrusion Detection System

We are connecting Intrusion Detection Systems to form a collaborative network. Here each IDS can choose its collaborative peers. The supporting peers will have varied expertise levels of detecting the intrusions. Our system has following features:

- An algorithm for intrusion detection systems which is effective, fair and incentivisable which helps in managing their acquaintances from which they can ask opinions about intrusions;
- A trust management model to reduce the negative impact of low expertise IDSes, dishonest IDSes and discover compromised ones
- Detection of Malicious insider activity
- System should be scalable and highly elastic in terms of trust evaluation, network size, and assessing the intrusion.

Links between IDSes indicate their collaborative relationships. Each node maintains a list of acquaintances whom it trusts the most and collaborates with. Nodes communicate by means of intrusion evaluation requests and corresponding feedback. There are two types of requests:

Intrusion Consultation and Evaluation Requests Whenever a suspicious behavior is detected by IDS and the expertise level still remains insufficient to make a decision, it sends other requests to the other friend IDSes for consultation. Feedback from the friend IDS's is cumulated and aggregated. A final conclusion is developed using those feedbacks. The information provided to friend IDS depends on the trust level of each friend.

Fake Test Messages Nodes in the Collaborative Intrusion Detection Network use these fake test messages for finding out the trust of IDS. These messages are "fake" consultation requests, which are formulated in a manner that makes them difficult to be distinguished from real consultation message requests [3].

The content of these messages depends on various factors. Some of them can be:

(1) Type of Attack
(2) Number of Peers
(3) Topology of Network etc.

A behavioral graph is made for describing the activities. The testing node has prior information about the true result of the fake testing message. It then uses the received feedback to derive at a trust value for the other nodes in the network. This can be done using standard machine learning and regression techniques. This will help us in identifying malicious nodes. IDSes use different metrics to rank and rate alerts. Let's assume there exists a function H, which maps Intrusion Detection System's alert ranking to a [0, 1] interval where 0 denotes minimal level of traffic and 1 highly dangerous intrusions. H follows more severe than partial order relationship which means if an alert a_j is more severe than alert a_i then H preserves that relationship by having $H(a_j) > H(a_i)$. The satisfaction of feedback is find out using these factors:

1. The received answer ($a \in [0, 1]$),
2. The difficulty level of the test message ($d \in [0, 1]$) and
3. The expected answer ($r \in [0, 1]$).

More the value of d, more difficult it would be to answer the request correctly. The difficulty of a test message can be determined by the age of signatures. The difficulty level is low for the messages generated from old signature; medium difficult for messages generated using new signature and high difficult for malicious traffic taken from previous attacks.

We can measure the quality of feedback using a function $Sat(r, a, d)$ ($\in [0, 1]$) for representing the level of satisfaction in the received answer which depends on the distance from the expected answer and difficulty of the message. It can be written as follows:

$$Sat(r,a,d) = \begin{cases} 1 - \left(\frac{a-r}{\max(c_1 r, 1-r)}\right)^{d/c_2} & a > r \\ 1 - \left(\frac{c_1(r-a)}{\max(c_1 r, 1-r)}\right)^{d/c_2} & a \leq r \end{cases} \tag{1}$$

where c_1 is used penalizing the wrong estimates. c_1 value is always >1 to show that the estimates which are lower than exact answer are penalized strongly than those that are higher. $c_2 \in R^+$ controls the satisfaction sensitivity, where large values means more sensitive to the distance between the correct and received answers. This equation makes sure that levels with low difficulty are more severe in their penalty than incorrect answers [3].

Our interest lies in finding out the distribution of the satisfaction levels in the answers provided by each peer IDS. We use this information for estimating the satisfaction level. We use a beta distribution with binary satisfaction level (satisfied, —satisfied). Here we are using Dirichlet distribution [4] for solving our problem which involves multi-valued satisfaction levels. This kind of distribution is most suited for our trust management model since the trust is updated based on the history of interactions.

We consider X as a discrete random variable which signifies the satisfaction level peer's feedback. X is chosen such that $X = \{x_1, x_2, \ldots, x_k\}$ ($x_i \in [0, 1]$, $x_{i+1} > x_i$) are the different levels of satisfaction. The trustworthiness of a given IDS or peer would be an input to our Game theoretic model for IDS.

$$T^{uv} = E[Y^{uv}] = \sum_{i=1}^{k} w_i E[p_i^{uv}] = \frac{1}{\gamma_0^{uv}} \sum_{i=1}^{k} w_i \gamma_i^{uv} \tag{2}$$

where,

- p_i^{uv}, denotes the probability that peer v provides answers to the requests sent by peer u with satisfaction level x_i.
- γ_i^{uv} is the cumulated evidence that v has replied to u with satisfaction level x_i.
- w_i, an associated weight of each satisfaction level x_i
- Y^{uv} be the random variable denoting the weighted average of the probability of each satisfaction level in p^{uv}.

5.2 Game Theoretical Model for Intrusion Detection System

Prevention of intrusion is just set of interaction between the IDS which protects a target system (TS) and the user. This situation can be studied in detail using Game Theory. A game (stage game) is constructed to model the interaction between user and IDS. It can be considered as an infinitely repeated game. The solutions to the stage game and to the repeated game are then given and interpreted. Using this model we can predict user intentions and preconditions for an attack and thus we can prevent any insider intrusions and outsider attacks.

We model the interactions using a two player stochastic game between the user and the IDS. A N-node intrusion detection network is considered in our model. Let the n nodes be denoted by N = {1, 2 ,…, N}. N_u^d will represent the set of neighbors for peer u with maximum distance d ∈ $R^{+,}$ i.e., N_u^d = {i ∈ N: dist(i, u) <= d, i <= u}, where dist: N N –> R^+ is a distance function measuring the distance between two nodes. N1 u will be same as N-u because it will contain all nodes except u. There is symmetric information flow in the network. R_u^v represents the set of resources demanded by v from u for its full satisfaction. Whereas the minimum acceptable set from u to v is represented by mv u. Let p_v^u ∈ R^+ be the set which u actually allocates to v. This parameter is decided by u and is private to u and v. Therefore, to satisfy v this should lie over the interval [m_u^v, r_u^v].

We assume that we are aware of trust values between each peer and its neighbors, as its a distributed trust management system. Let T^{uv} ∈ [0, 1] represent this trust value from u to v. p_{uv} parameter is dependent on this trust value T^{uv} perceived by u. Each of the peer will try to maximize its effort in order to satisfy its neighbors but having a capacity constraint C_u, determined by its own resource capacity such as bandwidth, CPU, memory, etc. Following relation will hold true:

$$\sum_{v \in \mathcal{N}_u^d} p_{uv} \leqslant C_u, \quad \text{for all } u \in \mathcal{N}. \tag{3}$$

In this model, an utility function NS_{uv} is defined to model the satisfaction level from a peer to its neighbors. It is defined as follows:

$$S_{uv} = \frac{\ln(\alpha \frac{p_{uv} - m_{vu}}{r_{vu} m_{vu}} + 1)}{\ln(\alpha + 1)} \tag{4}$$

where, a ∈ (0, 1), a system parameter to control satisfaction curve, ln(a + 1), the normalization factor, ln chosen because of its property of proportional fairness. We will set the net satisfaction S as Trust (which we got from Dirichlet Solution) multiplied by Satisfaction level of neighbors.

$$S_{uv} = NS_{uv} \times T^{uv} \tag{5}$$

Let U^u: $R_+^{L(u,d)}$ → R_+ be the peer u's aggregated altruistic utility, where L(u, d) = card(N_u^d), the cardinality of the set N_u^d. Let the payoff function, U_u, for u be given by:

$$U_u = \sum_{v \in \mathcal{N}_u^d} w_{uv} S_{uv} w_{uv} = T_v^u p_{vu} \tag{6}$$

where, w_{uv}, weight given on v's satisfaction level S_{uv}. More the trust, more is the weight. In this model, every peer $u \in N$ tries to maximize U_u within its resource capacity. To find optimum value, following function can be devised:

$$\begin{aligned} &\max_{\{p_{uv}, v \in \mathcal{N}_u^d\}} && \sum_{v \in \mathcal{N}_u^d} w_{uv} S_{uv} \\ &\text{s.t.} && \sum_{v \in \mathcal{N}_u^d} p_{uv} \leqslant C_u \\ & && m_{vu} \leqslant p_{uv} \leqslant r_{uv} \forall v \in \mathcal{N}_u^d \end{aligned} \tag{7}$$

where, S_{uv} and w_{uv} have been previously defined in Eqs. 2 and 3 respectively. As we observed earlier every peer needs to find an optimal value and thus an optimization problem (OP) to solve. This problem is a concave one in which the function is a concave function in p_{uv} constrained by the cardinality of the set N_u^d. We assume the size of whole network is large and peers within a radius d can only communicate with each other and thus N independent optimization problems are there. Therefore game can be modeled by triplet (N, A_u, U_u), where Au is action set for peer u for $u \in N$, U_u is the payoff function defined in Eq. 3 and N is the size of network [2]. Action here means allocation of resources. Action set, $A_u = \{p_u \in R^{L(u,d)+} - \sum v \in N_u^d\ p_{uv} <= C_u\} \cap \{p_u \in R^{L(u,d)+} - m_u^v <= p_{uv} <= r_{vu}, v \in N_u^d\}$. Following condition shows that action set is non-empty.

$$C_u \geqslant \sum_{v \in \mathcal{N}_u^d} m_{vu} \tag{8}$$

Lagrange relaxation can be used to solve for Nash equilibrium. Lagrangian of peer u's optimization problem based on three lagrangian multipliers can be used to devise a relaxed game model [2]. Using this relaxed model we can try to solve for Nash Equilibrium. First order KKT condition can then be applied to this optimization problem.

Since our system is comprised by taking the positive effects of Game Theory and Trust Management models, it should theoretically perform better than the available Intrusion Detection Systems. We can also refer from the results of Zhu et al. [1] and Fung et al. [13] for more introspection on the implementation details.

5.3 Extended Game Theoretical Model for Intrusion Detection System

The model proposed using game theoretic and trust management models can further be decomposed into smaller manageable components so we can figure out strategies individually for each sub model using Markov decision process along with game theoretic methods. Subgames can be defined using nearly isolated clustered state and MDP can be defined for states which has meaningful actions for only one player. Using the strategies from each sub model or components we can find the overall best-response for each player. The computation time is said to be reduced significantly using a such a decomposition method [1]. Even different attacks can be detected using Markov decision process.

We propose to use reinforcement learning to enhance the game theoretic model for IDS. Markov decision process is being used to solve reinforcement learning. In MDP, if the current state is some state s, an action available to s is then chosen by a decision maker a. The process randomly moves to a new state s′ with corresponding state transition function P_a (s, s′) and reward R_a (s, s′). Therefore, s′, the next state depends on decision maker a and the current state s. The state transitions satisfies the Markov property [2], which states that "effects of an action taken in a state depend only on that state and not on the prior history".

The MDP for our system is defined using filtering variables:

- S = State corresponding to one attack vector
- A = Actions which are either reporting attack or remaining silent
- P = Probability of transition is specific to the current state that is the attack vector.
- R = Reward is directly proportional to the trustworthiness factor taken from Dirichlet distribution
- γ = this depends on the Nash equilibrium stabilization time.

One of the most critical part in any MDP is the decision maker's policy [16]. This policy is defined by a function ∏ that allots an action ∏(s) to a state s. Policy function ∏ should be chosen in such a way to maximize sum cumulative function of the rewards. We need to take care of following criteria:

- The trustworthiness of the IDS
- The Nash equilibrium state
- Different attack attributes
- Rewards for moving from one state to another

Using the knowledge of the reward function R and the state transition function P, we try to devise a policy to maximize the discounted rewards.

Calculation of optimal policy can be done using standard set of algorithms which requires two arrays indexed by state: V, set of real values; ∏, the set of actions. The algorithm results to give ∏ the solution and V(s), the discounted sum of the rewards that can be gained by following that solution from state s [17].

The algorithm is a two-step one, these two steps are repeated for all states until the result becomes constant and no further changes are observed. They are defined as follows:

$$\pi(s) := \arg\max_a \left\{ \sum_{s'} P_a(s,s')(R_a(s,s') + \gamma V(s')) \right\} \tag{9}$$

$$V(s) := \sum_{s'} P_\pi(s,s')(R_\pi(s,s') + \gamma V(s')) \tag{10}$$

These steps can be done state by state or even all states at once or maybe even specifically more to certain states. The algorithm will converge to a optimal solution unless certain set is excluded all together. To resolve this a further function is defined which corresponds taking the action a and then continuing optimally:

$$Q(s,a) = \sum_{s'} P_a(s,s')(R_a(s,s') + \gamma(s')) \tag{11}$$

This above function is an unknown one but it learns based on (s, a) pairs with their outcomes s′. Thus, there is an array Q and "earns" to update it. This is known as Q-learning. Markov decision processes can be solved even without explicit specification of the transition probabilities using reinforcement learning. In this a simulator aids in accessing transition probabilities, which is generally restarted many times from a uniformly random initial state. As previously mentioned we propose a model by breaking it down to smaller sub module and solve it using MDP to find strategies for the game between the attacker and the IDS. Reinforcement learning enhances the MDP and thus helps in solving for an optimal state (Nash equilibrium). Thus, here we are able to extend upon the game theoretic approaches previously along with a Dirichlet based management system to make the system even more robust.

6 Analysis

We proposed two new models for Intrusion Detection Network and analyzed the previous available models.

Dirichlet Based Collaborative Intrusion Detection System is based using Dirichlet distribution as a trust distributing mechanism. Two types of requests are sent: Intrusion Consultation and Evaluation Requests and Fake Test Message request. A satisfaction function is formed using received feedback, expected answer and difficulty level of request. This helps in formulating a trust value for each peer. This method is generally fast but the accuracy is less than other systems.

Game Theoretical Model for Intrusion Detection System is made using the concept of Nash Equilibrium. Lagrange Multipliers were used to find the Nash Equilibrium state. The optimal strategy involves considering both the gameplay of individual player. This model can only be used when a single attacker is attacking the system. It also considers that the attacker always play using extreme rules that is he can't change from an attacker to a normal peer.

Game Theoretic Dirichlet Based Collaborative Intrusion Detection System uses the concept of Dirichlet Distribution based trust model along with Nash Equilibrium. We use the trust value which we get from Dirichlet Based Message Request and use them for formulating the Nash Equilibrium's Lagrange multipliers. This method has better accuracy than both the above methods and it takes into consideration that the attacker can change roles in between.

Table 1 Comparisons of different methodologies

Type of IDS	Features
Dirichlet based collaborative intrusion detection system	IDS built using Dirichlet distribution for distributing trust. Faster than conventional systems. Can handle only one attack at a time. Accuracy is less
Game theoretical model for collaborative intrusion detection system	IDS built using game theory and Nash equilibrium. Also handles one attack at a time. Considers attackers to attack in extreme fashion
Game theoretic Dirichlet based collaborative Intrusion detection system	IDS uses Dirichlet distribution along with Nash equilibrium. Trust derived from Dirichlet distribution and fed to Nash equilibrium. removes the flaw of game theoretical IDS and has better accuracy than trust based
Extended game theoretic Dirichlet based collaborative intrusion detection system	Uses game theory, Dirichlet distribution and Markov decision process. Better than all other alternatives as It considers all attack scenarios with high accuracy and power to counter simultaneous attacks. Slower than other systems

Extended Game Theoretic Dirichlet Based Collaborative Intrusion Detection System is basically the advanced version of the approach explained previously. It uses Markov Decision Process for modelling the Intrusion Detection system. Submodels are generated and reinforcement learning techniques are applied. This is the best model among all the proposed models. It even takes into consideration multi-attacker scenario. Time complexity of the system is high but it reduces the false positives which is a significant advantage. Due to the aforementioned features we can use it in high secure environments.

The summary of the analysis is done in following Table 1.

7 Conclusion

In this paper, we present an extended game theoretic approach along with Dirichlet based trust management model for an intrusion detection system. IDS are an important part of any software system and so also a hard task to make it efficient and robust. We first start off with a trust management model for Collaborative Intrusion Detection System. This model is based on Dirichlet density functions and it takes care of evaluating uncertainty in estimating future behavior of peers in IDS or between IDSes in an IDN. Measurement of this uncertainty aids to deploy an adaptive message exchange rate which then helps in making the system scalable. Along with the "forgetting factor", it is robust against some common attacks and threats. We then next moved on to game theoretic approaches and showed the existence of Nash equilibrium in the system and its uniqueness. This takes care of the free rider problem in the IDS. Then finally we move on to MDP and reinforcement learning which helps in breaking the model into smaller sub models to find the optimal strategy for the game. All these techniques and methodologies when put together gives us a secure and robust Collaborative Intrusion Detection System which enhances the security of the network and the overall system.

References

1. Yegneswaran, V., Barford, P., Jha, S.: Global intrusion detection in the DOMINO overlay system. In: Proceedings of Network and Distributed System Security Symposium (NDSS04) (2004)
2. Zhu, Q., et al.: A game-theoretical approach to incentive design in collaborative intrusion detection networks. In: International Conference on Game Theory for Networks, 2009. GameNets' 09. IEEE (2009)
3. Fung, C.J., et al.: Dirichlet-based trust management for effective collaborative intrusion detection networks. IEEE Trans. Network Service Manage. **8.2**, 79–91 (2011)
4. Zhu, Q., Tamer, B.: Dynamic policy-based IDS configuration. In: Proceedings of the 48th IEEE Conference on Decision and Control, 2009 held jointly with the 2009 28th Chinese Control Conference. CDC/CCC 2009. IEEE (2009)
5. Fung, C., Baysal, O., Zhang, J., Aib, I., Boutaba, R.: Trust management for host-based collaborative intrusion detection. In: 19th IFIP/IEEE International Workshop on Distributed Systems (2008)
6. Fung, C., Zhang, J., Aib, I., Boutaba, R.: Robust and scalable trust management for collaborative intrusion detection. In: 11th IFIP/IEEE International Symposium on Integrated Network Management (IM09), to appear (2009)
7. Karel, B1., Martin, R.: Trust-based solution for robust self-configuration of distributed intrusion detection systems. In: The 20th European Conference on Artificial Intelligence, ECAI (2012)
8. Alpcan, T., Basar, T.: A game theoretic approach to decision and analysis in network intrusion detection. In: Proceedings of the 42nd IEEE Conference on Decision and Control, Dec 2003
9. Alpcan, T., Basar, T.: A game theoretic analysis of intrusion detection in access control systems. In: 43rd IEEE Conference on Decision and Control, 2004. CDC, vol. 2. IEEE (2004)
10. Alpcan, T., Basar, T.: An intrusion detection game with limited observations. In: Proceedings of the 12th International Symposium on Dynamic Games and Applications (2006)
11. Nguyen, K.C., Alpcan, T., Basar, T.: Fictitious play with imperfect observations for network intrusion detection. In: Preprints of the 13th International Symposium Dynamic Games and Applications (ISDGA 2008). Wroclaw, Poland (2008)
12. Liu, H.M.Y., Comaniciu, C.: A Bayesian game approach for intrusion detection in wireless ad hoc networks. Valuetools, Oct 2006
13. Alazzawe, A., Asad N., Bayaraktar, M.M.: Game Theory and Intrusion Detection Systems. (2006)
14. Lye, K., Wing, J.M.: Game strategies in network security. Int. J. Inf. Secur. **4**(1–2), 71–86 (2005)
15. Alpcan, T., Basar, T.: An intrusion detection game with limited observations. In: Proceedings of the 12th Internationl Symposium on Dynamic Games and Applications (2006)
16. Wikipedia Contributors: Markov decision process. Wikipedia. The Free Encyclopedia. Wikipedia, The Free Encyclopedia, 7 April 2015. Web.16 April 2015
17. Wikipedia Contributors: Reinforcement learning. Wikipedia, The Free Encyclopedia. Wikipedia. The Free Encyclopedia, 9 April 2015. Web. 16 April 2015

Implementation of ECDSA Using Sponge Based Hash Function

M. Lavanya and V. Natarajan

Abstract Elliptic Curve Cryptography (ECC) is a public key cryptographic technique. Here the encryption and decryption are done in finite field either in prime mode or in binary mode. Goal of this work is to design a light weight and fast message authentication algorithm. The Digital signature Algorithm used in ECC i.e. Elliptic curve digital signature algorithm (ECDSA) uses SHA-1 as the algorithm for generating the hash code. In this paper we propose a technique of using Sponge hash for generating the hash code and signing the message with the newly generated hash code. This approach reduces the bytes per cycle time of the algorithm used in generating the hash code for authentication. when the bytes/cycle time is reduced then the energy consumption will also be reduced and the computation time is also reduced when used in resource constrained environments. *abstract* environment.

Keywords Elliptic curve cryptography · Sponge function · Hash code · Digital signature algorithm

1 Introduction

A Cryptographic hash function H is a process which takes an input of arbitrary length {0, 1}* and produces an output of fixed length $\{0, 1\}^n$ [1]. The output generated by the hash function is called as the message digest or hash code or hash value or imprint or digital finger print [2, 3]. The Cryptographic hash functions are classified as keyed and unkeyed functions, keyed functions are called as Message authentication codes (MAC) and unkeyed are called as Modification detection

M. Lavanya (✉) · V. Natarajan
MIT Campus, Anna University, Chennai, India
e-mail: drop2lavi@gmail.com

V. Natarajan
e-mail: natraj@mitindia.edu

M. Senthilkumar et al. (eds.), *Computational Intelligence, Cyber Security and Computational Models*, Advances in Intelligent Systems and Computing 412, DOI 10.1007/978-981-10-0251-9_33

codes (MDC) [4]. Generally hash functions are built using Merkle-Damgard construction or Sponge construction [5]. The main strategy of Merkle-Damgard construction is that, when a compression function is designed to be collision resistant the resulting hash function is said to be collision resistant. Sponge construction was introduced with SHA-3 [5] hash function which uses Keccak [6] family of sponge functions. Hash Functions find application in many fields including Digital signature, MAC, HMAC (Hash Message Authentication Codes), Kerberos authentication, Key derivation, One time password, PGP and SSL/TLS. Various attacks on hash functions [7–10] reveals that the construction of hash function does not make it at the expected security level. Security of cryptographic hash function is measured by three criteria collision resistance, pre-image resistance and second pre-image resistance [11]. Random oracle [12] is a mathematical function mapping each possible query to a random response from its output domain. It is a theoretical construction that satisfies all known security criteria for hash function. Random oracles are used as an ideal replacement for cryptographic hash functions in schemes where randomness assumptions are needed Iterated hash functions have state and inner collisions hence cannot be a good candidate for random oracle model [13], so finite state sponges are used for constructing hash code which ensures compact security. The input to a sponge function is a variable length string and produced an infinite output string. Sponge functions are used in various modes [14] like n-bit Hash function, One way function, MAC function, Random access stream cipher, Deterministic random bit generator (DRBG), Key derivation function. In this paper sponge functions are used for generating the hash code, which will be used by Elliptic curve Digital Signature Algorithm (ECDSA). ECDSA is elliptic curve analogue of Digital signature algorithm. ECDSA was first developed for ANSI by the Accredited standards committee on financial services X9. In ECDSA the public and private key pairs are generated based on certain domain parameters. These domain parameters are common to the group of users in communication for a period of time. There are different methods for the generation and management of domain parameters [15]. Domain parameters once generated are used for various purposes (Signature, Key generation). ECDSA contains a pair of keys for generation and verification, the key pairs are generated for a specific set of domain parameters. The public and private key of ECDSA are used only for the generation and verification of the signature. A new secret random number is also generated during the signature generation process. This random number is protected from unauthorized change.

The paper is organized as follows, Sect. 2 gives an over view of the sponge functions and the elliptic curve digital signature algorithm. Sect. 3 gives the is G proposed work describing the MBLAKE2b construction and message digest generation using sponge function. This generated digest is used in signature generation and verification using ECDSA, Sect. 4 describes the implementation setup and results and Sect. 5 concludes the work.

2 Related work

Security plays a vital role in all fields of communication in these days. There are many security services [11] like confidentiality, Integrity, Authentication, Non-repudiation, Availability and Data freshness. In the current world there are many cryptographic algorithms providing this services using different ciphers like conventional symmetric ciphers and asymmetric ciphers. There are number of algorithms for providing confidentiality in resource constrained environments like WSN, RFID etc. Some of the recently proposed cryptographic algorithms are PRESENT [16], HIGHT [17] and CURUPRIA-2 [18] which gives better performance than skipjack, the block cipher used in TinySec [19]. Most of the architectures develop message authentication by deploying some MAC algorithm derived from conventional block ciphers like CBC-MAC [20] which is strictly sequential. Some of the MAC Algorithms use Carter-wegman [20] design technique, the algorithms in this list are GMAC that uses GHASH [21] some other algorithms uses the ALRED design. SHARK, SQUARE, PELICIAN, MARVIN uses ALRED construction [21–25]. In this work we use the concept of sponge functions for generating the hash codes.

2.1 *Sponge Function*

The sponge is constructed by a simple iterated construction with a function f with variable length input and arbitrary length output. The output is constructed based on a fixed length transformation or permutation operating on ‘b’ bits. Sponge contains various operating states. The state is initialized to 0. State operates on ‘b’ bits, $b = r + c$. Input message is padded and cut into blocks of r bits. In the construction of sponge function there are two phases the absorption and squeezing. In absorption phase the r bit input is XORed into the first r bits of the state and then the fixed permutation function f is applied. This is repeated until all the input blocks are absorbed. The absorption phase is followed by the squeezing phase, where the first r bits of the state are returned as output block, interleaved with the permutation function f. The choice of the number of output blocks gives the requested no of bits (l). Finally the output is truncated to first l bits. The last c bits of the state are never affected directly by the input or the output, it determines the attainable security level of the construction. There are many instances of sponge functions like Keccak, Quark, Photon, Spongent etc., The sponge construction is shown in Fig. 1. There are many modes of operation of the sponge function, the hash function mode is considered in this paper, where the sponge function is used as a n-bit hash function by simply truncating the output.

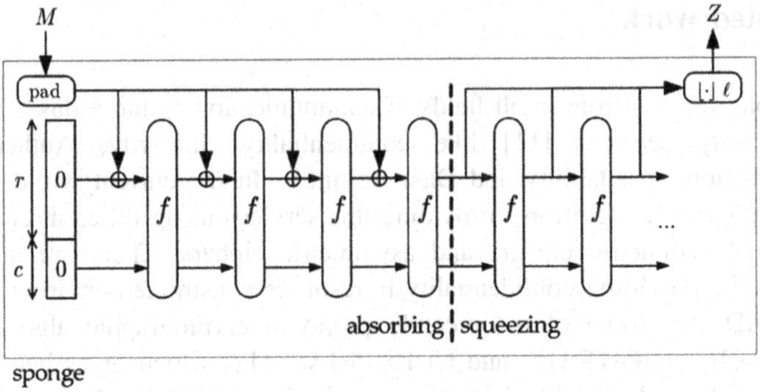

Fig. 1 Sponge construction

Algorithm Absorption Phase

Assume the state is initialized to zero and the message is padded and split into blocks of r bits each , f is the fixed permutation function. The number of bits in the state is b(b=r+c).

1. Initialize the state to 0, i.e., S= 0(b bits)
2. Let P be the length of the message after padding
3. for i =0; i ¡ p do
3.1 S = S ⊕ (P || 0(b-r))
3.2 S= f(S)
3.3 end for
4. return S

Algorithm Squeezing Phase

The message digest of desired length is generated in this phase 1.Truncate the state S to r bits. let it be Z

2. While Z < l do
2.1 S = f(S)
2.2 Concatenate Z and Length of S
2.3 end While
3. return length of Z
4. truncate Z to l bits

2.2 Elliptic Curve Digital Signature Algorithm

Elliptic Curve Digital Signatures are used for authentication. Selecting the correct domain parameter avoids the Pollard's rho or pholling hellman attacks. Let E be the elliptic curve, it is necessary that the number of F_q rational point on E be divisible by large prime number 'n'. where $n > 2^{160}$. The domain parameters of elliptic curves are

1. a, b from Fq
2. Compute N $= \#E(Fq)$
3. Verify N is divisible by large prime number n and $n > 2^{160}$
4. Select an arbitrary point $'G'$, $G \in E(Fq)$, repeat until $G \neq O$

ECDSA Key pair Generation The key generation depends on the domain parameters. The user selects a random number d, such that $d \in (1, n-1)$. Q is computed as $d^{*}G$, Where G is the point on Elliptic curve called the generator point. Now Q is the public key and d is the private key.

ECDSA signature Generation and Verification To sign a message the user should know all the domain parameters i.e $D = q, Fr, a, b, G, n, h$ and the public and the private keys Q and d.

Generation

1. Select a random integer k, such that $1 \leq k \leq n-1$
2. Compute k * G which will be of the form (x, y) convert x to integer.
3. Find $r = x \bmod n$, if $r = 0$ go to 1
4. Compute $k^{-1} \bmod n$
5. Find SHA-1(m) = e
6. $S = k^{-1}(e + dr) \bmod n$ if s = 0 go to 1 thus the generated signature is S and r, which is sent to the receiver.

Verification

1. Verify r and S in [1, n − 1]
2. compute SHA-1(m)
3. $w = s^{-1} \bmod n$
4. u = e w mod n, and v = rw mod n
5. compute $X = uG + vQ$
6. if X = 0 reject the signature, otherwise convert x coordinate to integer and compute v = x mod n.
7. accept of v = r. the receiver accepts the signature only if v = r.

3 Proposed Method

3.1 MBLAKE2b—A Modified Version of BLAKE2b

The hash function BLAKE [26, 27] is one of the second round finalist in SHA-3 competition conducted by NIST. There are two flavors of BLAKE; 32bit version (BLAKE2s) and 64 bit version (BLAKE2b [28]). The iteration method incorporated in BLAKE is modified version of the Merkle Damgard paradigm. It provides resistance to long message second pre-image attacks, its internal structure is built such that local collisions are impossible. The compression function used in BLAKE is modified version of the compression function used in cryptographic algorithm chacha [29]. In the proposed method we have modified the compression function of BLAKE2b [30], where original BLAKE compression function uses four inputs; a chain value, a message block, a salt, and a counter. In MBLAKE2b the constants and the salt value are not used so that the memory storage is reduced, the message block is not included in the compression function, instead it is initialized in the state along with the chain value. A round is a transformation of a state v, State is a 4 × 4 matrix initialized with the chain values of the BLAKE2b initial vector IV and the message. There are 12 rounds and 8 G functions for each round which is used for computing the diagonal step and the column step of the states. The advantage of the G functions is the diagonal step and the column step can be performed in parallel there by reducing the hardware requirement, and reducing the computation time of the algorithm. The G function used in modified MBLAKE2b is shown in the Fig. 2 and the column steps and diagonal steps are shown in Fig. 3. The state is represented as

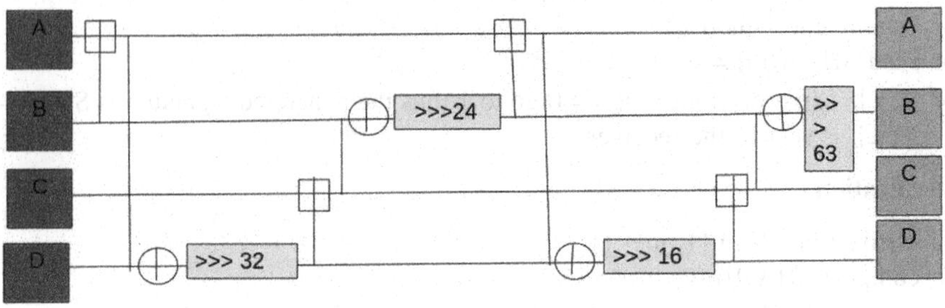

Fig. 2 Modified BLAKE2b;G function of each round

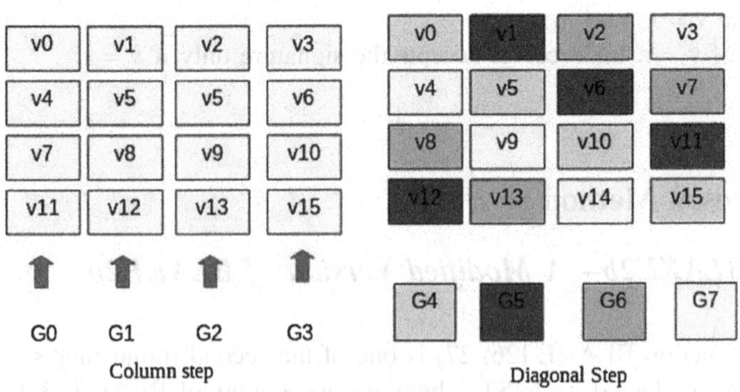

Fig. 3 Column step and diagonal step of MBLAKE2b

$$\text{State v} = \begin{Bmatrix} v0 & v1 & v2 & v3 \\ v4 & v5 & v6 & v7 \\ v8 & v9 & v10 & v11 \\ v12 & v13 & v14 & v15 \end{Bmatrix}$$

The G functions are; G0(v0,v4,v8,v12), G1(v1,v5,v9,v13), G2(v2,v6,v10,v14), G3(v3,v7,v11,v15), G4(v0,v5,v10,v15), G5(v1,v6,v11,v12), G6(v2,v7,v8,v13), G7(v3,v4,v9,v14).

MBLAKE2b version of BLAKE optimizes it in its speed. MBLAKE2b is faster than MD5 and also provides security similar to SHA-3 [31]. MBALKE2b uses 32 % RAM compared to BLAKE2b. Hence it is more suitable for resource constrained environments. In this work MBLAKE2b is used for generating the hash code. Both the compressed version and full version of MBLAKE2b are implemented and compared. Secure hash Algorithm (SHA-1) is also implemented. The running time of all these algorithms are compared. The elliptic curve digital signature algorithm(ECDSA) uses SHA-1 for generating a portion of its signature. Here SHA-1 is replaced by the sponge hash function using MBLAKEba as its compression function and their performance are compared. The generated sponge digest is used by the ECDSA algorithm for generating and verifying the signature. This novel idea of using a sponge based hash code in elliptic curve digital signature algorithm reduces the cpu utilization time.

3.2 Implementation Setup and Results

SHA-1 algorithm is implemented using GMP programming. The MBLAKE2b sponge hash is implemented in c, with both the full and the reduced version. The programs are implemented in Intel Pentium(R) dual Core processor with Ubuntu OS, with 3 GB memory The running time of both versions (normal and reduced) of MBLAKE2b are given in the Table 1.

Digital signature generation and verification are implemented in C with three different hash code generation algorithms. The message length is assumed to be fixed (512 bit block) and same for all the three cases and the computation time values are tabulated in Table 2. The signature generation and verification is also analyzed by varying the size of the message. When the message length is varied the Computation time for generation and verification increases proportionally when SHA-1 algorithm is used. In the case of sponge hash the computation time is maintained with some variation. The comparison is shown in Fig. 4.

Table 1 Comparison of computation time of different hash functions

Algorithm	Computation time (ms)
Sponge hash	0.7286
Reduced sponge hash	0.5
Sha-1	2.074

Table 2 Comparison of ECDSA with different message digests

Algorithm	Computation time (ms)
Sha-1	10.3815
Sponge hash	7.8127
Reduced sponge hash	5.64

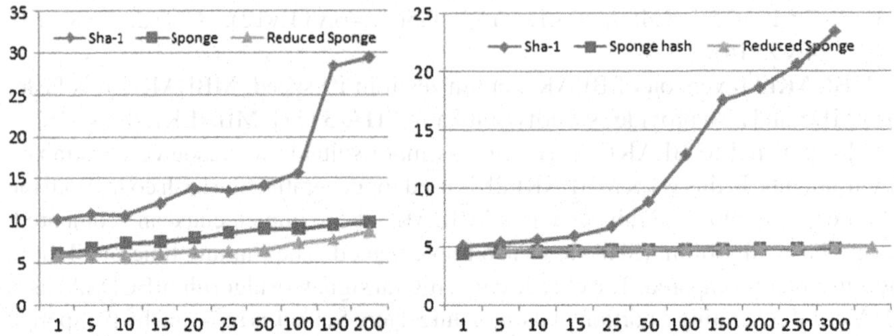

Fig. 4 **a** ECDSA signature generation. **b** ECDSA signature verification

4 Security Analysis

SHA-1, the most famous hash function used in traditional ECDSA is on the way out in present scenario. In Nov 2013 Microsoft announced that they wouldn't accept SHA-1 certificates after 2016. The traditional attacks on hash functions include output collision, second preimage, preimage, length extension and correlation immunity [14]. All these attacks concentrate on the capacity 'c' of the sponge function. When the value of 'c' is smaller the sponge function becomes more vulnerable because of the fact that the sponge functions operate on finite state. Most of the existing hash function follows Merkle-Damgard construction, where the compression function contains any block cipher with a loop to the next encryption and message digest is calculated from the last chaining value [26, 32]. According to this construction if a the message digest has provable collision resistance if the compression function is collision resistant [33]. The message digest here is the truncated value from the final chaining. The drawback of truncation process is the reduction proof for collision resistance is no longer valid. Some time we might be choosing the same digest value, if the value is smaller than the maximum. This is overcome in the Sponge function case, any digest length is supported and the resistance of the hash function is limited by the inner state.

Next coming to the security of ECDSA, Elliptic curve cryptography have high per bit security compared to other asymmetric key algorithms. For example, 160 bit key for ECC is equivalent to 1024 bit key of RSA [34]. Another important parameter of efficiency in ECC is choosing the domain parameters, NIST and

SEGC proposed recommended curves and domain parameters [34, 35]. Choice of curve with field size greater than 160 is recommended for cryptographic applications. When ECDSA is considered, the possible attacks can be on ECDLP, on Hash function and other attacks. Concentrating on the attacks on the hash function employed in ECDSA, the traditional ECDSA can be compromised if the SHA-1 algorithm is not preimage resistant or if it is not collision resistant. If SHA-1 is not preimage resistant an adversary can easily forge the signature. If it is not collision resistant, then the sender may repudiate signature The solution is the use of variable length hash function. Hence message digest based on sponge function is the best candidate.

5 Conclusion

Elliptic curve cryptography has relatively small key size compared with other asymmetric algorithms. This reduced key length results in smaller hardware area, shorter running time and saves energy. The usage of cryptographic sponge function with ECDSA provides an added advantage for application in resource constrained environments. The computation time of the algorithm for generating and verifying the digital signature is reduced when compared with the traditional algorithm. Hence such an algorithm finds application in resource constrained environments like Sensor networks, RFID etc.

References

1. Naor, M., Yung, M.: Universal one-way hash functions and their cryptographic applications. In: Proceedings of the Twenty First Annual ACM Symposium on Theory of Computing, pp. 33–43. ACM Press (1989)
2. Lucks, S.: Design principles for iterated hash functions. Cryptology ePrint Archive, Report 2004/253, 2004. http://eprint.iacr.org
3. Goldwasser, Micali, S., Rivest. R.: A digital signature scheme secure against adaptive chosen-message attacks. SIAM J. Comput. **17**(2), 281–308 (1998)
4. AlAhmad, M.A., Alshaikhli, I.F.: Broad View of Cryptographic Hash Functions. Addison-Wesley, Harlow, England (1999)
5. Damgrd, I.: A design principle for hash functions. In: Brassard, G. (ed.) Advances in Cryptology—CRYPTO ™89, 9th Annual International Cryptology Conference, Santa Barbara, California, USA, August 20–24, 1989, Proceedings, vol. 435. Lecture Notes in Computer Science, pp. 416–427. Springer (1990)
6. AlAhmad, M.A., Alshaikhli, I.F.: sha-3 fips. Addison-Wesley, Harlow, England (1999)
7. Joux, A.: Multicollisions in iterated hash functions application to cascaded constructions. In: Franklin M. (ed.) Advances in Cryptology—Crypto 2004. LNCS, no. 3152, pp. 306–316. Springer-Verlag (2004)
8. Kelsey, J., Schneier, B.: Second preimages on n-bit hash functions for much less than 2n work. In: Cramer, R. (ed.) Advances in Cryptology—Eurocrypt ™2005. LNCS, no. 3494, pp. 474–490. Springer-Verlag (2005)

9. Coron, J., Dodis, Y., Malinaud, C., Puniya, P.: Merkle-Damgard revisited: how to construct a hash ard function. In: Shoup, V. (ed.) Advances in Cryptology—Crypto 2005. LNCS, no. 3621, pp. 430–448. Springer-Verlag (2005)
10. Kohno, T., Kelsey, J.: Herding hash functions and the nostradamus attack. In: Vaudenay, S. (ed.) Advances in Cryptology—Eurocrypt ™2006, pp. 222–232. LNCS, no. 4004, Springer-Verlag (2006)
11. Menezes, A.J., van Oorschot, P.C., Vanstone, S.A.: Handbook of Applied Cryptography. CRC Press (1997)
12. Bellare, M., Rogaway, P.: Random oracles are practical: a paradigm for designing efficient protocols, In: ACM, (ed.), ACM Conference on Computer and Communications Security, pp. 62–73 (1993)
13. Bertoni, G., Daemen, J., Peeters, M., Van Assche, G.: Sponge Functions. ECRYPT Hash Function Workshop (2007)
14. Bertoni, G., Daemen, J., Peeters, M., Van Assche, G.: Cryptographic Sponge Functions. January 2011. http://sponge.noekeon.org/
15. Kopka, H., Daly, P.W.: fips-186, Nist. Addison-Wesley, Harlow, England (1999)
16. Bogdanov, A.: Present: An ultra-lightweight block cipher. In: Cryptographic Hardware and Embedded Systems—CHES ™2007. Heidelberg, Germany, Springer (2007)
17. Hong, D.: HIGHT: A new block cipher suitable for low-resource device. In: CHES. [S.l.: s.n.]. pp. 46–59. Lecture Notes in Computer Science (2006)
18. Simplicio, M.: The Curupira-2 block cipher for constrained platforms: Specification and benchmarking. In: Proceedings of the 1st International Workshop on Privacy in Location-Based Applications—13th European Symposium on Research in Computer Security (ESORICS ™2008). [S.l.]: CEUR-WS, 2008. v. 397
19. Karlof, C., Sastry, N., Wagner, D.: Tinysec: a link layer security architecture for wireless sensor networks. In: 2nd International Conference on Embedded Networked Sensor Systems—SenSys ™2004, pp. 162–175. ACM, Baltimore, USA (2004). ISBN 1-58113-879-2
20. Iwata, T., Kurosawa, K.: OMAC One-key CBC MAC. In: Fast Software Encryption—FSE ™2003. Lecture Notes in Computer Science, vol. 2887, pp. 129–153. Heidelberg, Germany, Springer (2003)
21. Mc Grew, D., Vigea, J.: The Galois/Counter Mode of Operation (GCM), May 2005. Submission to NIST Modes of Operation Process
22. Damen, J., Rijmen, V.: A new MAC construction Alred and a specific instance Alpha-MAC. In: FSE. [S.l.: s.n.], pp. 1–17 (2005)
23. Damen, J., Knudsen, L.R., Rijimen, V.: The block cipher Square. In: Fast Software Encryption—FSE ™97. Lecture Notes in Computer Science, vol. 1267, pp. 149–165. Springer, Haifa, Israel (1997)
24. Rijimen, V.: The cipher shark. In: Fast Software Encryption—FSE ™96. [S.l.]. Lecture Notes in Computer Science, vol. 1039, pp. 99–111. Springer (1996)
25. Damen, J., Rijmen, V.: The Pelican MAC Function 2005. Cryptology ePrint Archive, Report 2005/088
26. Merkle, R.: One way hash functions and DES. In: Brassard, G. (ed.), Advances in Cryptology—Crypto ™89, pp. 428–446. LNCS, no. 435. Springer-Verlag (1989)
27. Aumasson, J.P., Henzen, L., Meier, W., Phan, R.C.W.: SHA-3 proposal BLAKE. Submission to the SHA-3 Competition (2008)
28. AlAhmad, M.A., Alshaikhli, I.F.: Keccak. Addison-Wesley, Harlow, England (1999)
29. Aumasson, J.P., Neves, S., Wilcox-OHearn, Z., Winnerlein, C.: BLAKE2: simpler, smaller, fast as MD5
30. AlAhmad, M.A., Alshaikhli, I.F.: Lyra: Password-Based Key Derivation with Tunable Memory and Processing Costs. JCEN—J. Cryptograph. Eng. ISSN:2190-8508. Springer, pp. 1–15 (2014)
31. Kopka, H., Daly, P.W.: 10. IPS -202 draft SHA3 :Standards based on the fixed permutation function, 3rd edn. Addison-Wesley, Harlow, England (1999)

32. Rivest, R.: The MD4 message digest algorithm. In: Menezes, A., Vanstone, S. (eds.), Advances in Cryptology—Crypto ™90, pp. 303–311. LNCS, no. 537. Springer-Verlag (1991)
33. Wegman, M., Carter, J.: New hash functions and their use in authentication and set equality. J. Comput. Syst. Sci. **22**, 265–279 (1981)
34. National Institute of Standards and Technology: Recommended Elliptic Curves for Federal Government Use, Aug 1999
35. Standards for Efficient Cryptography Group (SECG): Recommended Elliptic Curve Domain Parameters. SEC 2, Sept 2000

Contrast-Enhanced Visual Cryptography Schemes Based on Perfect Reconstruction of White Pixels and Additional Basis Matrix

Thomas Monoth and P. Babu Anto

Abstract The existing pixel patterns for the visual cryptography scheme are based on the Perfect Reconstruction of Black Pixels (PRBP). Mathematically in PRBP the white pixels are represented by 0 and the black pixel by 1. In the usual binary image, the number of white pixels is much larger than the number of black pixels. Therefore, the perfect reconstructions of black pixels in visual cryptography schemes can decrease the contrast. Here, a visual cryptography scheme which is focused on the Perfect Reconstruction of White Pixels (PRWP) and hence can provide better clarity is presented. As in the case of all existing binary image file formats, PRWP represents white pixel by 1 and black pixel by 0. The visual cryptography scheme with PRWP can improve the clarity of reconstructed images. But by analysing the experimental results, we know that the number of black pixels in the reconstructed image is very less compared to the original image. Therefore increasing the black pixels in the reconstructed image can improve the contrast. In order to increase the black pixels use Additional Basis Matrix (ABM) to represent the new pixel pattern.

Keywords Visual cryptography · Perfect reconstruction of white pixels · Additional basis matrix · Visual secret sharing

1 Introduction

In 1995, Naor and Shamir introduced a very interesting and simple cryptographic method called visual cryptography to protect secrets [1]. Basically, visual cryptography has two important features. The first feature is its perfect secrecy and the

T. Monoth (✉)
Department of Computer Science, Mary Matha Arts & Science College, Mananthavady 670645, Kerala, India
e-mail: tmonoth@yahoo.com

P. Babu Anto
Department of Information Technology, Kannur University, Kannur 670567, Kerala, India

M. Senthilkumar et al. (eds.), *Computational Intelligence, Cyber Security and Computational Models*, Advances in Intelligent Systems and Computing 412, DOI 10.1007/978-981-10-0251-9_34

second is its decryption method which requires neither complex decryption algorithms nor the aid of computers. It uses only human visual system to identify the secret from the stacked image of some authorized set of shares. Therefore, visual cryptography is a very convenient way to protect secrets when computers or other decryption devices are not available. The simple decryption method is the reason that attracts many researchers to make further detailed enquiries in this research area. Nowadays, many related methods concerning the theory and the applications of visual cryptography are proposed.

An extended visual cryptography scheme (EVCS) was proposed by Ateniese et al. [2]. The image size invariant visual cryptography was proposed by Ito et al. [3]. There are also some other studies which focus on the methods without pixel expansion [4–6]. In 1996, Naor and Shamir proposed an alternative VCS model for improving the contrast in [7]. The other research works done by different authors are found in [8–11].

2 Visual Cryptography Scheme with PRWP

2.1 The Model

Let $P = \{1,\ldots, n\}$ be a set of elements called participants, and let 2^P denote all the subsets of P. Let $\Gamma_{Qual} \subseteq 2^P$ and $\Gamma_{Forb} \subseteq 2^P$, where $\Gamma_{Qual} \cap \Gamma_{Forb} = \emptyset$. Here the members of Γ_{Qual} are referred to as qualified sets and the members of Γ_{Forb} are called forbidden sets. The pair $(\Gamma_{Qual}, \Gamma_{Forb})$ is called the access structure of the scheme [12]. A $(\Gamma_{Qual}, \Gamma_{Forb})$ VCS with PRWP with relative difference $\alpha(m)$, contrast $\beta(m)$ and threshold $1 \le t_X \le m$ is realized using the $n \times m$ basis matrices MS^0 and MS^1 if the following two conditions hold:

1. If $X = \{i_1, i_2, \ldots, i_p\} \in \Gamma_{Qual}$ is a qualified set, then the "or" V of rows $\{i_1, i_2,\ldots, i_p\}$ of MS^0 satisfies $\omega_H(V) \ge t_X - \beta(m)$; whereas, for MS^1 it results that $\omega_H(V) \le t_X$.
2. If $X = \{i_1, i_2, \ldots, i_p\} \in \Gamma_{Forb}$ is not a qualified set then the two $p \times m$ matrices obtained by restricting MS^0 and MS^1 to rows $\{i_1, i_2,\ldots, i_p\}$ are equal up to a column permutation.

The collections C_0 and C_1 are obtained by permuting the columns of the corresponding matrix (MS^0 for C_0 and MS^1 for C_1) in all possible ways.

Formula 1 (Relative Difference)

Let $\omega_H(MS^0)$ and $\omega_H(MS^1)$ be the *hamming weight* corresponding to the basis matrices MS^0 and MS^1. Then relative difference $\alpha(m)$ is defined as:

$$\alpha(m) = (\omega_H(MS^0) - \omega_H(MS^1))/m$$

Formula 2(Contrast)

Let α(m) be the relative difference and m be the pixel expansion. The formula to compute contrast in different VCS with PRWP is:

$$\beta(m) = \alpha(m) \cdot m, \quad \beta(m) \geq 1$$

2.2 *The Construction of 2-out-of-2 VCS with PRWP*

The basic idea of visual cryptography scheme with PRWP can be explained by 2-out-of-2 VCS. The pixel layouts for the scheme are as shown in Table 1.

The basis matrices, MS^0 and MS^1 are:

$$MS^0 = \begin{bmatrix} 1 & 0 \\ 0 & 1 \end{bmatrix} \quad MS^1 = \begin{bmatrix} 1 & 0 \\ 1 & 0 \end{bmatrix}$$

The relative difference α(m) and contrast β(m) can be computed as:

$$\alpha(m) = 1/2, \quad \beta(m) = 1$$

The matrices C_0 and C_1 are:

$$C_0 = \left\{ \begin{bmatrix} 1 & 0 \\ 0 & 1 \end{bmatrix}, \begin{bmatrix} 0 & 1 \\ 1 & 0 \end{bmatrix} \right\} \quad \text{and } C_1 = \left\{ \begin{bmatrix} 1 & 0 \\ 1 & 0 \end{bmatrix}, \begin{bmatrix} 0 & 1 \\ 0 & 1 \end{bmatrix} \right\}$$

While observing the basis matrices MS^0 and MS^1, S^0 of VCS becomes MS^1 of PRWP scheme and S^1 becomes MS^0. Therefore, in VCS with PRWP scheme, to share a white pixel, the dealer randomly selects one of the matrices in C_1, and to share a black pixel, the dealer randomly selects one of the matrices in C_0 of Noar and Shamir scheme. From the results, both the relative difference and contrast of VCS with PRWP are equal to that of Noar and Shamir scheme.

Table 1 The pixel layout for 2-out-of-2 VCS with PRWP

Original pixel	Pixel value	Share1	Share2	Share1 + Share2
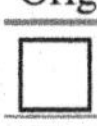	1			
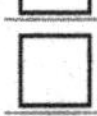	1			
	0			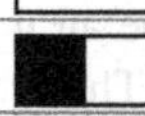
	0			

Fig. 1 The 2-out-of-2 VCS with PRWP of SI: SI (**a**), S_1 (**b**), S_2 (**c**) and $S_1 + S_2$ (**d**)

2.3 The 2-out-of-2 VCS with PRWP

The 2-*out-of-2* VCS with PRWP applied to Secret Images (SI) is shown in Fig. 1.

3 The Proposed Method

The visual cryptography scheme with PRWP can improve the clarity of reconstructed images [13]. But by analysing the results, we know that the number of black pixels in the reconstructed image is very less compared to the original image. Therefore increasing the black pixels in the reconstructed image can improve the contrast. In order to increase the black pixels use ABM to represent the new pixel pattern. The ABM for VCS with PRWP is represented by AS^1. The matrix AS^1 is used to share black pixels in the secret image. The AS^1 can be defined by an *n* x *m* Boolean matrix, $AS^1 = [as^1_{ij}]$, where

$$as^1_{ij} = 1 \Leftrightarrow \text{the jth subpixel in the ith share is white.}$$

$$as^1_{ij} = 0 \Leftrightarrow \text{the jth subpixel in the ith share is black.}$$

Formula 3 (*Additional Relative Difference for PRWP with ABM*)

Let $[s^1_{ij}]$ be an *n* x *m* basis matrix and $[as^1_{ij}]$ be an additional basis matrix of the same order. Then,

$$\alpha(m)^* = (\alpha_2 + \alpha(m))/2$$

where

$$\alpha_2 = (\omega_H(\mathrm{MS}^0) - \omega_H(\mathrm{AS}^1))/\mathrm{m},\ \alpha(m) = (\omega_H(\mathrm{MS}^0) - \omega_H(MS^1))/\mathrm{m}$$

where $\omega_H(MS^0)$ is the *hamming weight* (the number of ones) of the *m*-vector *V* of any *k* of the *n* rows in MS^0 and $\omega_H(\mathrm{AS}^1)$ is the *hamming weight* of the *m*-vector *V* of any *k* of the *n* rows in the additional basis matrix AS^1.

Formula 4 (*Additional Contrast for PRWP with ABM*)

Let α(m)* be the additional relative difference for PRWP and *m* be the pixel expansion. The formula to compute contrast in different VCS with PRWP and ABM is

$$\beta(\mathrm{m})^* = \alpha(\mathrm{m})^* \cdot \mathrm{m}, \quad \beta(\mathrm{m})^* \geq 1$$

3.1 The Construction of 2-out-of-2 VCS Based on PRWP and ABM

The visual cryptography scheme based on PRWP with ABM is illustrated by a 2-out-of-2 VCS with 4-subpixel layout. By increasing the number of pixel patterns for black pixels, the contrast of the reconstructed image can be improved without adding any computational complexity. The AS^1 for 2-out-of-2 VCS can be designed according to new pixel layout as

$$\mathrm{AS}^1 = \begin{bmatrix} 1 & 0 & 0 & 0 \\ 1 & 0 & 0 & 0 \end{bmatrix}$$

The matrices C_0 and C_1 can be designed as:

$$\mathrm{C}_0 = \left\{\pi\begin{bmatrix} 1 & 0 & 0 & 1 \\ 0 & 1 & 1 & 0 \end{bmatrix}\right\} \mathrm{C}_1 = \left\{\pi\begin{bmatrix} 1 & 0 & 1 & 0 \\ 1 & 0 & 1 & 0 \end{bmatrix}, \pi\begin{bmatrix} 1 & 0 & 0 & 0 \\ 1 & 0 & 0 & 0 \end{bmatrix}\right\}$$

The α(m)* and β(m)* are calculated as

$$\alpha(\mathrm{m})^* = 5/8,\ \beta(\mathrm{m})^* = 2.5$$

3.2 The Experimental Results

For evaluating the feasibility of the scheme, the experiment is conducted using 2-out-of-2 VCS.

Fig. 2 The 2-out-of-2 VCS with PRWP and ABM of SI: SI_1 (**a**), S_1 (**b**), S_2 (**c**) and $S_1 + S_2$ (**d**)

Table 2 The contrast and relative difference of VCS with PRWP and VCS with PRWP and ABM

VCS	VCS with PRWP		VCS with PRWP and ABM	
	α(m)	β(m)	α(m)*	β(m)*
2-out-of-2	½	2	$^{5}/_{8}$	2.5
2-out-of-3	$^{1}/_{3}$	1	½	1.5
3-out-of-3	$^{1}/_{4}$	1	$^{3}/_{8}$	1.5
3-out-of-6	$^{1}/_{12}$	1	$^{1}/_{6}$	2

The 2-out-of-2 VCS with PRWP and ABM

The Fig. 2 represents 2-out-of-2 VCS of secret images (SI) using PRWP with ABM.

Analysis of Experimental Results

For the analysis of experimental results, first compare the relative difference (α(m)) and contrast (β(m)) of VCS with PRWP and VCS with PRWP and ABM.

The results in the Table 2 shows that the relative difference and contrast of the VCS with PRWP and ABM method are better compared to those of the VCS with PRWP.

Consider the VCS with PRWP and VCS with PRWP and ABM scheme based on pixel by pixel approach with the help of graphs (Fig. 3).

By analyzing the graphs one can see that the number of white pixels is greater than that of black pixels in the secret images. But the black pixels are reduced and white pixels are increased significantly in the reconstructed secret images by using VCS with PRWP. Therefore, to enhance the contrast of reconstructed secret images requires increasing the black pixels and decreasing the white pixels. This is achieved in VCS with PRWP and ABM. Finally investigate the clarity of the

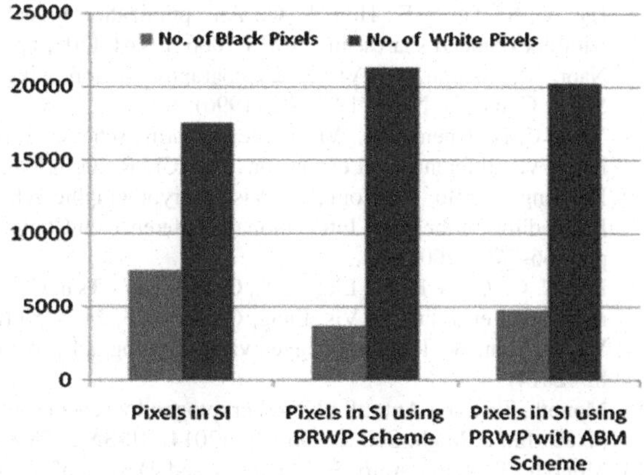

Fig. 3 The graphical representation of 2-out-of-2 VCS based on PRWP and VCS based PRWP and ABM using the secret image (SI)

reconstructed images in VCS with PRWP and VCS with PRWP and ABM using 2-out-of-2 scheme. From the Figs. 1 and 2, it is clear that the VCS with PRWP and ABM scheme achieve contrast-enhanced images than VCS with PRWP scheme.

4 Conclusions

This paper presents new methods for contrast-enhanced visual cryptography schemes based on PRWP and ABM. These methods are explained and implemented with examples. The contrast of the presented visual cryptography schemes based on PRWP with ABM and VCS with PRWP are compared here. Using these methods, some experiments were also conducted based on different VCS. These results are analysed by using tables and graphs and are also compared with the features of existing visual cryptography schemes.

References

1. Naor, M., Shamir, A.: Visual cryptography. In: Advances in Cryptology-Eurocrypt'94, LNCS 950, pp. 1–12. (1995)
2. Ateniese, G., Blundo, C., De Santis, A., Stinson, D.R.: Extended schemes for visual cryptography. Theoret. Comput. Sci. **250**, 1–16 (1996)
3. Ito, R., Kuwakado, H., Tanaka, H.: Image size invariant visual cryptography. IEICE Trans. **E82-A**(10), 2172–2177 (1999)
4. Yang, C.N., Chen, T.S.: Aspect ratio invariant visual secret sharing schemes with minimum pixel expansion. Pattern Recogn. Lett. **26**(2), 193–206 (2005)
5. Fang, L., Yu, B.: Research on pixel expansion of (2, n) visual threshold scheme. In: Proceedings of the IEEE International Symposium on Pervasive Computing and Applications, pp. 856–860. (2006)

6. Hsu, C.-S., Tu, S.-F., Hou, Y.-C.: An optimization model for visual cryptography schemes with unexpanded shares. In: ISMIS 2006, LNAI 4203, pp. 58–67. (2006)
7. Naor, M., Shamir, A.: Visual cryptography II: improving the contrast via the cover base. Secur. Commun. Netw. 197–202 (1996)
8. Yang, C.-N., Chen, T.-S.: Visual secret sharing scheme: improving the contrast of a recovered image via different pixel expansions. In: ICIAR 2006, LNCS 4141, pp. 468–479. (2006)
9. Kezheng, L., Bo, F., Hong, Z.: Visual cryptographic scheme with high image quality. In: Proceedings of the IEEE International Conference on Computational Intelligence and Security, pp. 366–370. (2008)
10. Lee, C.-C., Chen, H.-H., Liu, H.-T., Chen, G.-W., Tsai, C.-S.: A new visual cryptography with multi-level encoding. J. Vis. Lang. Comput. **25**, 243–250 (2014)
11. Wu, X., Sun, W.: Improved tagged visual cryptography by random grids. Sig. Process. **97**, 64–82 (2014)
12. Monoth, T., Babu Anto, P.: Visual cryptography schemes with perfect reconstruction of white pixels. Int. J. Math. Sci. **3**, 706–709 (2014). (ISSN:2278-8697)
13. Monoth, T., Babu Anto, P.: Analysis and Design of Tamperproof and Contrast-Enhanced Secret Sharing Based on Visual Cryptography Schemes, Ph.D Thesis, Kannur University, Kerala, India (2012). http://hdl.handle.net/10603/6102

Hash Based Two Gateway Payment Protocol Ensuring Accountability with Dynamic ID-Verifier for Digital Goods Providers

Venkatasamy Sureshkumar, R. Anitha and N. Rajamanickam

Abstract Mobile web payment for online shopping is done using only single gateway. If a customer wants to use balances of two different bank accounts in a single transaction then the customer has to transfer this fund to any one of the bank account and then he/she can start the transaction. This is a time consuming process. In this situation, the facility of making the payment via two gateways for a single transaction is more convenient to the customer. Our work is mainly devoted to design an efficient payment protocol that can be used in portable payment devices for making the payment via two gateways for digital goods providers. The proposed scheme is analyzed using an automated tool CPSA, based on different security features and computation overhead. Moreover, the computation overhead of this protocol is less than that of the other schemes.

Keywords Accountability · Anonymity · Payment protocol · Payment gateway

1 Introduction

Nowadays, there is a tremendous growth in online shopping. Online shopping is the best suitable mode for buying digital goods. Digital goods refer to items that are produced, stored and consumed in electronic form [1]. The suppliers of digital goods are known as digital goods providers. Compared to desktop computers, some portable payment devices like iPads, tablets, smartphones, etc. have lower power,

V. Sureshkumar (✉) · R. Anitha · N. Rajamanickam
Department of Applied Mathematics and Computational Sciences,
PSG College of Technology, Coimbatore 641004, India
e-mail: sand@mca.psgtech.ac.in

R. Anitha
e-mail: anitha_nadarajan@mail.psgtech.ac.in

N. Rajamanickam
e-mail: nrm@amc.psgtech.ac.in

M. Senthilkumar et al. (eds.), *Computational Intelligence, Cyber Security and Computational Models*, Advances in Intelligent Systems and Computing 412, DOI 10.1007/978-981-10-0251-9_35

limited storage, and less computational capacities. High computational operations such as encryption and decryption using public/private-key are difficult to perform in portable devices. Therefore, usage of Public Key Infrastructure (PKI) in the mobile payment device is not advisable. So it is highly desirable to use symmetric key cryptography in the design of the mobile payment protocol.

One of the important property, a payment protocol should satisfy is accountability. Accountability is the ability to trace an action between parties engaging in payment protocol and then hold them responsible for their transactions. Formally "Accountability is the property whereby the association of a unique originator with an object or action can be proved to a third party" [2]. Without accountability, a payment protocol may lead to disputes, so one cannot use this protocol with an untrusted party.

If a customer has two cards from two different banks and the price of the product is more than that of the amount he/she has in any of his/her single card but the total balance of these two cards is enough for the transaction then, the customer is unable to make payment for the required products in the conventional payment system. Here the customer has the total amount on two different cards and he/she may like to make the payment via two different gateways. Timed market policy described in [3] is a security policy for this scenario. To the best of our knowledge, so for no payment protocol has been proposed for this purpose.

Our protocol uses two gateways for payment, and the customer can use two cards in a single transaction. Our protocol uses dynamic identity verifier that is useful to provide customer anonymity. Existing protocols use the mechanism of bulk posting of the customers ID from the issuer bank that can be compromised, and its impact will affect the future sessions.

The rest of the paper is organized as follows. Section 2 discusses the related work. The four phases of the proposed protocol are described in Sect. 3. Section 4 presents the analysis of the proposed protocol against the accountability property using the automated tool CPSA. Section 5 compares the proposed protocol with the existing payment protocols. Finally, Sect. 6 concludes this paper with future work.

2 Related Work

This section describes the existing payment protocols and their working mechanisms in brief. The SET protocol [4] is a popular protocol for online payment using credit cards. This protocol requires the public key certificates of all the parties engaged in the protocol. The SET protocol has five phases namely payment initialization, purchase order, authorization, capture payment and card inquiry. In this protocol, the customer's bank details for payment are hidden from the merchant and purchase order information are hidden from the bank. Bellare et al. [5] proposed a family of protocols—iKP (i = 1, 2, 3) for secure electronic payment over the internet. These protocols are designed on the basis of public key cryptography. The 1KP, 2KP and 3KP protocols differ in the aspect of the number of parties having their own public key pairs. The level of security directly depends on the number of parties possessing the key pairs. The engaging parties of iKP are customer, merchant and payment gateway. Kungpisdan [6] proposed a symmetric-key based secure payment protocol. This protocol does not require to disclose the secret information (card details) during the transaction. This protocol is based on lightweight cryptosystem as it requires lower computation at each engaging parties. The five parties involved in this protocol are merchant, client, acquirer, issuer and payment gateway. This protocol has two sub-protocols, namely merchant registration protocol and payment protocol. The merchant registration protocol is executed by the client in order to register himself/herself. After the registration, the client starts to make payment with the merchant by executing the payment protocol. Isaac and Camara [7] proposed a secure payment protocol for restricted connectivity scenarios in M-Commerce. This secure payment protocol is based on public key cryptosystem, and it is applicable in restrictive connectivity scenarios. Tan Soo Fun et al. [8] proposed a lightweight and private mobile payment protocol by using Mobile Network Operator (MNO). This is a secure mobile payment protocol employs symmetric key operations and also it provides customer anonymity. Isaac and Zeadally [9] proposed an anonymous secure payment protocol in a payment gateway centric model for mobile environments. In this protocol, the client cannot contact the merchant directly. Every communication between client and merchant is only through the payment gateway. This protocol ensures customer anonymity.

[4, 5, 7] are public key based protocols and they are not suitable for making payment in the wireless network. Even though [6, 8, 9] are designed in the Symmetric Key Infrastructure (SKI) and it is suitable to make payment in the wireless network, there is no provision in those protocols for making payment via two gateway system. In the above said protocols, the anonymity property is achieved by the customer by creating user IDs in bulk from the Issuer bank. Our protocol has the provision of making payment via two gateways and creates dynamic IDs at the end of each session.

3 Gateway Payment System

Mobile web payment by the customer needs an interface between merchants website and financial institutions. This interface is called payment gateway. The payment gateway connects unsecured internet and the secured network of financial institutions (banks) [10]. There are four participants in this payment procedure: issuer, acquirer, merchant, and customer. Issuer is the bank that provides an account and card for the customer. Acquirer is the bank that provides an account for the merchant and process authorizations related to payments [11]. The payment process is shown in Fig. 1. In this context, the issuer and acquirer are in private financial institutions network and uses secured channel. In order to contact acquirer, the merchant uses the payment gateway, from the public internet in the unsecured channel. All communications are done only through the payment gateway.

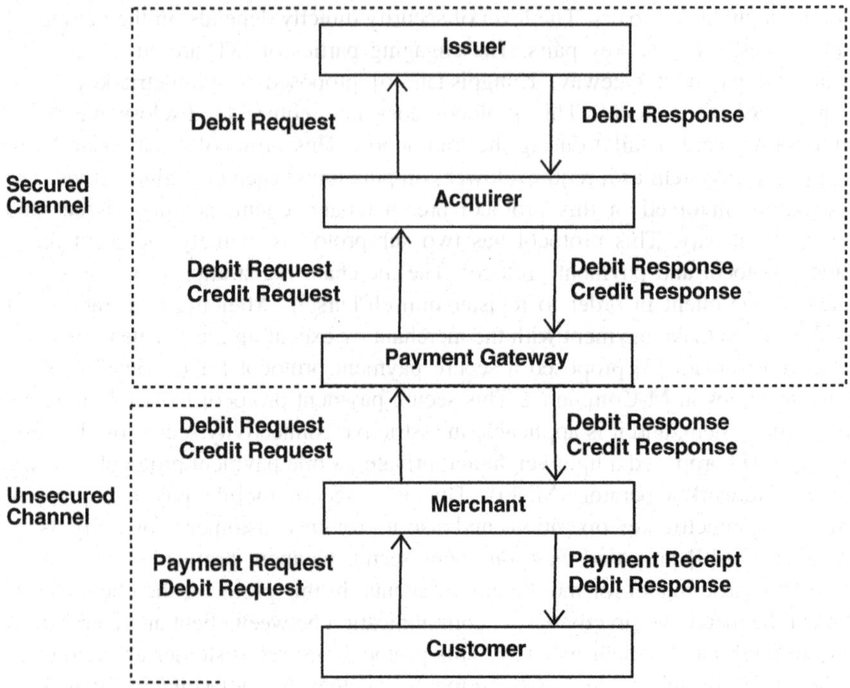

Fig. 1 Payment process

In this section, we propose a protocol to make payment via two gateway system for a single transaction. This article tries to develop a secure wireless payment protocol with advanced features.

3.1 Initial Assumptions

To make the proposed protocol suitable for the real world, some assumptions are listed out.

- Customer browses merchant website, finds out the product with respective price, and decides to make payment in two fold via two gateways.
- Customer registers with the two gateways separately, obtains the secrets, initial IDs and generates the symmetric session keys using appropriate key generation protocol.
- The session IDs are well established (well identified the by the parties at the appropriate sessions).
- The gateways are used to separate the banks in which the merchant and customer have their own accounts. Also, there exist an internal network among the gateway and the banks that can not have been attacked by the intruder.

- As a one-time process, the customer personally (offline) register at issuer bank. The customer submits his Account Information (AI) to prove his/her identity to the bank. After ascertaining the identity of the customer, the bank installs payment application software on the customer's device. The banker stores customer's AI along with his initial dynamic identity for the customer verification in its database. This identity remains to be dynamic and will frequently be updated for the purpose of maintaining customer anonymity. Similarly merchant should have done the registration at the acquirer bank.

3.2 *Two Gateway Payment Protocol*

In this section, we propose a new protocol that can be used to execute an online transaction between a customer and a merchant in the case when the customer wants to make the payment through two gateway system. Some notations used in the protocol are detailed in Table 1. The main actions within the proposed payment system are: Initialization of payment transaction, payment via the gateway G_1, payment via the gateway G_2 and dynamic identity update.

The proposed protocol is described below in four phases.

Table 1 Notations used in the protocol

Notations	Description
C	Customer
M	Merchant
G_1	First gateway
G_2	Second gateway
$TIDS_{req}$	Requisition of transaction IDs
amt_1	Partial amount to be paid via gateway G_1
amt_2	Partial amount to be paid via gateway G_2
OD	Order description about the goods purchased
TID_1	Transaction ID corresponds to the payment via gateway G_1
TID_2	Transaction ID corresponds to the payment via gateway G_2
K_{XY}	Symmetric session key shared between the two parties X and Y
χ	Commitment scheme with secret value r
IDC_{G_1}	Dynamic session ID of the customer associated with the gateway G_1
IDC_{G_2}	Dynamic session ID of the customer associated with the gateway G_2
$\{m\}_{K_{XY}}$	Encryption of the message m using the key K_{XY}
$h(m)$	Hash value of the message m
$h(m, K_{XY})$	The keyed hash value of the message m using the key K_{XY}

Phase 1: Payment Agreement (Initialization)

Customer browses for items, select items to be purchased from the Vshop and get an order that contain the list of items to be purchased. Before starting the purchase, the customer and the merchant agree upon the order description amount. The customer sends a request to the merchant for the transaction IDs associated with the specified partial amounts amt_1, amt_2 and along with this request customer also sends the details of the two gateways G_1, G_2. After receiving this request from the customer, merchant fixes the transaction IDs TID_1, TID_2 and associate them with the partial amounts amt_1, amt_2 via the two gateways G_1, G_2 in his database.

$$C \rightarrow M : m_1 = \{OD, TIDS_{req}, amt_1, G_1, amt_2, G_2\}$$
$$M \rightarrow C : m_2 = \{TID_1, TID_2\}_{K_{MC}}$$

At the end of this phase, the customer makes an agreement with the merchant to purchase the goods in two payments via two gateways and the merchant agrees to receive the amount in the specified pattern and provides the transaction identities accordingly.

Phase 2: Payment Via Gateway G_1

$$C \rightarrow M : m_3 = \{G_1, IDC_{G_1}, \{TID_1, amt_1, T_1, h(TID_1, amt_1, K_{MC})\}_{K_{CG_1}}\}_{K_{MC}}$$
$$M \rightarrow G_1 : m_4 = \{IDC_{G_1}, TID_1, amt_1, \{TID_1, amt_1, T_1, h(TID_1, amt_1, K_{MC})\}_{K_{CG_1}}\}_{K_{MG_1}}$$
$$G_1 \rightarrow M : m_5 = \{TID_1, amt_1, T_2, h(TID_1, amt_1, T_2, K_{CG_1}), h(TID_1, amt_1, K_{MC})\}_{K_{MG_1}}$$
$$M \rightarrow C : m_6 = \{TID_1, amt_1, T_2, h(TID_1, amt_1, T_2, K_{CG_1}), \chi(product, r), h(TID_1, amt_1, T_2, h(TID_1, amt_1, T_2, K_{CG_1}), \chi(product, r), K_{MG_1})\}_{K_{MC}}$$

This phase is initiated by the customer C by providing the debit request $\{TID_1, amt_1, T_1, h(TID_1, amt_1, K_{MC})\}_{K_{CG_1}}$ to the merchant M which can be encashed via the gateway G_1 while sending the partial payment information m_3 which includes the transaction identity TID_1, the partial amount amt_1 and the time stamp T_1 and is encrypted using the symmetric key between the customer and merchant. After receiving the payment information, merchant decrypts the message m_3 and extracts the debit request from it, which can not be understood by the merchant as it is encrypted using the key between C and G_1. Merchant sends the debit request along with the customer's dynamic identity IDC_{G_1} and the partial amount amt_1 by encrypting with the shared symmetric key between the merchant and the gateway G_1. After receiving this credit request message m_4 from the merchant, the gateway G_1 decrypts the message and verifies the partial amount amt_1 in both m_4 and debit request. The debit request includes keyed hash of transaction identity and the partial amount using the key between M and C, which can not be understood by the bank. This hash value is useful to the merchant that is sent by the customer via the gateway G_1. The gateway G_1 sends the message m_5 to the merchant that includes the credit response to the merchant and the debit response to the customer with the time stamp T_2. After receiving this message from the gateway G_1, merchant prepares the encrypted message m_6 using the key between M and G_1, which includes the transaction identity, the partial amount, the time stamp T_2, the debit response sent by the bank, the committed goods and the entire message that is hashed using the session key shared between the gateway G_1 and M. This hash value is useful for the customer later for the confirmation of the partial goods received for the paid partial amount.

Phase 3: Payment Via Gateway G_2

$$C \rightarrow M : m_7 = \{G_2, IDC_{G_2}, \{TID_2, amt_2, T_3, h(TID_2, amt_2, K_{MC})\}_{K_{CG_2}}\}_{K_{MC}}$$
$$M \rightarrow G_2 : m_8 = \{IDC_{G_2}, TID_2, amt_2, \{TID_2, amt_2, T_3, h(TID_2, amt_2, K_{MC})\}_{K_{CG_2}}\}_{K_{MG_2}}$$
$$G_2 \rightarrow M : m_9 = \{TID_2, amt_2, T_4, h(TID_2, amt_2, T_4, K_{CG_2}), h(TID_2, amt_2, K_{MC})\}_{K_{MG_2}}$$
$$M \rightarrow C : m_{10} = \{TID_2, amt_2, T_4, h(TID_2, amt_2, T_4, K_{CG_2}), r, h(TID_2, amt_2, T_4, h(TID_2, amt_2, T_4, K_{CG_2}), r, K_{MG_2})\}_{K_{MC}}$$

This phase is very similar to the phase 2, except in m_{10}. In the message m_{10}, the merchant sends the payment receipt along with the secret value r associated with the committed value in the message m_6. Using this secret value r, the customer can extract the digital goods from the committed value in the message m_6. This ensures the digital goods exchange feature.

Phase 4: ID verifier updation and Payment confirmation

The issuer bank assigns several session IDs to the customer because of the trust relationship between both of them. These session IDs are known only to the customer and the issuer bank and are used to prevent the merchant from knowing the identity of the customer. If these session IDs are generated in bulk, there is a possibility for an attacker to compromise the future session IDs, and hence they are to be frequently updated for maintaining freshness.

$$C \rightarrow G_1 : m_{11} = \{updateID, (T_2 - T_1), success/failure\}_{K_{CG_1}}$$
$$G_1 \rightarrow C : m_{12} = \{result\}$$
$$C \rightarrow G_2 : m_{13} = \{updateID, (T_4 - T_3), success/failure\}_{K_{CG_2}}$$
$$G_2 \rightarrow C : m_{14} = \{result\}$$

The identity updation process is carried out as follows

$$ID_{CG_1}(new) = h(ID_{CG_1}(old), (T_2 - T_1)) \text{ and}$$
$$ID_{CG_2}(new) = h(ID_{CG_2}(old), (T_4 - T_3))$$

this will be done on both customer and gateway sides.

4 Accountability Analysis of the Proposed Protocol Using CPSA

In this section, the informal analysis of the accountability property of proposed protocol is carried out.

4.1 Accountability

Accountability requires that if a protocol execution fails or does not achieve its goal, then it is possible to determine which participant misbehaved or did not follow the protocol. In simple words, the verifier (Judge) should able to associate the message with the creator. In order to make a particular party is accountable for his sent message, it is sufficient to prove that the message is created by that party, and it is not a forwarded message.

4.2 CPSA

The Cryptographic Protocol Shapes Analyzer (CPSA) is an automated tool working on the basis of the formal method strand space model [12–14]. CPSA tries to find out all possible different skeletons. The skeleton describes the set of possible executions for a cryptographic protocol. It is also called the shape of the protocol. Many of the existing protocols have only finite number of shapes. Using these (after excluding the isomorphic) shapes, it is easy to determine the authentication and secrecy properties. Based on Dolev-Yao model, the shape analysis is carried out. The CPSA program accepts a sequence of problem descriptions and provides the steps for solving each problem [15]. For each input problem, CPSA is given some initial behavior, and it discovers what shapes are compatible with it. The red colour nodes represents unrealized nodes, blue colour nodes represents realized nodes and black colour nodes represents sending nodes. The initial behavior is from the point of view of one participant. For given one participant's view, the analysis discloses, what the other participants would have done, given the participant's view.

4.3 Scenarios

The following scenarios details the output of CPSA for the proposed protocol.

Scenario 1 CPSA describes an initial skeleton in the merchant's view with a strand of height 4, shown in Fig. 2a. The two unrealized nodes are receiving nodes and the

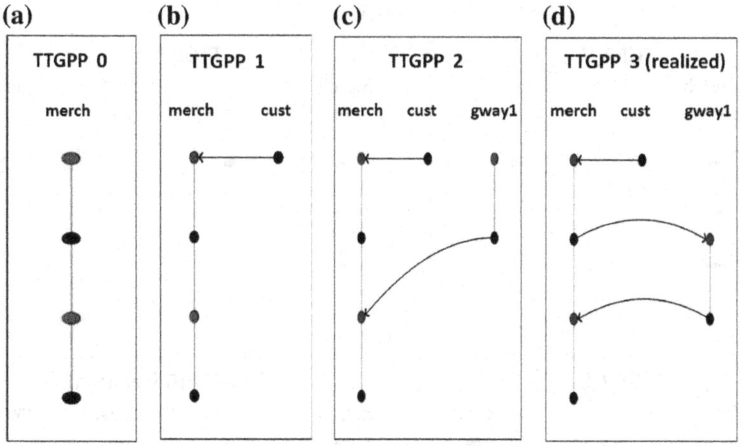

Fig. 2 From initial skeleton to realized skeleton in the merchant's view

remaining two nodes are sending nodes. The first node receives the message m_3 from the customer which is realized in Fig. 2b. At the third node, the merchant receives the message m_5, and it is realized by adding the gateway strand, shown in Fig. 2c. The gateway strand has two nodes in which the first node is an unrealized node, and it is realized by constructing an edge from the existing merchant strand as shown in Fig. 2d. In this way, CPSA finds a realized skeleton in Fig. 2d.

Scenario 2 In this scenario, CPSA depicts the same set of skeletons from scenario 1 with the added intruder strand. The figures which are shown in Fig. 3a–d details that there is no edge connecting the intruder's strand and the legitimate user's strand in each skeleton. This implies that there is no attack exists while exchanging the messages between the parties.

Scenario 3 This scenario is produced in CPSA in the payment gateway G_1 view. The initial skeleton produced by CPSA is shown in Fig. 4a, which has an unrealized node in it. The unrealized node is realized by adding the merchant's strand of height 2, shown in Fig. 4b, in which the first node is unrealized. the unrealized node in the added merchant strand is realized by adding the customer strand of height 1 and hence the skeleton is realized in Fig. 4c.

Scenario 4 This scenario is created in CPSA in the customer point of view. The customer strand of height 2 is introduced in the initial skeleton shown in Fig. 5a. In this initial skeleton, the second node receives the message and hence it is a critical node. This critical node is realized by adding merchant strand of height 4 in Fig. 5b. The added strand has two critical nodes, and they are realized as stated in scenario 1 and depicted in the Fig. 5c–e. The completely realized skeleton in the customer's view is shown in Fig. 5e.

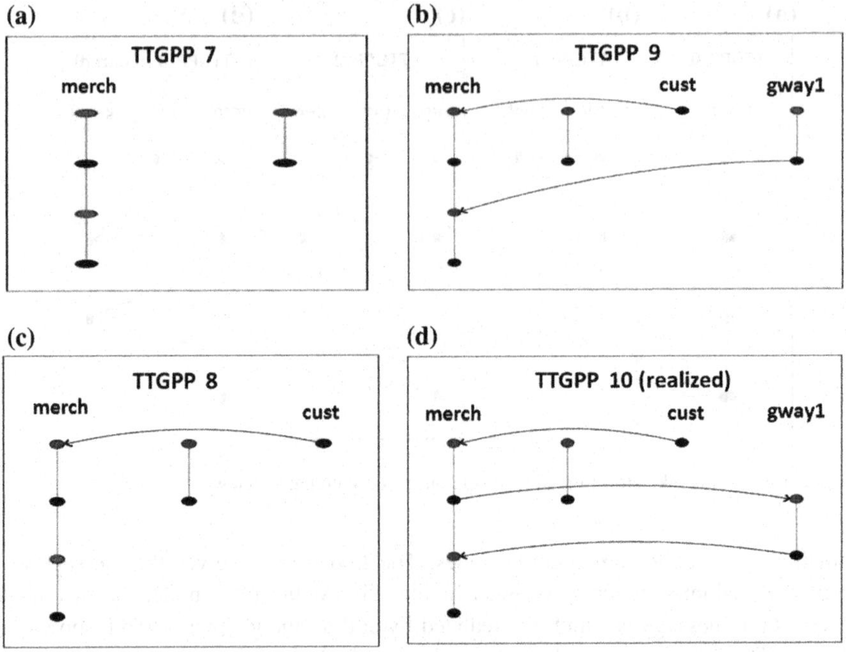

Fig. 3 From initial skeleton to realized skeleton with an intruder strand in the merchant's view

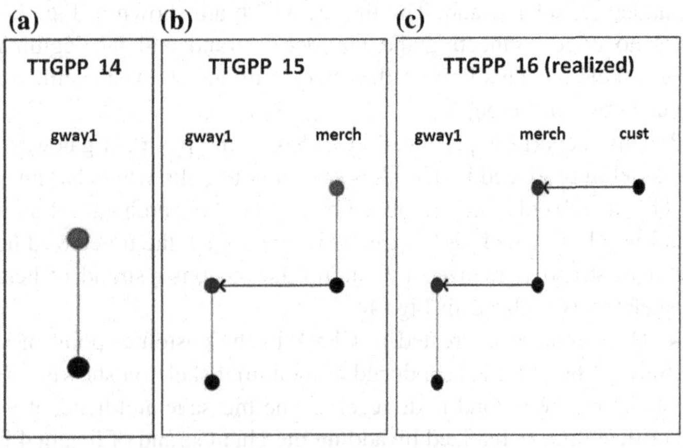

Fig. 4 From initial skeleton to realized skeleton in the gateway G_1 view

Scenario 5 This scenario describes the same set of skeletons from scenario 4 with the added intruder strand. The figures which are shown in Fig. 6a–e details that there is no edge connecting the intruder's strand and the legitimate user's strand in

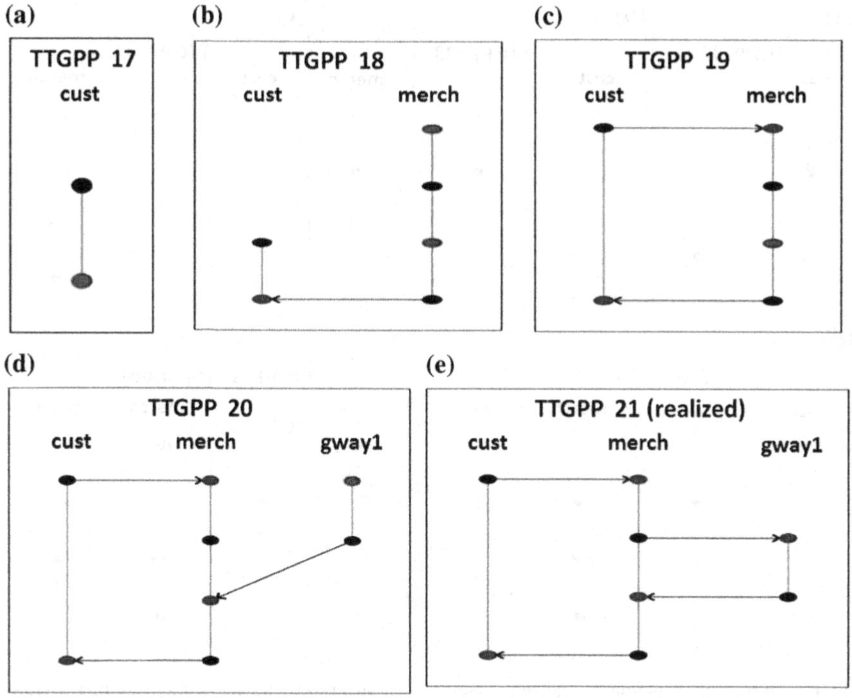

Fig. 5 From initial skeleton to realized skeleton in the customer's view

each skeleton. This implies that there is no attack exists while exchanging the messages between the parties.

4.4 *Verification of Accountability Property Using CPSA*

The accountability issue appears to have only in the payment phases. Phase 2 and phase 3 are identical to each other and hence it is sufficient to carry out the analysis only for phase 2. The possible accountability issues are primarily focused on the association of

- the customer with his sent debit request
- the merchant with his sent credit request
- the gateway with his sent credit response
- the merchant with his sent payment receipt and digital goods

Argument: 1 (The message m_3 can only be created by the customer) The message m_3 is an encrypted message using the shared key between the customer and merchant that contains the encrypted payment information to the gateway using

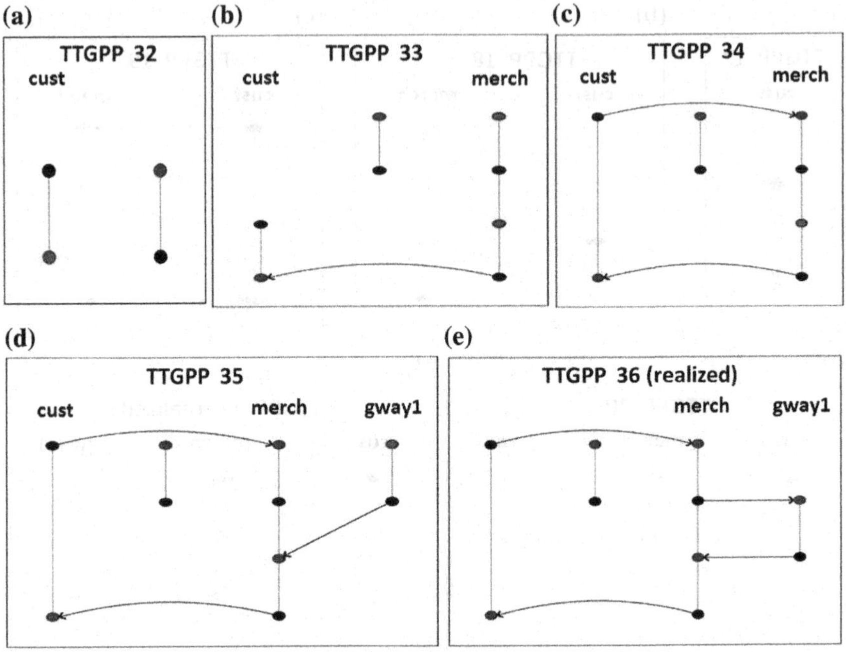

Fig. 6 From initial skeleton to realized skeleton with an intruder strand in the customer's view

the shared key between the gateway and customer. Hence, the construction of the message requires knowledge of both the keys K_{MC} and K_{CG_1}. This implies that the message m_3 can have been created only by the customer. Figures 2d and 4c show the existence of unique customer in the merchant and gateway point of view. Moreover the realized skeleton in Fig. 5e shows that the message m_3 is not a forwarded message.

Argument: 2 (The message m_4 can only be created by the merchant) The message m_4 is an encrypted message using the shared key between the merchant and the gateway which contains the payment information to the gateway that contains the keyed hash value $h(TID_1, amt_1, K_{MC})$ which cannot have been created by the gateway because the key K_{MC} is known only to the merchant and the customer. Thus, this message should only be created by the merchant. The unique existence of the merchant strand is detailed in the scenario 3 and 4 with realized skeletons in Figs. 4c and 5e. Moreover the realized skeleton in Fig. 2d shows that the message m_4 is not a forwarded message.

Argument: 3 (The message m_5 can only be created by the gateway G_1) The message m_5 is an encrypted message using the shared key between the gateway G_1 and the merchant which contains the keyed hash value $h(TID_1, amt_1, T_2, K_{CG_1})$ which can not have been created by the merchant due to the occurrence of the key K_{CG_1}. Hence, this message can have only been created by the gateway G_1. The unique existence of the gateway G_1 is shown in Fig. 2d in the scenario 1. Moreover the realized skeleton in Fig. 5e shows that the message m_5 is not a forwarded message.

Argument: 4 (The message m_6 can only be created by the merchant) The message m_6 is an encrypted message using the key shared between the customer and the merchant and it contains the keyed hash value $h(TID_1, amt_1, T_2, h(TID_1, amt_1, T_2, K_{CG_1}), \chi(product, r), K_{MG_1})$ which can not have been created by the customer due to the usage of the key K_{MG_1}. Thus, the message m_6 can only be created by the merchant. The unique existence of the merchant strand is shown in Fig. 5e. Moreover the realized skeleton in Fig. 5e shows that the message m_6 is not a forwarded message.

Based on these arguments, it is very clear that the each message sent can be associated with the concerned sender. Thus, this protocol satisfies the accountability property. Moreover, as in phase 4, the ID verifiers are updated at the end of each success/failure transactions and hence the customer anonymity is preserved.

5 Comparative Analysis

In this section, the performance of the proposed protocol with the existing protocols is compared in the aspects of computational overhead and the advanced features they possess.

5.1 Computational Complexity

The proposed payment protocol is compared with some existing payment protocols in the perspective of computational complexity. The comparison is done on the basis of the number of cryptographic operations carried out by each party. The cryptographic operations are categorized based on [6]. Table 2 demonstrates the numbers of cryptographic operations by the involved parties of each protocol.

The comparative analysis in Table 2 reveals that the proposed payment has reduced the number of cryptographic operations applied to the existing payment protocol.[1] As SET and iKP protocols are designed in PKI, they require encryption and signature verification which need high computational task. Compared with [6, 8, 9] protocols, the proposed payment protocol does not apply more symmetric key encryption and hash functions, but only employs the symmetric key generation that further requires lower computational work. Furthermore, merchant interacts only with the payment gateway (not with the acquirer and issuer banks). It reduces the additional communications and maintains the secrecy of confidential

[1]As the existing payment systems are designed only for single gateway, the comparison is made on the basis of single gateway.

Table 2 Comparative analysis of computational complexity

Cryptographic operations	SET [4] (1997)	iKP [5] (2000)	KSL [6] (2004)	LMPP [8] (2008)	PCMS [9] (2012)	Proposed
Public-key encryptions	3	1	0	0	0	0
Public-key decryptions	3	1	0	0	0	0
Signature generations	5	3	0	0	0	0
Signature verifications	5	7	0	0	0	0
Symmetric-key encryptions/decryptions	3	0	11	11	10	7
Hash functions	5	7	2	3	3	1
Keyed-hash functions	0	1	5	0	3	3
Key generation	0	0	4	6	9	6

information. In the nutshell, the proposed payment protocol has improved the performance of existing payment protocol by reducing the number of cryptographic operations.

5.2 *Protocol Features*

The performance of the proposed protocol is compared with that of the existing protocols in Table 3 in the aspect of the advanced features possessed by the protocols. Compared with the existing protocols, the proposed protocol has the facility of making payment through two gateways and the proposed protocol has the provision for digital goods exchange in the lightweight cryptosystem.

Table 3 Comparative analysis with existing payment protocols

Protocols	Accountability	Anonymity	Two gateway	Goods exchange	Infrastructure
SET [4]	Yes	No	No	No	PKI
iKP [5]	Yes	Yes	No	No	PKI
KSL [6]	Yes	Yes	No	No	SKI
LMPP [8]	Yes	Yes	No	No	SKI
PCMS [9]	Yes	Yes	No	No	SKI
Proposed protocol	Yes	Yes	Yes	Yes	SKI

6 Conclusion

This work provides a suitable payment protocol for making the payment via two gateways. Correctness of the protocol is obtained by using the automated tool CPSA. The proposed protocol is compared with the existing payment protocols in the aspect of computational overhead and advanced features. Comparative analysis shows that the proposed protocol is well suited for the portable payment system. For the future work, the implementation of the proposed protocol is to be done using suitable virtual test bed. The formal proof for the correctness of accountability and anonymity property are to be provided using the formal method strand space. The protocol is to be enhanced by including suitable sub-protocols in such a way that it achieves atomicity property.

References

1. Bhasker, B.: Electronic Commerce: Framework, Technologies and Applications. Tata McGraw-Hill Education (2013)
2. Kailar, R.: Reasoning about accountability in protocols for electronic commerce. In: Proceedings of the IEEE Symposium on Security and Privacy, pp. 236–250. Oakland, CA (May 1995)
3. Rajamanickam, N., Nadarajan, R.: Implementing real-time transactional security property using timed edit automata. In: Proceedings of the Sixth International Conference on Security of Information and Networks, pp. 429–432. Aksaray, Turkey (2013)
4. SET Secure Electronic Transaction Specification. Book 3: Formal Protocol Definition. SET Secure Electronic Transaction LLC (1997)
5. Bellare, M., Garay, J., Hauser, R., Herzberg, A., Krawczyk, H., Steiner, M., Tsudik, G., Van Herreweghen, E., Waidner, M., et al.: Design, implementation, and deployment of the ikp secure electronic payment system. IEEE J. Sel. Areas Commun. **18**(4), 611–627 (2000)
6. Kungpisdan, S., Srinivasan, B., Le, P.D.: A secure account-based mobile payment protocol. In: Information Technology: Coding and Computing, 2004. Proceedings. ITCC 2004. International Conference on, vol. 1, pp. 35–39. IEEE (2004)
7. Isaac, J.T., Camara, J.S.: A secure payment protocol for restricted connectivity scenarios in m-commerce. In: E-Commerce and Web Technologies, pp. 1–10. Springer (2007)
8. Fun, T.S., Beng, L.Y., Likoh, J., Roslan, R.: A lightweight and private mobile payment protocol by using mobile network operator. In: International Conference on Computer and Communication Engineering. ICCCE 2008, pp. 162–166. IEEE (2008)
9. Isaac, J.T., Zeadally, S.: An anonymous secure payment protocol in a payment gateway centric model. Procedia Comput. Sci. **10**, 758–765 (2012)
10. Waidner, B.B., Mueller, A., Pedersen, T., Schunter, M.: Architecture of Payment Gateway. SEMPER Consortium (1996)
11. Hwang, J.J., Hsueh, S.C.: Greater protection for credit card holders: a revised set protocol. Computer Stand. Interfaces **19**(1), 1–8 (1998)
12. Fabrega, F., Herzog, J.C., Guttman, J.D.: Strand spaces: why is a security protocol correct? In: Security and Privacy, 1998. Proceedings. 1998 IEEE Symposium on, pp. 160–171. IEEE (1998)
13. Sureshkumar, V., Anitha, R.: Analysis of electronic voting protocol using strand space model. In: Recent Trends in Computer Networks and Distributed Systems Security, pp. 416–427. Springer (2014)

14. Sureshkumar, V., Ramalingam, A., Anandhi, S.: Analysis of accountability property in payment systems using strand space model. In: Security in Computing and Communications, pp. 424–437. Springer (2015)
15. Ramsdell, J.D., Guttman, J.D.: Cpsa Primer. The MITRE Corporation, Bedford, MA, USA 1 (2009)

Decrypting Shared Encrypted Data Files Stored in a Cloud Using Dynamic Key Aggregation

Maria James, Chungath Srinivasan, K.V. Lakshmy and M. Sethumadhavan

Abstract Cloud file sharing is a topic that has been in top flight for a long period of time and even in near future. Many methods like access control lists, attribute based encryption, identity based encryption and proxy re-encryption were introduced for sharing data stored in a cloud. But their drawbacks lead to the introduction of key aggregation based file sharing where the content provider generates an aggregate key based on files to be shared and distribute it among the users. Even though this mechanism is very much effective, it has the drawback that if once a user gets an aggregate key corresponding to a set of files from the data owner, then the data owner cannot revoke the permission for accessing the updated files to the user, who already possess the aggregate key. Our scheme in this paper provides an extension to this existing work which guarantees a dynamic generation of the aggregate key and thereby will not allow any unauthorised access to the updated files in the cloud.

Keywords Encrypted data · Cloud security · Key aggregation · Bilinear pairing

1 Introduction

Data sharing has always existed since the beginning of life. Humans love to share information with one another whether it is one to one or in a closed group or with second level of acquaintances or with public. With the evolution of computers, people moved from hard copies to soft copies which made data theft easier. Saving digitally has its own advantage as well as disadvantages. The same theory applies to cloud also. Now sharing became more easy when cloud storage came to new

M. James (✉) · C. Srinivasan · K.V. Lakshmy · M. Sethumadhavan
TIFAC-CORE in Cyber Security, Amrita Vishwa Vidyapeetham, Coimbatore 641112, India
e-mail: jms.maria@gmail.com

C. Srinivasan
e-mail: chungathsrinivasan@gmail.com

M. Senthilkumar et al. (eds.), *Computational Intelligence, Cyber Security and Computational Models*, Advances in Intelligent Systems and Computing 412, DOI 10.1007/978-981-10-0251-9_36

evolution. But with this evolution in data storage, security and sharing of cloud data has also raised to a new level.

Taking a case in the medical field investors are moving towards such companies that take up patients data to the cloud. There are immense possibilities in this field because of the huge data involved. Even from the birth of a child there will be numerous test conducted and a dozen scans and reports which any individual take up during normal course of his life. People migrate to different places and taking these reports to various hospitals and doctors is inefficient and tiresome and numerous duplicate reports are generated which adds up the cost. Taking the entire data into the cloud gives any doctor added advantage of understanding the history of the patient and doing the right medication. The patient can also have the privilege to share the report with the set of people and clinics and even deny access in future which adds to the privacy of the patient.

Since the data is distributed easily in cloud, the attack on cloud storage is also wide. Since the data may be present on multiple servers the data is at a risk of unauthorized physical access of data. For example, the data is made as multiple copies and is moved to different location and this leads to an increase in the unauthorized data recovery attacks. Unauthorized view to the data can be prevented by encrypting the data which converts the data into a form that cannot be easily understood by an attacker.

Also large numbers of users are given access to the data. So there is a chance that the attackers belongs to the group of users who has access to the data or the attacker can bribe or blackmail a user who is inside the trusted set of users and thus gain access to the data. It is also impractical to monitor which data has transferred to which location and also to check if the data has reached the correct destination.

In this paper we provide an extension to the existing work of Chu [1], which guarantees a dynamic generation of the aggregate key and thereby will not allow any unauthorised access to the updated files in the cloud.

The paper is organized as follows: Sect. 2 describes the prerequisites for better understanding of the scheme, Sect. 3 gives a survey on the similar works and problems in the literature. In Sect. 4 we define our problem and describe Cheng-Kang Chu's key aggregate cryptosystem for data sharing in a cloud. Section 5 describes the proposed dynamic key aggregate cryptosystem.

2 Bilinear Pairing

A pairing is a bilinear map defined over elliptic curve [2, 3, 4] subgroups. Let $\mathbb{G}_1$ and $\mathbb{G}_2$ be two groups of the same prime order q (usually cyclic). Let $P \in \mathbb{G}_1$ chosen to be an arbitrary generator of $\mathbb{G}_1$.

A mapping $e : \mathbb{G}_1 \times \mathbb{G}_1 \rightarrow \mathbb{G}_2$ satisfying the following properties is called a bilinear map.

1. *Bilinearity*: $e(aR, bS) = e(R, S)^{ab}$, $\forall R, S \in G_1$ and $a, b \in \mathbb{Z}$.
   In general for $R, S, T \in \mathbb{G}_1$

$$\begin{aligned} e(RS, T) &= e(R, T) \cdot e(S, T) \\ e(R, ST) &= e(R, S) \cdot e(R, T). \end{aligned}$$

2. *Non-degeneracy*: If P is a generator of $\mathbb{G}_1$, then $e(P, P)$ is a generator of $\mathbb{G}_2$. That is $e(P, P) \neq 1$.
3. *Computability*: There exists an efficient algorithm to compute $e(R, S)\, \forall R, S \in \mathbb{G}_1$.

The following properties of a bilinear pairing can be easily verified. For all $S, T \in \mathbb{G}_1$:

1. $e(S, \mathcal{O}) = 1$ and $e(\mathcal{O}, S) = 1$.
2. $e(S, -T) = e(-S, T) = e(S, T)^{-1}$.
3. $e(aS, bT) = e(S, T)^{ab}$ for all $a, b \in \mathbb{Z}$.
4. $e(S, T) = e(T, S)$.
5. If $e(S, R) = 1$ for all $R \in \mathbb{G}_1$, then $S = \mathcal{O}$.

$\mathbb{G}_1$, $\mathbb{G}_2$ are the groups that are isomorphic to one another as they have the same order and are cyclic. They are different groups in the sense that of representing the elements and computation of operations are performed differently.

3 Literature Survey

This section describes the possible solutions on data sharing [5, 6] in secure cloud storage and explains the pros and cons of each method.

3.1 Attribute Based Encryption

One of the method used for securely storing data in cloud server is using Attribute Based Encryption (ABE) [7, 8]. In the scheme defined by Tu et al. [8] each user is identified by a set of attributes. The encrypted file stored in cloud is decrypted by the user who get the corresponding secret key. The secret key is securely transmitted to the user who satisfies the access control policies set by the data owner. When the access right to a particular user is revoked the entire ciphertext is re-encrypted in the cloud. Thus the user will not be able to access the data again through the use of old key. The problem faced by the earlier scheme is that, once the users access rights are revoked the ciphertext need to be re-encrypted which leads to heavy overhead even though the computation is done at cloud. Li et al. [7]

extends the ABE scheme in the use for personal health records (PHR). The method has drawbacks when certain files have to be shared among multiple roles. This leads to collision problem and leads to privacy issues.

3.2 Proxy Re-encryption

Proxy re-encryption is another emerging technique used for sharing files stored in a cloud [9, 10]. In this method a semi-trusted proxy is used in order to re-encrypt the data. The data encrypted by the data owner is first send to the semi-trusted proxy. The proxy re-encrypts the data using the data owners public key converting it into a file that can be decrypted by the secret key of the demanding party. The proxy gets no idea about the data that it converts and thus the data remains secure. We demonstrate a basic Proxy Re-encryption scheme in Fig. 1.

Inspite of the above listed advantages the scheme suffers from the disadvantage of occurrence of collusion attacks. If the user and the proxy collude, the user can get access to the private key of the entire group. Also too many encryption as well as decryption occurs at the side of proxy which may lead to denial of service attack.

3.3 Key Aggregate Cryptosystem

One of the most recent efficient method to share data stored in a cloud to various users is by the use of key aggregate cryptosystem [1, 11, 12]. The methods that include sharing of files may be by sharing the key to the user if all the files are encrypted by the same key. But this leads to security issues if the data owner wishes to share only some of the files. So the next possible method is to encrypt each file with a separate secret key. When sharing the files the decryption keys corresponding to the encrypted files is securely transmitted to the user. This leads to large amount of transmission of secure data. But the key aggregate cryptosystem proposed by Chu et al. in [1] solves all the above mentioned problems. The system encrypts each file with different keys and when a user demands for a particular set of files, the data owner computes an aggregate key which integrates the power of the individual decryption keys corresponding to the files in the set.

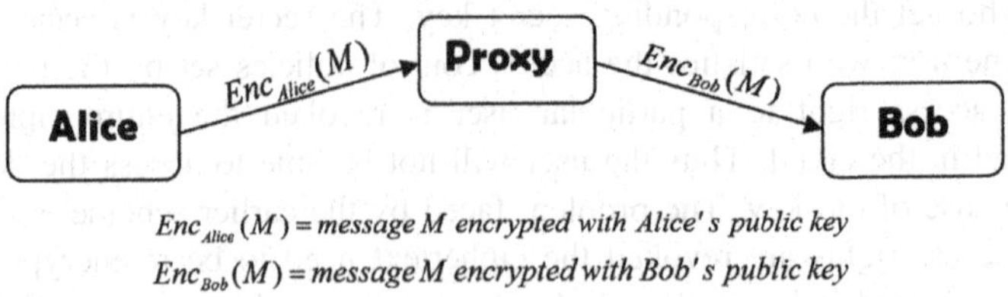

Fig. 1 Proxy re-encryption scheme

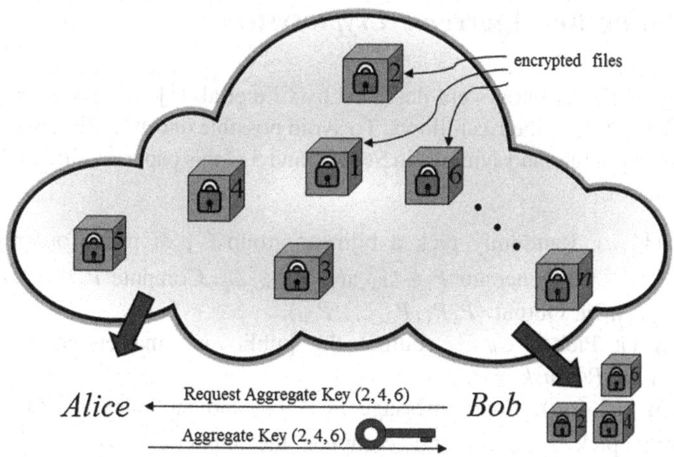

Fig. 2 Key aggregate cryptosystem

Two major drawbacks exists for this scheme. First drawback is that the data owner should predefine the number of classes that is the number of files that will be used by him. To extend the predefined number of classes is a tedious work. The second major drawback of the system is that once the data owner gives an aggregate key corresponding to a set of files to a user and later at some point of time modifies some of these files and wishes to revoke the access right of the same files to the previous user is not possible. Once a user is in hold of an aggregate key for a set of files, he will have access to any files in that set at any point of time. Hence this existing system enables for an unauthorized access of data and in this paper we suggest a solution to this problem by dynamically updating the aggregate key. The Fig. 2 gives a block diagram for key aggregate cryptosystem scheme, where Alice is the data owner and Bob is an end user requesting the aggregate key for the set of files with indices (2, 4, 6).

4 Problem Definition

The aim of the paper is to suggest a new version to the existing key aggregate cryptosystem proposed in [1]. In some scenarios this existing scheme enables an unauthorized access to the data files stored in the cloud. Our scheme helps to share encrypted data stored in a cloud by providing a dynamic aggregate key to the end users. So whenever a user requests the same set of files, the aggregated key generated would be different. This method would be extremely useful when the data owner shares an aggregated key for a set of files to a user and later wants to modify these files and wish not to grant access to these files for the same user. In other words this updated scheme prevents from unauthorized access of data files from the cloud.

4.1 Existing Key Aggregate Cryptosystem

The key aggregate cryptosystem designed by Chu et al. [1] consists of five phases. These phases are described as follows. To avoid possible uncertainities, we choose to use the same variables and symbols in Sects. 1 and 5 of this paper, similar to that given in [1].

1. **Setup** $(1^\lambda, n)$: Randomly pick a bilinear group $\mathbb{G}_1$ of prime order p where $2^\lambda \leq p \leq 2^{\lambda+1}$, a generator $P \in \mathbb{G}_1$ and $\alpha \in_R \mathbb{Z}_p$. Compute $P_i = P^{\alpha^i} \in \mathbb{G}_1$ for $i = 1, 2, \ldots, 2n$. Output $(P, P_1, P_2, \ldots, P_{2n})$.
2. **Keygen** (): Pick $\gamma \in_R \mathbb{Z}_p$, output the public and master-secret key pair: $(pk = v = P^\gamma,\ msk = \gamma)$.
3. **Encrypt** (pk, i, m): For a message $m \in \mathbb{G}_2$ and an index $i \in \{1, 2, \ldots, n\}$, randomly pick $t \in_R \mathbb{Z}_p$ and compute the ciphertext as $C = (P^t,\ (vP_i)^t,\ m.e(P_1, P_n)^t) = (c_1, c_2, c_3)$. The calculated aggregate key is sent over a secure channel.
4. **Extract** $(msk = \gamma,\ S)$: For the set S of file indices j's, the aggregate key is computed as $K_S = \prod_{j \in S} P^\gamma_{n+1-j}$. Since S does not include 0, $P_{n+1-j} = P^{\alpha^{n+1-j}}$ can always be retrieved from $(P, P_1, P_2, \ldots, P_{2n})$.
5. **Decrpt** (K_s, S, i, C): If $i \notin S$, output null. Otherwise, return the message

$$m = c_3 \cdot \frac{e\left(K_s \cdot \prod_{j \in S, j \neq i} P_{n+1-j+i}, c_1\right)}{e\left(\prod_{j \in S} P_{n+1-j}, c_2\right)}$$

The correctness can be verified as follows:

$$\begin{aligned}
& c_3 \cdot \frac{e\left(K_S \cdot \prod_{j \in S, j \neq i} P_{n+1-j+i}, c_1\right)}{e\left(\prod_{j \in S} P_{n+1-j}, c_2\right)} \\
&= c_3 \cdot \frac{e\left(\prod_{j \in S} P^\gamma_{n+1-j} \cdot \prod_{j \in S, j \neq i} P_{n+1-j+i},\ P^t\right)}{e(\prod_{j \in S} P_{n+1-j}, (vP_i)^t)} \\
&= c_3 \cdot \frac{\prod_{j \in S} e\left(P^\gamma_{n+1-j}, P^t\right) \prod_{j \in S, j \neq i} e\left(P_{n+1-j+i}, P^t\right)}{\prod_{j \in S} e\left(P_{n+1-j}, v^t\right) \prod_{j \in S} e\left(P_{n+1-j}, P^t_i\right)} \\
&= c_3 \cdot \frac{e\left(\prod_{j \in S, j \neq i} P_{n+1-j+i},\ P^t\right)}{e\left(\prod_{j \in S} P_{n+1-j},\ P^t_i\right)} \\
&= \frac{m \cdot e(P_1, P_n)^t}{e(P_1, P^t_n)} \\
&= m
\end{aligned}$$

5 Dynamic Key Aggregation

The modified key aggregate cryptosystem consist of five phases. They are as described below:

1. **Setup** $(1^{\lambda}, n)$: Randomly pick a bilinear group $\mathbb{G}_1$ of prime order p where $2^{\lambda} \leq p \leq 2^{\lambda+1}$, a generator $P \in \mathbb{G}_1$ and $\alpha \in_R \mathbb{Z}_p$. Compute $P_i = P^{\alpha^i} \in \mathbb{G}_1$ for $i = 1, 2, \ldots, 2n$. Output $(P, P_1, P_2, \ldots, P_{2n})$.
2. **Keygen** (): Pick $\gamma \in_R \mathbb{Z}_p$, output the public and master-secret key pair: $(pk = v = P^{\gamma}, msk = \gamma)$.
3. **Encrypt** (pk, i, m): For a message $m \in \mathbb{G}_2$ and an index $i \in \{1, 2, \ldots, n\}$, randomly pick $t \in_R \mathbb{Z}_p$ and compute the ciphertext as
$$C = (c_{1i}, c_{2i}, c_{3i}) = (P^t, (vP_i)^t, m.P^{\gamma.t}.e(P_1, P_n)^t)$$
4. **Extract** $(msk = \gamma,\ S,\ u_i : i \in S)$: Let S be the set of indices of the data files requested by the user. The data owner calculates the aggregate key
$$K_S = \prod_{j \in S} P^{\gamma}_{n+1-j}$$
and $u_i = c^{\gamma}_{1i}$ for each $i \in S$ and sends $(K_S,\ u_i : i \in S)$ over a secure channel to the user.
5. **Decrypt** (K_S, S, i, u_i, C): If $i \notin S$, output null. Otherwise, compute
$$m = c_3 \cdot \frac{e\left(K_s \cdot \prod_{j \in S, j \neq i} P_{n+1-j+i}, c_1\right)}{u_i \cdot e\left(\prod_{j \in S} P_{n+1-j}, c_2\right)}$$

The correctness of the above scheme for decrypting the file i can be verified as follows:

$$\begin{aligned}
& c_3 \cdot \frac{e\left(K_S \cdot \prod_{j \in S, j \neq i} P_{n+1-j+i}, c_1\right)}{u_i \cdot e\left(\prod_{j \in S} P_{n+1-j}, c_2\right)} \\
&= c_3 \cdot \frac{e\left(\prod_{j \in S} P^{\gamma}_{n+1-j} \cdot \prod_{j \in S, j \neq i} P_{n+1-j+i},\ P^t\right)}{u_i \cdot e\left(\prod_{j \in S} P_{n+1-j}, (vP_i)^t\right)} \\
&= c_3 \cdot \frac{\prod_{j \in S} e\left(P^{\gamma}_{n+1-j}, P^t\right) \prod_{j \in S, j \neq i} e\left(P_{n+1-j+i}, P^t\right)}{u_i \cdot \prod_{j \in S} e\left(P_{n+1-j}, v^t\right) \prod_{j \in S} e\left(P_{n+1-j}, P^t_i\right)} \\
&= c_3 \cdot \frac{e\left(\prod_{j \in S, j \neq i} P_{n+1-j+i},\ P^t\right)}{u_i \cdot e\left(\prod_{j \in S} P_{n+1-j},\ P^t_i\right)} \\
&= \frac{m \cdot P^{\gamma t} \cdot e(P_1, P_n)^t}{u_i \cdot e(P_1, P^t_n)} \\
&= m
\end{aligned}$$

6 Conclusion

The proposed system in this paper is an extension of the paper by Cheng-Kang Chu who describes a system which provides an efficient way in which the encrypted data stored in a cloud is shared among various users by sending just a single key for a set of files. The system is more efficient than any other data sharing scheme as this method involves sharing of just one key. Our method extends this work by ensuring that if once a key for a set of files is shared to the user it cannot be reused by the same or different user due to dynamic key aggregation.

Usually in cloud storage large number of ciphertexts is to be accomodated. So a solution to extend the number of ciphertext classes still remains.

References

1. Chu, C.K., Chow, S.S.M., Tzeng, W.G., Zhou, J.Y., Deng, R.H.: Key-aggregate cryptosystem for scalable data sharing in cloud storage. IEEE Trans. Parallel Distrib. Syst. **25**(2), 468–477 (2014)
2. Silverman, J.H.: Advanced Topics in the Arithmetic of Elliptic Curves, vol. 151. Springer Science & Business Media (1994)
3. Brezing, F., Weng, A.: Elliptic curves suitable for pairing based cryptography. Des. Codes Crypt. **37**(1), 133–141 (2005)
4. Menezes, A.J., Okamoto, T., Vanstone, S.A.: Reducing elliptic curve logarithms to logarithms in a finite field. IEEE Trans. Inf. Theor. **39**(5), pp. 1639–1646 (1993)
5. Feldman, L., Patel, D., Ortmann, L., Robinson, K., Popovic, T.: Educating for the future: another important benefit of data sharing. The Lancet **379**(9829), 1877–1878 (2012)
6. Tran, D.H., Nguyen, H.L., Zhao, W., Ng, W.K.: Towards security in sharing data on cloud-based social networks. In: Information, Communications and Signal Processing (ICICS) 2011 8th International Conference on, IEEE, pp. 1–5. (2011)
7. Li, M., Yu, S., Zheng, Y., Ren, K., Lou, W.: Scalable and secure sharing of personal health records in cloud computing using attribute-based encryption. IEEE Trans. Parall. Distrib. Syst. **24**(1), 131–143 (2013)
8. Tu, S., Niu, S.Z., Li, H., Xiao-ming, Y., Li, M.J.: Fine-grained access control and revocation for sharing data on clouds. In: Parallel and Distributed Processing Symposium Workshops & Ph.D. Forum (IPDPSW), 2012 IEEE 26th International. IEEE, pp. 2146–2155. (2012)
9. Ateniese, G., Fu, K., Green, M., Hohenberger, S.: Improved proxy re-encryption schemes with applications to secure distributed storage. ACM Trans. Inf. Syst. Secur. (TISSEC) **9**(1), 1–30 (2006)
10. Liu, Q., Wang, G., Wu, J.: Clock-based proxy re-encryption scheme in unreliable clouds. In: Parallel Processing Workshops (ICPPW), 2012 41st International Conference on, IEEE, pp. 304–305. (2012)
11. Guo, F., Mu, Y., Chen, Z.: Identity-based encryption: how to decrypt multiple ciphertext using a single decryption key. In: Pairing-Based Cryptography–Pairing 2007, pp. 392–406. Springer (2007)
12. Guo, F., Mu, Y., Chen, Z., Xu, L.: Multi-identity single-key decryption without random oracles. In: Information Security and Cryptology, pp. 384–398. Springer (2008)

A Lock and Key Share (2, *m*, *n*) Random Grid Visual Secret Sharing Scheme with XOR and OR Decryptions

T. Farzin Ahammed and Sulaiman Mundekkattil

Abstract Security of secret sharing schemes has become a significant issue as secret communication is now being widely used. Random grid based visual secret sharing (RG-VSS) schemes has drawn greater attention taking into consideration its own specific advantages. A (2, n) RG-VSS produces n shares with equal priorities and any of those two shares can be combined to restore the original image. The proposed system boosts the security of (2, n) RG-VSS by integrating a lock and key share concept. A lock and key share (2, m, n) scheme generates m lock and n key shares, and retrieval of original secret is possible if and only if one share is selected from the set of m lock shares and other from the set of n key shares. Both OR-based and XOR-based decryptions are implemented. Further, necessary theoretical proofs are supplied for the correctness of the implementation.

Keywords Random-grids · Lock and key share · Visual cryptography · XOR · OR

1 Introduction

Recently visual cryptography (VC), one of the most fascinating techniques to perform secret sharing gained more recognition. Naor and Shamir [1] were the inventers in accomplishing the (k, n) visual cryptography (VC) technique, which involves slicing an image into n different shares. The secret image could be revealed if and only if, at least k shares were stacked, where $n \geq k$. Kafri and Keren [2] initially presented the encryption of binary secret images by random grids (RG), each of which generated into two meaningless random grids (shares), with the same size of the original secret image. The decryption process is the same as that of traditional

T. Farzin Ahammed (✉) · S. Mundekkattil
Department of Computer Science and Engineering, TKM College of Engineering, Kollam, Kerala, India
e-mail: farzinahammed@gmail.com

S. Mundekkattil
e-mail: sulaimancse@gmail.com

M. Senthilkumar et al. (eds.), *Computational Intelligence, Cyber Security and Computational Models*, Advances in Intelligent Systems and Computing 412, DOI 10.1007/978-981-10-0251-9_37

VC. The image encryption by random grids removes the disadvantages of codebook based VCS. VC-based VSS always require a codebook to support encryption. The designated codebook significantly intensifies the data size and cuts transmission rate. Even though the RG-based VSS schemes beat VC-based ones by predominantly avoiding pixel expansion, the former had been less studied than the latter until Shyu [3] and Chen and Tsao [4, 5] made additional research. XOR based VSS schemes were investigated extensively in [6, 7]. User-friendly schemes random-grid based VCS were also implemented by Chen and Tsao [8]. Collusion attacks in RGVCS have been discussed by Lee and Chen [9] whereas the visual quality improvement strategies of various schemes are discussed in [10–12]. All the existing visual secret sharing schemes assign equal weightage to the shares. The proposed lock and key share (2, *m, n*) RG-VSS introduces the concept of two sets of shares by incorporating a security model. This scheme is implemented with XOR and OR decryptions and the results are compared. The formal proof validates the precision and security while the experimental results show that the proposed scheme does work.

The rest of the paper is organized as follows. The review of the basic VSS scheme by random grids is given in the next section. The proposed lock and key share (2, *m, n*) RG-VSS scheme is presented in Sect. 3. Section 4 analyses the performance of our scheme. Section 5 demonstrates the experimental results. Section 6 provides the comparative analysis of OR and XOR decryptions. The conclusions and future works are included in Sect. 6.

2 Related Works

Kafri and Keren [2] proposed (2, 2) RG-VSS containing three different algorithms to encode a binary image. A binary secret image S can be encoded into two different random grids R1 and R2 such a way that stacking reveals the secret image. Kafri and Keren's [2] popular algorithm is presented as Algorithm 1, where "$\in_r$ {0, 1}" denotes randomly choosing a number from {0, 1} with same probability and "¬" denotes the complement operation (e.g. ¬0 = 1 and ¬1 = 0).

Algorithm 1.

Input: A secret binary image S.

Output: Two random grids (R_1, R_2).

Step 1: Generate a random grid R_1 where $R_1(i, j) \in_r \{0, 1\}$.

Step 2: For each pixel S (i, j), compute

$$R_2(i,j) = \begin{cases} R_1(i,j) & \text{If } S(i,j) = 0 \\ \neg R_1(i,j) & \text{otherwise} \end{cases}$$

Step 3: Output (R1, R2).

3 The Proposed Lock and Key Share (2, *m*, *n*) RG-VSS

The proposed lock and key share (2, *m*, *n*) RG-VSS introduces the concept of two different set of prioritized shares. Instead of giving each share equal priority the model classifies shares in such a way that two sets with *m* lock shares and *n* key shares are defined. The decryption operation is successful only when out of the two shares combined, one share is from lock shares and another is from key shares. The proposed lock and key share (2, *m*, *n*) RG-VSS formally stated as Algorithm 2, which takes one secret image S as input and output two sets of shares −*m* lock shares and *n* key shares. Decryption using two shares from either *m* lock shares or *n* key shares will gave a transparent image.

Algorithm 2.

Input: A secret binary image S.

Output: *m* lock shares and *n* key shares.

Step 2.1: For each pixel S (i, j) do the following:

Step 2.2: Use algorithm 1 to encode pixel S (i, j) so that two bits t_1 and t_2 are obtained.

Step 2.3: Generate *m* lock share bits by duplicating t_1 bit $m - 1$ times.

Step 2.4: The *m* lock share bits are distributed into *m* lock shares (L_1, L_2, ….., L_m).

Step 2.5: Generate *n* key share bits by duplicating t_2 bit $n - 1$ times.

Step 2.6: The *n* key share bits are distributed into *n* key shares (K_1, K_2, …..,K_n).

Step 2.7: Repeat Steps 2.2 through 2.6 until all pixels S (i,j) of secret image S are encoded.

Decryption involves XOR or OR operation between two shares; one being lock share and other key share only recovers the secret. Figure 1 gives an intuitive idea about Algorithm 2.

4 Performance Analysis

This section includes basic definitions from Shyu [3], Chen and Tsao [4].

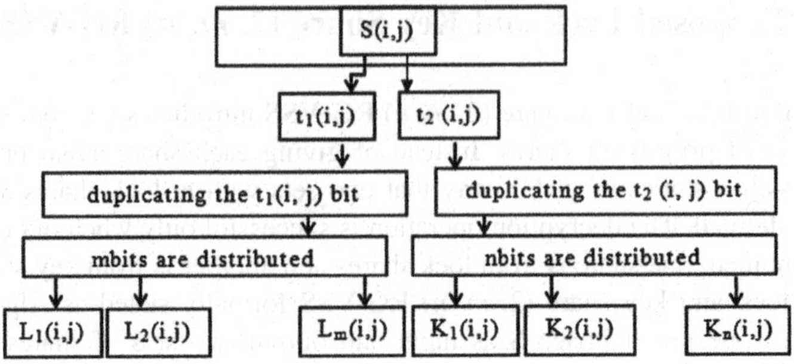

Fig. 1 Encoding process of proposed lock and key share (2, *m, n*) RG-VSS

4.1 Definitions

Definition 1 Average light transmission, Shyu [3]. In a binary image *X*, for a particular pixel *x*, the light transmission, $l[x]$. l [x] is represented by the probability of pixel *x* is transparent. l [x] for a transparent pixel is 1 and for an opaque pixel is 0. The average light transmission of an image *X* is defined as

$$L(X) = \frac{1}{hw}\sum_{i=1}^{h}\sum_{j=1}^{w} l[X[\mathrm{i},\mathrm{j}]] \tag{1}$$

Definition 2 Contrast, Shyu [3]. Visual quality of the reconstructed image *B* for the original secret image *A* is measured by the parameter, contrast α calculated as

$$\alpha = \frac{L[B[A(0)]] - L[B[A(1)]]}{1 + L[B[A(1)]]} \tag{2}$$

In which A(0) denotes all transparent pixels in A and B[A(0)] denotes all pixels encoded from A(0). A(1) and B[A(1)] denotes respective definitions for opaque pixels.

Definition 3 Contrast condition, Chen and Tsao [4]. The reconstructed image B displays the content of the original secret *A* if $\alpha > 0$. If L[B[A(0)]] = L[B[A(1)]] then it reveals no information about A.

4.2 Proof of Correctness of Proposed Model

Lemma 1 *Each share generated by proposed model whether it is lock share or it key share displays no information about the secret, i.e.* L_1, L_2...L_m *and* K_1, K_2...K_n *reveals nothing about the secret.*

Proof The proposed model generates two type of shares, *m* lock share and *n* key shares. Lock shares are clearly a random grid, means it is assigned by t_1 which is random distribution of transparent and opaque pixels (0 and 1 resp.).
The light transmissions are:

$$L[LSi[A(0)]] = L[LSi[A(1)]]$$

Key shares are generated by duplicating the t_2 bit. There *n* random grids will be the t_2 bit random grid which also reveals no information about the secret because the t_2 bit grid is random.

Since L[KSi[A(0)]] = L[KSi[A(1)]], each share does not reveal any information about secret image.

Lemma 2 *While using OR decryption, the recovered image is meaningless when either any of the two lock shares or any of the two key shares are used for decryption.*

Proof Since each lock share is the t_1 bit random grid, the light transmission of recovered image by any two lock shares is calculated by

$$L[LSi(A(0))\,\mathrm{OR}\,LSj(A(0))] = L[A](0)].\,\mathrm{Because}\,0\,\mathrm{OR}\,0 = 0.$$
$$L[LSi(A(1))\mathrm{OR}\,LSj(A(1))] = L[A(1)].\,\mathrm{Because}\,1\,\mathrm{OR}\,1 = 1.$$

Therefore, L[LS(A(0))] = L[LS(A(1))].

Since L[LS(A(0))] = L[LS(A(1))], α will be 0. Therefore when any two lock shares are used, recovered image is meaningless. Similarly for any two key shares also.

Lemma 3 *While using XOR decryption, the recovered image is transparent when either any of the two lock shares or any of the two key shares are used for decryption.*

Proof Since each lock share is the t_1 bit random grid, the light transmission of recovered image by any two lock shares is calculated by

$$L[LSi(A(0))\,\mathrm{OR}\,LSj(A(0))]= L[A(0)].\,\mathrm{Because}\,0\,\mathrm{OR}\,0 = 0.$$
$$L[LSi(A(1))\,\mathrm{OR}\,LSj(A(1))]= L[A(0)].\,\mathrm{Because}\,1\mathrm{OR}1 = 0.$$

Therefore, L[LS(A)] = 1, plies all pixels are white.

Therefore when any two lock shares are used, recovered image is transparent. Similarly for any two key shares also.

4.3 Contrast of Decoded Image of Algorithm 2

Theorem 1 *While using OR decryption, the contrast of the recovered image which is generated by the any one lock share and any one key share is ½.*

Proof Suppose that if S(i, j) is 0.

If t_1 bit is 0, then L[LS(A(0)) OR KS(A(0))] = L[A(0)]. Because 0 OR 0 = 0.

If t_1 bit is 1, then L[LS(A(1)) OR KS(A(1))] = L[A(1)]. Because 1 OR 1 = 1.

Step 2 of Algorithm 1 implies that if (i, j) = 0, then corresponding lock share pixel and key share pixel are the same.

Step 1 of Algorithm 1 implies that pixels are probabilistically distributed resulting half of the pixels are white and other half of the pixels are black.

So if S(i, j) = 0, then resultant pixel = $1/2 * L[A(0)] + 1/2 * L[A(1)] = 1/2$.
Suppose that if S(i, j) = 1,

If t_1 bit is 0, then L[LS(A(0)) OR KS(A(1))] = L[A(1)]. Because 0 OR 1 = 1.

If t_1 bit is 1, then L[LS(A(1)) OR KS(A(0))] = L[(A(1)]. Because 1 OR 0 = 1.

Step 2 of Algorithm 1 implies that if (i, j) = 1, then corresponding lock share pixel and key share pixel are complementary.

So if S(i, j) = 1, then resultant pixel = L[A(1)] = 0. Using (2) α will be (½ − 0)/(1 + 0) = ½.

Theorem 2 *While using XOR decryption, the contrast of the recovered image which is generated by the any one lock share and any one key share is* 1.

Proof Suppose that if S(i, j) is 0.

Ift_1 bit is 0, then L[LS(A(0)) XOR KS(A(0))] = L[A(0)]. Because 0 XOR 0 = 0.

If t_1 bit is 1, then L[LS(A(1)) XOR KS(A(1))] = L[A(1)]. Because 1 XOR 1 = 0.

Step 2 of Algorithm 1 implies that if (i, j) = 0, then corresponding lock share pixel and key share pixel are the same.

Step 1 of Algorithm 1 implies that pixels are probabilistically distributed resulting half of the pixels are white and other half of the pixels are black.

So if S(i, j) = 0, then L[A(0)] = 1.

Suppose that if S(i, j) = 1,

If t1 bit is 0, then L[LS(A(0)) XOR KS(A(1))] = L[A(1)]. Because 0 XOR 1 = 1.
If t1 bit is 1, then L[LS(A(1)) XOR KS(A(0))] = L[A(1)]. Because 1 XOR 0 = 1.

Step 2 of Algorithm 1 implies that if (i, j) = 1, then corresponding lock share pixel and key share pixel are complementary.

So if S(i, j) = 1, then resultant pixel = L[A(1)] = 0. Using (2) α will be (1 − 0)/(1 + 0) = 1.

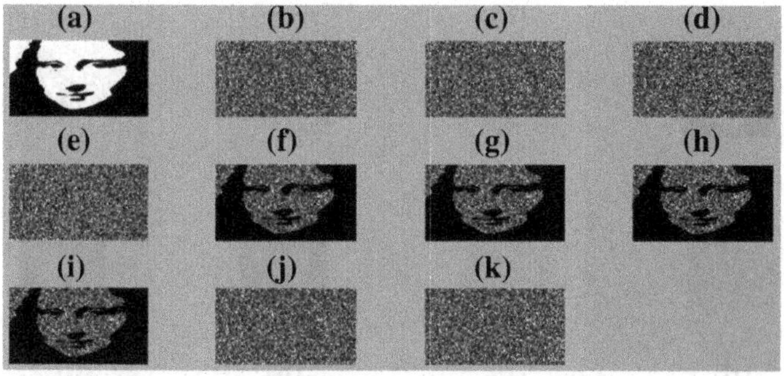

Fig. 2 The experimental results of (2, 2, 2) lock and key share OR based RG-VCS of all images of size 200 × 200. **a** S, **b** LS_1, **c** LS_2, **d** KS_1, **e** KS_2, **f** LS_1 OR KS_1, **g** LS_1 OR KS_2, **h** LS_2 OR KS_1, **i** LS_2 OR KS_2, **j** LS_1 OR LS_2, **k** KS_1 OR KS_2

5 Experimental Results

Images in Fig. 2 are the experimental results for the proposed model of (2, 2, 2) lock and key OR based RG-VSS, where image (a) is the secret image, (b) and (c) are lock shares and images (d) and (e) are the key shares. (f) and (g) generated when the lock share (b) decrypted with key shares (c) and (d) respectively. Similarly (h) and (i) generated with lock share (c). When two lock shares (b) and (c) are used for decryption, (j) is obtained which is a meaningless image as expected. Similarly (i) is obtained with key shares (d) and (e). Images in Fig. 3 are the experimental results for the proposed model of (2, 3, 2) lock and key OR based RG-VSS, where image (a) is the secret image, (b), (c) and (d) are lock shares and (e) and (f) are key shares. (g) and (h) generated when the lock share (b) decrypted with key shares (c) and (d) respectively. Similarly (i) and (j) generated with lock share (c), (k) and (l) with lock share (d) respectively. Meaningless images are (m) (n) and (o) are revealed when any two lock shares are used for decryption. (p) is revealed when two keys shares are used. Similarly, images in Figs. 4 and 5 are the results when XOR operation is used for decryption (Table 1).

6 Comparison and Discussion

Lock and key share (2, *m*, *n*) RG VSS scheme implemented with two different kinds of decryption provide different results. With OR decryption constant time (O(1)) is needed for decryption as decryptions involves merely stacking the result. Merely stacking the result implies binary OR operation which takes constant time. By Theorem 1 using OR decryption, the contrast of the recovered image which is generated by the any one lock share and any one key share is ½. XOR decryption is done based on XOR decrypting device which is proportional to number of shares

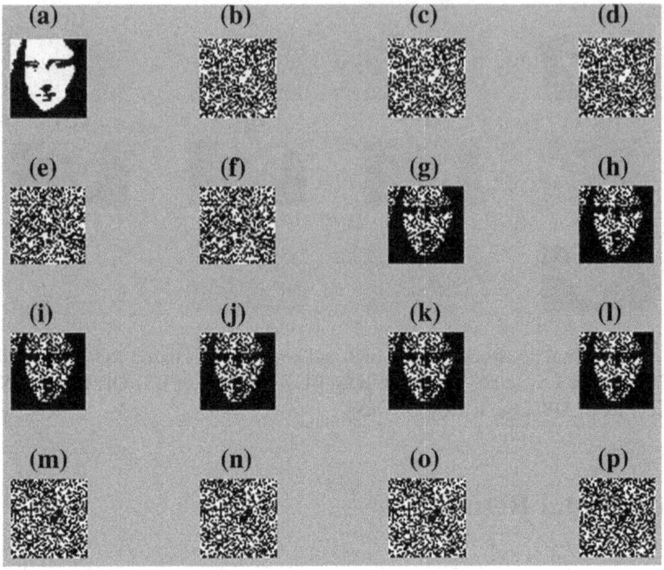

Fig. 3 The experimental results of (2, 3, 2) lock and key share OR based RG-VCS of all images of size 200 × 200. **a** S, **b** LS_1, **c** LS_2, **d** LS_3, **e** KS_1, **f** KS_2, **g** LS_1 OR KS_1, **h** LS_1 OR KS_2, **i** LS_1 OR KS_3, **j** LS_2 OR KS_1, **k** LS_2 OR KS_2, **l** LS_2 OR KS_3, **m** LS_1 OR LS_2, **n** LS_1 OR LS_3, **o** LS_2 OR LS_3, **p** KS_1 OR KS_2

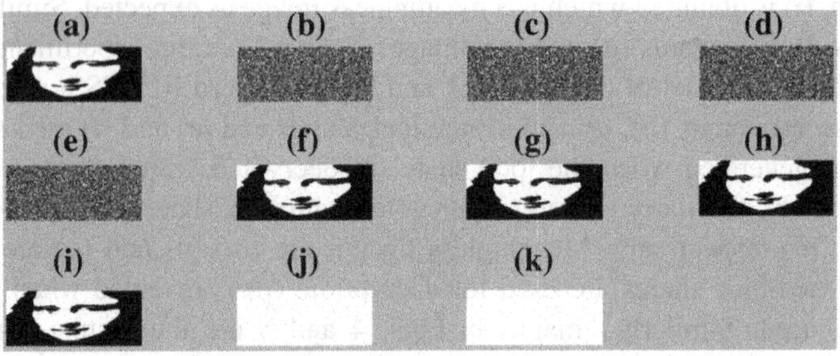

Fig. 4 The experimental results of (2, 2, 2) lock and key share XOR based RG-VCS of all images of size 200 × 200. **a** S, **b** LS_1, **c** LS_2, **d** KS_1, **e** KS_2, **f** LS_1 OR KS_1, **g** LS_1 OR KS_2, **h** LS_2 OR KS_1, **i** LS_2 OR KS_2, **j** LS_1 OR LS_2, **k** KS_1 OR KS_2

XORed. With XOR decryption O(t) is the time needed where t being the number of shares XORed. Since the proposed model is (2, *m*, *n*) RG VSS scheme, the value of t will be 2. XOR decryption results in perfect reconstruction of the secret. By Theorem 2 using XOR decryption, the contrast of the recovered image which is generated by the any one lock share and any one key share is 1. Since both types of decryptions are available, it provides flexibility for users to choose according to the requirement. So there is a trade-off between contrast of the revealed image and time complexity of the decryption, considering XOR and OR decryptions.

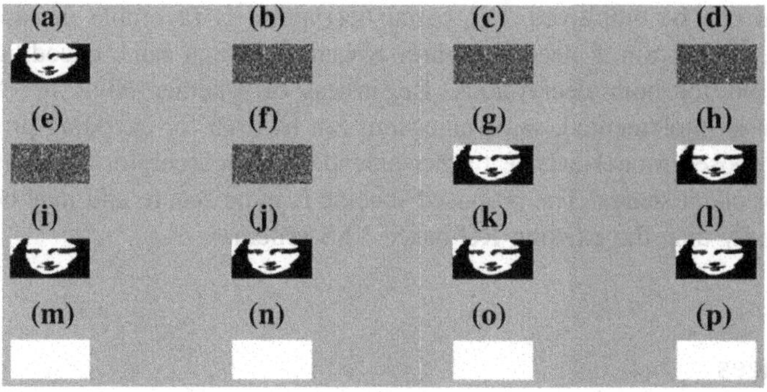

Fig. 5 The experimental results of (2, 3, 2) lock and key share XOR based RG-VCS of all images of size 200 × 200. **a** S, **b** LS_1, **c** LS_2, **d** LS_3, **e** KS_1, **f** KS_2, **g** LS_1 OR KS_1, **h** LS_1 OR KS_2, **i** LS_1 OR KS_3, **j** LS_2 OR KS_1, **k** LS_2 OR KS_2, **l** LS_2 OR KS_3, **m** LS_1 OR LS_2, **n** LS_1 OR LS_3, **o** LS_2 OR LS_3, **p** KS_1 OR KS_2

Table 1 Comparison table

Parameters for comparison	Lock and key share (2, *m*, *n*) random grid VSS	
	With OR decryption	With XOR decryption
Time complexity of decryption	0(1)	0(t)
Contrast of the reveled secret	½	1

7 Conclusion and Future Work

This paper firstly introduces the concept of lock and key shares in RG-VSS. By inheriting the traditional (2, 2) RG-VCS, the proposed lock and key share (2, m, n) RG-VSS can be constructed. Existing schemes do not deal with security of share images. Existing (2, n) random grid VCS proposed in [4, 5] do not consider the case when the intruder able to capture some of the shares. Proposed work considers this case and a secure encoding model is proposed. Also the schemes in [4, 5] never consider splitting of the shares and maintain security. One of the advantages is that incrementing the security of traditional (2, n) RG-VCS. Consider the case of a (2, 1, n) lock and share VCS. The security of a traditional (2, n) scheme is increased in such a way that by protecting single lock share, an intruder cannot recover anything about the secret even if he acquires *n* key shares. Second advantage is that providing a two-tier priority share model. In applications in which we require two sets of users, in which within a set security is needed and secret is only visible if any one of the share images from each set is needed, this scheme can be deployed. The significance of the proposed scheme is that there is a mandatory requirement that out of two shares, one share must be from lock shares and other from key shares. Third advantage is that according to the requirement of reconstructed secret, XOR or OR decryption can be employed. If perfect reconstruction is needed XOR based decryption can be employed. OR based decryption is favorable in cases where immediate decryption is needed. Fourth advantage is that same encoding process can be used for both decryptions. Regardless of whether XOR or OR based decryption is implemented, same algorithm can be used for encoding process.

The proposed model can be further extended to incorporate meaningful cover images for client shares. The proposed scheme is more secure and may be applied more broadly than the existing RG-based VSS schemes.

References

1. Naor, M., Shamir, A.: Visual cryptography. In: Proceedings of Advances in Cryptology: Eurocrypt94, Lecture Notes in Computer Science, vol. 950, pp. 1–12. Springer, Berlin (1995)
2. Kafri, O., Keren, E.: Encryption of pictures and shapes by random grids. Opt. Lett. **12**(6), 377–379 (1987)
3. Shyu, S.J.: Image encryption by random grids. Pattern Recog. **40**(3), 1014–1031 (2007)
4. Chen, T., Tsao, K.: Threshold visual secret sharing by random grids. J. Syst. Softw. **84**, 1197–1208 (2011)
5. Chen, T.H., Tsao, K.H.: Visual secret sharing by random grids revisited. Pattern Recog. **42**, 2203–2217 (2009)
6. Yang, C.N., Wang, D.S.: Property analysis of XOR-based visual cryptography. IEEE Trans. Circuits Syst. Video Technol. **24**, 189–197 (2014)
7. Liu, F,. Wu, C.K.: Optimal XOR based (2, n)-visual cryptography schemes. IACR Cryptol. 545 (2010). ePrint Archive
8. Chen, T., Tsao, K.: User-friendly random-grid-based visual secret sharing. IEEE Trans. Circuits Syst. Video Technol. **21**(11), 1693–1703 (2011)
9. Lee, Y., Chen, T.: Insight into collusion attacks in random-grid-based visual secret sharing. Sig. Process. **92**, 727–736 (2012)
10. Wu, X., Sun, W.: Improving the visual quality of random grid-based visual secret sharing. Signal Process. **93**, 977–995 (2013)
11. Liu, F., Guo, T., Wu, C., Qian, L.: Improving the visual quality of size invariant visual cryptography scheme. J. Vis. Commun. Image Represent. **23**, 331–342 (2012)
12. Guo, T., Liu, F., Wu, C.: Threshold visual secret sharing by random grids with improved contrast. J. Vis. Commun. Image Represent. **23**, 331–342 (2012)

Multilevel Multimedia Security by Integrating Visual Cryptography and Steganography Techniques

M. Mary Shanthi Rani, G. Germine Mary and K. Rosemary Euphrasia

Abstract Multimedia security facilitates the protection of information in multiple forms of digital data such as text, image, audio and video. Many approaches are available for protecting digital data; these include Cryptography, Steganography, and Watermarking etc. A novel approach is proposed for multimedia data security by integrating Steganography and Visual Cryptography (VC) with the goal of improving security, reliability and efficiency. The proposed technique consists of two phases. The first phase is used to hide the message dynamically in an Cover Image1 by changing the number of bits hidden in RGB channels based on the indicator value. VC schemes conceal the Cover Image2 into two or more images which are called shares. In the second phase two shares are created from a Cover Image2 and the stego image created in the first phase is hidden in these two shares. The shares are safe as they reveal nothing about the multimedia content. The Cover Image2, stego image and the hidden message can be recovered from the shares without involving any complex computation. Experimental results show that the new scheme is simple and retrieves many multimedia contents and accomplishes a high level of security.

Keywords Steganography · Visual cryptography · HVS · Multimedia security

M. Mary Shanthi Rani (✉)
Department of Computer Science and Applications, Gandhigram Rural Institute-Deemed University, Gandhigram, Tamil Nadu, India
e-mail: drmaryshanthi@gmail.com

G. Germine Mary · K. Rosemary Euphrasia
Department of Computer Science, Fatima College, Madurai, Tamil Nadu, India
e-mail: germinemary@yahoo.co.in

K. Rosemary Euphrasia
e-mail: rmeuph@yahoo.com

M. Senthilkumar et al. (eds.), *Computational Intelligence, Cyber Security and Computational Models*, Advances in Intelligent Systems and Computing 412, DOI 10.1007/978-981-10-0251-9_38

1 Introduction

Safety of the information transmitted across the internet is an imperative issue, as it is affected by many serious problems such as hacking, duplications and malicious usage of digital information. Some of the objects transmitted online may be vital secret images, and in such cases the senders have to take information security issues into consideration. Being a type of secret sharing scheme steganography and VC can be used in a number of applications.

Steganography is the art of "concealed writing". It is used for secure communication between two parties. In steganography some medium such as image, audio or video is used as cover medium to hide secret information. The cover object with hidden message is called stego object. Least Significant Bit substitution (LSB) is one of the most conventional techniques for hiding considerably large secret message without introducing much visible distortions [1]. Even though it is imperceptible to Human Visual System (HVS), many steganalysis tools are available to detect the presence of message in the stego medium. To prevent the deduction by steganalysis, a method is suggested, in which the stego image is concealed inside the shares of a Cover Image2 created using VC technique.

VC is a technique in which a Cover Image is split up into n distinct meaningless images called shares [2]. Each of the shares looks like a group of random pixels and of course looks meaningless by itself. Naturally, any single share does not disclose anything about the Cover Image. The Cover Image can be decrypted by stacking together all the n shares. The shares generated in VC techniques are meaningless and look like random dots. This is an added advantage to store any secret information because it will not make any visual distortions after hiding the secret information. The above said advantages of steganography and VC are taken into consideration in the proposed method.

The proposed technique consists of two phases. The first phase is used to hide the message dynamically in an image by changing the number of bits hidden in RGB channels based on the indicator value. In the second phase the stego image created in the first phase is hidden in the two shares created from a Cover Image2 using VC. The Cover Image2, stego image and the hidden message can be recovered from the shares without involving any complex computation. The secret message is embedded into an image, which is then embedded inside the shares thereby fully safeguarding the inbuilt-secret message. The secret text can only be interpreted when all the shares are transmitted to the receiver. Thereby, the overall efficiency is increased and security is strengthened. This paper is organized as follows. Section 2 reviews the related work in steganography and visual cryptography, Sect. 3 presents the proposed scheme and Sect. 4 reports the experimental results and discussions. Finally, conclusions appear in Sect. 5.

2 Related Works

VC and Steganography are the current area of research which has lot of scope. Many research works are presented in VC and Steganography separately. VC is a special encryption technique to hide information in images in the form n shares in such a way that it can be decrypted by the HVS when overlapped. It is impossible to retrieve the secret information from any n − 1 shares. The original image can be revealed only when all the shares are simultaneously available. This concept of VC is explored to preserve the privacy of digital data [3]. Steganography is the art and science of writing hidden messages in such a way that no one, apart from the sender and intended recipient, suspects the existence of the message. Images are the most popular cover objects used for Steganography as it has large space for information hiding and high redundancy in representation [4].

The multi-share crypt-stego authentication system proposed by Deshmukh and Sonavane [5] uses both VC and Steganography. In this method there are two main components encoder and decoder. At the encoder side, key image is the input to VC stage which divides key image into multiple shares. The output of VC stage is forwarded to Steganography stage for hiding shares into Cover Image. At the decoder side, stego image is the input to reverse steganography stage for extraction of hidden shares. All output shares are forwarded to VC stage which overlaps all shares to generate key image. Then the output of decoder stage is checked with certain threshold to decide whether access is allowed or is denied for user. Ambar et al. [6] suggested a method for multilevel data security by combining VC and Steganography. In this method an encrypted secret message in the form of an image is split into 2 VC shares and is hidden in 2 different Cover Images. From the cover images the shares are retrieved and secret message is revealed.

In the above works the secret image is converted to VC shares and is hidden in cover images using Steganography. These stego images are susceptible to Steganalyis. Moreover more number of cover images is needed to send a single secret image. In order to overcome the problem of susceptibility and to restrict the number of cover image to one a novel method is suggested in the proposed study.

3 Proposed System

Cryptography and Steganography are the two most commonly used techniques used for data security. Steganography only hides the message where as cryptography scrambles the message. In order to cope with the shortcomings of these techniques and to enhance the security a multilevel multimedia security technique is proposed by integrating Steganography and VC. The proposed system consists of three phases. The architecture of the recommended system is shown in Fig. 1.

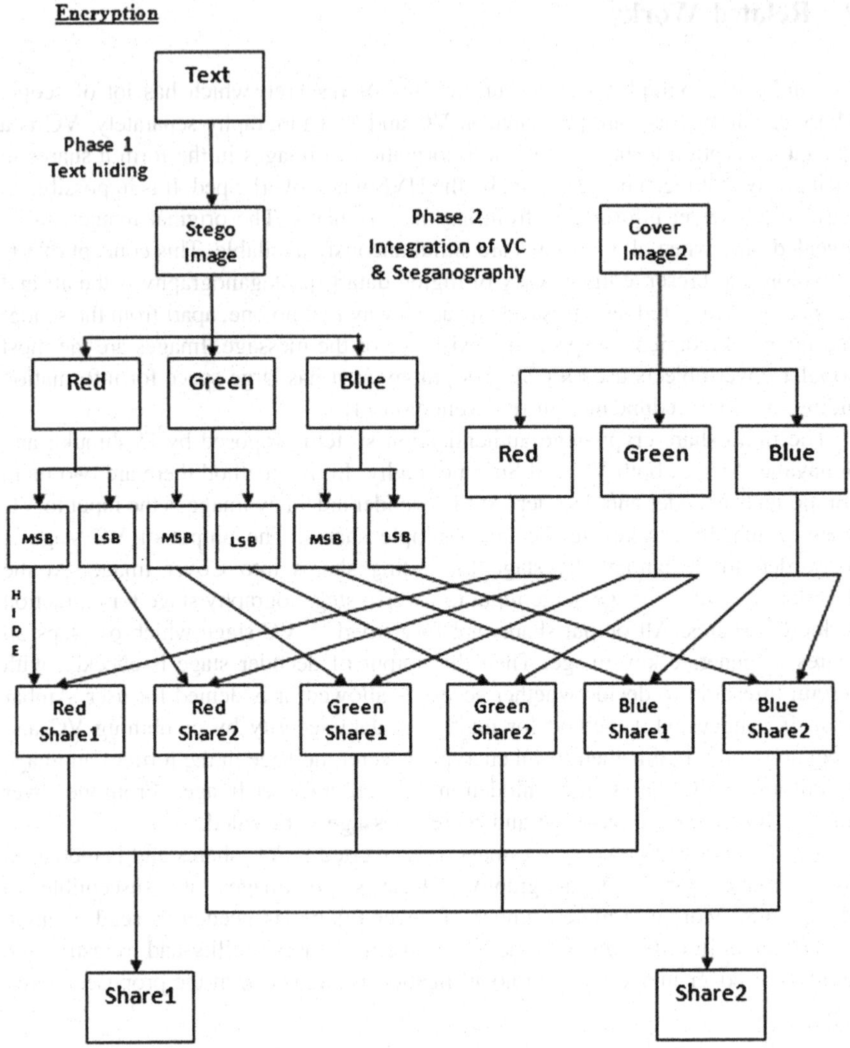

Fig. 1 Proposed system architecture

3.1 *Phase 1: Stego Image Creation*

In this phase the secret message is embedded into random pixels of the Cover Image1 and the steps are described below.

Step 1: Read the secret message and convert them into bytes.
Step 2: Read the Cover Image1 and split into RGB channels.

Step 3: Select one of the color channel using Pseudo Random Number Generator (PRNG)
Step 4: Hide 4 bits of the secret message in a pixel based on the Indicator value.

Step 3 and Step 4 involves random selection of channel and an indicator to hide data and is explained below. Before hiding secret data in each pixel one of the color channels of the pixel will be selected at random as shown in Table 1.

After selecting the channel at random, the 3 MSBs of the selected color channel is used as an indicator to decide whether to hide/not to hide the data in the current pixel and the number bits to be hidden in each color channel as shown in Table 2.

If the three most significant bits are same, either 0 0 0 or 1 1 1, then no data will be hidden in that pixel. Otherwise 1 or 2 least significant bits of the three components will be replaced by the secret message bits as given by the Eqs. (1) and (2).

$$\left.\begin{array}{l} R(x,y) = R(x,y) - (R(x,y)\,\%2) \mid 1 \text{ bit of secret message} \\ G(x,y) = G(x,y) - (G(x,y)\,\%2) \mid 1 \text{ bit of secret message} \\ B(x,y) = B(x,y) - (B(x,y)\,\%2) \mid 1 \text{ bit of secret message} \end{array}\right] \quad (1)$$

$$\left.\begin{array}{l} R(x,y) = R(x,y) - (R(x,y)\,\%4) \mid 2 \text{ bits of secret message} \\ G(x,y) = G(x,y) - (G(x,y)\,\%4) \mid 2 \text{ bits of secret message} \\ B(x,y) = B(x,y) - (B(x,y)\,\%4) \mid 2 \text{ bits of secret message} \end{array}\right] \quad (2)$$

Table 1 PRNG for channel selection

Rand. no.	Channel selected
0	RED
1	GREEN
2	BLUE

Table 2 Indicator and bits hidden

Indicator (3 MSB bits)	Channel		
	Red	Green	Blue
0 0 0	No data hidden		
0 0 1	2 Bits	1 Bit	1 Bit
0 1 0	1 Bit	2 Bits	1 Bit
0 1 1	1 Bit	1 Bit	2 Bits
1 0 0	1 Bit	1 Bit	2 Bits
1 0 1	1 Bit	2 Bits	1 Bit
1 1 0	2 Bits	1 Bit	1 Bit
1 1 1	No data hidden		

3.2 Phase 2: Hiding stego image in VC shares

In this phase the stego image created in phase 1 is embedded into the VC shares of Cover Image2 and the steps are described below.

Step 1: Read the stego image and read each pixel value
Step 2: Separate the 8 bits of each color component into 2 nibbles
Step 3: Read the Cover Image2 and create 2 shares for each color channel
Step 4: Hide the first nibble (MSB) in share1 and second nibble (LSB) in share2

Step 3 and Step 4 which involves creation of VC shares and hiding of stego image simultaneously is explained below. The Cover Image2 is split up into 3 color channels (RGB) and two shares are created depending on the intensity of pixel values (whether it is greater than or less than 128) of each color channel. It expands each pixel into a two 2 × 2 block (B1 and B2) to which a color is assigned as shown in Fig. 2. This shows the blocks created for Red channel. Similarly blocks are created for Blue and Green channels. The fourth pixel of B1 is replaced with first nibble and B2 is replaced with second nibble of stego image. B1 of all pixels form share1 and B2 form share2.

3.3 Phase 3: Extraction Process

The process of decryption is very simple. Neither the stego image nor the Cover Image2 is needed to recover the multimedia content. Cover Image2 can be revealed by overlapping the two shares without any mathematical operations. The stego image can be recovered by tracing, extracting and combining the values of the fourth pixel of every 2 × 2 block in the 2 shares. From the recovered stego image secret message can be extracted. Thus, this multi-level stego-vc system helps to transfer the messages securely which is really hard to crack.

Fig. 2 Block creation for red channel

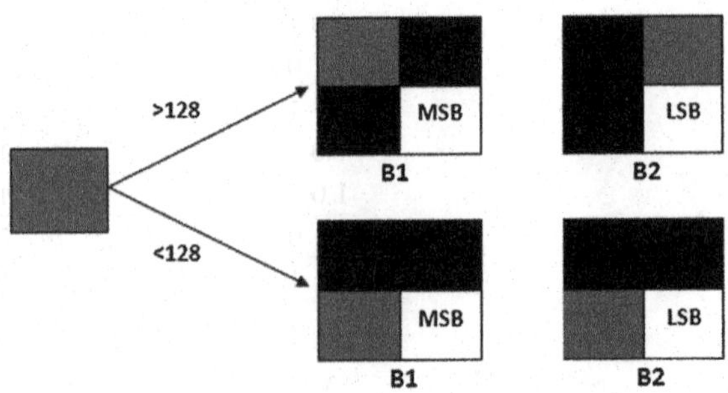

4 Results and Analysis

The proposed system makes use of the advantages of VC and Steganography in hiding multimedia data. The algorithm written in Java language is tested with sample data and the result is shown in Figs. 3, 4 and 5.

Some of the salient features of this work is analyzed and listed below.

(1) **Imperceptibility and Peak Signal to Noise Ratio (PSNR)**

Phase 1 of the suggested method has been tested by hiding a text file in a Cover Image1 and is evaluated by calculating PSNR value. The PSNR in decibels is computed between two images. This ratio is often used as a quality measurement between the original and the reconstructed image. The higher is the PSNR, better is the quality of the reconstructed image. $PSNR = 20\log_{10}\left(\frac{MAX_f}{\sqrt{MSE}}\right)$ MAXf is the maximum signal value that exists in the Cover Image.

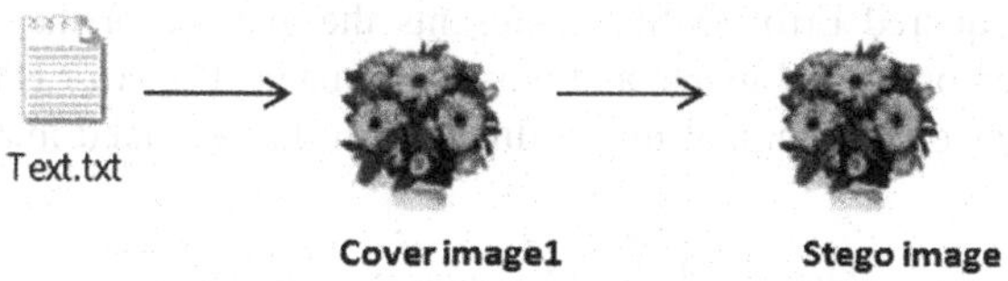

Fig. 3 Result of phase 1-hiding text file in cover image1

Fig. 4 Result of phase 2-hiding stego image in blocks of cover image2

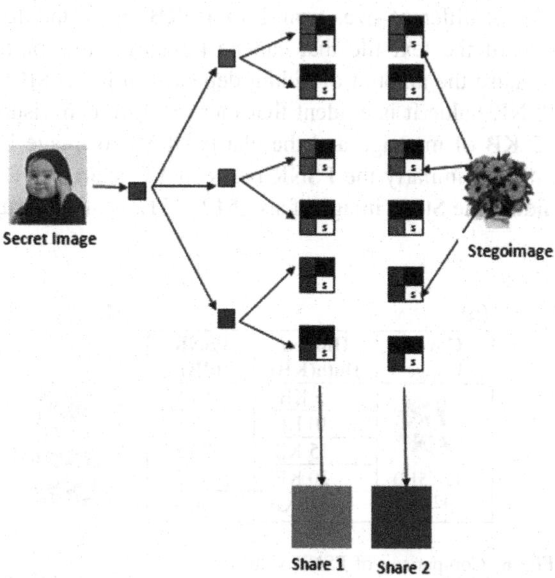

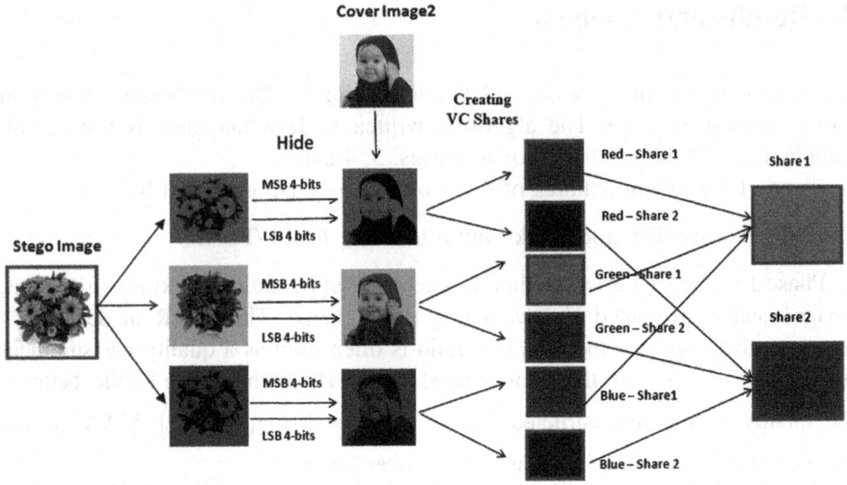

Fig. 5 Experimental result

The Mean Squared Error (MSE) represents the average of the squares of the "errors" between our actual image and our stego image. The error is the amount by which the values of the original image differ from the degraded image

$$MSE = \frac{1}{mn}\sum_{0}^{m-1}\sum_{0}^{n-1}\|f(i,j) - g(i,j)\|^2$$

The PSNR between Cover Image1 and Stego image is calculated by hiding text files of different sizes from 5 to 45 KB and is tabulated in Fig. 6a. The maximum size of the text file that can be hidden in this particular image is only 45 KB, because the amount of hiding depends on the MSB values of the pixels. From the PSNR value it is evident that there is no much visual distortion even after hiding 45 KB of message and the clarity of Stego image is almost same as the original image. Similarly the PSNR between VC shares of Cover Image2 before and after hiding the Stego image of size 512 × 512 is also calculated and is shown in Fig. 6b.

(a)

Cover Image	Hidden Data(KB)	PSNR (dB)
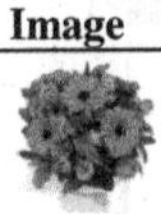 (512× 512) 428 KB	5 KB	54.55
	10 KB	53.78
	15 KB	53.52
	20 KB	53.23
	44.9KB	50.98

(b)

Cover Share (1024 X 1024)	Stego Image (512 X 512)	Stego Share (1024 X 1024)	PSNR
			32.91
			32.98

Fig. 6 Comparison of PSNR values

(2) **Resistance to Steganalysis**

Even though the changes made in the Cover Image due to data hiding is imperceptible to HVS, many steganalysis tools are available to detect the presence of secret message in the stego medium. Deduction by steganalysis can be evaded using VC technique by concealing the stego image inside the shares of Cover Image2 created. The produced shares are always meaningless even after hiding the stego image as shown in Fig. 6b. This system ensures that hackers could not perceive any clues about the Cover Image2 from the shares created.

(3) **Multilevel Security**

It provides three levels of security to the information being transmitted.

- Dynamic and Random hiding of the secret message in an image in phase 1
- Embedding the stego image in VC shares in phase 2
- Meaningless and clueless shares of the Cover Image2 created in phase 2

Because of this the intruders cannot easily break the system, even if they realize the existence of a secret data.

(4) **Multimedia Security**

This system facilitates the user to hide multiple data in different formats such as text and image simultaneously. In this method a stego image, an image with secret message, is hidden in the shares of the Cover Image2. So the receiver can extract the Cover Image2, Stego image and secret message from the received shares. Thus this algorithm provides facility for sending more than one form of data secretly.

(5) **Message Integrity**

A security algorithm is considered as an efficient one if the receiver is able to extract the exact message that was hidden and sent. In this suggested algorithm the Secret Message is hidden in the spatial domain of the image and no transformations are applied and the message retrieved in the extraction process is exactly the same as the hidden message. Thus this method assures data integrity.

(6) **Robust and simple**

This method is very simple because the data are hidden just by replacing the least significant bits and it needs less resource. No complex decryption algorithm required for extraction as we use VC. The results of the algorithm confirm the robustness of the system.

5 Conclusion

In the previous work [6] VC shares are first created and they are hidden inside 2 cover images using steganography. The stego images look like normal image and the steganalysis tools can easily detect the presence of hidden secret. But in the proposed system stego images are hidden inside the VC shares which are meaningless and cannot be deducted by steganalysis tools. VC and Steganography play a major role in data security. Since this proposed method incorporates these two technologies for multilevel multimedia security, highly confidential message can be transmitted across the network confidently. The algorithm is tested with text and image files. The results prove that the quality of the recovered image and message is same as that of the original image and message. Any digital data can be transferred in a secured way using this scheme. More amount of information can be transmitted by increasing the number of VC shares. In this proposed system, since the stego image is hidden in the spatial domain of the VC shares it is prone to transform domain attacks. To overcome, this work can be extended to hide the data in the frequency domain of the shares.

References

1. Mandal, J.K., Debashis, D.: Color image steganography based on pixel value differencing in spatial domain. Int. J. Inf. Sci. Tech. (IJIST) **2**(4), 83–93 (2012)
2. Naor, M., Shamir, A.: Visual Cryptography. In: Advances in Cryptology-EUROCRYPT'94, pp. 1–12 (1995)
3. Mary Shanthi Rani, M., G.G, Mary.: MSKS for data hiding and retrieval using visual cryptography. Int. J. Comput. Appl. **108**(4), 41–46
4. Mary Shanthi Rani, M., Euphrasia, K.R.: Randomized hiding of an encrypted image with hidden data in RGB spatial domain. In: Proceedings of the National Conference on Recent Advances in Computer Science and Applications, pp. 148–151 (2015). ISBN 978-93-84743-57-4
5. Deshmukh, O., Sonavane, S.: Multi-share crypt-stego authentication system. Int. J. Comput. Sci. Mobile Comput. **2**(2), 80–90 (2013)
6. Ambar, R., et.al.: Survey paper on multilevel security using VC and BPCS. Int. J. Comput. Sci. Inf. Technol. Res. **3**(1), 58–62 (2015)

K Out of N Secret Sharing Scheme with Steganography and Authentication

P. Mohamed Fathimal and P. Arockia Jansi Rani

Abstract With communication, preservation and processing of information becoming digital, security of such transactions is a matter of prime concern. This has resulted in the development of encryption and cryptography. The degree of security of confidential data is measured by the involvement of one or more people —at times security is guaranteed when a single hand is involved, while at others secrets kept in parts by many ensures security. This has given rise to secret sharing that has emerged as a hot area of research in computer science in the past decade. Existing schemes of secret sharing suffer from pixel expansion and degradation in visual quality of the recovered secret and cover images. This paper aims to develop a steganography and authenticated image sharing (SAIS) scheme for gray and color images with meaningful shares and the authentication ability to detect the manipulation of shadow images. The main features of this scheme is the use of simple Boolean and arithmetic operations and thus the computational complexity is reduced from $O(n\log^2 n)$ to $O(n)$.

Keywords Threshold secret sharing scheme · Access control · Steganography · Authentication · Secret image sharing · XOR · Pixel expansion

1 Introduction

In the intricately connected world of our times, all documents (text, images and multimedia) have to be stored in a database for future access and sharing. With the emergence of high-speed networks and cloud computing infrastructure, it has become common practice to process and handle highly confidential information as images, and to share them with other users. For example, in medical applications

P. Mohamed Fathimal (✉) · P. Arockia Jansi Rani
Department of Computer Science and Engineering, Manonmaniam Sundaranar University, Tirunelveli, Tamil Nadu, India
e-mail: fatnazir@gmail.com

M. Senthilkumar et al. (eds.), *Computational Intelligence, Cyber Security and Computational Models*, Advances in Intelligent Systems and Computing 412, DOI 10.1007/978-981-10-0251-9_39

such as tele-diagnosis and tele-consultation information exchange is required over a highly secure network. For such an exchange, medical records are computerized at clinics and hospitals in the form of EMR (Electronic Medical Records). Another usage is to create a single patient record system, which is shared among hospitals to avoid multiple entries and help build a consistent medical history of patients.

On the flipside of such record keeping is that the confidential images might be destroyed intentionally or accidentally if they are held in a single place or a single person. Therefore, protection of integrity and confidentiality of the images is of paramount importance in such exchanges. The second issue is the need to keep invaders unaware of not only the content of the secret image itself but also of the very fact that an image is being transferred.

A (k, n) threshold secret sharing scheme is a cryptographic technique designed to distribute a secret S for n participants in such a way that a set of k or more participants can recover the secret S, and a set of k − 1 or fewer participants cannot obtain any information about S.

To make the shares free from modifications, a (k, n) secret image sharing scheme was designed using steganography so that shadow images looks like a cover image. A (k, n) steganography and authenticated image sharing (SAIS) scheme has the authentication ability to detect the manipulation of shadow images. Thus a (k, n) SAIS Scheme has two additional properties -Steganography and authentication ability.

The remainder of this paper is organized as follows. Section 2 discusses the related literature. Section 3 explains the discussion of the proposed scheme. Section 4 discusses the Experimental Results. Finally, Sect. 5 concludes the paper.

2 Related Literature Review

The research on secret image sharing scheme is riven in two directions. One-based on the visual cryptography and another one-based on Lagrange's Polynomial Interpolation.

The visual cryptography introduced by Naor and Shamir [1] allows visual information (pictures, text, etc.) to be encrypted in such a way that secret can be restored without the aid of a computing device. Kang [2] developed an encryption method to construct color Extended Visual Cryptographic scheme (EVC) with VIP synchronization and error diffusion for visual quality improvement. VIPs are pixels on the encrypted shares that have color values of the original images, VIPs at random positions in subpixels degrades the visual quality. VIP synchronization prevents the color and contrast of original images from degradation even with matrix permutation. The other one polynomial-based secret image sharing proposed by Shamir [3] and Blakely [4] is to hide secret pixels as constant terms in (k − 1) degree polynomials using Lagrange Interpolation. A (k, n) secret image sharing scheme that can retrieve the lossless secret image with large embedding capacity based on Sudoku was developed by Chang [5]. After revealing the secret image from any t of n shadow images, the authorized participants can restore the original host image from the shadow images. The reversibility of this scheme allows authorized participants to reconstruct the distorted host image to the original one after retrieving the secret data. Some researchers developed a (k, n) SAIS schemes and these schemes are briefly discussed below.

2.1 (k, n) Shamir Secret Sharing Scheme

In Shamir's secret sharing scheme, there are n participants and a mutually trusted dealer. Let p be a large prime and GF(p) denote Galois Field of order p. The scheme constructs a (k − 1)—degree polynomial y = f(x) = $(a_0 + a_1 x\, a_2 x^2 + \cdots a_{k-1}$ $x^{k-1})$ mod z where $a_1 \ldots .a_{k-1}$ are randomly chosen from GF(p), and a_0 = f(0) = s is the shared secret. Then s (i) = f (xi) mod p for i = 1, 2... n. is computed and distributed to the corresponding participants over a secure channel. With the knowledge of t pairs of (xi, s(i)), the shared secret is calculated as

$$\text{s} = \text{f}(0) = \prod_{j=1 j!=i}^{k} \frac{(-x_i)}{x_i - x_j)} \bmod p^{\infty} \tag{1}$$

To reduce the size of the shadow images by 1/k times the size of the secret image, Lin and Thien [6] used all coefficients in a (k − 1)-degree polynomial. Ahmad et al. [7] used both steganography and shared secret scheme to protect patient medical report or electronic patient record (EPR) and the patient medical image.

Lin and Tsai [8] proposed a (k, n)-SAIS scheme in which every k secret pixelsare embedded into the coefficients $(a_0\, a_1 \ldots a_{k-1})$ of a (k − 1) degree Shamir's polynomial f(x). The output of f(x) is an eight-bit tuple $s_1\, s_2\, s_3\, s_4\, s_5\, s_6\, s_7\, s_8$ and one parity bit p for authentication (even or odd parity) depending on the binary parity sequence generated by the secret key. This scheme reconstructs the original image with slight distortion. Further, this scheme has weak authentication mechanism.

Yang et al. [9] developed an improved version of Lin and Tsai in which the modulus value p is set to the Galois Field GF (2^8) for the restoration of the secret image without distortion. To avoid the authentication weakness, it used the Hash Based Message Authentication Code (HMAC). High computational complexity and reduced visual quality of the stego images are the drawbacks of this scheme.

Chang et al. [10] developed an SAIS in which the Chinese Remainder Theorem (CRT) has been used to improve the authentication ability. This scheme compromises the visual quality of the stego images as more number of bits are embedded into stego blocks with the high authentication ability.

Eslami et al. [11] reduced the computational complexity from Shamir's O (nlog^{2}n) to O(n) by employing a cellular automata (CA) method to construct a (k, n)-SAIS scheme. This scheme provides double authentication to verify whether a stego-image is tampered.

Yang et al.'s (k, n)-SAIS [12] scheme is based on symmetric bivariate polynomial f(x, y) of degree $(k-1)$. In this scheme, every participant can authenticate the other stego images using symmetric property f(x, y) = 2f(y, x). If every shadow gains two votes from other participants, then the stego images are authenticated successfully.

3 Proposed Work

This section describes the proposed secret sharing scheme in detail. The scheme consists of two processes- Sharing and Recovery. Sharing Process has four phases-Initialization, Pre-processing, Partitioning and Embedding. The RecoveryProcess has two phases—Regeneration and Authentication Phase and Post Processing.

3.1 Sharing Process

Initialization Phase In this scheme, with the knowledge of any k shares, the dealer can reconstruct the secret image. To improve the security, participant authentication requires unique key(x, f(x)) to validate their shares. So when the hackers steal the share from any participant, they cannot recover the original image unless they provide the correct unique key value(x, f(x)). Thus this phase ensures unique key value for each participant. But this value for authentication of share is generated only in the dealer side using Lagrange interpolation scheme.

Every participant unique value is generated using the following procedure.

Step 1: Generate $f(x) = \left(a_0 + a_1x + a_2x^2 \ldots + a_{k-1}x^{k-2}\right) \bmod z$ where the coefficients a_0 a_1 … a_{k-1} are random numbers in the range of [1, 255].

Step 2: The dealer then computes $y(x) = (x, f(x))$. i.e. $y(1) = (1, f(1)), y2 = (2, f(2)) \ldots y(n) = (1, f(n))$.

This authentication unique id is a pair of two integers with the condition that $x \neq 0$. If any k-1 of participants gathers, they can reconstruct the coefficients.

PreProcessing Phase For a secret image I of size $r \times c \times d$, each pixel of this image I is subjected to bitwise XOR operation. For example a pixel w and the main key Mkey generated is represented in Fig. 1.

The least significant bit *w*8 is extracted from each pixel and successive 8 bits are converted to the decimal value and stored in image *S*. The size of *Mkey* is $r \times c \times d$ and the size of *S* is $r \times c/8 \times d$. *Mkey* will be the main key stored in the database. This pre-processing phase helps to reduce the size of secret image by eight times.

w_1	w_2	w_3	w_4	w_5	w_6	w_7	w_8

$w_1 \oplus w_5$	$w_2 \oplus w_6$	$w_3 \oplus w_7$	$w_4 \oplus w_8$	$w_5 \oplus w_7$	$w_6 \oplus w_8$	$w_7 \oplus_8$	*0*

Fig. 1 An example pixel and the main key generated in initialization phase

Partitioning Phase This phase describes the procedure of sharing the image S among n participants such that combination of unique k shares from k participants will generate the secret image without distortion. Each participant share is generated by the following algorithm.

A random primary key image *Pkey* of size $r \times c/8 \times d$ is generated and is xor-ed with the image S to produce the resultant image *R* of size $r \times c/8 \times d$. Each pixel in the image *R* obtained in step 1 is subtracted from the modulus *k* (threshold number of shares). The resultant *R*1 is then XOR ed with the original image *R* to obtain *R*2. The *Skey* generated in this step acts as the Master share and is stored in the database. The *R*2 calculated is then divided by $k - 1$ to generate share for each participant. Here *k* is the threshold number of participants. This share generated is then right shifted circularly by c times. The value c is obtained by multiplying participant number x and the coefficient $a_{x \bmod k-1}$, which is used in the initialization phase. Similarly, *Skey* generated in step 5 is circularly right shifted by a_0 and stored in the database as Share 1.

Algorithm 1

Input:

a. *Number of Shares n,*
b. *Number of shares required to recover the shares k,*
c. *Secret Image S of size* $r \times c/8 \times d$
d. *set of n unique ids y(1),….y(n)(for each participant) and coefficients* $a_0 a_1 \ldots a_{k-1}$ *of polynomial f(x)*

Output:

N shares S (1)...S (n) of size $r \times c/8 \times d$

Step1: Generate random image Pkey of size $r \times c/8 \times d$.

Step 2:XOR the random image with secret image.

R=bitxor(S, Pkey)

Step 3: Subtract R from it modulus k-1.

R1=R- (R mod (k-1))

Step 4: XOR the output generated in step 1 and step2.

$R2 = R \oplus R1$

Step 5: Secondary Key is produced by XOR ing the R2 with the primary key generated in step 1.

$Skey = Pkey \oplus R2$

Step 6: Each cell value in R1 obtained in step 2 is divided by k-1(threshold number of shares)

R3=R1/ (k-1)

Step 7: In order to aid authentication and to avoid fabrication of shares,image R3 is left shiftedcircularly by c times to generate the share foreach participant.

$$c_x = (a_{x \bmod k-1} * x) \bmod 255$$

$$S(x) = rightcircularshift\ (R3, c_x) \text{ where } 2 \le x \ge n$$

Step 8:

$$c_1 = (a_0) \bmod 255$$

$$S(1) = rightcircularshift\ (Skey, c_1)$$

$v_{11}...v_{18}$	$v_{21}...v_{28}$	$v_{31}..v_{38}$	$v_{41}..v_{48}$
$v_{51}..v_{58}$	$v_{61}..v_{68}$	$v_{71}..v_{78}$	$v_{81}..v_{88}$

$v_{11}...v_{17}w_{11}$	$v_{21}...v_{27}w_{12}$	$v_{31}...v_{37}w_{13}$	$v_{41}...v_{47}w_{14}$
$v_{51}...v_{57}w_{15}$	$v_{61}...v_{67}w_{16}$	$v_{71}...v_{77}w_{17}$	$v_{81}...v_{87}w_{18}$

Fig. 2 Stego block and the modified stego block of the proposed scheme

Embedding Phase In this phase, a cover image of size $r \times c \times d$ is selected for each participant. Then shares generated in partitioning phase are embedded in the LSB of each pixel in the cover image. The shared pixel *w is represented* in binary as $w_1\ w_2\ w_3\ w_4\ w_5\ w_6\ w_7\ w_8$ and binary form of eight pixels in the cover image $v_1\ v_2\ v_3\ v_4\ v_5\ v_6\ v_7\ v_8$ and the modified block after replacing shared pixel bits $\mathbf{v_{18}\ v_{28}\ v_{38}\ v_{48}\ v_{58}\ v_{68}\ v_{78}\ v_{88}}$ of the cover image pixels is represented in Fig. 2.

This share-embedded cover image and the authentication id $y(x) = (x, f(x))$ where $1 \le x \ge n$ is distributed to each of n participants.

3.2 Recovery Process

The recovery process has two phases.

Regeneration and Authentication Phase In this phase, the original image is recovered by combining shares from $k - 1$ participants and one share from database (*Skey*) as described in *algorithm* 2. First, the LSB of the cover image will be extracted and successive eight bits in each row are combined and converted to decimal form and stored as *SS*(x) of size $r \times c/8 \times d$. Then, the authentication id (x, $f(x)$) of the $k - 1$ participants will be collected and the polynomial f(x) is solved to find the coefficients. Each extracted share *SS*(x) is left-shifted circularly by the resultant of coefficient multiplied by x *mod* 255 to calculate *SSS*(x). Check whether the shares *SSS*(x) are all equal to assure that cover image is free from modification. If all the shares are equal, the master share *S* (1) from the database is then left-shifted circularly by a_0 and stored in *SSS* (1). Then the participants shares *SSS* (2)… *SSS*(k) are added and XORed with *SSS*(1). The output *RR* is then expanded to the size of $r \times c \times d$ by converting each pixel into binary form and store each bit as one pixel in the resultant image *RRR*.

Algorithm 2

Input:
1. *k-1 number of participant cover images of size r×c×d,*
2. *Share from the database of size r×c/8×d*
3. *set of k-1 unique ids y(1),….y(k-1)(for each participant)*

Output: Secret Image I of size r×c×d

Step1: Generate shares SS(x) from LSB of k-1 coverimages where 1 ≤ x ≥ k-1.

Step 2: with the knowledge of k-1 pairs of (x, f(x)), determine the (k-2) degree polynomial f(x) and the coefficients are calculated using the following equation.

$s = f(0) = \prod_{j=1 j!=i}^{k} \frac{(-xl)}{(xl-xj)} mod\ p$

Step 3: Authenticate shares and check the integrity of the cover image by calculating

$$c_x = (a_{x mod\ k-1} * x)\ mod\ 255$$

$SSS(x) = leftcircularshift\ (SS(x), c_x) where\ 2 \le x \ge k.$

if SSS(2)==SSS(3)…==SSS(k-1), the shares are not modified and sharefrom database can be accessed.

$c_1 = (a_0)\ mod\ 255$

$$SSS(1) = leftcircularshift\ (S(1), c_1)$$

Step 4: calculate RR

RR=SSS (1) ⊕ (SSS (2)+SSS (3)..+ SSS (k))

Step 5: Size of RR is r×c/8×d. To expand it into the size of r×c×d, each pixel of RR is represented in the binary form and each bit is stored as one pixel in RRR.

Post Processing Phase: The resultant *RRR* is subjected to post processing using *Mkey* to recover the secret image. Let the pixel in the secret image be $w_1w_2w_3w_4w_5w_6w_7\ w_8$. The image *RRR* obtained in Regeneration and Authentication Phase has 8th bit $w8$ *in each pixel.* To recover the remaining 7 bits, the main key *Mkey* stored in database that has bits of secret image pixel is XORed with the pixels of *RRR* as shown in Fig. 3.

Pixel of the recovered image is extracted as below.

$$w_8 = RRR\,(1)$$
$$w_7 = RRR\,(1) \oplus M_7 = w_8 \oplus w_7 \oplus w_8 = w_7$$
$$w_6 = M_6 \oplus w8 = (w6 \oplus w8) \oplus w8 = w8$$
$$w_5 = M_6 \oplus M_5 = w_6 \oplus w_5 \oplus w_6 = w_5$$
$$w_1w_2w_3w_4 = w_5w_6w_7w_8 \oplus M_1M_2M_3M_4$$

Thus, the secret image can be recovered without any loss.

M_1 ($w_1 \oplus w_5$)	M_2 ($w_2 \oplus w_6$)	M_3 ($w_3 \oplus w_7$)	M_4 ($w_4 \oplus w_8$)	M_5 ($w_5 \oplus w_7$)	M_6 ($w_6 \oplus w_8$)	M_7 ($w_7 \oplus w_8$)	M_8 0

w_1 ($M_1 \oplus w_5$)	w_2 ($M_2 \oplus w_6$)	w_3 ($M_3 \oplus w_7$)	w_4 ($M_4 \oplus w_8$)	w_5 ($M_6 \oplus M_5$)	w_6 ($M_6 \oplus w_8$)	w_7 (RRR (1) $\oplus M_7$)	w_8 RRR (1)

Fig. 3 Main key and the regenerated pixel of the secret image

4 Experimental Results and Analysis

In this section, the feasibility of the proposed scheme is demonstrated with the results of some simulations. The method described in this paper is implemented in *Matlab* 10.0 running on *Windows*8. Experiments performed on *i*5 *Processor* with 4 *GB* of memory.

To test the visual quality of the image and to compare the performance of the system with the previous scheme, the same gray test image *Jet F*16 with size 256 * 256 has been taken as the secret image while *Leena, Pepper, Baboon, Boat images* of size 512 * 512 are selected as the cover (stego) images. The values set for k is 3 and n is 5. So for this (3, 5) proposed sharing scheme, the average PSNR for the cover image is 54.96 dB.

The criteria for the visual quality of the image is the peak signal to noise ratio which is defined as

$$MSE = \frac{1}{mn}\sum_{1}^{m}\sum_{1}^{n}[I_{ij} - I'_{ij}]^2 \tag{2}$$

$$PSNR = 20 * \log_{10}(\max_f / sqrt(MSE)) \tag{3}$$

Legend

I original image of size $m \times n$

I' recovered image of size $m \times n$

$\mathbf{max}_f$ maximum intensity value that exists in the original image (255)

The higher the PSNR value, better the quality of the reconstructed image [13–16]. This scheme can be applied for color image. For a color test image *Jet F*16 with size 256 * 256 as the secret image and *Lena, Pepper, Baboon, Boat and Gold Hill images* of size 256 * 256 with their PSNR for a (3,5) proposed scheme is shown in Fig. 4.

The PSNR for the Lena, Baboon and Pepper images for different schemes are compared with the proposed scheme is shown in Table 1. It shows that visual quality of the share embedded cover images and the recovered image are better than the existing schemes.

15 color test images are used and the PSNR of the some stego images for different k and n values are obtained in Table 2. From the table it is observed that average PSNR is 51 dB for cover images with no pixel expansion and 58 dB for cover images with pixel expansion of 4.

To check the integrity of the stego image, the detection ratio is defined in C-N yang al.'s (k, n) scheme (2010) as *DR = NTDP/NTP* where *NTP* is the number of tampered pixels and the *NTPD* is the number of tampered pixels detected. In the proposed scheme, if any stego image is modified, the share extracted from that stego image will not be equal to the unmodified stego images and the number of

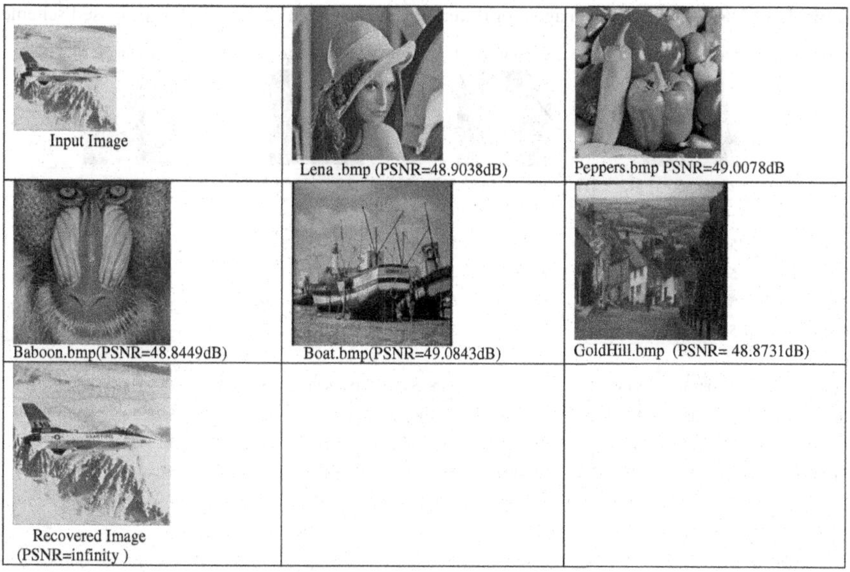

Fig. 4 Experimental results of the proposed scheme for color images with no expansion in cover images

Table 1 Comparison of PSNR (dB) values for the proposed (2,3) scheme with the existing schemes

(2,3) Schemes		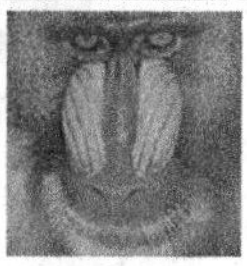	
Lin et al.'s scheme	43.82	43.81	43.78
Yang et al.'s scheme	46.11	46.14	46.14
Chang et al.'s scheme	42.70	42.30	42.09
Eslami et al.'s scheme	47.10	47.08	47.17
C-N yang et al.'s scheme	52.91	52.90	52.91
Proposed scheme	58.38	58.40	58.45

tampered pixels are detected by comparing with other shares. To evaluate the detection ratio, previous schemes added the pepper image in the left corner of the stego images. Similarly the pepper image is added in the stego images and the detection ratio obtained is 0.6 for the proposed scheme whereas DR = 0 for Lin and Tsai's Scheme, 0.5 for Yang et al. scheme, 0.97, 0 for Eslami's Scheme (Fig. 5).

Thus, this scheme prevents the fake stego images generated by dishonest participants or invaders by comparing with other stego images and the tampered pixels are detected with the ratio of 0.6.

Table 2 PSNR (dB) of cover images for different test images in (2,3) and (3,4) proposed scheme

Secret image	Stego image 1 		Stego image 2 		Stego image 3 		Stego image 4 	
(256 * 256)	256 * 256	512 * 512	256 * 256	512 * 512	256 * 256	512 * 512	256 * 256	512 * 512
(2,3)								
Im1.bmp	52.5643	58.4053	52.3976	58.4001	52.3755	58.3591	–	–
Im2.bmp	51.2825	57.2257	51.1316	57.1662	51.1597	57.1379	–	–
Im3.bmp	54.6619	60.5640	54.5579	60.5744	54.5632	60.5478	–	–
Im4.bmp	51.2825	59.6173	51.1316	59.5316	51.1597	59.4848	–	–
(3,4)								
Im1.bmp	49.0843	54.9294	48.8449	54.8854	48.8626	54.8579	48.9038	54.8762
Im2.bmp	51.2825	57.2257	51.1316	57.1662	51.1597	57.1379	51.2539	57.1585
Im3.bmp	52.9842	58.8695	52.8186	58.8664	52.8419	58.8359	52.9422	58.8538
Im4.bmp	52.5107	58.4825	52.3493	58.4136	2.3885	58.3762	52.4940	58.3720

Fake stego image
Authentication of the proposedscheme with DR=0.6

Fig. 5 Fake stego image and the result of authentication (DR = 0.6)

Table 3 shows the comparison of the proposed scheme with the existing schemes. From the table it is observed that PSNR is high for the proposed scheme with no authentication bits in the stego images and detection ratio of 0.6 in the proposed scheme. When the cover image of size 4 times that of the secret image is used, the average PSNR value obtained for the cover image is 58 dB which is higher than the other existing schemes. When pixel expansion is 1, i.e., the cover image and the secret image sizes are equal, the average PSNR of the cover image is 51 dB. For computational complexity in the recovery process, authentication phase requires calculations for solving a polynomial one time (O(1)) and circular shift operation for k shares (O(k)). The regeneration phase requires addition of k − 1 shares and one Boolean XOR operation. The addition of shares being a matrix with size n x m is equivalent to doing m summations of length n. Each length n

Table 3 Comparison of the proposed Scheme with the Existing Scheme

	Lin et al.'s scheme	Yang et al.'s scheme	Chang et al.'s scheme	Eslami et al.'s scheme	C-N Yang et al.'s scheme	Proposed scheme	
						Pixel expansion = 4	Pixel expansion = 1
Meaningful shadow image	Yes	Yes	Yes	Yes	Yes	Yes	Yes
Lossless secret image	Yes	Yes	Yes	Yes	Yes	Yes	Yes
Expansion of stego-image	4	4	4	4	4	4	1
Quality of shadow image	43 dB	46 dB	42 dB	47 dB	52 dB	58 dB	51 dB
authentication bits in a stego-block	1	1	4	0	0	0	0
DR	0	0.5	0.97	0.997	0.997	0.6	0.6
Computational complexity	O(nlog^{2}n)	O(nlog^{2}n	O(nlog^{2}n)	O(n)	O(nlog^{2}n)	O(n)	O(n)

summations requires O(n) operations, it would have m * O(n). However, multiplicative constants are ignored for big-O notation and the resultant complexity will be O(n).

5 Conclusion

In this paper, a novel (k, n) steganography and authenticated image sharing scheme is proposed. This scheme generates meaningful shares by embedding the secret image in the cover images. This scheme reduces the computational complexity of Shamir's scheme from $(n\log^2 n)$ to O(n). Further, this scheme has uniform stego image size equal to the secret image irrespective of k values and thus avoids pixel expansion problem. Experimental results show that the visual quality of the stego images is high and size of the cover image is reduced when compared to existing schemes. This scheme prevents fake stego images and thus ensures authentication. This scheme can further be enhanced with the sharing of multiple secret images.

References

1. Noar, N., Shamir, A.: Visual Cryptography, Advances In Cryptology: Eurocrypt'94. Springer, Berlin (1995)
2. Kang, I.K., Arce, G.R., Lee, H.K.: Color extended visual cryptography using error diffusion. IEEE Trans. Image Process. (2011)
3. Shamir, A.: How to Share a Secret. Commun. ACM **22**(11), 612–613 (1979)
4. Blakley, G.R.: Safeguarding cryptographic keys. In: Proceedings of FIPS National Computer Conference, vol. 48, pp. 313–317 (1979)
5. Chang, C.-C., Lin, P.-Y., Wang, Z.H, Li, M.C.: A sudoku-based secret image sharing scheme with reversibility. J. Commun. **5**(1) (2010)
6. Thien, C.C., Lin, J.C.: Secret image sharing. Comput. Graph. **26**, 765 (2002)
7. Ahmad, T., Studiawan, H., Ahmad, H.S., Ijtihadie, R.M., Wibisono, W.: Shared secret-based Steganography for protecting medical data. In: International Conference on Computer, Control, Informatics and Its Applications, (2014)
8. Lin, C., Tsai, W.: Secret image sharing with steganography and authentication. J. Syst. Softw. 405–414 (2004)
9. Yang, C.N., Chen, T.C., Yu, K.H., Wang, C.C.: Improvements of image sharing with steganography and authentication.. J. Syst. Softw. 1070 (2007)
10. Chang, C., Hsieh, Y., Lin, C.: Sharing secrets in stego images with authentication. Pattern Recogn. 3130–3137 (2008)
11. Eslami, Z., Razzaghi, S.H., Ahmadabadi, J.Z.: Secret image sharing based on cellular automata and steganography. Pattern Recogn. **43**, 397 (2010)
12. Yang, C.-N., Ouyang, J.-F., Harn, L.: Steganography and authentication in image sharing without parity bits. Opt. Commun. 1725–1735 (2012)
13. Fathimal, P.M., Rani, P.A.J.: K out of N secret sharing scheme for gray and color images. In: IEEE International Conference on Electrical, Computer and Communication Technologies, March 2015

14. Fathimal, P.M., Rani, P.A.J.: Bidirectional serpentine scan based error diffusion technique for color image visual cryptography. Int. J. Sci. Eng. Technol. Res. (2014)
15. Fathimal, P.M., Rani, P.A.J.: (N, N) Secret color image sharing scheme with dynamic group. Int. J. Comput. Netw. Inf. Secur. (2015)
16. Fathimal, P.M., Rani, P.A.J.: Design of block based visual secret sharing scheme for color images. Int. J. Appl. Eng. Res. (2015)

A Centralized Trust Computation (CTC) Model for Secure Group Formation in Military Based Mobile Ad Hoc Networks Using Stereotypes

S. Sivagurunathan and K. Prathapchandran

Abstract Self-organized, distributed nature of mobile nodes, frequent changes in topology and shared wireless medium and other characteristics make MANET differ from other networks and encourage military communications that can run on the top of MANET. At the same time, such characteristics pose security threats in military communication. As trust computations play a vital role in decision making and suitable for resource constrained devices, in this paper we propose a Centralized Trust Computation model (CTC) to ensure authentication, based on own experiences, recommendations, social and sense making trusts of team members evaluated by commanders that create an efficient and secure teams based on stereotypes model hence the adversaries who intentionally harm the mission and selfish members due to lack of resources can be identified and isolated from the network by the way authentication can be achieved and secure group can be formed.

Keywords Mobile ad hoc networks · Authentication · Security · Trust · Stereotypes · Military

1 Introduction

Mobile ad hoc networks is simply called as MANET which is capable of self-configuration, self-healing mobile nodes that are connected via shared wireless mediums that form their own infrastructure therefore no pre-defined infrastructure. Mobile nodes are moving unrestricted hence network topologies always change.

S. Sivagurunathan (✉) · K. Prathapchandran
Department of Computer Science and Applications, Gandhigram Rural Institute – Deemed University, Gandhigram, Dindigul 624 302, Tamil Nadu, India
e-mail: svgrnth@gmail.com

K. Prathapchandran
e-mail: kprathapchandran@gmail.com

M. Senthilkumar et al. (eds.), *Computational Intelligence, Cyber Security and Computational Models*, Advances in Intelligent Systems and Computing 412, DOI 10.1007/978-981-10-0251-9_40

Each mobile device is in a situation to forward the traffic that unrelated to its own use hence they act as routers as well as ordinary mobile nodes [1]. This distinct nature offers information-sharing at all levels of military environments, assisting enhanced situational awareness, a clearer understanding of leader's goals, and the ability for mobile soldiers to self-synchronize. In addition, Self-configuration and self-healing nature facilitates and reduces the need for manual configuration and intervention in battlefield. Due to ad hoc nature mobile soldiers have the ability to operate with or without connectivity to a centralized network. When we compare with commercial networks, military networks differ in two ways such as Military is subject to obvious threats that are not subject to restriction protocols and the failures of a technology to execute can cost in terms of loss of life. Mobile soldiers in the battlefield are operating on the basic assumption that all the fighters are fully cooperating in achieving mission. But achieving such cooperation is difficult due to the open nature of MANET. Among the security requirements such as confidentiality, integrity, non-repudiation authentication, and availability, authentication is important because communicating entities must be assured with the identity of each other and these identities should be authorized and recognized before the communication begins [2]. Trust is a word which is originally derived from the social sciences. Trust is defined as "one entity (trustor) is willing to depend on another entity (trustee) [3]. Stereotypes are defined as expectations made about a group of people and are applied to individuals regardless of their personal characteristics because of their affiliation with a certain group [4]. Sensemaking describes what people do so as to decide how to act in the conditions they encounter [5, 6].

2 Related Work

There exists an extensive range of literature dealing with security and trust both in general MANET and also MANET based group communication like military scenarios. The following are the key literature available. The authors [7] proposed a security model to maximize the network performance and minimize trust bias based on the social and quality of trust with intimacy, healthiness, energy and cooperativeness into account. The authors [8], proposed a light weight trust based security model to ensure authentication and detect the malicious nodes. Abort from the above the authors [9–12] proposed trust based secure models for mobile ad hoc networks based on either direct or in direct trust and both trust establishment mechanisms. Our work differs from the existing work that discussed above and we make use of sense making concept as one of the core factor in trust evaluation.

3 The Proposed Model

3.1 Model Design and Assumptions

We assume a pure military based MANET environment without a centralized administration. We assume that mobile devices such as handheld radio, man pack radio, laptops, cameras and Personal Digital Assistants (PDA) are carried by dismounted soldiers and each armed soldier vary in their speed, energy and social characteristics thus reflecting the heterogeneous architecture. We assume each soldier walking speed range from (0 to M) meter/second where 0 denotes the upper range and M denotes the maximum range and walking randomly towards to achieve a mission. At the time of initial network deployment the entire team is trustworthy and authentic and equipped with well all resources. We place any number of the armed soldiers (S) under a team (T) and it is controlled by a commander (C) depends on the situations in emergency and it's all depends on the mission coordinator and likewise all the teams in the military environment can be formed and all teams are small scale in size. We assume that command will pass from upper level to lower level means commander to team members and a team members can interact with the commander but the probability is low when compare the previous one. Then, the armed commanders are well defined in terms of their security, resources and social things that means commanders never compromise and having high energy, processing, and memory and bandwidth devices and they do not change their position; they never move from one team to another due to military constrains. A solider can move from one team to another without the knowledge of commander due to disconnection, mobility or failure of networks and also soldiers can interact with their own teammates only.

We also assume armed soldiers are often behaving maliciously or selfishly affected by their inherent nature as well as environmental or operational or social conditions. We also assume military operations are executed with the support of network operations; for instance a commander wants to order a soldier to do a particular task then he makes use of underlying networking operations such as control and data packet forwarding as a core factor for military commands. To execute commands from a commander to soldiers and soldiers interact with the commander, an underlying MANET routing protocols are needed hence in that case we make use of AODV routing protocol. Every commanders and soldiers maintain a trust table where all the trust related information is stored. The trust table structure of both soldier and commander shown in Tables 1 and 2. We also assume that soldiers trust value as well as overall trust as a continuous real numbers in the range 0–1 with representation of 1 means completely trusted soldier, 0.5 means partially trusted soldier and 0 means adversaries.

Table 1 Trust table maintained by each soldiers

Soldiers ID	CP	DP	DT	CID

Where, *ID* soldier's identity, *CP* control packet, *DP* data packet, *DT* direct trust, *CID* commander ID

Table 2 Trust table maintained by each commander about their own team members

Soldier ID	DT	InDT	ST	SMT	OT	D	TUT

Where, *ID* soldier's identity, *DT* direct trust, *InDT* indirect trust, *ST* social trust, *SMT* sense making trust, *OT* overall trust, *D* decision, *TUT* trust update time

3.2 Centralized Trust Based Computation (CTC)

The CTC model consists of the following phases.

Phase 1: Trust Computation (Soldiers to Soldiers and Commander to Soldiers) As mentioned earlier, initially all the soldiers are cooperating well and trustworthy. Due to inherent nature of the network, soldiers may behave selfishly or misbehave so as to result in poor mission performance. Therefore a commander in a situation to execute the trust computation model. The trust computation model is described as follows, initially all the soldiers broadcast the Hello packets instead of initiating route discovery process or checking their own routing table for desired route. So that every soldier and commander ensure their one hop neighboring soldier's as well as commander ultimately only one hop neighbors respond to the Hello packets because they are in same communication range. Hence every soldier and commander concludes how many soldiers are staying as one hop neighbors. After that every soldier and commander execute the trust evaluation mechanism based on direct experience. The following Eq. 1 shows the direct trust computation of soldiers amongst soldiers and commander to soldiers.

$$\left.\begin{aligned} \text{S2SDT sisj(n)} &= \mu 1(\text{CP sisj(n)}) + \mu 2(\text{DP sisj(n)}) \\ \text{C2SDT ctsj(n)} &= \mu 1(\text{CP ctsj(n)} + \mu 2(\text{DP ctsj(n)}) \end{aligned}\right\} \quad \begin{aligned} &\text{where i, j, n, t} = 1, 2, 3\ldots \\ &\text{i} \neq \text{j},\ \text{j} \neq \text{t}\ \mu_1 + \mu_2 = 1 \end{aligned} \tag{1}$$

where S2SDT represents the direct trust evaluation between the soldiers, C2SDT represents the direct trust evaluated by commander to solider. Si and ct denotes the evaluating soldier and the commander, Sj denotes evaluated soldier by Si and ct. CP denotes control packet (forwarding or responding ratio) and DP denotes data packets forwarding ratio over time (t) with n number of interactions.

As AODV is used as an underlying routing protocol, control packets such as route request (RREQ), route reply (RPLY) packets are used in route discovery process and route error (RERR) and HELLO packets are used in route maintenance process. While evaluating trust these packets are also considered because they are providing significant contribution towards the routing operations by the way achieve successful mission execution. Though adversaries can also utilize such packets but utilizing probability of such packets are relatively low compared with honest soldiers. Hence ratio of Control Packet forwarding or responding is calculated over the period of time based on the Eq. 2 with n number of interactions.

$$\left.\begin{aligned} CP_{sisj}(n) &= \mu_1 RREQ_{sisj}(n) + \mu_2 RPLY_{sisj}(n) + \mu_3 RERR_{sisj}(n) + \mu_4 HELLO_{sisj}(n) \\ CP_{ctsj}(n) &= \mu_1 RREQ_{ctsj}(n) + \mu_2 RPLY_{ctsj}(n) + \mu_3 RERR_{ctsj}(n) + \mu_4 HELLO_{ctsj}(n) \end{aligned}\right\} \\ \text{where } i, j, n, t = 1, 2, 3\ldots, i \neq j, j \neq t, \mu 1 + \mu 2 + \mu 3 + \mu 4 = 1 \tag{2}$$

The Data Packet (DP) forwarding ratio of each soldier is calculated based on the Eq. 3.

$$\left.\begin{aligned} DP\ sisj(n) &= \sum[NDF\ sisj(n)/NDR\ sisj(n)] \\ DP\ ctsj(n) &= \sum[NDF\ ctsj\ (n)/NDR\ ctsj\ (n)] \end{aligned}\right\} \quad \begin{aligned} &\text{where } i, j, n, t = 1, 2, 3\ldots, \\ &i \neq j, j \neq t, \end{aligned} \tag{3}$$

Likewise every soldier and commander could calculate the trust value of all their own members' and update their trust table. Each node can monitor its neighboring nodes' forwarding behavior by using passive acknowledgment [13].

In this phase, each solider only focusing on direct trust and they never look for indirect trust and other trusts because it always leads to additional overhead in terms of communication bandwidth and processing since military devices may be light weight hence direct trust is only consider for soldiers.

Phase 2: Trust Computation (Commander to Soldiers) Once phase 1 is executed, all the soldiers and commanders are able to know their own team members trust values. In phase 2, a commander will evaluate the trust value of each of its soldiers by using indirect or recommendation, social trust and sense making trust and direct trust which are computed in phase 1. The reasons behind to consider all level of trust is, sometimes direct trust is not enough to assess the trustworthiness of a soldier because a soldier A can interact with soldier B and he can obtain good impression in providing services with B at the same time, Solider A may not get good impression with solider C due to unexpected situations in that case a direct trust is often insufficient or even non-existent [14]. Beyond the indirect trust, social trust can also make impact on trustworthiness for example each soldier has different social aspects in terms of honesty, friendship, privacy and cooperativeness with their companion personally and each manned devices have different capability in terms of their energy, cooperativeness, honesty and closeness. For instance a device with high capacity may evaluate the trustworthiness of other devices more strictly than a device with low capacity [15]. With this idea in mind we derived the social trust and applied that in trust evaluation. The reason for considering sense making trust is to know how the soldiers understand the mission, how do deal with the present scenario and execute the mission accordingly. From the above knowledge, we derived the indirect or recommendation trust of soldiers by making interaction with other soldiers who has the experience of evaluating soldiers in phase 1. The following Eq. 4 illustrates the indirect trust evaluation by a commander with respect to each soldier.

$$\begin{aligned} &\text{C2SinDTCtSj (n)} = \sum \mu i\{\text{C2SDTCtSi (n)} * \text{S2SDTsjsi (n)}\} \\ &\text{where n, i, j, t} = 1, 2, 3\ldots, j \neq t \text{ and } \mu 1 + \mu 2 + \mu 3 + \cdots + \mu n = 1 \end{aligned} \tag{4}$$

Next the commander initiates the social trust calculation of each soldier based on the following Eq. 5

$$\begin{aligned} \text{C2SSoTCtSj(n)} = \; & \mu 1 \text{Coop\_Ctsj}(n) + \mu 2 \, \text{Eng\_Ctsj}(n) \\ & + \mu 3 \, \text{Clo\_Ctsj}(n) + \mu 4 \, \text{Hon\_Ctsj}(n) \\ & \text{where } n, j, t = 1, 2, 3.., \mu 1 + \mu 2 + \mu 3 + \mu 4 = 1 \text{ and } j \neq t \end{aligned} \tag{5}$$

Finally, sense making trust will be derived from how a solider is responding for a particular assigned mission by whether he understands the mission clearly, understand the present situation, find a way to accomplish a mission and execute a mission accordingly. All these things happen only if the soldier is responding to a task that is assigned by the commander otherwise we can assume that the solider did not understand the mission clearly and he may be in trouble in making communication or he deliberately wants to avoid the mission. The following Eq. 6 is used to illustrate the sense making trust; it derived from receiving response from the soldiers.

$$\text{C2SSmTCtSj(n)} = \sum \underset{N=1}{\overset{n}{\text{ARCtsj}}} / \underset{N=1}{\overset{n}{\text{TRCtsj}}} \quad \text{where} \quad t, j = 1, 2, 3 \quad \text{and} \quad j \neq t \tag{6}$$

Phase 3: Trust Aggregation and updation Trust aggregation phases will execute once the phase 2 is completed. This phase is used to aggregate all levels of trust such as direct, indirect, social and sense making trusts so as to complete the overall trust of a particular soldier that will help to take decision on a particular soldier based on the stereotypes model. The following Eq. 7 is used to compute the overall trust of a particular soldier by a commander.

$$\begin{aligned} \text{C2SOT CtSj(n)} = \; & \mu 1(\text{C2SDT ctsj(n)}) + \mu 2(\text{C2SinDTctSj(n)}) + \mu 3(\text{C2SSoTCtSj(n)}) \\ & + \mu 4(\text{C2SSmT CtSj(n)}, \\ & \text{where } \mu 1 + \mu 2 + \mu 3 + \mu 4 = 1, j \neq t\ , n, t, j = 1, 2, 3\ldots \end{aligned} \tag{7}$$

Table 3 Interpretation of trust based on stereotypes

Group	Trust nalue	Soldiers security level	Teams various security level
1	If overall trust < (0.0-T1)	Untrusted soldier	Become unsecure team
2	If overall trust > (T2-1)	Trusted soldier	Become secure team
3	If overall trust == (T1-T2)	Partially trusted solider	Become neutral team

From the above equation; C2SOT CtSj(n) denotes the overall trust of a particular soldier evaluated by the commander over n interactions and μ denotes the weighting factor. Trust is updated according to the satisfaction degree of commander on a particular mission. If the mission is executed according to what the commander expects, there is no need of trust update. Otherwise trust update is executed by the commander. Commander's satisfaction degree depends on all levels of trusts obtained by each soldier in their team.

Phase 4: Secure group formation based on stereotypes As soon as the phase 3 is completed, now the commander of a team can know about the overall trust value of its group members and the trust value is used to make decision on them. According to the stereotypes model [16], three types of groups can be formed under a team; secure, unsecure and partially secure teams and based on the definition of stereotypes, we define the group as a community of soldiers who show some common properties in terms of their trust level and behave similarly in certain aspects. Because the common property is shared by all the soldiers in the group, then the group act a collective entity to represents its members. The overall trust is used to classify the teams into various levels such as secure, unsecure and neutral. The following Table 3. Illustrates the various trust levels of soldiers and reflection in their team.

3.3 *Soldier's Dynamic Adaptation*

Due to the armed soldier's movement, disconnection and link failure, mission fails often but soldiers are in situation to complete the mission successfully so as to maintain the consistency of mission. To complete the mission successfully, each soldier should follow the dynamic adaptation algorithm that is described as follows. When a soldier wants to join team A from team B, it first broadcasts its "request" message along with overall trust and its current team commander's ID say team A to team B with small Time To Live (TTL) value. Based on the trust value possess by a soldier a commander will take a decision based on the stereotypes model. However intimation about the new soldier can know by all other soldiers only when phase 1 executes because a military scenario possess highly confidential information since the worth of information is higher than number of soldiers involved in the mission.

4 Experimental Results and Discussion

The proposed model is implemented in Network Simulator 3 (NS3). It is a discrete event simulator tool. Our study area is 500 m × 1000 m for simulation. The movement of soldier's is restricted to a maximum of 2 m/s because a human being walking speed range from 0 to 2.5 m/s and we utilize the Random Waypoint mobility model. The following Table 4 Illustrates the simulation parameters that we have set for the evaluation of the proposed trust model.

Experiment 1 Detection of untrusted Soldiers with varying number of adversaries' nodes under CTC. It effectively detects the misbehaving nodes though the number of adversaries' nodes increases. The sample datasets of the entire nodes with the malicious nodes being 20 % of the overall number has obtained and based on the datasets we have drawn the plots to visualize or identify the untrusted nodes. Similarly the simulation was performed for varying percentage of malicious nodes with the proposed trust model. In Fig. 1 the detection ratio against the percentage of adversaries is shown. The Figs. 2 and 3 show the detection of malicious activities for the dataset with 20 and 60 % of adversaries. As mentioned earlier, though the number of malicious nodes increases our trust model effectively detects the untrusted nodes. For example whenever we increase Malicious nodes to 40 and 60 %, the detection ratio also increases by 40 and 50 % respectively. Hence we can easily detect the untrusted nodes and make decision and apply stereotypes model and place the nodes accordingly.

Table 4 Simulation parameters

System parameters	Values utilized
No. of mobile nodes	50
% of black hole node	20, 60 %
Mobility model	RWPM
Simulation duration	100 s
Time interval	0.5 s
Simulation size	500 m × 1000 m
Routing protocol	AODV
Data rate	3072 bps
Packet size	64 Bytes
Wi-Fi ad hoc	802.11 b
Data traffic	UDP
Max node speed	20 m/s
Node pause	0 s
Transmission range	7.5 dbm
Threshold value 1	>=5.5
Threshold value 2	<=5.5. >4.0

under 60 % of malicious

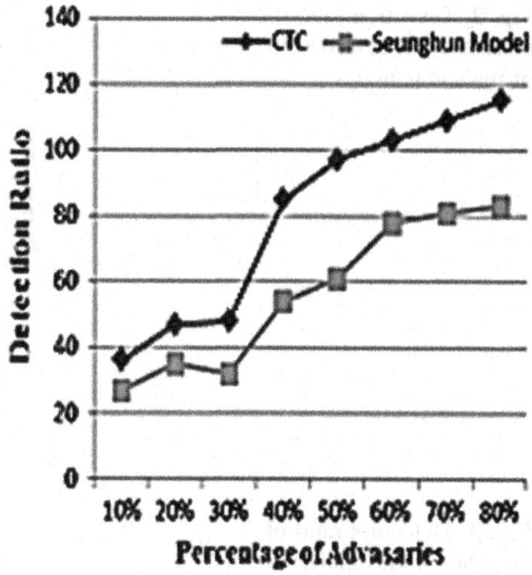

Fig. 1 Detection ratio of adversaries

Experiment 2 The Fig. 1 shows the detection ratio between CTC and [12] whenever the number of adversaries has increase in CTC model detection also dramatically increased when compare with [12] model.

Experiment 3 In this experiment the performance metrics such as packet dropping ratio, packet delivery ratio and end to end delay are analyzed (Figs. 2 and 3).

Packet dropping ratio This metric is calculated by the difference between the total number of packets actually sent and the total number of packets actually received during the simulation. Hence the Fig. 4 clearly shows that packet dropping ratio of CTC model is relatively low compared with [12]. During the simulation the adversaries could be isolated by using CTC model so that it shows better results.

Packet delivery ratio This metric analyses the packet delivery ratio of each node as well as for the overall network. It is measured by the number of packets actually received divided by number of packets actually sent. The Fig. 5 depicts the packet delivery ratio of CTC is very high over [12] because as mentioned earlier, adversaries are isolated from the network, as they are not involved in mission.

End to End delay It is measured by the average time taken by a packet from the source to the destination. Hence it is calculated by difference between the arrival time and sending time of packets from the source to the destination and the results will be divided by total number of connections between the sources to the destinations for each communication. The Fig. 6 compares the end to end delay of CTC model with [12]. From the above performance metrics, it is clearly understood that the CTC model is better compared with [12].

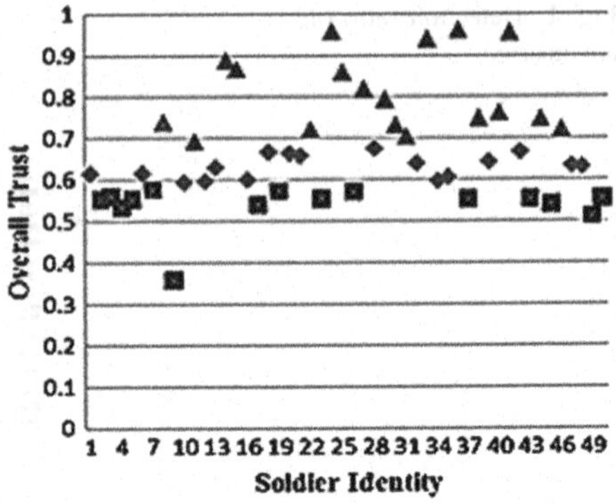

Fig. 2 Detection ratio of untrusted nodes under 20 % of malicious nodes

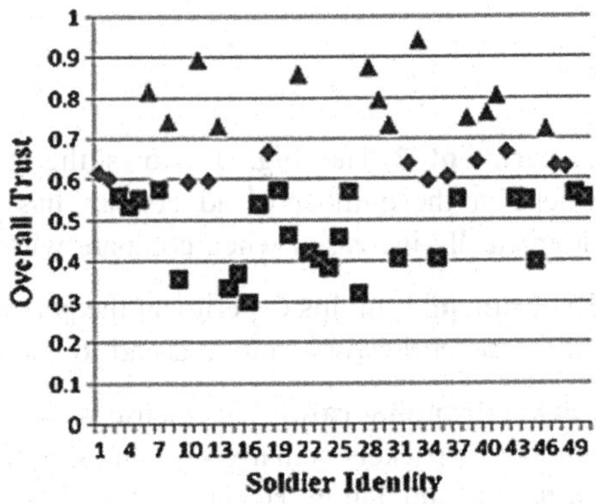

Fig. 3 Detection ratio of untrusted nodes under 60 % of malicious nodes

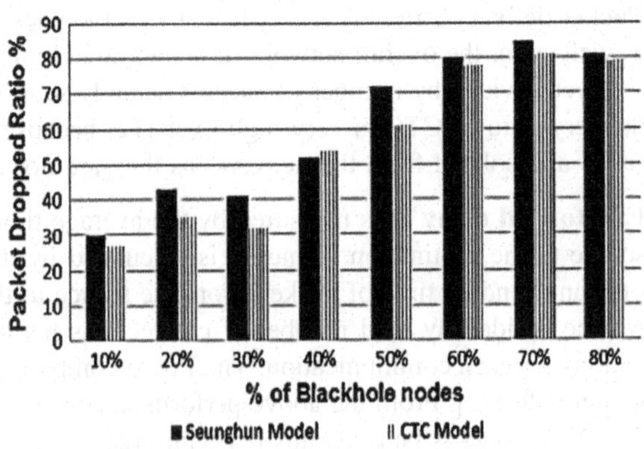

Fig. 4 Packet dropped ratio

Fig. 5 Packet delivery ratio

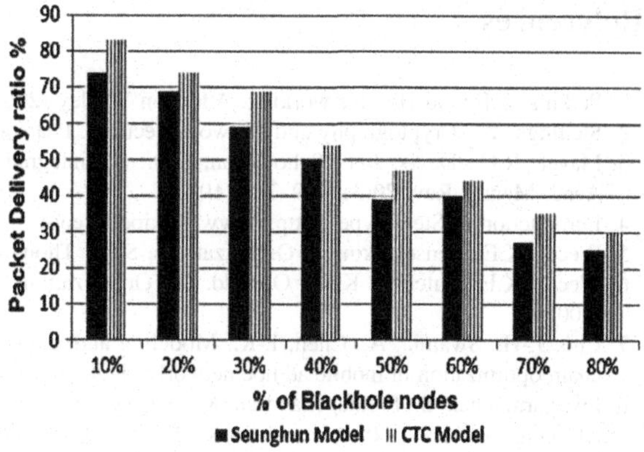

Fig. 6 End to end delay

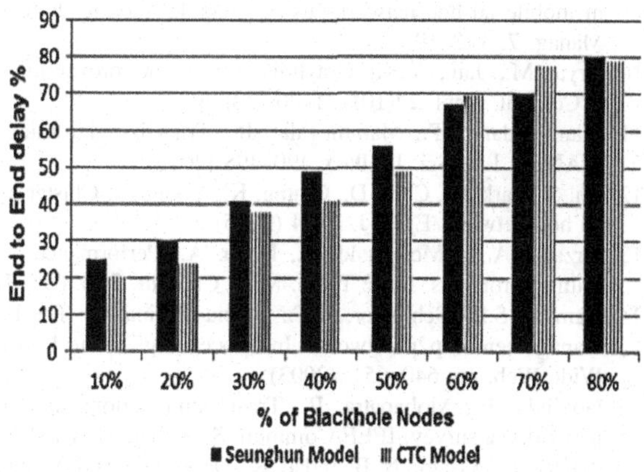

5 Conclusion

In this paper we have proposed a Centralized Trust Computation model (CTC) to identify untrusted nodes in order to ensure the authentication and secure team formation for military scenario using stereotypes. To evaluate the trustworthiness of participating solider by a commander, we make use of direct, indirect, social and sense making trust values since the proposed model give maximum effort to evaluate a trustworthiness of a particular soldier. We assign the overall trust evaluation task to Commanders since it is highly configured and it reflects the real time scenario without much overhead. Moreover we avoid armed soldiers to calculate the overall trust so those devices' energy can be saved and overhead is avoided, hence they involve themselves in a mission for a long time.

Acknowledgments This research work is supported by University Grant Commission, India, through a Major Research Project, Grant (UGC.F No: 42-128/2013 (SR)).

References

1. Perkins, C.E.: Ad Hoc Networking. Addision Wesley (2001)
2. Stallings, W.: Cryptography and Network Security. Pearson Education (2003)
3. Mayer, R.C., Davis, J.H., Schoorman, F.D.: An integrative model of organizational trust. Acad. Manag. Rev. **20**(3), 709–734 (1995)
4. Introduction to Stereotypes. http://www.Wikipedia.com
5. Wecik, K.E.: Sensemaking in Organizations. Sage, Thousand Oka (1995)
6. Wecik, K.E., Sutcliffe, K.M., Obsfeld, D.: Organizations and the Process of Sensemaking. (2005)
7. Cho, J.-H., Swamia, A., Chen, I.-R.: Modeling and analysis of trust management with trust chain optimization in mobile ad hoc networks. J. Netw. Comput. Appl. **35**, 1002–1012 (2012)
8. Sivagurunathan, S., Prathapchandran, K.: A light weight trust based security model for mobile ad hoc networks. In: Proceedings of the International Conference on Electrical, Instrumentations and Communication Engineering-Recent Trends and Research Issues (2015)
9. Velloso, P.B., Laufer, R.P., de Cunha, D.O., Duarte, O.C.M.B., Pujolle, G.: Trust management in mobile ad hoc networks using scalable maturity based model. IEEE Trans. Netw. Serv. Manag. **7**, 3 (2010)
10. Arya, M., Jain, Y.K.: Grayhole attack and prevention in mobile ad hoc network. Int. J. Comput. Appl. **27**(10), 21–26 (2011)
11. Manikandan, S.P., Manimegalai, R.: Trust based routing to mitigate black hole attack in MANET. Life Sci. J. **10**(4), 490–498 (2013)
12. Jin, S., Park, C., Choi, D., Chung, K., Yoon, H.: Cluster-based trust evaluation scheme in an ad hoc network. ETRI J. **27**, 4 (2005)
13. Pirzada, A.A., McDonald, C., Datta, A.: Performance comparison of trust based reactive routing protocols. IEEE Trans. Mob. Comput. **5**(6), 695–710 (2006)
14. Kamvar, S.D., Schlosser, M.T., Garcia-Molina, H.: The Eigen trust algorithm for reputation management in p2p networks. In: Proceedings of the 12th International Conference on World Wide Web, pp. 640–651. (2003)
15. Govinda, K., Mohapatra, P.: Trust computations and trust dynamics in mobile ad hoc networks: a survey. IEEE Commun. Surv. Tutorials **14**, 3 (2012)
16. Ameza, F., Assam., N, Beghdad, R.: Defending AODV routing protocol against the black hole attack. Int. J. Comput. Sci. Inf. Secur. **8**(2), 112–117 (2010)

Cost Effective Rekeying Approach for Dynamic Membership Changes in Group Key Management

Raja Lavanya, K. Sundarakantham and S. Mercy Shalinie

Abstract Security is an important requirement in reliable group communication over open networks in order to prevent intruder attack. A common secret key called group key is generated for encrypting the group information. A distributed key management methodology based on dynamic decentralized group key agreement protocol is required to handle this issue. Rekeying or new group key generation is based on membership driven or time driven. Rekeying is needed whenever a new single member comes to the group or an existing member goes out. Individual rekeying operations in a large group of users leads to the increase the rate of message exchanges which leads to performance degradation. This paper investigates communication rounds, computation complexity and storage of keys in existing key management schemes and introduces an enhanced decentralized Sub Group Key Management (SGKM) approach which is efficient for rekeying. In the proposed work, a group is fragmented into sub groups and the sub groups are managed with one encryption key and multi decryption key protocol by the respective manager. The performance of proposed algorithm is analyzed under different join and leave possibilities.

Keywords Group key agreement · Dynamic membership · Dynamic rekeying · Authentication · Reliable group communication

---

R. Lavanya (✉)
Department of Information Technology, Thiagarajar College of Engineering, Madurai, Tamil Nadu, India
e-mail: rlit@tce.edu

K. Sundarakantham · S.M. Shalinie
Department of Computer Science and Engineering, Thiagarajar College of Engineering, Madurai, Tamil Nadu, India
e-mail: kskcse@tce.edu

S.M. Shalinie
e-mail: shalinie@tce.edu

M. Senthilkumar et al. (eds.), *Computational Intelligence, Cyber Security and Computational Models*, Advances in Intelligent Systems and Computing 412, DOI 10.1007/978-981-10-0251-9_41

1 Introduction

Group Key management is a primary security service for group communication. It can be defined as a set of processes and mechanisms involved in the group key generation, rekeying operation and key relationship management [1]. For managing the group, whenever group membership changes occurs, a new group key has to be generated and delivered to the members in order to achieve maximum security. Thus the new member is prevented in accessing the old data and the departed member is prevented to access new data. Many researchers have proposed number of dynamic group key management approaches for multicast namely, Simple Key Distribution Center (SKDC), Logical Key Hierarchy (LKH) [2], Master-Key-Encryption-based Multiple Group Key Management (MKE-MGKM) [3], One-way Function Tree (OFT) [4] and Shared Key Derivation (SKD) [5]. In this paper, a cost effective SGKM approach is proposed which significantly reduces the above said complexities.

In SGKM, network is organized as a collection of sub groups and each sub group is managed by one manager. The proposed scheme facilitates both unicast and multicast transmission for key delivery. The sub group manager has the responsibility of managing the group key. The communication between root and sub group manager is done via multicast whereas the group key is securely delivered to the members using unicast. The manager generates the group encryption key and allows the members to use their secret mac id as their unique decryption keys. During the occurrence of membership changes, the group encryption key alone is rekeyed and the decryption keys are kept as same. SGKM provides protection against inside attack and outside attack and ensures the security requirements namely, forward secrecy, backward secrecy, group secrecy, confidentiality, authentication and integrity. So it is very difficult to cryptanalyze.

2 Related Works

Based on the literature survey, according to topological structure, the group key management schemes can be classified into three categories namely, centralized key management scheme, decentralized key management scheme and distributed key management scheme. In centralized key management scheme, a central server is responsible for managing the whole group which includes generating the group key, delivering the group key to the members and rekeying due to join and leave operation [6]. The rekeying cost is high because, more intermediate keys are rekeyed. Also server is still in a security bottleneck as the compromised server causes the failure of whole network. In decentralized key management scheme, the network is organized as a collection of subgroups and each subgroup is managed by the respective heads of the subgroup [7–9]. This scheme significantly reduces the rekeying cost because the head of the subgroup takes care of key management.

In distributed key management scheme, all the members in the network involve in the key management. Each member must contribute for the generation of group key by sharing the part of its key material [10]. Many Key agreement schemes use one encryption key and multi decryption key protocol [11, 12] for key management. In this scheme, rekeying cost is high since all the members again contribute to the generation of the new group key.

Three existing key management schemes are considered. One is OFT which is centralized key management scheme. The second one is Iolus which is a decentralized key management scheme and the final one is SKD which is the distributed key management scheme.

2.1 *Existing Key Management Schemes*

2.1.1 One-Way Function Tree (OFT)

Sherman and McGrew proposed that in OFT, the balanced binary tree is used for implementing the key management where bottom-up approach is used for key generation [4]. The OFT tree is solely created and managed by the root. The root node sends a notification about a member entering or exiting the group, so that the modified members will refresh themselves and compute the new key. Because of its bottom up approach, the complexity depends on the height (h) of the tree. During rekeying the whole path from the modified node to the root is updated. For a join operation, the new member involves in the decryption of the new keys and existing members involves in both key decryption and key derivation. Meanwhile, the server does key encryption, key derivation and random key generation. For a leave operation, the server performs key encryption, key derivation.

2.1.2 Iolus

Mittra proposed that, in Iolus, a large group is divided into many smaller subgroups and each subgroup is managed by a sub group controller which eliminates the single point failure [13]. Two main objects namely Group Security Controller for top-level subgroup and Group Security Agent for controlling each subgroup are involved. The communication between the subgroups is achieved by security intermediaries through bridging. For a data transmission between groups, decryption and re-encryption is performed. Iolus can be used to build protocols that provide an independent group key management service for other security aware multicast applications.

2.1.3 Shared Key Derivation (SKD)

Lin et al. proposed that Shared key derivation [5] is a combination of centralized and distributed key management [12]. The key derivation is performed by the members and the server does not generate, encrypt and distribute the new keys. Initially the server distributes the group key to the members. Later when a member joins or leaves the group, the new group key is generated by each member by using old key. This ensures backward secrecy.

2.2 *Cost on Existing Schemes*

The following operations can be carried out for calculating communication, computation and storage costs.

2.2.1 Communication Cost

Communication cost means the number of keys transmitted by a node when rekeying. OFT and SKD depends on the height of the tree.

2.2.2 Computation Cost

Computation cost refers to the number of operations involved in encryption, decryption, key derivation and Random key generation during rekeying. Here also the number of operations depend on the height of the tree.

2.2.3 Storage Cost

Storage cost is the cost required for storing the keys which depends on the size of the key. In OFT and SKD, the storage cost at server depends on the number of users in the group and keys for members.

2.3 *Individual Rekeying*

For every single member join or single member leave, a new group generation is performed. For a large group of members, the group key generation for single membership changes has the following drawbacks.

First, more number of message exchanges in the group key generation and distribution which leads to performance degradation. It consumes more bandwidth for real time group communication.

Second, a member of the group may need a huge amount of memory to temporarily hold the dynamic rekeys and messages before they are encrypted. The reason for this issue is the increase in the dynamic rekeys message delivery delay and the increase in the rate of join and leave request.

3 System Model

The primary functions of SGKM are sub group formation, group key generation and key refreshment.

3.1 Formation of Sub Group

In a highly dynamic networked environment, different sub groups are formed which significantly reduce the cost involved in rekeying process. For this decentralized approach, the subgroups are managed by their respective sub group manager. The manager generates the group encryption key and allows the members to use their secret mac id as their unique decryption keys. This eliminates the single point vulnerability problem as well as one–affect –N issue. In SGKM, the sub group manager is responsible for generating the common group key and updating the group key when the member join or leave the group.

3.1.1 Selection Procedure for Sub Group Manager

The sub group is constructed by selecting the largest weighted node as a manager in that level. For that, each node shares the MAC ID with its neighbors which are in the close geographical area. The weight estimation is based on the mac id. For constructing the sub group in user level, the maximum hop count between all the nodes is one. After selection of the sub group manager for all groups, next election is for choosing the head of different subgroup managers. The largest weighted subgroup manager is elected as head and the same process is continued till the root is found. The above process is illustrated in Fig. 1.

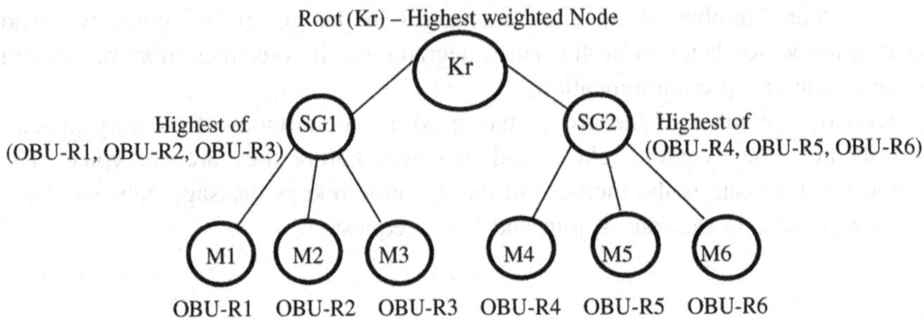

Fig. 1 Selection procedure for sub group manager

4 Proposed Approach

Rekeying Process:

In SGKM, the entire group is not involved in the rekeying process because of the subgroup management. A node's request is granted to join or leave from the sub group, the new group key should be generated by the subgroup manager.

4.1 Rekeying Due to Join Operation

For readers' convenience, rekeying process due to join and leave in SGKM is discussed based on the following example of a key tree, as illustrated in Fig. 2. In

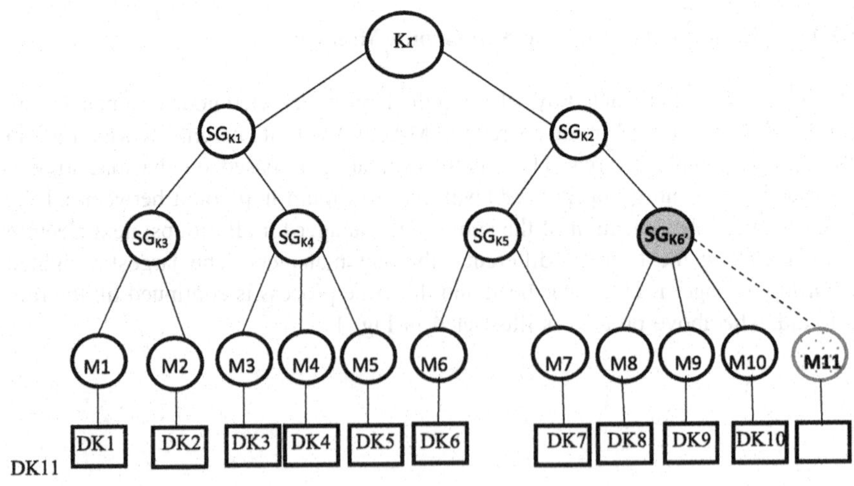

Fig. 2 Rekeying due to join and leave of M13

this example, Kr is the Root of the tree having six intermediate nodes as subgroup managers and 10 members.

When the new member M11 wants to join in a subgroup, first, it sends a join request to its subgroup manager SG6. The manger SG6 is responsible for generating a new group key and allowing M11 to use its decryption key. So it records the request and forwards it to the root through higher level managers.

The root looks up its table to confirm whether the request is new or duplicate/replay. If it is a duplicate, i.e., the join request is from the same node or existing member, the root does not accept until it makes sure that this new node leaves from the old group. This can be used to prevent inside attacker. If the request is from a new node, the root authenticates the node and authentication is sent back to the group manager SG6. Now SG6 generates the new key SG_{K6}'.

Note that, the new node M11 is unable to access the old key SG_{K6}. Thus the proposed scheme ensures the backward secrecy. In SGKM, only one key is generated during rekeying, hence it achieves the better performance.

```
Procedure join (member Mi, subgroup manager SGi, root Kr)
{
Mi sends join request to SGi
SGi sends Mi request to Kr
Kr verifies for authentication
If (Mi = valid)
Sends yes to SGi
else
rejects the request
then
if (SGi receives acceptance)
Generates new group key (qk) and send notification to Mi
else
return null.
}
```

4.2 Rekeying Due to Leave Operation

4.2.1 Leaving of Member

Suppose the member M11 wants to leave from the subgroup, first M11 sends the leave request to its manager SG6′. SG6′ forwards the request to Kr. Kr looks its table to confirm that the requested node is the existing member of the group. After getting the confirmation, Kr sends the authentication message to SG6′. Meantime, Kr removes the identity of M11 from the table. Then SG6′ removes the member M11 and generates the new group key SG_{k6}'' for that group and notify to all its members. So the member who left from the group is unable to use SG_{k6}''. This ensures the forward secrecy.

In leave operation, only one key is generated. Hence the proposed SGKM is efficient.

Procedure leave (member Mi, subgroup manager SGi, root Kr)
{
Mi sends leave request to SGi
SGi sends Mi request to Kr
Kr verifies for authentication
If (Mi = valid)
Sends yes to SGi
else
rejects the request
then
if (SGi receives acceptance)
release Mi and Generates new group key (qk) and send notification to existing Mi
else
return null.
}

4.2.2 Leaving of Sub Group Manager

Suppose, the subgroup manager SG6 wants to leave from the group, it has to send the leaving request to Kr. After receiving the acceptance from Kr, SG6 will leave from the group. Then the next higher weighed node SG6′ becomes the manager of that subgroup. SG6′ sends this update information to Kr and generates the new group key K_6' for that subgroup and send to all its members. Thus the left manager SG6 is unable to involve in the communication between SG6′ and its members.

Procedure leave (subgroup manager SGi, root Kr)
{
SGi sends leave request to Kr
Kr verifies for authentication
If (SGi = valid)
Sends yes to SGi
else
rejects the request
then
if (SGi receives acceptance)
next higher weighted is selected as new SGi'
else
return null.
}

In SGKM, whenever changes occur in the subgroups due to join or leave, only the common group key which is used for encryption alone is updated however, the decryption keys are unchanged. When compared to the existing schemes mentioned in Sect. 2.1, the number of keys for updating, number of encryption and decryption are significantly reduced in SGKM.

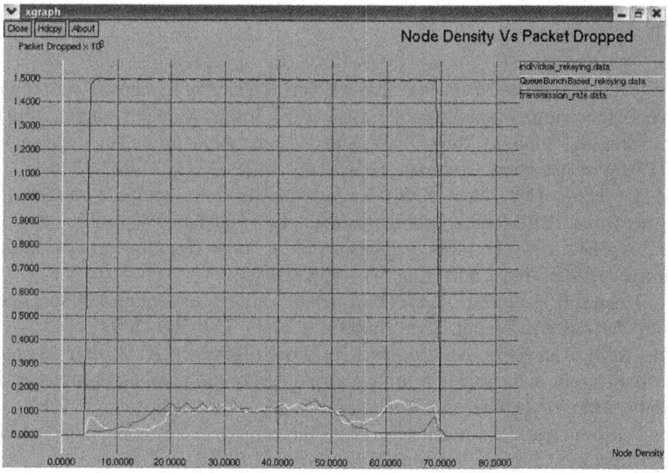

Fig. 3 Node density vs packets dropped ratio of OFT and SKD

5 Performance Analysis

In this section, the cost effectiveness of the proposed scheme SGKM is analyzed and compared with OFT and SKD. In SGKM, only one key generation and one key encryption is needed for a join and leave. When the existing members and new member receive the encrypted data, they have to perform single decryption only. Thus the reduction in computation cost, communication cost and storage cost is achieved (Fig. 3).

6 Conclusion

In the proposed scheme during rekeying, one common encryption key alone is transmitted, hence communication cost is one. Regarding computation cost, one key encryption and one key generation is performed at sub group manager in both join and leave process. Meanwhile, single key decryption is performed by members. This results in the lower computation cost. In SGKM, the storage cost is equivalent to the number of sub groups, therefore the proposed scheme experiences lower storage cost. So, SGKM is having better performance in terms of communication, computation and storage cost. The simulation results show that the proposed SGKM is efficient and achieves better performance when compared to existing schemes OFT and SKD.

References

1. Park, Y.H., Je, D.H., Park, M.H., Seo, S.W.: Efficient rekeying framework for secure multicast with diverse-subscription-period mobile users. IEEE Trans. Mobile Comput. **13**(4) (2014)
2. Zhou, J., Song, M., Song, J., Zhou, X.-W, Sun, L.: Autonomic group key management in deep space DTN. Wireless Peers Commun. (2013). doi 10.1007/s11277-013-1505-1
3. Park, M.H., Jeong, H.Y., Seo, S.W.: Key management for multiple multicast groups in wireless networks. IEEE Trans. Mobile Comput. **12**(9) (2013)
4. Sherman, A., McGrew, D.: Key establishment in large dynamic groups using one-way function trees. IEEE Trans. Soft. Eng. 444–458 (2003)
5. Lin, J.C., Huang, K.H., Lai, F., Lee, H.C.: Secure and efficient group key management with shared key derivation. Comput. Stand Interfaces **31**(1), 192–208 (2009)
6. Naranjo, J.A.M., Antequera, N., Casado, L.G., López-Ramos, J.A.: A suite of algorithms for key distribution and authentication in centralized secure multicast environments. J. Comput. Appl. Math. **236**, 3042–3051 (2012)
7. Dandon, L., Runtong, Z., Chuanchen, W.: Efficient group key management scheme with hierarchy structure. Chin. J. Electron. **21**(2), 249–253 (2012)
8. Juhyung, S., Jun, S.L., Seung, W.S.: Topological key hierarchy for energy-efficient group key management in wireless sensor networks. Wirel. Pers. Commun. **52**(2), 359–382 (2012)
9. John, S.P., Samuel, P.: A predictive cluster technique for effective key management in mobile ad hoc networks. Inf. Secur. J. **20**(4), 250–260 (2011)
10. Xiao, P., Jingsha, H., Yingfang, F.: Distributed group key management in wireless mesh networks. Int. J. Secur. Appl. **6**(2), 478–480 (2012)
11. Liao, J., Hui, X.L., Qing, Q.P., Yi, L., Yu, M.W.: A public ley encryption scheme with one-encryption and multi-decryption. Chin. J. Comput. **35**(5), 1059–1067 (2012)
12. Zhou, J., Sun, L., Zhou, X., Song, J.: High performance group merging/splitting scheme for group key management. Wirel. Pers. Commun. **75**, 1529–1545 (2014). doi:10.1007/s11277-013-1436-x
13. Mittra, S.: Iolus: A framework for scalable secure multicasting. ACM SIGCOMM Comput. Commun. Rev. **27**(4), 288 (1997)

A Multipath Routing Protocol Which Preserves Security and Anonymity of Data in Mobile Ad Hoc Networks

J. Jaisooraj, M.K. Sulaiman and Manu J. Pillai

Abstract Mobile Ad Hoc Networks (MANETs) have become a likely candidate for a wide range of threats. Each and every routing protocol in MANETs should preserve anonymity of mobile devices and data, along with unlinkability of data packets. With an aim of improving routing efficiency, many routing protocols either partially sacrificed anonymity or disregarded unlinkability. Here, we propose a multipath routing protocol that will enhance security as well as efficiency of routing in MANETs. Anonymity and unlinkability are completely preserved through the concepts of keyed hash chain and per-hop data packet appearance alteration. Also, the routing efficiency is enhanced through the formulation of a route selection parameter called Link Stability Metric (LSM) and the deployment of an elaborate route maintenance phase. The proposed system has been simulated and tested, and the results indicate that security as well as routing efficiency is enhanced considerably.

Keywords MANETs · Anonymous routing · Keyed hash chain · Link stability metric etc.

J. Jaisooraj (✉) · M.K. Sulaiman · M.J. Pillai
Department of Computer Science and Engineering, TKM College of Engineering, Kollam, Kerala, India
e-mail: jaisooraj11@gmail.com

M.K. Sulaiman
e-mail: sulaimancse@gmail.com

M.J. Pillai
e-mail: manujpillai@gmail.com

M. Senthilkumar et al. (eds.), *Computational Intelligence, Cyber Security and Computational Models*, Advances in Intelligent Systems and Computing 412, DOI 10.1007/978-981-10-0251-9_42

1 Introduction

The usage of mobile applications and smartphones has reached such scaling heights that we seldom come across people who do not carry a mobile phone. This, along with an increased availability of wireless technologies like Wi-Fi on smartphones has forced Mobile Ad Hoc Networks (MANETs) to increase its channel capacity. Also, the openness and cooperative nature of MANETs have made them a likely candidate for a wide range of threats. Thus, a routing protocol designed for MANETs should consider the security of data packets being transmitted as important as the efficiency of underlying routing algorithm.

The security of data in MANETs is characterized primarily by two important factors- anonymity and unlinkability. Pfitzmann and Hansen have accurately defined anonymity as "the state of being not identifiable within a set of subjects, the anonymity set" [1]. Their expertise in the field also provided us with a classy definition for unlinkability as "the notion of a third party being unable to distinguish whether any two or more items of interest are related" [1]. A detailed survey of the existing routing protocols indicates that each of them gave more priority to either the routing part or the security part. Routing part or the security part, whichever it may be, is partially sacrificed owing to the fact that mobile devices have limited power and processing capabilities. These are done to avoid the additional delays caused by expensive cryptographic operations. Hence, many of the existing routing protocols either partially sacrificed anonymity [2, 3] or disregarded unlinkability [3]. Also, the survey shows TARo (Trusted Anonymous Routing) protocol to be the most secure among existing anonymous routing protocols [4]. But in TARo also, the efficiency of underlying routing algorithm is compromised for enhancing the security. Thus, we can say that the result of survey did not include a routing protocol which gave equal priorities to both the routing part as well as the security part. In this paper we aim to add a novel protocol, which is secure as well as efficient, to the list of routing protocols in MANETs.

Here, we propose a Secure Efficient On-Demand Routing (SEODR) protocol which gives equal priorities to both the security of data being transmitted as well as the efficiency of routing algorithm lying underneath. SEODR establishes security of data packets by preserving anonymity through the usage of a keyed hash chain, and unlinkability through per-hop data packet appearance alteration. Also, a Link Stability Metric (LSM) has been formulated based on three of the most important factors in a dynamic networking environment (as in MANETs) namely—signal strength, mobility of nodes, and energy consumption of nodes. LSM periodically measures the stability of links and thereby establishes the current best routes at different instances of time. SEODR further increases the efficiency of routing algorithm by incorporating a detailed route maintenance phase which facilitates route switching whenever the performance of an active route falls below a threshold.

The remainder of this paper is organized with related work being presented in Sect. 2, security considerations in Sect. 3, proposed system design in Sect. 4, results and discussions in Sect. 5, and conclusions and future work in Sect. 6.

2 Related Work

An extensive survey conducted on anonymous routing in MANETs flashed light on some of the most prominent anonymous routing protocols. The existing protocols preserve anonymity through the usage of onion routing, pseudonyms, and invisible implicit addressing. The method of implicit addressing makes use of a trapdoor message encrypted by either the public key or shared key of the receiver. The node which possesses the key for decrypting the trapdoor message will be the intended receiver. On the other hand, techniques like per-hop packet appearance alteration [2, 5, 6], traffic mixing [7], and dummy packet injection [2] are employed to achieve unlinkability.

With the aim of achieving an increased performance, the existing anonymous routing protocols tend to partially sacrifice either anonymity [2, 3] or unlinkability [3]. In the protocols like Discount-ANODR and MASK, the cost of implicit addressing is cut off by using real identity of receiver in the route discovery phase. Then, there is another important anonymous routing protocol named SDAR which uses encrypted real identities of nodes and is known only to the sender and the receiver [8]. Thus anonymity is partially sacrificed in the sense that anonymity of intermediate nodes are guaranteed only to the observers, but not to the sender and the receiver. So we can say that Discount-ANODR, MASK, and SDAR belong to the category of protocols which sacrifice security for the sake of routing efficiency.

The most efficient among the existing anonymous routing protocols is the Trusted Anonymous Routing (TARo) protocol which preserves a very high degree of anonymity and unlinkability. Also, the unwanted cryptographic operations performed by intermediate nodes are avoided in TARo. But the main drawback of TARo is that even though it achieves a higher level of security, the techniques for improving the efficiency of underlying routing algorithm are not considered. Therefore as already mentioned, our proposed routing protocol aims to provide equal considerations to security as well as routing efficiency.

3 Security Considerations

The openness of wireless networks (and hence MANETs) make them exposed to a wider range of threats. A MANET operates without the aid of a fixed networking infrastructure. This makes unauthorized network monitoring in MANETs much easier compared to the infrastructure based networks. The option of deploying a centralized Public Key Infrastructure (PKI) is also not available, once again due to the above reason. Another challenge is the resource constraints of mobile devices; they suffer from a limited battery capacity as well as limited buffer space. This limits the usage of secure (yet computationally expensive) cryptographic operations like public key encryption and decryption. This section gives a brief description regarding the attacks that can affect routing in MANETs and also states the measures taken by SEODR to thwart these attacks.

Attacks like identity spoofing, link spoofing, replay attack, man-in-the-middle attack, wormhole attack, Sybil attack, routing table overflow attack etc. are bound to affect MANET routing protocols at any point of time [9]. The main aim of such attackers is to trace the route and take control of sensitive information being transmitted. These attacks are launched by attackers after closely monitoring the network traffic. The following are the main traffic monitoring techniques that are being considered here [10]:

- Message coding analysis: Attackers use the concept of pattern matching to trace messages that do not change their coding for a long time.
- Message length analysis: Here, the attackers make use of length of a message while it is being transmitted through the network.

In the design of proposed system, we assume that an attacker can hamper the anonymity condition by revealing identities of sender, receiver, and en-route nodes. Also, an attacker can disturb the unlinkability condition by linking packets from the same communication flow. Our assumption also includes the presence of external as well as internal adversaries. External adversaries are mobile nodes that can eavesdrop, record, alter and inject packets to initiate various attacks [11]. Internal adversaries refer to the compromised en-route nodes that can boast of the necessary cryptographic secret required to reveal and generate legitimate messages [11]. The proposed routing protocol (i.e. SEODR) attempts to preserve anonymity of data, sender, receiver, and en-route nodes using the concept of a keyed hash chain [12]. Message coding analysis as well as message length analysis is prevented using the concept of per-hop data packet appearance alteration [2, 5, 6]. The appearance of data packets is changed frequently, thereby not allowing an attacker to figure out whether the data packets belong to same flow or not. This ensures unlinkability of data packets. Another important fact to be considered is that a node which uses long term identities is likely to be traced by an attacker, thereby compromising location privacy. Hence, SEODR uses short term pseudonyms for nodes to avoid such a condition.

4 SEODR Design

Secure Efficient On-Demand Routing (SEODR) protocol also has been designed as a fully distributed on-demand multipath routing protocol. The routing process is on-demand, i.e. a route discovery is initiated only when a request arrives from an upper layer and a data packet needs to be delivered to an intended destination. It sets multiple anonymous routes to the destination as a part of routing process. Also, SEODR has been designed with an aim of providing equal priorities to both the security part as well as the efficiency part.

SEODR takes certain measures to ensure the security of data packets that are being transmitted through the network. SEODR does not expose the real and long term identifiers of nodes to the routing process. Instead, it assigns new short term pseudonyms to nodes and uses that for routing. This is done to ensure location privacy of nodes. Another important security feature incorporated with SEODR is that it uses both end-to-end and hop-by-hop encryption of data packets. End-to-end encryption is aimed at increasing the security of data packets, whereas hop-by-hop encryption is aimed at preserving the unlinkability of data flows [4]. It also uses the concept of a keyed hash chain to preserve anonymity of sender, receiver, enroute nodes, and data packets [13]. Shared keys between sender and enroute nodes, that are required for keyed hash chain, are exchanged using Diffie–Hellman mechanism. These security measures are absent in most of the routing protocols that are primarily aimed at increasing the efficiency of routing algorithm.

In order to enhance the efficiency of routing algorithm, the concept of node disjoint routing is used. As a result, only the routes that do not have common nodes are chosen from the multiple anonymous routes established. These node disjoint routes are ranked according to the decreasing order of a metric called the Link Stability Metric (LSM). Also, these routes are maintained by the periodic transmission of a control packet. Whenever the LSM of current best route goes below a threshold, control is immediately shifted to the next best route. This eliminates the overhead of route rediscovery associated with most of the secure routing protocols.

SEODR successfully carries out the task of providing equal priorities to security and routing efficiency in four steps: route discovery, best route selection, data transmission, and route maintenance. Of these, route discovery and data transmission are aimed at providing the desired security of data packets by preserving anonymity and unlinkability whereas best route selection and route maintenance are aimed at providing the desired efficiency of routing algorithm by eliminating overheads. The various notations used here are shown in the following table (Table 1) [4].

4.1 Route Discovery

Each node maintains a destination table which contains a list of destinations and corresponding pre-shared secrets and keyed hash chains. For the efficient working of SEODR, we assume that there are only a limited number of destinations for each node. A node also maintains an active session table (to maintain the list of active routing sessions), a forwarding table, and a routing table. Route Discovery in SEODR is completed in two steps- route request and route reply. Route Request and Route Reply uses RREQ and RREP messages respectively to accomplish their own specific tasks.

Table 1 Notation table

Notation	Parameter
$\mathbf{N_x}$	Node **X**, where $\mathbf{N_s}$, $\mathbf{N_d}$ represent source and destination
F_{type}	message flag, *type* = *RREQ*, *RREP*, *DATA* or *RERR*
$[\cdot]_K$	Symmetric encryption using key *K*
$H(\cdot)$	One-way hash function
$[A\|B]$	Concatenation of content A and B
K_{sd}	Source-destination shared key
K^i_{sd}	*i*th element of the source-destination key chain
PS_x	Pseudonym of node *x*. $PS_x = H(DH^p_x)$
DH^s_A, DH^p_A	Diffie–Hellman secret and public key generated by node A

Route Request

An RREQ message of the following format is broadcasted into the network by a sender (Ns) whenever it wishes to send a message to a particular destination (pseudonyms are as given in Table 1) [4]:

$$(\text{FRREQ},\ K^i\text{sd},\ [\text{PSs}||\text{padding}]K^{i+1}\text{sd},\ DH^p\text{s},\ \text{PSi},\ \text{ttl})$$

The major advantage of this method is that the true identities of both source and destination are never used for verification. Instead, it is done by using a keyed hash chain (Fig. 1) [4]. The hash operations have been performed using SHA1 algorithm. A node can check whether it is the intended receiver by searching for K^isd in its destination key chains. If found, the node tries to decrypt the trapdoor message using K^{i+1}sd. The message can then be verified by comparing H(DH ^{P} $_s$) and PSs. If both are same, the route request is validated and the protocol enters route reply phase. Also, an intermediate node has to check the active sessions table to see whether the session id is already present. If it is already present, the node which forwarded that RREQ is put into relay node list and the packet is dropped. On the other hand, if the node does not find K^isd in its destination key chains, it replaces PSi with its own pseudonym, decrements the value of ttl (time-to-live) by one, and forwards the modified message.

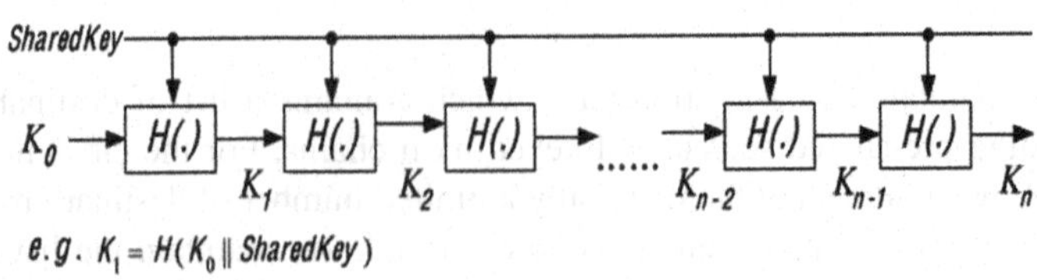

Fig. 1 Keyed hash chain

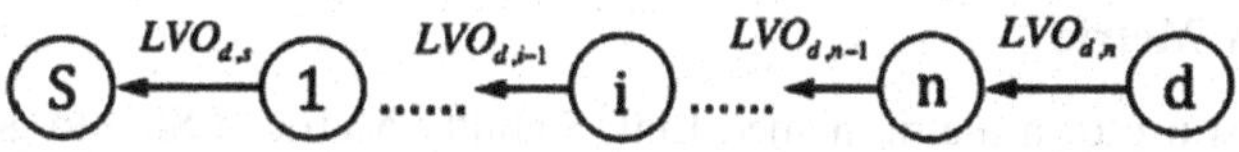

$$LVO_{d,s} = DH_1^P \,||\, [PS_s \,||\, LVO_{d,1}]_{K_{s1}} \qquad LVO_{d,n-1} = DH_n^P \,||\, [PS_{n-1} \,||\, LVO_{d,n}]_{K_{sn}}$$

$$LVO_{d,i-1} = DH_i^P \,||\, [PS_{i-1} \,||\, LVO_{d,i}]_{K_{si}} \qquad LVO_{d,n} = DH_d^P \,||\, [PS_n \,||\, PS_d \,||\, padding]_{K_{sd}}$$

Fig. 2 Link verification onion (LVO)

Route Reply

This phase makes use of a Link Verification Onion (Fig. 2) to preserve anonymity of both data and mobile nodes [4]. On receiving a valid route request message, the destination node (Nd) creates an RREP message of the following format (pseudonyms are as given in Table 1 [4]:

$$(\text{FRREP},\ K^i\text{sd},\ [\text{PSd}||\text{padding}]K^{i+1}\text{sd},\ \text{PSi},\ \text{LVOd},\text{i})$$

On receiving an RREP message, a node Ni computes the shared key and creates a new LVO as shown in Fig. 2. Then, PSi is replaced with the last hop pseudonym and the modified message is forwarded. In order to discover multiple routes, RREP message is forwarded to all nodes in the relay node list. Following that, Ni computes uplink (H(LVOd, i)) and downlink (H(LVOd, i − 1) and is stored in the forwarding table. This denotes the end of route discovery and all related records can be removed from the active sessions table.

The source node verifies the route by sequentially deriving the shared keys. Then H($DH^{P}{}_{i}$) and the encry pted pseudonym PSi are checked to see whether they are matching. Finally, the route is said to be verified if PSd decrypted from [PSd||padding]$K^{i}{}_{sd}$ is the same as that of PSd obtained from the core of LVO. A route identifier is assigned to all the valid routes and is calculated as Rid = H (LVOd, s). The list of routes and their corresponding identifiers are then entered into the routing table.

4.2 Best Route Selection

Node Disjoint Routing

SEODR makes use of the concept of node disjoint routing. This is a concept that is absent in most of the secure anonymous routing protocols. Here, only the routes that do not share common nodes are considered as valid. This avoids the condition of a common node carrying the traffic of more than one routes. Thus, it enhances the whole network life time significantly. Also, this is the step that helps in balancing the overhead caused by expensive cryptographic operations used for maintaining anonymity and unlinkability.

Link Stability Metric

SEODR makes use of a metric named Link Stability Metric (LSM) to govern the process of best route selection. LSM characterizes the stability of links present in each route. In a dynamic networking environment, like that in MANETs, link stability depends on factors like signal strength (S), mobility of nodes (M), and the energy consumption of nodes (E).

Signal strength (S) can be calculated based on the parameters like St (strength of transmitted signal), Gt (antenna gains at transmitter), Gr (antenna gains at receiver), and d (distance between transmitter and receiver) as follows:

$$S = St\ Gt\ Gr(\lambda/4\pi d)^{n} \quad (1)$$

where n can take the values from 2 (free-space propagation) to 5(strong/attenuation).

Mobility Factor (M) is calculated based on a parameter called Link Expiration Time (LET) which in turn is based on position of nodes as well as the relative velocity between nodes [14]. Consider (Xi, Yi) and (Xj, Yj) to be the position coordinates of nodes i and j respectively. Also, let (Vi, Θi) and (Vj, Θj) denote (velocity, direction) of nodes i and j respectively. Then M can be calculated as follows [14]:

$$M = -(ab + cd)/(a + c^2) \quad (2)$$

where,

$$a = V_i \mathrm{Cos}\,\Theta_i - V_j \mathrm{Cos}\Theta_j \quad (3)$$

$$b = X_i - X_j \quad (4)$$

$$c = V_i \mathrm{Sin}\,\Theta_i - V_j \mathrm{Sin}\,\Theta_j \quad (5)$$

$$d = Y_i - Y_j \quad (6)$$

Each node calculates its own energy consumption periodically (say, in every T seconds) and estimates the Drain Rate (DR) [14]. Energy consumption (E) is calculated as follows:

$$DR = a\,DR_{old} + (1 - a)DR_{new} \quad (7)$$

where a takes any value in the range of 0 to 1, so as to give more weightage to the updated value, DRnew.

Finally, LSM can be formulated as follows:

$$LSM = (S \times M)/E \tag{8}$$

Route Selection

LSM is calculated for each links in each of the node disjoint routes established. The minimum value of LSM is found out for each route. Then, maximum among those minimum LSM values is figured out. The route containing the above mentioned maximum of minimum LSM values is considered the best. Other routes are sorted accordingly.

4.3 Data Transmission

Now that we have anonymously discovered the routes and also chosen the best route among them, next step is to securely transmit data through the selected best route. In SEODR, the source builds a cryptographic onion for each data packet. Data is encrypted using shared keys of nodes appearing in the sequence [4]. When a node receives the data packet, it peels off one layer of data onion using its corresponding shared key. After that, the modified message is forwarded. This process continues until the data packet reaches the intended destination. For a forward path consisting of nodes- Ns, Na, Nb, Nc, Nd; the cryptographic onion will be of the following form:

$$[[[[Data]K_{sd}]K_{sc}]K_{sb}]K_{sa}$$

4.4 Route Maintenance

To further enhance the efficiency of routing, SEODR periodically keeps track of performance all the established routes. This is achieved through periodic transmission of a Packet Update (PU) message. Route switching is initiated when one among the following two situations arises:

- LSM of a particular route falls below a threshold. In such a case, the control is immediately shifted to the next best route.
- LSM of the current best route falls below LSM of another route. Here also, route switching is performed.

Such an elaborate route maintenance (which is absent in most of the secure routing protocols) in SEODR helps in significantly reducing the control overhead associated with the initiation of a route rediscovery. This is another enhancement that helps SEODR to balance the high cost cryptographic operations associated with maintenance of anonymity and unlinkability.

5 Results and Discussions

The proposed routing protocol, SEODR, has been simulated and tested using Scalable Wireless Ad Hoc Network Simulator (SWANS). Simulations involving 100 nodes were performed for 10 min duration. Mobility of nodes was given values from 0 to 10 m/s to test various cases. Additionally, the packet size was set as 512 bytes and data rate as 4 packets/s.

Even though computationally intensive cryptographic operations are used, SEODR shows an increased packet delivery ratio and a reduced control overhead when compared to the other anonymous routing protocols. This is mainly because of the fact that overhead associated with cryptographic operations is balanced by incorporating the concept of node disjoint routing and an elaborate route maintenance phase. Also as mentioned in Sect. 3, a variety of attacks are effectively handled by SEODR. Thus, we can say that SEODR achieves a higher routing efficiency without compromising on security requirements. Here, comparisons are made with respect to the performance of other important anonymous routing protocols. Figure 3 shows the variations in packet delivery ratio and control overhead when plotted against mobility of nodes.

6 Conclusions and Future Work

In this paper, we have proposed a Secure Efficient On-Demand Routing (SEODR) protocol for MANETs which is aimed at providing equal priorities to both the security part as well as the routing part. The results indicate that, the usage of the concept of node disjoint routing and an elaborate route maintenance phase have caused an increased packet delivery ratio as well a reduced control overhead. In

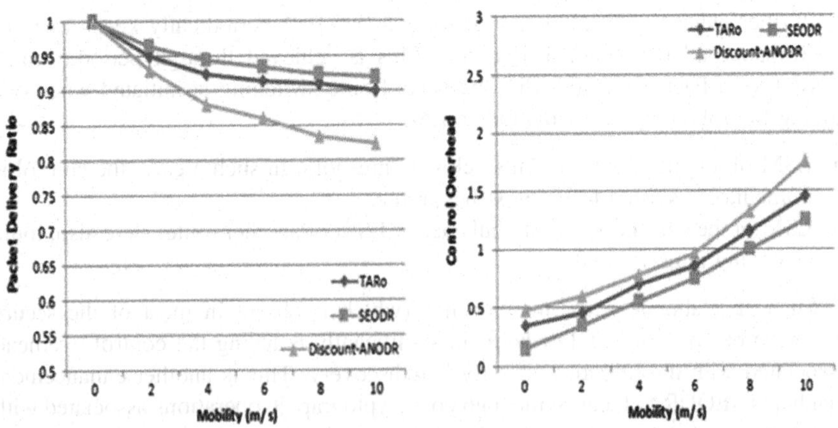

Fig. 3 Packet delivery ratio versus mobility and control overhead versus mobility

future, we aim to further increase the security by concentrating on the data being transmitted. Also, we aim to fuzzify the route selection parameter so as to further enhance the efficiency of underlying routing algorithm.

References

1. Pfitzmann, A., Hansen, M.: Anonymity, unlinkability, undetectability, unobservability, pseudonymity, and identity management—A Consolidated Proposal for Terminology. Tech. Rep. Feb 2008
2. Yanchao Zhang, W.L., Liu, W., Fang, Y.: MASK: anonymous on-demand routing in mobile ad hoc networks. Trans. Wirel. Commun. IEEE **21**, 2376–2385 (2006)
3. Yang, L., Jakobsson, M., Wetzel, S.: Discount anonymous on demand routing for mobile ad hoc networks. In: SECURECOMM, vol. 6 (2006)
4. Chen, J.T., Boreli, R., Sivaraman, V.: TARo: trusted anonymous routing for MANETs. In: Sixth IEEE/IFIP International Symposium on Trusted Computing and Communications (TrustCom-10). IEEE Computer Society, pp. 756–762 (2010)
5. Seys, S., Preneel, B.: ARM: anonymous routing protocol for mobile ad hoc networks. Int. J. Wirel. Mob. Comput. **3**, 145–155 (2009)
6. Boukerche, A., Ren, Y.: ARMA: an efficient secure ad hoc routing protocol. In: GLOBECOM, pp. 1268–1272 (2007)
7. Ghaderi, J., Srikant, R.: Towards a theory of anonymous networking. In: Proceedings of the 29th conference on Information communications, INFOCOM'10. IEEE Press, pp. 686–694 (2010)
8. Boukerche, A., El-Khatib, K., Xu, L., Korba, L.: SDAR: a secure distributed anonymous routing protocol for wireless and mobile ad hoc networks. In: Annual IEEE Conference on Local Computer Networks, pp. 618–624 (2004)
9. Deng, H., Li, W., Agrawal, D.P.: Routing security in wireless ad hoc networks. IEEE Commun. Mag. **40**(10) (2002)
10. Rennhard, M., Plattner, B., Rafaeli, S., Mathy, L., Hutchison, D.: An architecture for an anonymity network. In: IEEE International Workshops on Enabling Technologies, p. 165 (2001)
11. Zhou, D.: Security Issues in Ad Hoc Networks. CRC Press, Inc (2003)
12. Lamport, L.: Password authentication with insecure communication. Commun. ACM **24**, 770–772 (1981)
13. Pfitzmann, A., Waidner, M.: Networks without user observability design options. Eurocrypt **85**, 245–253 (1986)
14. Upadhyaya, S., Gandhi, C.: Node Disjoint Multipath Routing Considering Link and Node Stability Protocol. IJCSI, Jan 2010

An Efficient Continuous Auditing Methodology for Outsourced Data Storage in Cloud Computing

Esther Daniel and N.A. Vasanthi

Abstract Inspite of enormous advancement in cloud data storage and computation technologies, data security remains challenging. This paper presents an efficient auditing method for integrity verification of outsourced data. The proposed scheme combines scheduling with data integrity verification mechanisms. The experimental results on computation and communication cost have also been obtained. The result shows that the proposed scheme is efficient and intuitive choice for cloud storage.

Keywords Cloud computing · Security · Data storage · Integrity · Auditing

1 Introduction

A tremendous amount of data is produced every day by people all over the world. With expansion of affordable smart phones, tablets and other devices equipped with camera and recorder, the volume of this data is growing rapidly. Cloud storage is an important service of cloud computing, which allows data accumulators to move data from their local computing systems to the cloud [1]. It provides low-cost, scalable, location-independent platform to store and manage user's data. Therefore naturally more and more data accumulators start to use cloud storage services. Businesses needed the power of the cloud storage to compete with the latest era of information but are reluctant to move to cloud as they expect powerful privacy, security, and control. Since the data are stored in the cloud, data owners need to be assured, that their data are not accessed or even misused by unauthorized users. Another issue is that data can be lost in cloud. Data loss can occur in any infras-

E. Daniel (✉)
Karunya University, Coimbatore, India
e-mail: estherdaniell@gmail.com

N.A. Vasanthi
Dr. NGP Institute of Technology, Coimbatore, India
e-mail: vasanti.au@gmail.com

M. Senthilkumar et al. (eds.), *Computational Intelligence, Cyber Security and Computational Models*, Advances in Intelligent Systems and Computing 412, DOI 10.1007/978-981-10-0251-9_43

tructure, no matter what kind of reliable measures the cloud service providers (CSP) would take [2]. Therefore, owners need to be convinced that the data are correctly stored in the cloud. If any problems regarding to owner's data occur, the owner has to be notified in time. This paper is organized into following sections. Section 2 reviews few of the existing techniques given in the literature and Sect. 3 briefs an efficient continuous auditing methodology. Sections 4 and 5 presents the algorithm of the proposed auditing method and analyses its efficiency. Finally Sect. 6 gives the conclusion and future work of this paper.

2 Related Works

In recent years, there has been an exhaustive research on a topic of integrity auditing of cloud data storage. The Provable Data Possession (PDP) settings were considered to be more competent and practical to use. The PDP method proposed by Ateniese et al. [3] was the first scheme to provide probabilistic proof by sampling random blocks and public verifiability at the same time. They have also proposed two variants of PDP schemes [4, 5] for better efficiency and to support partial dynamic operations. Juels et al. [6] developed a POR based approach which best suited for verification of stored encrypted files but a heavy preprocessing so Compact POR scheme [7] when compared with PDP and POR assured the cloud servers possess the target files and guarantees full recovery. An efficient POR scheme [8] based of polynomial commitment for reducing the communication cost was introduced. DPDP [9] methods supported full data dynamics with the rank based skip list data structure but with no public verifiability Wang et al. [10, 11] proposed a scheme that can support both data dynamics and public verifiability entrusted to a TPA with Merkle hash tree to verify updates enabling Batch auditing efficient. Privacy preserving [11] was carried out by using random masking technique to prevent any part of original file being extracted. The protocol supports public verifiability without help of a third-party auditor increasing unnecessary computation and communication cost at the client side. Although this protocol shows interesting approach to data integrity checking and it achieves batch auditing, it doesn't support dynamic auditing. Similar masking technique was also used in IPDP [12]. Furthermore both IPDP and Wang's schemes incur heavy computation cost on TPA. Grounded on these research findings a need for efficient TPA which schedules and audits the files on the remote servers with reduced communication and computation cost is evident. So we propose an efficient and secure auditing protocol based on scheduled and continuous auditing ensures reduced communication and computation cost with increased assurance level of data file remotely stored.

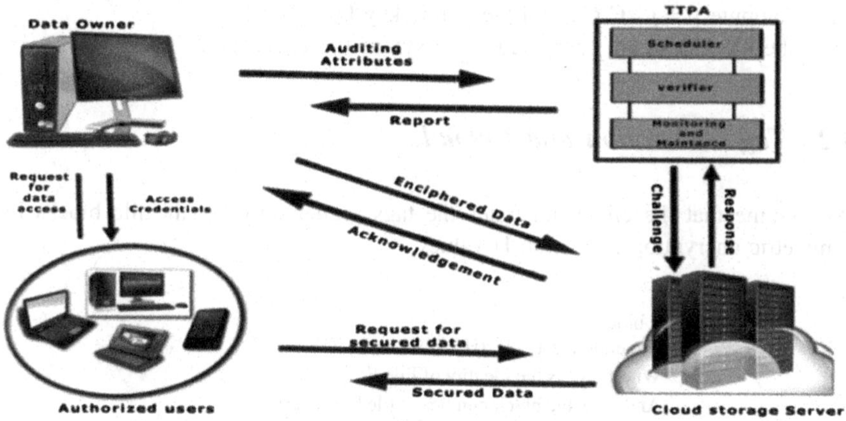

Fig. 1 Efficient continuous auditing framework for cloud storage

3 Continuous Auditing System Model

We consider a cloud storage auditing system as shown in Fig. 1 with entities as Data Owners (DO), Cloud Storage Servers (CSS), Trusted Third Party Auditor (TTPA), and Authorized User (AU)

4 The Proposed Algorithm

Continuous Integrity Auditing Scheme is using cryptography together with bilinearity property of the bilinear pairing and comprises of the following five algorithms.

4.1 Preprocessing and Key Generation

We consider file F, which has m data components as F = (F1, F2……Fm). Private data components need to be encrypted with their corresponding keys. Each data component Fk is divided into nk data blocks as Fk = (mk1; mk2……..mknk). By using the data fragment technique. The DO performs the following operations

(1) Select an implicit security parameter λ to govern the overall security of the scheme and choose the random large prime number 'p'
(2) Let 'g' be the generator of G1, a group of prime order p
(3) Choose a random number x ∈ Z_p, where x is the secret key S_k

(4) Compute $v = g^x \in G1$ and the public key $P_k = \{g,v\}$
(5) The file blocks are encrypted using symmetric encryption.

4.2 Tag Generation and Upload

We assume that the client generates the tags sequentially for the file blocks by symmetric encryption E of hash H values

1) **for each** block 'i'
    Generate tag $T_i = E_{sk}(H(id||i)H(m_i)) \in G1$
    Where 'id' is the identity of File F
    And 'i' is the block number of file blocks m

end
2) Upload the file blocks $\{m1......m_n\}$ to the CSP
3) The signature tags $\{T_1............T_n\}$ for the respective file blocks are to be send to CSP and TPA

4.3 Challenge Invoke

Based on the scheduling policy T of the data the TPA sends an auditing request to CSP. The TPA selects random data blocks to construct a Challenge set Q based on the abstract information M of the file blocks

$$\text{chal} = \{\text{file ID, total number of blocks, version number}(v_i), \\ \text{time variant random number}(r), \text{ time stamp}(ts)\}$$

The chal request is sent to the server.

4.4 Proof Checking

CSP receives the chal, and retrieves the data file blocks and tags. Sends the response P for the auditor for the challenge raised by the TPA

(1) Calculate $P = H\ (id||m_i||v_i||r||ts)$
(2) The TPA takes as inputs the P from the server, the secret hash key, the public tag key and the abstract information of the data M

Table 1 Scheduled continuous auditing parameters

Audit period 'T'	1 day/1 week/1 month
Number of data blocks	10,000 (up to)
Size of data blocks	100–1024 KB 1 MB–1 GB 1–2 GB
Data sensitivity level-trust value	1 (highly sensitive) 0.5 (medium) 0 (non sensitive)
Network bandwidth	1 Gbps

(3) TPA computes the ID hash values of all the challenged data blocks and puts together the challenge hash
(4) The proof is verified. It outputs the auditing result as 0 (False-Data Modified) or 1 (True-Data Secure).

4.5 Continuous Scheduled Auditing

The continuous auditing enables the clients' data file (DF) to be secure and also to detect and prevent any misbehavior either by the server or attackers at an early stage. The Table 1 gives the auditing parameters for the scheduled audit process.

```
1) If (DF=Sensitive)
        Perform challenge with T= 1 day
elseif (DF = Moderately Sensitive)
        Perform challenge with T=1 week
elseif (DF=Non Sensitive )
        Perform challenge with T=1 month
2) Perform verification for the proof sent by the server
3) Check if H (id||m_i||v_i||r||ts) == T_i||v_i||r||ts
4) If "verified" continue with the challenge
   else
   Notify user and server by sending the generated attack message
```

5 Experimental Result and Analysis

For the implementation of the proposed auditing scheme it has been used C/C++ programming language. The experiments were carried out in DevStack—the development environment of OpenStack. To validate the efficiency of our system the simulation of the auditing protocol was compared with privacy preserving protocol [11] and IPDP Protocol [12] the following results were analyzed. Figures 2, 3 and 4 reveals that the computation cost at the DO, CSP and the TTPA

Fig. 2 Computation cost (ms) at the DO

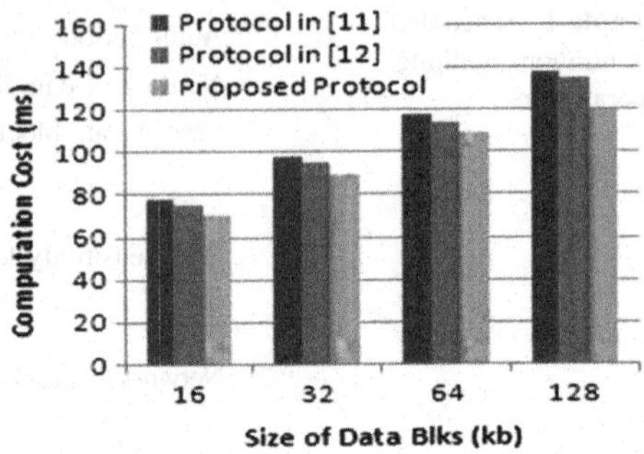

Fig. 3 Computation cost (ms) at the CSP

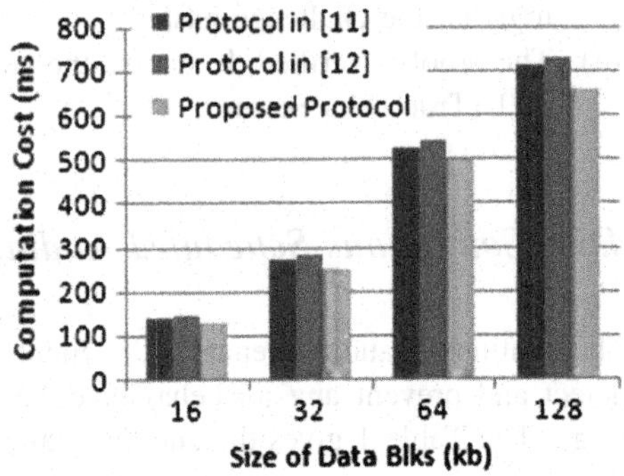

Fig. 4 Computation cost (ms) at the TTPA

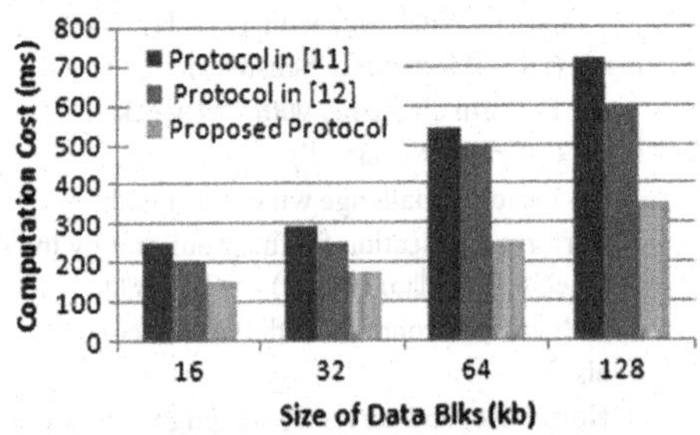

is reduced to a certain level as the data are not further divided into sectors as in existing system. As the DO categorizes the data to be stored as sensitive and non sensitive data the auditing cost is reduced to a great extent, the challenges are raised based on the sensitivity of the data and due to random sampling of the data block the communication cost to and from the Auditor gets reduced Fig. 5. The Scheduled Auditing reduces communication and computation cost with increased assurance level of data file remotely stored.

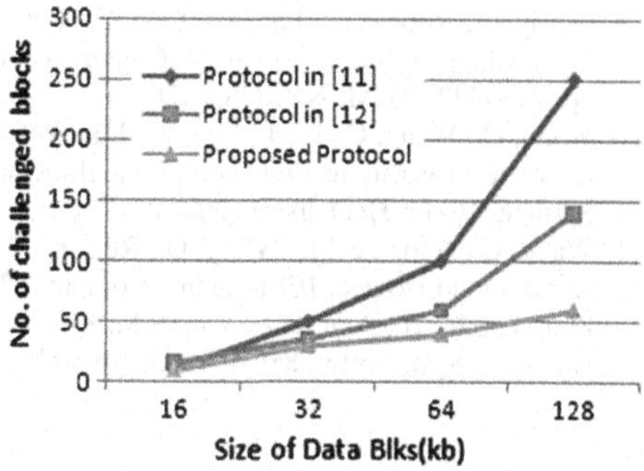

Fig. 5 Communication cost at the TTPA

6 Conclusion

Our proposed auditing protocol ensures privacy of the data by symmetric encryption along with bilinear pairing. The categorization of data sensitivity level along with scheduling of audit period for the data greatly improves the assurance level of the data integrity and as well as reduces the communication cost of the TTPA. The proposed protocol provides efficient auditing with reduced computation and communication cost when compared with privacy preserving protocol. Further this protocol can be improved to detect the misbehaviors if any by prediction models and a prevention mechanism will be studied.

References

1. Armbrust, M., Fox, A., Griffith, R., Joseph, A.D., Katz, R., Konwinski, A., Lee, G., Patterson, D., Rabkin, A., Stoica, I.: A View of Cloud Computing, pp. 50–58. ACM (2010)
2. Kher, V., Kim, Y.: Securing distributed storage: challenges, techniques, and systems. In: Proceedings of the ACM Workshop on Storage Security and Survivability, pp. 9–25. (2005)
3. Ateniese, G., Burns, R., Curtmola, R., Herring, J., Kissner, L., Peterson, Z., Song, D.: Provable data possession at untrusted stores. In: Proceedings of the 14th ACM Conference on Computer and Communications Security, pp. 598–609. (2007)
4. Ateniese, G., Burns, R., Curtmola, R., Herring, J., Khan, O., Kissner, L., Peterson, Z., Song, D.: Remote data checking using provable data possession. ACM Trans. Inf. Syst. Secur. (2011)
5. Ateniese, G., Pietro, R.D., Mancini, L.V., Tsudik, G.: Scalable and efficient provable data possession. In: Proceedings of the 4th International Conference on Security and Privacy in Communications Networks, pp. 1–10. Turkey (2008)
6. Juels, A., Kaliski Jr, B.S.: Pors: Proofs of retrievability for large files. In: Proceedings of the 14th ACM conference on Computer and Communications security, pp. 584–597. ACM (2007)
7. Shacham, H., Waters, B.: Compact proofs of retrievability. In: Proceedings of the 14th International Conference on the Theory and Application of Cryptology and Information Security, pp. 90–107. (2008)
8. Xu, Chang, E.-C.: Towards efficient proofs of retrievability. In: Proceedings of the 7th ACM Symposium on Information, Computer and Communications Security, pp. 79–80. NY, USA: ACM (2012)

9. Erway, C., Kupc̦u, A., Papamanthou, C., Tamassia, R.: Dynamic provable data possession. In: Proceedings of the 16th ACM Conference on Computer and Communications Security, pp. 213–222. ACM, NY, USA (2009)
10. Wang, Q., Wang, C., Li, J., Ren, K., Lou, W.: Enabling public verifiability and data dynamics for storage security in cloud computing. In: Computer Security, ESORICS 2009, pp. 355–370. Springer Berlin Heidelberg (2009)
11. Wang, C., Chow, S.M., Wang, Q., Ren, K., Lou, W.: Privacy-preserving public auditing for secure cloud storage. IEEE Trans. Comput. **62**, 362–375 (2013)
12. Zhu, Y., Hu, H., Ahn, G.J., Yau, S.S.: Efficient audit service outsourcing for data integrity in clouds. J. Syst. Softw. **85**(5), 1083–1095 (2012)

Part IV
Computational Models

A Study on Building Seamless Communication Among Vehicles in Vanet Using the Integrated Agent Communication Model (IACM)

N. Sudha Bhuvaneswari and K. Savitha

Abstract The main objective of this research paper is to propose an Integrated Agent Communication Model for Vehicular Ad hoc Networks. The model integrates five predefined mobile agents that aids in performing specific tasks to establish congestion free, co-operative, efficient channel utilization, energy conservative and optimized communication between nodes in vehicular ad hoc networks. This architecture significantly improves the functionality of the control unit to co-ordinate the agents and to balance the load of mobile agents that carries the information. The approach is to integrate all the agents in a single architecture and apply size reduction techniques to create light weight mobile agents for efficient performance and to improve the Quality of Service in VANET.

Keywords LPMA · MATLB · MSA · PCM · VAISTC4

1 Introduction

The world in which we live is built with technology, life has become crippled without technology. People have started becoming more dependent on gadgets and now and then in every minute or every hour a new gadget is being introduced in the market that has capabilities that out beat the already existing similar gadgets. With development of technology almost all the domains are surviving only with this technology and out of this the vehicle network is also one prime area that has seen a lot of development over the years with technological invention.

Vehicular Ad Hoc Network (VANET) has found to be an important area of research and many researches are on the go to improve this network. Numerous researches are going on in VANET highlighting the specific areas like passenger safety, vehicle safety, cost reduction in terms of routing, message broadcasting,

N. Sudha Bhuvaneswari (✉) · K. Savitha
School of IT & Science, Dr. G. R. Damodaran College of Science, Coimbatore, India
e-mail: sudhanarayan03@gmail.com

M. Senthilkumar et al. (eds.), *Computational Intelligence, Cyber Security and Computational Models*, Advances in Intelligent Systems and Computing 412, DOI 10.1007/978-981-10-0251-9_44

improving the Quality of Service (QoS) and security issues of vehicular ad hoc networks [1].

VANET not only deals with vehicle and passenger safety but the study and research on VANET has extended its hands into other supportive applications like cruising, automatic vehicle parking, speed controlling using sensors, message transmission between vehicles, message broadcasting and vehicle identification and prioritization [2].

Similarly while talking about VANET another major area of research that has attracted many researchers to explore new things is the Mobile Agent Technology. The mobile agent paradigm has found a prominent place in almost all areas of research and it is highly noted for its predominant features like platform independence, virtual execution environment and improved latency and bandwidth [3], The agent computing paradigm is best known for its role in decision making at times of uncertainty in dynamic changing environment. It has found to grow rapidly in these areas. The Intelligent Transportation System is one such perfect domain of traffic and transportation systems which highly dynamic and distributed and that is well suited for an agent-based approach [4].

The domain of traffic and transportation systems is well suited to an agent-based approach because of its geographically distributed nature and its alternating busy-idle operating characteristics [5]. The agent characteristics like autonomy, collaboration, and reactivity are more suited to apply on an Intelligent Transportation Systems. Agents are independent and they do not require any manual interrupts to handle them and this feature is very well suited to implement automated management of highway traffics [6].

The main objective of building an Integrated Agent Communication Model (IACM) for VANET is to efficiently create and handle a congestion free, co-operative, communicative, optimized, energy efficient Road Side Unit (RSU) quality service in VANET using an optimized mobile agent based approach. To achieve this various objectives to be met are: congestion avoidance, congestion detection, congestion control, handling high priority vehicle at congestion points, energy conservation, integration of agents for seamless communication and agent size shedding.

2 Research Methodology

The objective of congestion detection and congestion control demands identification of road status and handling of congestion on a busy lane. This uses an agent based approach VAISTC4 (Vibrant Ambient Intelligent System for Traffic Congestion Control in Coimbatore City) that uses sensor agents and effector agents along with a pre-processing unit to identify the type of vehicle and to identify road status and congestion.

The objective of congestion avoidance along with traffic intensity computation is handled using a middleware and MATLB (Mobile Agent for Traffic Load

Balancing) agent. MATLB architecture controls traffic in the existing VAISTC4 architecture with the help of SALSA middleware. Based on vehicle identity, road capacity, lane capacity, speed etc., it calculates the traffic intensity.

The current traffic intensity is calculated as follows:

$$\begin{aligned}\text{Speed Limit (X)} = {} & (\text{Length of Road} * \text{width of Road}) \\ & - (\text{WaitTime}\,(\text{Number of Railway Crossings} \\ & + \text{Number of Safety Installations} + \text{other installations}))\end{aligned} \tag{1}$$

$$\text{Traffic density (TD)} = \text{Number of vehicles (m)/mean Speed limit (x)} \tag{2}$$

$$\text{Lane density (LD)} = \text{Traffic Density (TD)/Number of Lanes (N)} \tag{3}$$

$$\text{Traffic Intensity (TI)} = \text{Number of Vehicles (n)/unit of time (t)} \tag{4}$$

$$\text{Traffic Intensity (TI)} = \text{Traffic Density (TD)} * \text{Mean Speed (u)} \tag{5}$$

where TI stands for Traffic Intensity, X-Speed Limit, u-Mean Speed, N-Number of Lanes and LD = Lane Density. Type of Roads include with their volume represented as integer Highway-300, Middleway-200, Motorway-150, Subway-75.

Using the above computed traffic intensity the risk factor in travelling is identified as acceptable, acceptable risk, unacceptable risk and tolerant risk.

The identification of emergency and high priority vehicles and automatic traffic speed control is achieved through the usage of PCM (Prioritization and Congestion Management) Agent along with the support of MATLB agent and Hall based sensors. This agent looks for emergency vehicles in a congested lane and if one is found, the lane is cleared immediately by activating the RFID tag to control the traffic signal.

Multihop Selfless Mobile Agent (MSA) is designed to identify selfish nodes and isolate them to build a co-operative vehicular network. MSA works in co-ordination with MATLB and PCM agent for identification of selfish nodes. Delivery ratio of each node is determined which is compared with a pre-estimated threshold value to identify selfish nodes.

$$\text{RPC}\,(i,j) = \text{RPC}\,(i,j) + 1 \tag{6}$$

$$\text{FPC}\,(i,j) = \text{FPC}\,(i,j) + 1 \tag{7}$$

$$\text{DR}\,(i,j) = \text{RPC}\,(i,j)/\text{FPC}\,(i,j) \tag{8}$$

These identified selfish nodes are then isolated, so that the other nodes in the network can work in co-operation to build an efficient communicative network.

A hybrid approach called MALP (Mobile Agent LEACH PEGASIS) integrates the application of mobile agents, MH-LEACH and PEGASIS to build an energy efficient, optimized communication vehicular network using the clustering approach. Efficient Energy Conservation (EEC) algorithm describes various stages of cluster formation, cluster head selection, cluster head chaining and performance is evaluated based on the factors like transmission delay, delivery ratio, energy consumption and conservation. Clustering of vehicles in the network and the cluster head selection is done with the help of the following distance and energy formula:

$$D = \mathrm{Sqrt}(x1 - x2)^2 + (y1 - y2)^2 \tag{9}$$

where the co-ordinates of node1 be (x1, y1) and the co-ordinate of node2 is (x2, y2).

The node with highest energy E is considered as the head of the cluster and computed as:

$$E = E_r/E_i * CH_p \tag{10}$$

$$E^1 = \sum E_r / \sum E_i CH_p \tag{11}$$

$$CH_p = NH_{net}/NN_{ne} \tag{12}$$

where E is the energy and E1 is the energy of current cluster head, Er is the residual energy, Ei is the initial energy, CHp is the proportion of number of cluster head nodes (NHnet) to the number of all nodes in the network (NNnet). Here CHp is used as constant to calculate the tolerance limit of the header and the tolerance limit is assumed in this work to be 5 % which is the default limit used by MH-LEACH protocol.

An integrated model for efficient and seamless communication between vehicular nodes is built that is capable of handling VANET issues related to safety, routing and broadcasting using the various mobile agents defined and discussed above like VAISTC4, MATLB, PCM, MSA and LPMA. Resource optimization of these mobile agents in terms of agent size is achieved through merging all these agents under a single model IACM. One important issue considered in this model is the agent size that keeps growing as the agent moves from node to node and by applying load shedding mechanism the size of the agent is reduced, simultaneously increasing bandwidth utilization. An Optimized Communicative Algorithm (OCA) is proposed that best describes the IACM model.

3 Integrated Agent Communication Model (IACM)

This integrated agent based architecture integrates many predefined mobile agents like VAISTC4, MATLB, PCM, MSA and LPMA and improves the functionality of the control unit to co-ordinate these agents and to balance the load of mobile agents that carries the information. The approach is to integrate all the agents in a single

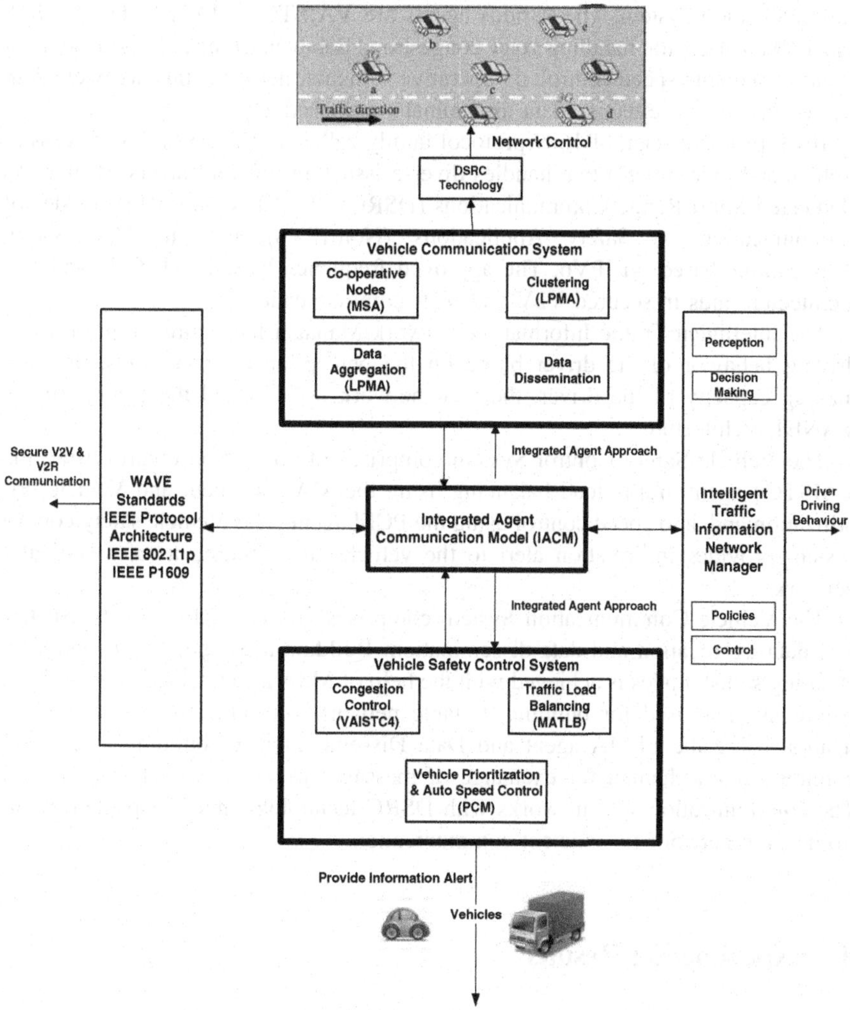

Fig. 1 Integrated agent communication model (IACM)

architecture and apply size reduction techniques to create light weight mobile agents for efficient performance and to improve the Quality of Service in VANET.

The Integrated Agent Communication Model is shown in Fig. 1. IACM is a model that uses agents to carry out safety and non-safety applications in VANET in order to build a safety, congestion free, auto speed control, vehicle prioritization, co-operative, stable network with seamless communication for efficient bandwidth utilization with energy conservation and optimization of message transfer. The various components that make up this model are:

The core component of the IACM model composes a set of agents that interfaces with various safety and non-safety applications of VANET to build an efficient communication system. The various agents are VAISTC4, MATLB, PCM, MSA and LPMA used for handling road congestion, Balancing traffic, prioritization of vehicles and auto speed control, co-operative vehicular network and energy efficient stable network for effective data dissemination respectively.

IEEE protocol stack of 1609 protocol family called WAVE (Wireless Access in Vehicular Environment) that handles diverse issues in different layers. It supports Dedicated Short Range Communications (DSRC). WAVE supports two modes of communication (i) Safety Applications (Non-IPV6) and (ii) Non Safety Applications based on IPV6. The approved frequency band is 5.0 GHz and this architecture aids in secured V2V and V2R communication.

The Intelligent Traffic Information Network Manager has a direct impact in the driving behavior of the driver based on the perception of road status, decision making capacity of the driver, the road network policies and the control of the VANET architecture.

The Vehicle Safety Control System comprises of congestion control using the VAISTC4 agent, traffic load balancing using the MATLB agent and Vehicle prioritization and auto speed control using the PCM agent. This Vehicle safety control system provides information alert to the vehicles for building a safer vehicular network.

The Vehicle Communication System comprises of co-operative nodes, clustering, data aggregation and data dissemination. Building a co-operative network by isolating selfish nodes is achieved with the help of MSA agent, Clustering and Data Aggregation is used for building a stable network with efficient bandwidth utilization using the LPMA agent and Data Dissemination for building a seamless communication channel for exchange of messages between the vehicular nodes. The communication system works with DSRC technology and is capable of controlling the network for efficient communication.

4 Experimental Results

The IACM communication model is simulated using MOVE, SUMO, NS2 and Netlogo. The simulation scenario considers a freeway node that is approximately 2000 * 2000 m long with 50 m distance between the roads apart. The scenario considers a busy road with houses, shopping malls and mansions on both sides of the road and this lane is one of the important lane that connects the source node taken for consideration with that of the destination node. There are other local roads that interact with the freeway or at times run in parallel. Here these sub-roads can be considered as alternate routes when the main freeway is congested. The total duration of the simulation under study is 10 min and during this 10 min the simulation run is conducted to identify traffic intensity, congestion detection, congestion avoidance and identifying emergency vehicles, giving priority to emergency vehicles. The performance of the model is studied in detail based on the

Table 1 Simulation parameters

Parameter	Value
Number of vehicles	100
Maximum speed	120 km/h
Area of simulation	2000 * 2000 m
Distance between streets	50 m
Agent packet size	512 Bytes
Timing of packet generation	1/s
Simulation time	900 s
Routing protocol	AODV
Transmission/communication range	250 m

above said factors using the Netlogo simulator by setting the road environment, vehicle count and other criteria.

In the following simulation the behavior of congestion along with the performance of the IACM model in message passing and handling congestion is studied in detail. The freeway considered for simulation is the normal busy route and initially all the vehicles take up this route to reach the destination understudy. If there is no congestion and the traffic intensity is tolerable, there are no issues until the road becomes congested. Once congestion starts occurring, the speed of vehicles has to be automatically reduced and the vehicles can take alternate routes till congestion is cleared. Therefore the IACM model has to immediately broadcast message about congestion at a particular point and by seeing this message the vehicles at the vicinity of congestion try to take alternate routes and other vehicles can take either an alternate path or cut down their speeds. Using this model the simulation study shows that within 6 min the congested area starts clearing and the entire congestion is cleared within 10 min. The simulation also shows the performance in terms of packet delivery ratio, energy conservation and bandwidth utilization (Table 1).

For optimal simulation of IACM, the influence of road traffic on the network and vice versa needs to be considered. For this purpose, simulators like MOVE, SUMO, NS2, Qualnet, Netlogo and Insight Maker are used. Traffic simulation is performed by the microscopic road traffic simulation package SUMO. Traffic scenarios can be configured by importing detailed road layouts from a Geographic Information System (GIS) and inserting traffic flows according to inductive loop measurements.

Congestion control after implementing MATLB is shown in Fig. 2. According to this simulation the traffic intensity is calculated and compared with the threshold and if heavy traffic is identified the control room automatically re-routes traffic so that the traffic intensity is kept on bay with the threshold.

The grid pattern of urban scenario to study the vehicle prioritization process is shown in Fig. 3. From the Fig. 3 it is inferred that the wait time of emergency vehicles is reduced by clearing the lane and the other plots regarding stopped cars, average speed and wait time of normal vehicles are reduced gradually to improve the performance of the vehicular traffic.

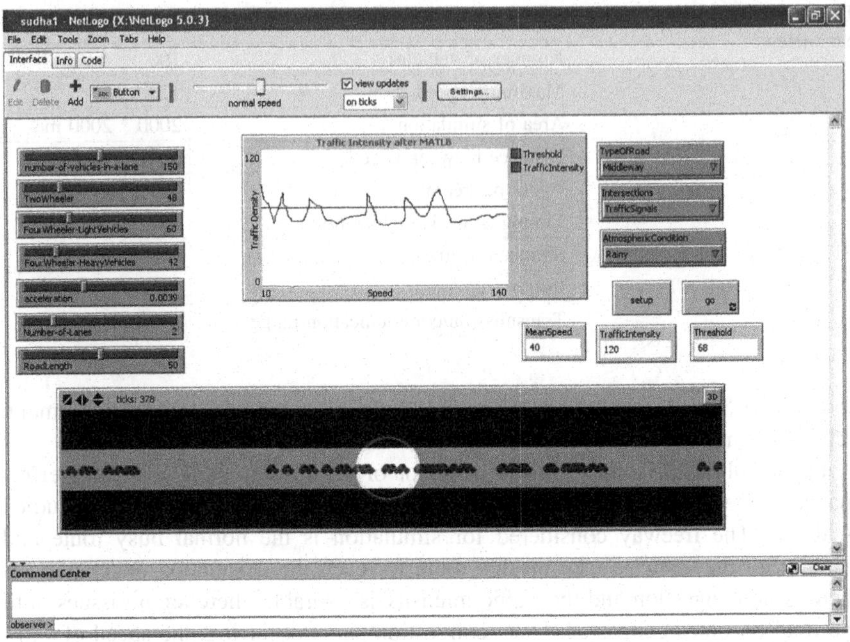

Fig. 2 Traffic intensity after using MATLB

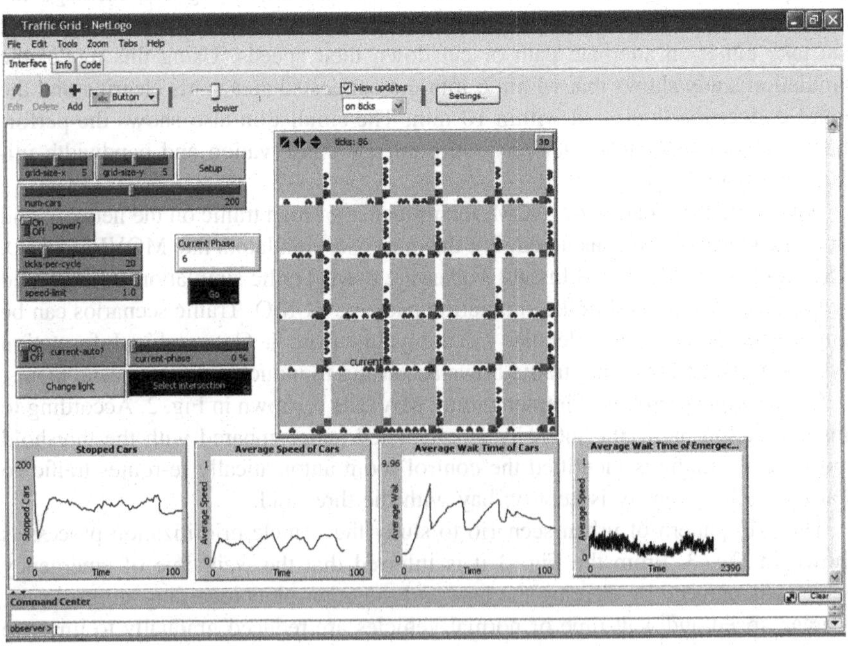

Fig. 3 Performance evaluation using PCM

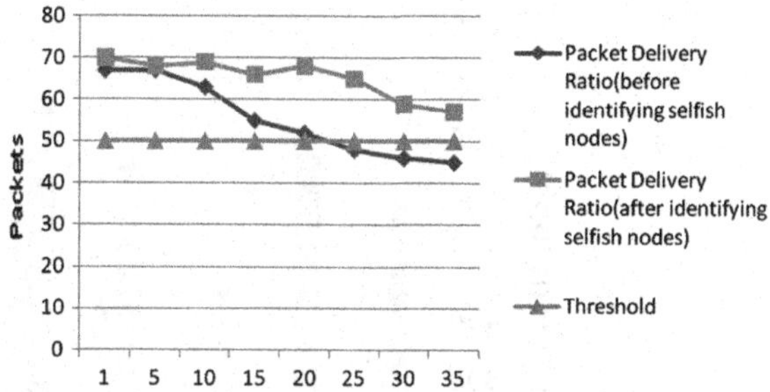

Fig. 4 Packet delivery ratio using MSA

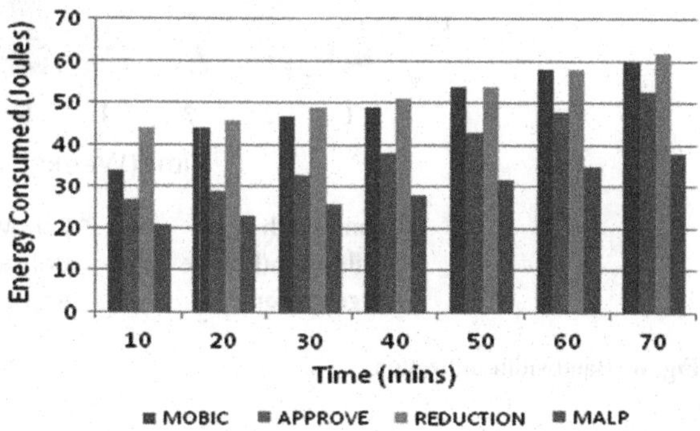

Fig. 5 Energy conservation

The performance of IACM model is shown in Fig. 4. From Fig. 4 it is observed that after using MSA agent for identification and isolation of selfish node, it is inferred from the figure that the packet delivery ratio is maintained above the threshold value without degrading the performance of the system.

Energy consumption of vehicular nodes is shown in Fig. 5. The graph makes a comparison of energy consumption between various clustering approached like MOBIC, APPROVE, REDUCTION and MALP. Based on the study it is inferred from the Fig. 5 that energy consumed on using MALP is less compared to other clustering approaches.

The bandwidth utilization of the model before performing data aggregation and mirroring using LPMA agent and after performing data aggregation and mirroring using LPMA agent is shown in Fig. 6. From Fig. 6 it is inferred that after using LPMA agent for data aggregation there is a drastic change in bandwidth utilization which is measured every week.

The simulated result of the IACM communication model is shown in Fig. 7. There are four graphs generated highlighting on the various aspects of communication like condition of received packets, packets received by transmitter, inter reception time by transmitter and handling emergency messages. From the graph of received packets lost/ok it is inferred that the number of correctly received packets or more than lost or corrupted messages.

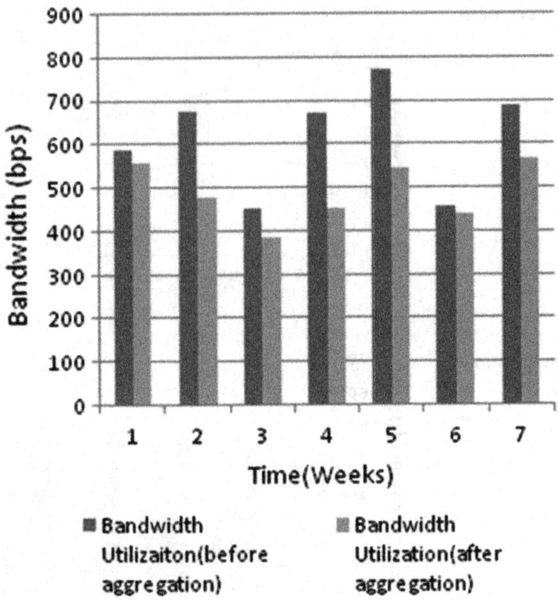

Fig. 6 Bandwidth utilization

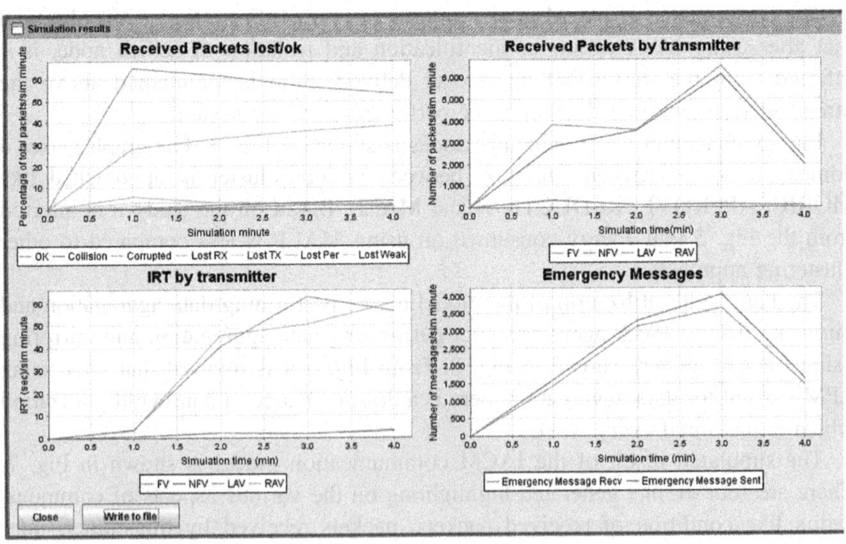

Fig. 7 Performance of integrated agent communication model (IACM)

5 Conclusion

The thesis entitled Study on Integrated Agent Communication Model (IACM) for Vehicular Ad hoc Networks using Mobile Agent proposes a communication model for exchange of messages using an integrated agent based approach. The IACM model focuses on studying communication between vehicular nodes in a vehicular ad hoc network. The problem of data dissemination between vehicular nodes is addressed using mobile agent technology which ensures the suitability of the solutions for deployment in real scenarios. This IACM model employs agents VAISTC4, MATLB, PCM, MSA and LPMA to achieve the task of effective seamless communication with quality service.

References

1. Pesel, R., Maslough, O.: Vehicular Ad-Hoc Networks (VANET) Applied to Intelligent Transportation Systems (ITS), Electronics and Telecommunications. Universite de Limoges (2012)
2. Kumar, R., Dave, M.: Mobile agent as an approach to improve QoS in vehicular ad hoc network. IJCA Spec. Issue Mob. Ad Hoc Netw. (MANETs) 2010
3. Kotz, D., Gray, R.S.: Mobile agents and the future of the internet. ACM Operating Syst. Rev. **33**(3), (1999)
4. Lange, O.: Programming and Deploying Java Mobile Agents with Aglets. Addison-Wesley (1998)
5. Wang, F.Y.: Towards a revolution in transportation operations: AI for complex systems. IEEE Intell. Syst. **23**(6), (2008)
6. Chen, B., Cheng, H.H.: A review of the applications of agent technology in traffic and transportation systems. IEEE Trans. Intell. Transp. Syst. **11**(2), (2010)

A Hybrid Approach for Data Hiding Through Chaos Theory and Reversible Integer Mapping

S.S.V. Nithin Kumar, Gunda Sai Charan, B. Karthikeyan, V. Vaithiyanathan and M. Rajasekhar Reddy

Abstract Steganography is the science which deals with hiding the message so that the intruder cannot even detect its existence. This paper proposes an efficient approach which embeds data into the cover image which is in the frequency domain. DCT is applied to the cover image using integer mapping and the encrypted secret is embedded into it using 3, 3, 2 LSB substitution. The encryption is achieved through chaos theory and ceaser cipher technique. The integer mapping is a method which transforms a given image into its DCT transform and transforms back into the spatial domain without any loss. The proposed method provides a high level of security since secret data is encrypted using chaos theory and embedded into DCT transformed cover image.

Keywords Chaos theory · Hash function · Data hiding · LSB replacement · Integer mapping

1 Introduction

Digitization of data has created a revolution in the field of communication by making exchange of data very convenient. Since the Internet which is used as a medium for exchange of data is an open source, security problems like modification and interception of data are very easy. Several approaches have been developed to make secure communication over the internet. Cryptography is one such technique in which secret information is converted into unrecognizable forms in such a way that only authorized user can transform it back to the original message. But transforming into unrecognizable forms may not meet the requirement in all situations. In some scenarios, hiding the existence of secret data may be needed. Steganography is one such technique which can achieve this requirement.

S.S.V. Nithin Kumar (✉) · G.S. Charan · B. Karthikeyan · V. Vaithiyanathan · M. Rajasekhar Reddy
School of Computing, SASTRA University, Thanjavur 613401, India
e-mail: nithinkumarssv@gmail.com

M. Senthilkumar et al. (eds.), *Computational Intelligence, Cyber Security and Computational Models*, Advances in Intelligent Systems and Computing 412, DOI 10.1007/978-981-10-0251-9_45

Steganography is a Greek word meaning covered writing, is an art and science of hiding secret data [1] into a cover object using the embedding algorithm. In this secret media can be plain text, image, audio or a video file. Cover object can be an image, audio or video. Steganography using image [2] as a cover object can be performed in spatial domain or in transform domain. In transformed domain steganography cover image is transformed into the frequency domain by applying Discrete Cosine Transform (DCT) [3, 4], Discrete Fourier Transform (DFT) or wavelet transform and then is used for embedding secret data. In Spatial domain, cover image is used as such, for hiding secret data. This paper deals with Image Steganography in the frequency domain. High Imperceptibility, high payload capacity and more robustness are the main goals of any steganography technique. Payload capacity refers to how many bits per pixel of a cover image are used for hiding secret data. Imperceptibility refers to the existence of the message is not detectable by intruding. Robustness is the capacity to resist attack from the intruder.

2 Literature Survey

Avinash et al. [5] proposed a method to hide the data in the frequency domain using pixel valued difference. Initially discrete wavelet transform is applied to obtain four sub bands. Then the pixel valued difference method is applied to all the four bands and the secret data is hidden in the cover image. Koikara et al. [6] devised a method which uses the techniques of Distributed Image Steganography (DIS) and Discrete Cosine Transform. DIS involves embedding the secret data into multiple carrier images. In this paper, set of gray scale images are converted into discrete cosine transform and then DIS is applied to obtain the stego image. Karthikeyan et al. [7] proposed a method in which thresholding is used to extract the features of an object. The threshold is calculated using the Tsall is entropy model. In this a non-extensive parameter q is used to distinguish long range correlation.

Bandopadyaya et al. [8] proposed a method to hide the secret data into cover image by encrypting it using chaos theory and then the encrypted data is embedded into the cover using 3, 3, 2 LSB replacement algorithm. Azia et al. [9] proposed an algorithm called cyclic chaos based steganographic algorithm in which secret data is embedded in a color image. In this cyclic chaos theory generates seeds for the pseudo random number generator which utilizes this information for finding channel and pixel positions on host image.

Karthikeyan et al. [10] proposed an image steganography in which image is embedded using modified Hill cipher method. In this a one dimensional matrix is formed from the secret image after applying encryption algorithm, then it is converted into binary form. In each cover image pixel last two LSB are replaced with secret data bits. Thenmozhi et al. [11] proposed an algorithm to hide the secret data into the cover which is in transformed domain. The cover image is converted into discrete wavelet transform. The secret data is encrypted using chaos theory. The encrypted message is embedded into the transformed cover image. Lin [12] introduced a technique called reversible integer mapping to convert an image into its DCT. This method prevents the rounding errors which occur when the stego image is transformed back into the spatial domain after secret data is embedded.

3 Proposed Scheme

3.1 Chaos Theory

Chaos theory [13] is a field of study in mathematics which is widely used in various domains of science. It mainly deals with the behavior of the dynamic systems which are highly sensitive to the initial conditions. It means that by giving a minor variation in the initial conditions, divergent outcomes are obtained which are highly unrelated to each other. In the proposed scheme, it uses a logistic map method in order required chaotic sequences. The chaotic maps [14] are defined by equation

$$X_{k+1} = \mu x_k (1 - x_k) \tag{1}$$

Usually $0 < \mu < 4$ and $0 < x_k < 1$. For this to generate chaotic maps $0 < x_k < 1$ $k = 0, 1, 2, 3 \ldots\ldots$ and $3.5699496 < \mu < 4$. The value of μ is chosen from this range because most uncorrelated chaotic sequences are generated if it is chosen this way.

3.2 Ceaser Cipher Technique

Ceaser cipher is a stream cipher technique in which plain text is encrypted character by character. In Ceaser cipher a fixed number called key is added to obtain the cipher text.

3.3 Reversible Integer Mapping to Change the Cover into DCT

The cover image is converted into DCT using Eq. (2). The resultant image has pixel values which contain decimal parts. When secret data are to be embedded into the image when it is in DCT it leads to rounding errors [15]. This is because after embedding the data in the frequency domain and when it is transformed back into spatial domain, in order to obtain stego image, pixel values must be rounded off since only integers are allowed in spatial domain. As a result the secret message cannot be exactly retrieved back. Hence this scheme uses a reversible integer mapping [16] method which avoids the rounding errors by establishing a mapping between integers of spatial domain and frequency domain. The reversible integer mapping [15] can be explained as given below.

$$y_k = \sum_{n=1}^{N} x_n \frac{1}{2} \cos\left(\frac{\pi}{N}\left(n - 1 + \frac{1}{2}\right)(k - 1)\right) \quad (2)$$

For k = 1….N

Let the given matrix be X = [x_{ij}] where 1 <= i <= M and 1 <= j <= N for an image of M × N. Let us assume that the notation <z> represents the rounded off value of z. The DCT of the given image is computed as Y = AX where Y is the matrix obtained after DCT and A is defined as $A = PS_N S_{N-1} \ldots S_0$ is called the factorization matrix, provided all its leading principle Here P is the permutation matrix which is given in Eq. (3) and each S_m is defined as $S_m = I + (e_m s_m)^T$ for m = 1, 2..,N, where I is the identity matrix of order N × N and e_m is the mth column vector of identity matrix and s_0, s_1,.. s_N are the row vectors of the matrix constructed through following steps:

1. By using the formula for DCT given in (2), construct a two dimensional matrix with the coefficients of DCT.
2. Convert the matrix such a way all its leading principle sub-matrices [12] are equal to 1.

The DCT matrix Y is computed as $Y = <P<S_N<S_{N-1}<\ldots<S_0X>\ldots>$. It means that at each mth step S_m matrix is multiplied and rounding is performed. The resultant DCT matrix consists of only integers which can be considered as the mapped values for corresponding integers of image in the spatial domain. After embedding the secret data, the stego image is obtained by applying the inverse DCT. This can be achieved as $X = <P<S_N^{-1}<S_{N-1}^{-1}<\ldots\ldots<S_0^{-1}X>\ldots>$.

Hence this scheme uses a reversible integer mapping method which avoids the rounding errors by establishing a mapping between integers of spatial domain and frequency domain.

$$P = \begin{bmatrix} 0 & 0 & 0 & 0 & 0 & 1 & 0 & 0 \\ 1 & 0 & 0 & 0 & 0 & 0 & 0 & 0 \\ 0 & 1 & 0 & 0 & 0 & 0 & 0 & 0 \\ 0 & 0 & 1 & 0 & 0 & 0 & 0 & 0 \\ 0 & 0 & 0 & 0 & 0 & 0 & 1 & 0 \\ 0 & 0 & 1 & 0 & 0 & 0 & 0 & 0 \\ 0 & 0 & 0 & 0 & 0 & 0 & 0 & 1 \\ 0 & 0 & 0 & 0 & 1 & 0 & 0 & 0 \end{bmatrix} \quad (3)$$

3.4 Embedding Technique

This paper uses the idea of 3, 3, 2 LSB replacement technique [17] for embedding the encrypted secret message. The encrypted secret message is embedded into the cover when it is in the DCT form. Each byte of the secret data is embedded into one pixel of the DCT applied cover image [18]. In the spatial domain each pixel of cover image has 8 bits for each red, green, blue components. In order to embed data in the DCT cover, each value is represented as 10 bit binary number. An extra bit is preserved for storing the sign of the DCT value. Secret bits are embedded in the least significant 5 bits of cover image pixel containing 10 bits using a hash function. The secret data is embedded into the cover in such a way that the first 3 bits are embedded in 5 least significant bits of red component and next 3 bits in 5LSBs of green component and last 2 bits in 5LSBs of blue component. Here only 2 bits are embedded in the blue component because the chromatic influence of blue color on human eye is more than red and blue. The hash function which is mentioned above is given as

$$k = p\%n \tag{4}$$

Here k-LSB bit position of the cover where the secret bit is to be embedded.

p Position of the bit in secret data which is to be embedded

n No of LSB bits in cover pixel

3.4.1 Steps for 3, 3, 2 LSB Replacement

1. Five LSB bits of each of the red, green, blue bytes of each pixel of the cover image are identified.
2. Embed eight bits of secret bits into its corresponding cover image pixel in the positions obtained by hash functions.
3. Repeat this procedure for all the secret data is embedded into cover image.
4. Finally, obtain a stego image.

Decoding Algorithm for 3, 3, 2 LSB replacement algorithm is as follows

1. From every pixel of cover image find LSB bits used for embedding the secret data bits.
2. Retrieve the secret bits from the cover image pixel in the order of 3, 3, 2 respectively.
3. Finally, obtain a secret image.

3.5 Illustration of Chaotic Encryption

In chaotic encryption [19] L × B matrix is constructed from cipher text obtained from Ceaser cipher encryption. Then divide L × B matrix into L parts of each size 1 × B.

Now generate chaotic sequence x_k, k = 0, 1, 2…N − 1 by considering some initial conditions for μ, x by using logistic map equation as explained previously. Then threshold is calculated by finding Arithmetic mean of chaotic sequence generated. Next by comparing each x_k with T, binary bit sequence B_k is generated in such a way that if $X_k \geq T$ then $B_k = 1$, else $B_k = 0$. Each part is encrypted by applying an XOR operation secret part and its corresponding Binary Bit Sequence generated.

3.5.1 Encryption Algorithm

1. Construct a matrix of size L × B with the cipher text obtained from the Ceaser cipher technique as its elements.
2. The matrix is divided into L parts of each size 1 × B.
3. For each part chaotic sequence [20] is generated by selecting some initial conditions for μ, x using the logistic map method.
4. Threshold value (T) is calculated by finding mean of generated chaotic sequence.
5. Binary bit sequence is generated by comparing T with each value of chaotic sequence in the following way:

   a. If $x_k \geq T$ then $B_k = 1$
   b. Else $B_k = 0$

6. Each part is encrypted by applying XOR between Binary bit sequence and secret part.
7. Repeat above steps for all the remaining parts.

For decoding pixels of the stego image used for embedding are identified and secret bits are retrieved from LSB bits. Matrix of size L × B is constructed from the obtained secret bits and is divided into L parts. Then for each part chaotic sequence is generated with the same initial values for x, μ as used in encryption. From the calculated chaotic sequence Binary Bit Sequence is calculated as explained in encryption algorithm section E. Apply XOR operation between Binary Bit Sequence and secret part. The above procedure is repeated for remaining parts.

3.5.2 Decryption Algorithm

1. Take input as stego image and retrieve the secret data from LSB bits of pixels of stego image.
2. Break the secret image into parts.
3. Generate chaotic sequence for each part by selecting same initial values for μ, x as used in encryption.
4. The threshold value is calculated by calculating the mean of chaotic sequence.
5. Develop Binary Bit sequence as explained above from the chaotic sequence generated.

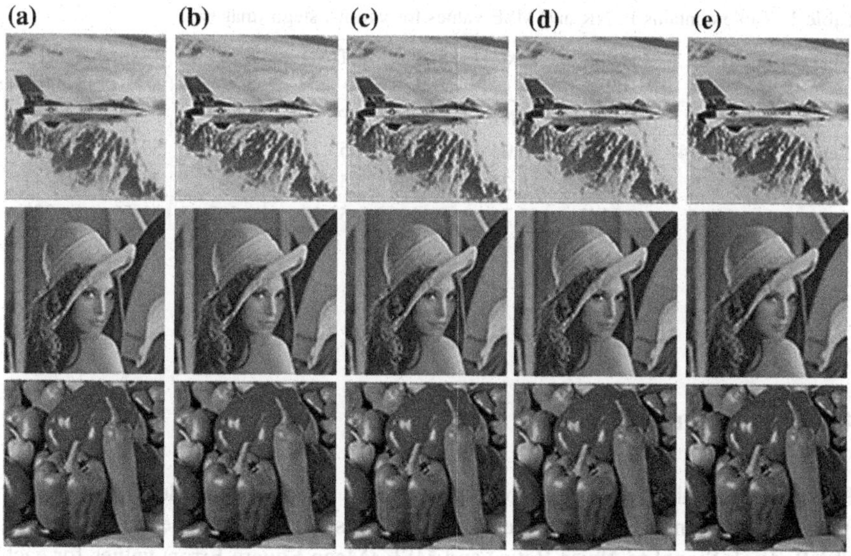

Fig. 1 **a** Cover images. **b** Stego image for secret text of size 100. **c** Stego image for secret text of size 200. **d** Stego image for secret text of size 300. **e** Stego image for secret text of size 500

6. Each part is decrypted by applying XOR between Binary Bit Sequence and secret data.
7. Repeat the above procedure for all the remaining parts (Fig. 1).

3.5.3 Architecture for Encoding

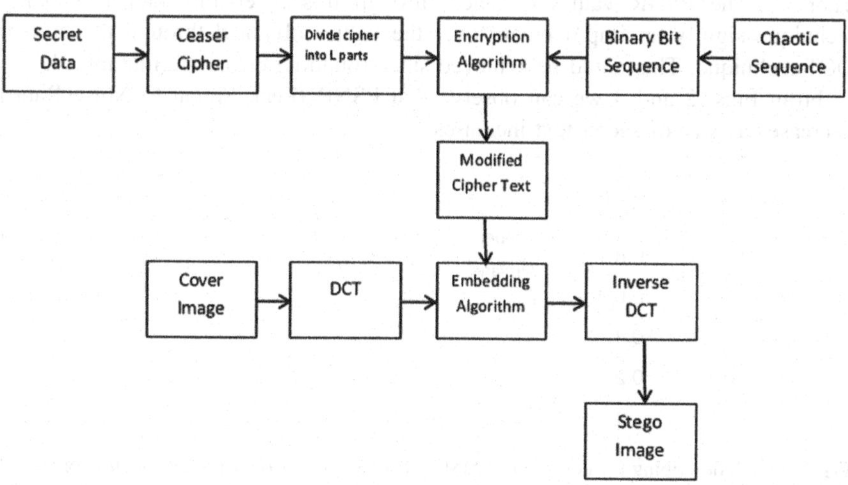

Table 1 Table contains PSNR and MSE values for various stego images

Image	Text	Size 100	Text	Size 200	Text	Size 300	Text	Size 500
Name	PSNR	MSE	PSNR	MSE	PSNR	MSE	PSNR	MSE
Lenna	65.474	0.0184	58.04	0.1020	53.71	0.276	53.87	0.266
Airplane	66.26	0.0154	58.26	0.0969	53.96	0.261	54.20	0.247
Pepper	66.45	0.0147	57.71	0.1100	53.38	0.298	53.66	0.279

See Table 1.

4 Experimental Results

Performance evaluation is done by considering 3 cover images, details of 3 cover images are given in Table 1. Performance measurement is done by calculating PSNR (Peak Signal to Noise Ratio) and MSE (Mean Square Error) values for each stego image.

$$PSNR = 10 \times \log_{10}\left(\frac{255}{MSE}\right) \tag{5}$$

$$MSE = \frac{1}{(L \times B)} \sum_{i=1}^{L} \sum_{j=1}^{B} (X_{ij} - Y_{ij})^2 \tag{6}$$

Where X_{ij} represents cover image pixel and Y_{ij} represents stego image pixel values. The values of PSNR and MSE for all stego images are stored in above Table 1. The PSNR values obtained through this reversible integer mapping technique signify an improvement over the values obtained through the normal DCT technique. This could be achieved since rounding errors are minimized.

From Figs. 2 and 3 we can observe that PSNR (Peak Signal to Noise Ratio) decreases as size of secret text increases.

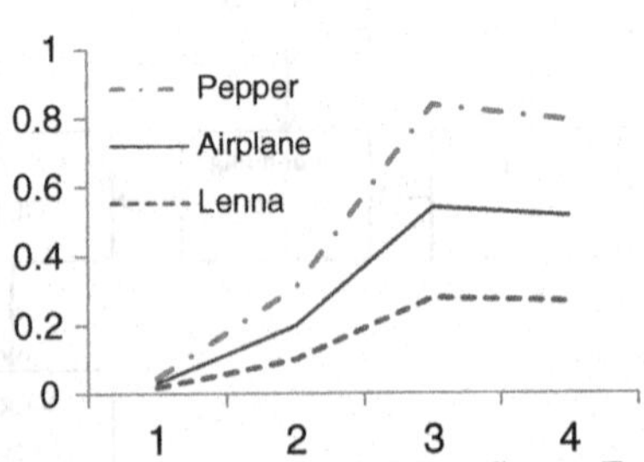

Fig. 2 Graph describing the deviation of MSE (Mean Square Error) with the variation of size of secret text

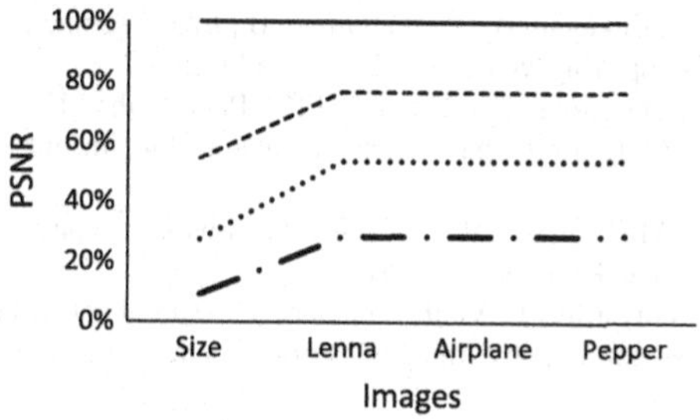

Fig. 3 Graph describing the deviation of PSNR values with the variation in size of secret text

5 Conclusion

In this paper reversible integer mapping technique is used to convert cover image into Discrete cosine transform by which rounding errors are minimized. Secret data is embedded into the DCT cover media because of which capturing the data from the cover image becomes very difficult for intruder. This technique is resistant to noise attack on the least significant bits because the embedding is achieved in the frequency domain. It is also resistant to blind attack [21], which can predict the existence of the payload based on the classification through a set of trained data, because even if the existence is determined the payload is still secured with chaos encryption. In this paper two levels of encryption of secret data is achieved by Ceaser cipher technique and chaos theory. Proposed algorithm achieved high security and performance which can be observed from high values of PSNR and MSE values achieved. Future work includes innovating new techniques to hide the data in the frequency domain obtained through reversible integer mapping.

References

1. Bender, W., Gruhl, D., Morimoto, N., Lu, A.: Techniques for data hiding. IBM Syst. J. **35**, 313–316 (1996)
2. Chang, C.C., Chen, T.S., Chung, L.Z.: A steganographic method based upon JPEG and quantization table modification. Inf. Sci. **141**, 123–138 (2011)
3. Zhao, Z.-J., Chen, H.-X., Chen, M.-S.: Integer Implementation of four-dimensional matrix DCT. Jilin Daxue Xuebao (Gongxueban)/J. Jilin University (Engineering and Technology Edition) **38**(3), 700–703 (2008)
4. Plonka, G.: A global method for invertible integer DCT and integer wavelet algorithms. Appl. Comput. Harmonic Anal. **16**(2), 90–110 (2004)
5. Gulve, A.K, Joshi, M.S.: An image steganography method hiding secret data into coefficients of integer wavelet transform using pixel value differencing approach. Math. Problems Eng. 684824 (2015)
6. Koikara, R., Deka, D.J., Gogoi, M., Das, R.: A novel distributed image steganography method based on block-DCT. Lecture Notes Electr. Eng. **315**, 423–435 (2014)

7. Vaithiyanathan, V., Karthikeyan, B., Venkatraman, B.: Image segmentation based on modified Tsallis entropy. Contemp. Eng. Sci. **7**(9–12), 523–529 (2014)
8. Bandyopadhyay, D., Dasgupta, K., Mandal, J.K., Paramartha, D.: A novel secure image steganography method based on Chaos theory in spatial domain. Int. J. Secur. Privacy Trust Manage. **3**(1) (2014)
9. Aziz, M., Tayarani N.M.H., Afsar, M.: A cyclic chaos-based cryptic-free algorithm for image steganography. Nonlinear Dynamics (2015)
10. Karthikeyan, B., Chakravarthy, J., Vaithiyanathan, V.: An enhanced Hill cipher approach for image encryption in steganography. Int. J. Electron. Secur. Digit. Forensics **5**(3–4), 178–187 (2013)
11. Thenmozhi, S., Chandrasekaran, M.: A novel technique for image steganography using nonlinear chaotic map. In: 7th International Conference on Intelligent Systems and Control, ISCO (2013)
12. Lin, Y.-K.: A data hiding scheme based upon DCT coefficient modification. Comput. Stand. Interf. **36**(5), 855–862 (2014)
13. Sudha, K.L., Prasad, M.: Chaos image encryption using pixel shuffling with Henon map. Proc. Elixir Elec. Engg **38**, 4492–4495 (2011)
14. Zaghbani, S., Rhouma, R.: Data hiding in spatial domain image using chaotic map. In: 5th International Conference on Modeling, Simulation and Applied Optimization, ICMSAO, 6552626 (2013)
15. Yusong, Y., Qingyun, S.: Reversible DCT mapping integers to integers and lossless image compression. Ruan Jian Xue Bao/J. Softw. **11**(5), 620–627 (2000)
16. Hao, P., Shi, Q.: Matrix factorizations for reversible integer mapping. IEEE Trans. Signal Process. **49**(10), 2314–2324 (2001)
17. Karthikeyan, B., Ramakrishnan, S., Vaithiyanathan, V., Sruthi, S., Gomathymeenakshi, M.: An improved steganography technique using LSB replacement on a scanned path image. Int. J. Network Secur. **16**(1), 14–18
18. Qian, Z., Zhang, X.: Lossless data hiding in JPEG bitstream. J. Syst. Softw. **85**(2), 309–313 (2012)
19. Fridrich, J., Du, R., Meng, L.: Steganalysis and LSB encoding in color images. In: Proceedings of ICME 2000, N. Y., USA
20. Zhang, Y., Zuo, F., Zhai, Z., Xiaobin, C.: A new image encryption algorithm based on multiple chaos system. In: Proceedings of the International Symposium on Electronic Commerce and Security (ISECS '08), 2008, pp. 347–350
21. Sravanthi, G.S., Sunitha Devi, B., Riyazoddin, S.M., Janga Reddy, M.: A spatial domain image steganography technique based on plane bit substitution method. Global J. Comput. Sci. Technol. (2012)

Fluid Queue Driven by an $M/E_2/1$ Queueing Model

K.V. Vijayashree and A. Anjuka

Abstract This paper deals with the stationary analysis of a fluid queue driven by an $M/E_2/1$ queueing model. The underlying system of differential difference equations that governs the process are solved using generating function methodology. Explicit expressions for the joint steady state probabilities of the state of the background queueing model and the content of the buffer are obtained in terms of a new generalisation of the modified Bessel function of the second kind. Numerical illustrations are added to depict the convergence of the joint probabilities with the steady state solutions of the underlying queueing models.

Keywords Generating function · Steady state probabilities · Modified Bessel function of second kind · Buffer content distribution

Mathematics Subject Classification 60K25 · 90B22

1 Introduction

A stochastic fluid flow model is an input-output system where the input is modelled as a continuous fluid that enters and leaves the storage devices called a buffer according to randomly varying rates. In these models, the fluid buffer is either filled or depleted or both at rates determined by the current state of the background Markov process. *Markov Modulated Fluid Queues* are particular class of fluid models useful for modelling many physical phenomenon and they often allow tractable analysis. Fluid queues finds a wide spread applicability in computer and communication systems [1, 2], manufacturing systems [3] etc.

K.V. Vijayashree (✉) · A. Anjuka
Department of Mathematics, Anna University, Chennai, India
e-mail: vkviji@annauniv.edu

A. Anjuka
e-mail: anjukaatlimuthu@gmail.com

M. Senthilkumar et al. (eds.), *Computational Intelligence, Cyber Security and Computational Models*, Advances in Intelligent Systems and Computing 412, DOI 10.1007/978-981-10-0251-9_46

In recent years, queueing theory has paid considerable attention to Markov Modulated Fluid Queues. Fluid modelling approach is appropriate in situations where the individual units have less impact on the overall behaviour of the system. For example, in ATM environment this modelling approach has been found to be quite effective because all the informations are transported using small fixed-sized cells that are statistically multiplexed and the interarrival time between cells at the time of generation is constant for several contiguous cells. Certain interesting real world applications of Markov Modulated Fluid Flow models can be found in [4–7].

Steady state behaviour of Markov driven fluid queues have been extensively studied in the literature [8–10]. Recently fluid models driven by an $M/M/1$ queue subject to various vacation strategies were analysed in steady state by Mao et al. [11, 12]. The results were further extended to fluid models driven by an $M/M/1/N$ queue with single and multiple exponential vacations [13, 14] and an $M/G/1$ queue with multiple exponential vacations using spectral method [15]. For a fluid models driven by an $M/M/c$ queue with working vacations, Xu et al. [16] obtained the expression for the buffer content distribution in Laplace domain using matrix analytic method. Further, for a MAP-modulated fluid model with multiple vacation, Baek et al. [17] presents the Laplace-Stieljes transform of the fluid level at an arbitrary point of time in steady state. However, in most of the literature relating to fluid queues driven by a vacation queueing models, the buffer content distribution is expressed in the Laplace domain. More recently, Vijayashree and Anjuka [18] presented an explicit expression for the buffer content distribution of a fluid queueing model modulated by an $M/M/1$ queue subject to catastrophes and subsequent repair. Ammar [19] obtain the explicit expression for the joint steady state probabilities of a fluid queue driven by an $M/M/1$ queue with multiple exponential vacation.

In this paper, we analyze a fluid queue modulated by a single server queueing model with Poisson arrival and Erlang type 2 service time distribution. The Erlang family of probability distributions provides far greater flexibility in modelling real life service patterns than does the exponential. The background queueing model considered in this paper is ideal for modelling many practical systems. For example, the mechanism of protein synthesis involves two major phases, namely, transcription (involving transfer of genetic information from DNA to mRNA) and translation (involving translation of the language of nucleic acid into that of proteins) wherein each phase can be assumed to follow exponential distribution with identical parameter. Similarly, in a hospital, the arriving patient undergoes two types of services, namely, enrolling the patients name at the reception counter to fix an appointment and meeting the doctor. Such situations where the arriving units gets served in two phases are quite common in real time situations.

This paper deals with the stationary analysis of the fluid model driven by an $M/E_2/1$ queueing model. Explicit analytical expression for the joint steady state probabilities of the number of customers in the background queueing model and the content of the buffer are obtained. The results are based on the new generalisation of the modified Bessel function of the second kind and its generating function introduced by Griffiths et al. [20].

2 Model Description

Consider a single server queueing model with Poisson arrival and Erlang distributed service times involving two phases of service, referred to as $M/E_2/1$ queueing model. Customers arrive according to a Poisson process at the rate of λ and the service time has an Erlang type-2 distribution with parameter μ. The Erlang distribution can be viewed as being made up of 2 exponential phases, each with mean $\frac{1}{2\mu}$. The phases are numbered backward that is, 2 is the first phase of service and 1 is the second phase (a customer leaving phase 1 actually leaves the system). The state transition diagram for the model under consideration is given in Fig. 1. The state space is given by

$$\Omega = \{(0) \bigcup (n,s), n = 1, 2, \cdots, s = 1, 2\}.$$

Consider a fluid queueing model driven by an $M/E_2/1$ queue. Let $N(t)$ denote the number of customers in the system at time t and $J(t)$ denote the phase of the customer undergoing service at time t. Further, let $\{C(t), t \geq 0\}$ represent the buffer content process where $C(t)$ denotes the content of the buffer at time t. During the busy period of the server, the fluid accumulates in an infinite capacity buffer at a constant rate $r > 0$. The buffer depletes the fluid during the idle periods of the server at a constant rate $r_0 < 0$ as long as the buffer is nonempty. Hence, the dynamics of the buffer content process is given by

$$\frac{dC(t)}{dt} = \begin{cases} 0, & N(t) = 0, C(t) = 0 \\ r_0, & N(t) = 0, C(t) > 0. \\ r, & N(t) > 0. \end{cases}$$

Under steady state conditions, let us define the joint probability distribution functions of the Markov process $\{(N(t), J(t), C(t)), t \geq 0\}$ as

$$F_0(x) = \lim_{t \to \infty} Pr\{N(t) = 0, C(t) \leq x\}, \ x > 0$$

$$F_{n,s}(x) = \lim_{t \to \infty} Pr\{N(t) = n, J(t) = s, C(t) \leq x\} \text{ for } x > 0, (n,s) \in \Omega \backslash (0).$$

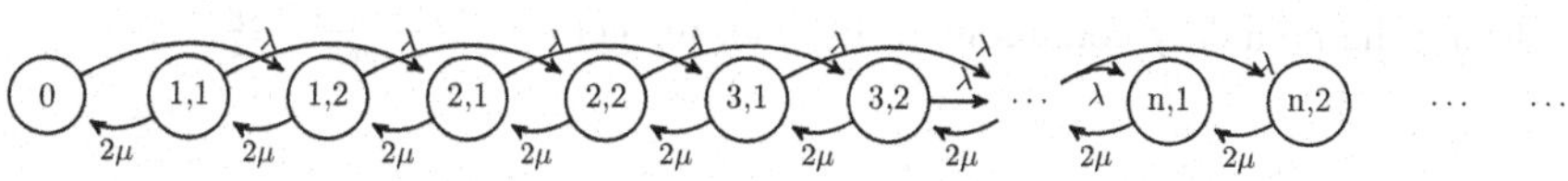

Fig. 1 State transition diagram

Clearly the 3-dimensional process $\{N(t), J(t), C(t)\}$ represent a fluid queue driven by an $M/E_2/1$ queue subject to the stability condition given by $d<0$ where

$$d = r_0\pi_0 + r\sum_{n=1}^{\infty}\sum_{s=1}^{2}\pi_{n,s} = r_0\pi_0 + r(1-\pi_0)$$

and $\pi_0 = \frac{\mu-\lambda}{\mu}$. Therefore $d = \frac{(r_0-r)(\mu-\lambda)+r\mu}{\mu}$. Note that the probability for the content of the buffer to be empty leads to

$$F_0(0) + \sum_{n=1}^{\infty}\sum_{s=1}^{2}F_{n,s}(0) = \frac{d}{r_0} \tag{1}$$

Using standard methods, the system of differential difference equations governing the fluid queue model are given by

$$r_0\frac{dF_0(x)}{dx} = -\lambda F_0(x) + 2\mu F_{1,1}(x), \tag{2}$$

$$r\frac{dF_{1,1}(x)}{dx} = -(\lambda+2\mu)F_{1,1}(x) + 2\mu F_{1,2}(x), \tag{3}$$

$$r\frac{dF_{1,2}(x)}{dx} = -(\lambda+2\mu)F_{1,2}(x) + 2\mu F_{2,1}(x) + \lambda F_0(x), \tag{4}$$

$$r\frac{dF_{n,1}(x)}{dx} = -(\lambda+2\mu)F_{n,1}(x) + 2\mu F_{n,2}(x) + \lambda F_{n-1,1}(x), \quad n\geq 2, \tag{5}$$

and $$r\frac{dF_{n,2}(x)}{dx} = -(\lambda+2\mu)F_{n,2}(x) + 2\mu F_{n+1,1}(x) + \lambda F_{n-1,2}(x), \quad n\geq 2. \tag{6}$$

with the boundary conditions

$$\begin{aligned} F_{n,s}(0) &= 0, \quad (n,s)\in \Omega\backslash(0) \text{ and} \\ F_0(0) &= a, \quad \text{for some constant } 0<a<1. \end{aligned} \tag{7}$$

Using the boundary conditions in Eq. (1), we get $a = \frac{(r_0-r)(\mu-\lambda)+r\mu}{r_0\mu}$.

3 Stationary Analysis

This section presents explicit expressions for the joint steady state probabilities of the background queueing model and the content of the buffer in terms of a new generalization of the modified Bessel function of the second kind. Define the generating function

$$G(y,x)=\sum_{n=1}^{\infty}\sum_{s=1}^{2}F_{n,s}(x)y^{2(n-1)+s}+\frac{r_0}{r}F_0(x)$$

with $G(y,0)=\frac{r_0 a}{r}$. To solve Eqs. (2)–(6), we multiply the second equation by y, the third equation by y^2, the fourth by $y^{(2n-1)}$, $n=2,3,\cdots$, and the fifth by y^{2n}, $n=2,3,\cdots$ and sum by column leading to

$$\begin{aligned}
r\left[\frac{r_0}{r}F_0(x)+\sum_{n=1}^{\infty}\sum_{s=1}^{2}F_{n,s}(x)y^{2(n-1)+s}\right] = \\
&-(\lambda+2\mu)\left[F_{1,1}(x)y+F_{1,2}(x)y^2+\sum_{n=2}^{\infty}F_{n,1}(x)y^{2n-1}\right.\\
&\left.+\sum_{n=2}^{\infty}F_{n,2}(x)y^{2n}\right]+\frac{2\mu}{y}\left[F_{1,1}(x)y+F_{1,2}(x)y^2+F_{2,1}(x)y^3\right.\\
&\left.+\sum_{n=2}^{\infty}F_{n+1,1}(x)y^{2n-1}+\sum_{n=2}^{\infty}F_{n,2}(x)y^{2n}\right]+\lambda y^2\left[F_0(x)\right.\\
&\left.+\sum_{n=2}^{\infty}F_{n-1,1}(x)y^{2n-3}+\sum_{n=2}^{\infty}F_{n-1,2}(x)y^{2n-2}\right]-\lambda F_0(x),
\end{aligned}$$

which on simplification leads to

$$\begin{aligned}
r\frac{\partial G(y,x)}{\partial x} &= -(\lambda+2\mu)\left[G(y,x)-\frac{r_0}{r}F_0(x)\right]+\frac{2\mu}{y}\left[G(y,x)-\frac{r_0}{r}F_0(x)\right]\\
&\quad+\lambda y^2\left[F_0(x)+G(y,x)-\frac{r_0}{r}F_0(x)\right]-\lambda F_0(x)\\
&=\left[-(\lambda+2\mu)+\frac{2\mu}{y}+\lambda y^2\right]G(y,x)+\frac{2\mu r_0}{r}\left(1-\frac{1}{y}\right)F_0(x)\\
&\quad+\lambda\left[-\left(1-\frac{r_0}{r}\right)+y^2\left(1-\frac{r_0}{r}\right)\right]F_0(x).
\end{aligned}$$

Therefore, we obtain the linear differential equation given by

$$\frac{\partial G(y,x)}{\partial x} = \frac{1}{r}\left[-(\lambda+2\mu) + \frac{2\mu}{y} + \lambda y^2\right] G(y,x) \\ + \frac{1}{r}\left[\frac{2\mu r_0}{r}\left(1-\frac{1}{y}\right) + \lambda\left(1-\frac{r_0}{r}\right)(y^2-1)\right] F_0(x). \tag{8}$$

Integrating the above equation yields

$$G(y,x) = \frac{a r_0}{r} exp\left[-\left(\frac{\lambda+2\mu}{r}\right)x\right] exp\left[-\frac{1}{r}\left(\lambda y^2 + \frac{2\mu}{y}\right)x\right] \\ + \frac{2\mu r_0}{r^2}\left(1-\frac{1}{y}\right)\int_0^x exp\left[-\left(\frac{\lambda+2\mu}{r}\right)(x-z)\right] exp\left[-\frac{1}{r}\left(\lambda y^2 + \frac{2\mu}{y}\right)(x-z)\right] F_0(z)dz \\ + \frac{\lambda}{r}\left(1-\frac{r_0}{r}\right)(y^2-1)\int_0^x exp\left[-\left(\frac{\lambda+2\mu}{r}\right)(x-z)\right] exp\left[-\frac{1}{r}\left(\lambda y^2 + \frac{2\mu}{y}\right)(x-z)\right] F_0(z)dz. \tag{9}$$

Here, we recall a new generalization of the modified Bessel function of second type [2] given by

$$\widetilde{I}_n^{k,s}(z) = \left(\frac{z}{2}\right)^{n+k-s} \sum_{r=0}^{\infty} \frac{\left(\frac{z}{2}\right)^{r(k+1)}}{(k(r+1)-s)!\Gamma(n+r+1)}, \tag{10}$$

where $z \in C$, $s \in 1,2,\cdots,k$, $n = 0, \pm 1, \pm 2, \cdots$, $k = 1,2.\cdots$. Its generating function takes the form

$$\exp\left[\frac{1}{r}\left(\lambda y^k + \frac{k\mu}{y}\right)z\right] = \sum_{n=-\infty}^{\infty}\sum_{s=1}^{k}(\beta y)^{k(n-1)+s}\widetilde{I}_n^{k,s}(\alpha z), \tag{11}$$

where $\alpha = 2\left[\frac{\lambda}{r}\left(\frac{k\mu}{r}\right)^k\right]^{\frac{1}{k+1}}$ and $\beta = \left(\frac{\lambda}{k\mu}\right)^{\frac{1}{k+1}}$. It is exactly this form of the generating function with appropriate power we need for the solution of the Erlang queue. Substituting Eq. (11) in the RHS of Eq. (9) leads to

$$
\begin{aligned}
G(y,x) =& \frac{ar_0}{r} exp\left[-\left(\frac{\lambda+2\mu}{r}\right)x\right] \sum_{n=-\infty}^{\infty}\sum_{s=1}^{2}(\beta y)^{2(n-1)+s}\widetilde{I}_n^{2,s}(\alpha z) \\
&+ \frac{2\mu r_0}{r^2}\int_0^x exp\left[-\left(\frac{\lambda+2\mu}{r}\right)(x-z)\right] \sum_{n=-\infty}^{\infty}\sum_{s=1}^{2}(\beta y)^{2(n-1)+s}\widetilde{I}_n^{2,s}(\alpha z)F_0(z)dz \\
&- \frac{2\mu r_0}{r^2}\int_0^x exp\left[-\left(\frac{\lambda+2\mu}{r}\right)(x-z)\right] \sum_{n=-\infty}^{\infty}\sum_{s=1}^{2}\beta^{2(n-1)+s}y^{2(n-1)+s-1}\widetilde{I}_n^{2,s}(\alpha z)F_0(z)dz \\
&+ \frac{\lambda}{r}\left(1-\frac{r_0}{r}\right)\int_0^x exp\left[-\left(\frac{\lambda+2\mu}{r}\right)(x-z)\right] \sum_{n=-\infty}^{\infty}\sum_{s=1}^{2}\beta^{2(n-1)+s}y^{2n+s}\widetilde{I}_n^{2,s}(\alpha z)F_0(z)dz \\
&- \frac{\lambda}{r}\left(1-\frac{r_0}{r}\right)\int_0^x exp\left[-\left(\frac{\lambda+2\mu}{r}\right)(x-z)\right] \sum_{n=-\infty}^{\infty}\sum_{s=1}^{2}(\beta y)^{2(n-1)+s}\widetilde{I}_n^{2,s}(\alpha z)F_0(z)dz.
\end{aligned} \tag{12}
$$

Comparing the coefficients of y^{2n-1} on both sides of Eq. (12) leads to

$$
\begin{aligned}
F_{n,1}(x) =& \frac{ar_0}{r} e^{\left[-\left(\frac{\lambda+2\mu}{r}\right)x\right]} \beta^{2n-1}\widetilde{I}_n^{2,1}(\alpha x) \\
&+ \frac{2\mu r_0}{r^2}\beta^{2n-1}\int_0^x e^{\left[-\left(\frac{\lambda+2\mu}{r}\right)(x-z)\right]}\widetilde{I}_n^{2,1}(\alpha(x-z))F_0(z)dz \\
&- \frac{2\mu r_0}{r^2}\beta^{2n}\int_0^x e^{\left[-\left(\frac{\lambda+2\mu}{r}\right)(x-z)\right]}\widetilde{I}_n^{2,2}(\alpha(x-z))F_0(z)dz \\
&+ \frac{\lambda}{r}\left(1-\frac{r_0}{r}\right)\beta^{2n-3}\int_0^x e^{\left[-\left(\frac{\lambda+2\mu}{r}\right)(x-z)\right]}\widetilde{I}_{n-1}^{2,1}(\alpha(x-z))F_0(z)dz \\
&- \frac{\lambda}{r}\left(1-\frac{r_0}{r}\right)\beta^{2n-1}\int_0^x e^{\left[-\left(\frac{\lambda+2\mu}{r}\right)(x-z)\right]}\widetilde{I}_n^{2,1}(\alpha(x-z))F_0(z)dz, \quad n \geq 1.
\end{aligned} \tag{13}
$$

Comparing the coefficients of y^{2n} on both sides of Eq. (12) leads to

$$
\begin{aligned}
F_{n,2}(x) =& \frac{ar_0}{r} e^{\left[-\left(\frac{\lambda+2\mu}{r}\right)x\right]} \beta^{2n}\widetilde{I}_n^{2,2}(\alpha x) \\
&+ \frac{2\mu r_0}{r^2}\beta^{2n}\int_0^x e^{\left[-\left(\frac{\lambda+2\mu}{r}\right)(x-z)\right]}\widetilde{I}_n^{2,2}(\alpha(x-z))F_0(z)dz \\
&- \frac{2\mu r_0}{r^2}\beta^{2n+1}\int_0^x e^{\left[-\left(\frac{\lambda+2\mu}{r}\right)(x-z)\right]}\widetilde{I}_{n+1}^{2,1}(\alpha(x-z))F_0(z)dz \\
&+ \frac{\lambda}{r}\left(1-\frac{r_0}{r}\right)\beta^{2n-1}\int_0^x e^{\left[-\left(\frac{\lambda+2\mu}{r}\right)(x-z)\right]}\widetilde{I}_{n-1}^{2,2}(\alpha(x-z))F_0(z)dz \\
&- \frac{\lambda}{r}\left(1-\frac{r_0}{r}\right)\beta^{2n}\int_0^x e^{\left[-\left(\frac{\lambda+2\mu}{r}\right)(x-z)\right]}\widetilde{I}_n^{2,2}(\alpha(x-z))F_0(z)dz \quad n \geq 1.
\end{aligned} \tag{14}
$$

Thus, $F_{n,1}(x)$ and $F_{n,2}(x)$ for all possible values of n are expressed in terms of $F_0(x)$. It still remains to determine $F_0(x)$. Toward this end, let $\hat{G}(y,s)$ and $\hat{F}_0(s)$

denote the Laplace transform of $G(y,x)$ and $F_0(x)$ respectively. Taking Laplace transform of Eq. (8) and after simplification, we get

$$\hat{G}(y,s) = \frac{\left\{\frac{ar_0}{r} + \frac{1}{r}\left[\frac{2\mu r_0}{r}\left(1-\frac{1}{y}\right) + \lambda\left(1-\frac{r_0}{r}\right)(y^2-1)\right]\hat{F}_0(s)\right\}ry}{-\lambda y^3 + (\lambda+2\mu+rs)y - 2\mu}. \tag{15}$$

Observe that the denominator of Eq. (15) is a cubic polynomial in y and hence it will have three zeros. However, by Rouche's theorem only one zero lies within the unit circle (say $y_0(s)$). It can be readily seen that

$$y_0(s) = \left(\sqrt{\frac{\mu^2}{\lambda^2} - \frac{(\lambda+2\mu+rs)^3}{27\lambda^3}} - \frac{\mu}{\lambda}\right)^{(1/3)} + \frac{(\lambda+2\mu+rs)}{3\lambda\left(\sqrt{\frac{\mu^2}{\lambda^2} - \frac{(\lambda+2\mu+rs)^3}{27\lambda^3}} - \frac{\mu}{\lambda}\right)^{(1/3)}}.$$

The above root also admits a series expansion of the form

$$y_0(s) = \frac{2\mu}{\lambda+2\mu+rs} + \sum_{n=1}^{\infty} \frac{\lambda^n}{(2\mu)^n n!}\frac{(3n)!}{(2n+1)!}\left(\frac{2\mu}{\lambda+2\mu+rs}\right)^{3n+1},$$

which on inversion yields

$$y_0(x) = \frac{2\mu}{r}e^{-\left(\frac{\lambda+2\mu}{r}\right)x} + \sum_{n=1}^{\infty}\frac{\lambda^n}{(2\mu)^n n!}\frac{(3n)!}{(2n+1)!}\left(\frac{2\mu}{r}\right)^{3n+1}\frac{x^{3n}}{(3n)!}e^{-\left(\frac{\lambda+2\mu}{r}\right)x}. \tag{16}$$

Note that $y_0(s)$ should also satisfy the numerator of $\hat{G}(y,s)$ and hence

$$\hat{F}_0(s) = \frac{\frac{ar_0}{r}}{\frac{1}{r}\left[\frac{2\mu r_0}{r}\left(\frac{1}{y_0(s)} - 1\right) + \lambda\left(\frac{r_0}{r} - 1\right)(y_0(s)^2 - 1)\right]}. \tag{17}$$

After certain simplification, Eq. (17) leads to

$$\hat{F}_0(s) = \frac{ar}{2\mu}\sum_{m=0}^{\infty}\sum_{n=0}^{m}[y_0(s)]^{n+1}[\delta(s)]^{m-n}, \tag{18}$$

with

$$\delta(s) = \frac{(r_0 - r)\lambda}{2\mu r_0}y_0(s)[1+y_0(s)].$$

Inversion of Eq. (18) yields

$$F_0(x) = \frac{ar}{2\mu}\left(\sum_{m=0}^{\infty}\sum_{n=0}^{m}[y_0(x)]^{*(n+1)} * [\delta(x)]^{*(m-n)}\right), \tag{19}$$

where $y_0(x)$ is given by Eq. (16) and

$$\delta(x) = \frac{(r_0 - r)\lambda}{2\mu r_0}\left(y_0(x) + [y_0(x)]^{*2}\right).$$

Here $*n$ denotes the n-fold convolution. Hence all the joint steady state probabilities are obtained explicitly in terms of new generalization of the modified Bessel function of the second kind.

Remark The stationary buffer content distribution of the fluid model under consideration is given by

$$F(x) = \lim_{t\to+\infty} P\{C(t) \le x\} = F_0(x) + \sum_{n=1}^{\infty}\sum_{s=1}^{2} F_{n,s}(x),$$
$$= \frac{ar_0}{r} + \left(1 - \frac{r_0}{r}\right)F_0(x)$$

where $F_0(x)$ is given by the Eq. (19).

4 Numerical Illustrations

This section illustrates the variations of the joint steady state probabilities of the buffer content and the state of the background queueing model for varying values of the parameters.

Figure 2 depicts the variations of $F_0(x)$ against the buffer content, x for $\lambda = 1, \mu = 2, r_0 = -3.5, r = 1$ and for $k = 1, 2$. As discussed earlier, the boundary conditions are given by $F_0(0) = a = 0.3571$. As x tends to infinity, $F_0(x)$ will converge to the corresponding steady state probabilities of the background queueing model, π_0. Therefore $F_0(x)$ starts with the constant a, increases with increase in x and converges to $\pi_0 = 0.5$ as x tends to infinity. Note that the value of π_0 is independent of the constant k.

Figures 3 and 4 depicts the variations of $F_{n,1}(x)$ for $n = 1, 2, 3, 4$ and $F_{n,2}(x)$ for $n = 1, 2, 3$ against the buffer content, x for $\lambda = 1, \mu = 2, k = 2, r_0 = -3.5, r = 1$ and different values of n. As x tends to infinity, $F_{n,1}(x)$ and $F_{n,2}(x)$ will converge to the corresponding steady state probabilities of the background queueing model.

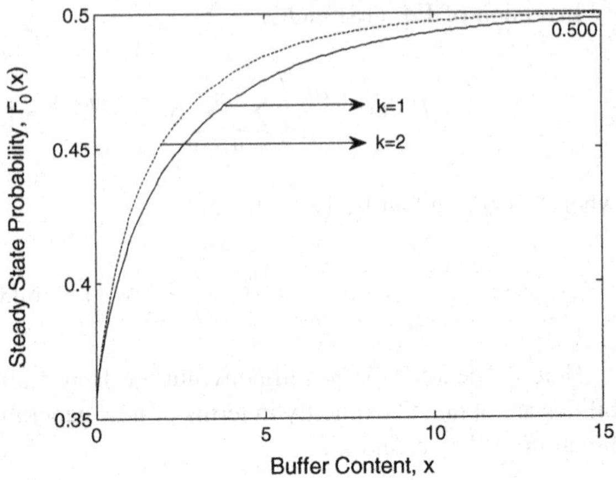

Fig. 2 Variations of $F_0(x)$ against x

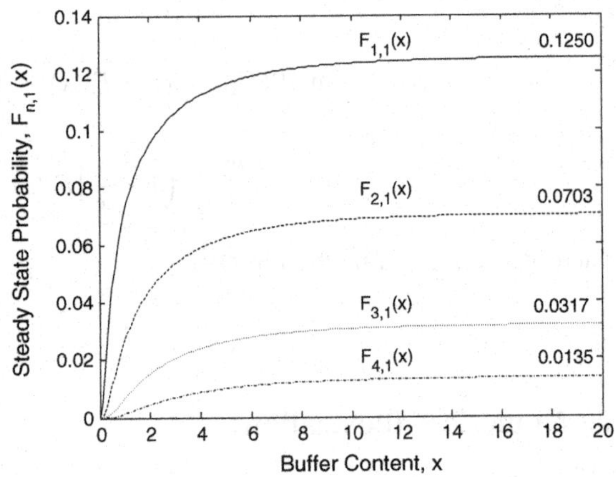

Fig. 3 Variations of $F_{n,1}(x)$ against x

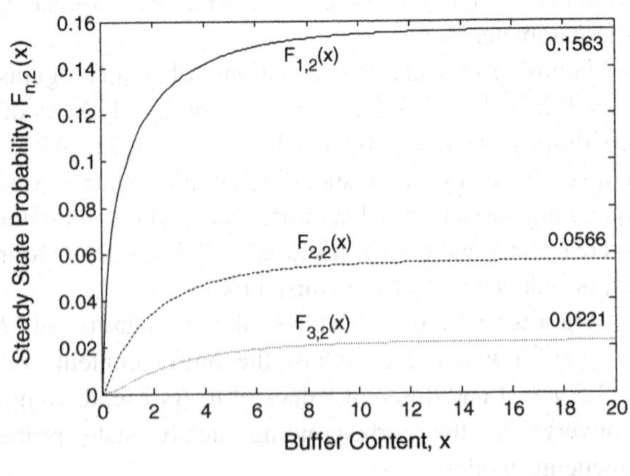

Fig. 4 Variations of $F_{n,2}(x)$ against x

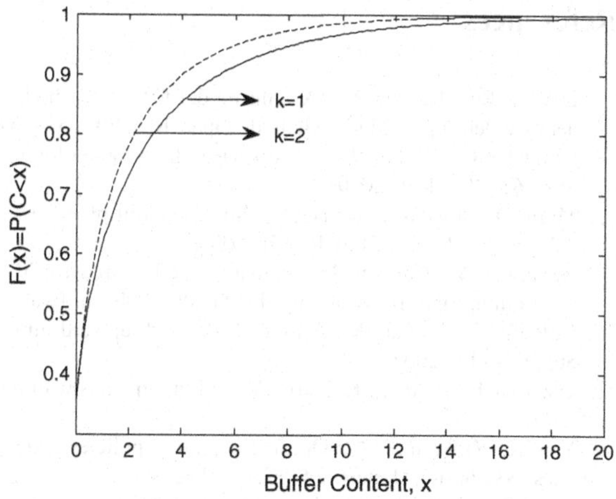

Fig. 5 Variations of $F(x)$ against x

Figure 5 depicts the variations of the buffer content distribution, $F(x)$ against the buffer content, x for $\lambda = 1, \mu = 2, r_0 = -3.5, r = 1$ and different values of k (namely $k = 1$ and 2). It is seen that $F(x)$ increases with increase in the value of the parameter k and converges to 1 as x tends to infinity. Thus all the joint steady state probabilities of the fluid queue driven by an $M/E_2/1$ queue are explicitly obtained under steady state and their corresponding behaviour is illustrated numerically for varying values of the parameters.

5 Conclusion

Markov Modulated Fluid Flows (MMFF) are a class of fluid models wherein the rates at which the content of the fluid varies is modulated by the Markov process evolving in the background. This paper studies a fluid model driven by an $M/E_2/1$ queueing model. The study of such models provide greater flexibility to the design and control of input and output rates of fluid flow thereby adapting the fluid models to wider application background. The governing system of infinite differential difference equations are explicitly solved using Laplace transform and generating function methodology. Closed form analytical solutions helps to gain a deeper insight into the model and other related performance measures. Numerical investigation are also carried out to visualize the behaviour of the joint state probabilities and the buffer content distribution against appropriate parameter values. The results can be extended to study fluid queues modulated by an $M/E_k/1$ queueing model.

References

1. Bekker, R., Mandjes, M.: A fluid model for a relay node in an ad noc network: the case of heavy-tailed input. Math. Methods Oper. Res. **70**, 357–384 (2009)
2. Latouche, G., Taylor, P.G.: A stochastic fluid model for an ad hoc mobile network. Queueing Syst. **63**, 109–129 (2009)
3. Mitra, D.: Stochastic theory of a fluid model of producers and consumers couple by a buffer. Adv. Appl. Probab. **20**, 646–676 (1988)
4. Aggarwal, V., Gautam, N., Kumara, S.R.T., Greaves, M.: Stochastic fluid flow models for determining optimal switching thresholds. Perform. Eval. **59**, 19–46 (2005)
5. Kulkarni, V.G., Yan, K.: A fluid model with upward jumps at the boundary. Queueing Syst. **56**, 103–117 (2007)
6. Tzenova, E., Adan, I., Kulkarni, V.G.: Fluid models with jumps. Stochastic Models. **21**, 37–55 (2005)
7. Yan, K., Kulkarni, V.G.: Optimal inventory policies under stochastic production and demand rates. Stochastic Models. **24**, 173–190 (2008)
8. Adan, I., Resing, J.: Simple analysis of a fluid queue driven by an *M*/*M*/1 queue. Queueing Syst. **22**, 171–174 (1996)
9. Silver Soares, A., Latouche, G.: Matrix analytic methods for fluid queues with finite buffers. Perform. Eval. **63**, 295–314 (2006)
10. Virtamo, J., Norros, I.: Fluid queue driven by an *M*/*M*/1 queue. Queueing Syst. **16**, 373–386 (1994)
11. Mao, B., Wang, F., Tian, N.: Fluid model driven by an *M*/*M*/1 queue with multiple exponential vacation. In: The 2nd International Conference on Advanced Computer Control, vol. 3, pp. 112–115 (2010)
12. Mao, B., Wang, F., Tian, N.: Fluid model driven by an *M*/*M*/1 queue with multiple exponential vacations and N-policy. J. Appl. Math. Comput. **38**, 119–131 (2012)
13. Mao, B., Wang, F., Tian, N.: Fluid model driven by an *M*/*M*/1/*N* queue with single exponential vacation. Int. J. Inf. Manag. Sci. **21**, 29–40 (2010)
14. Mao, B., Wang, F., Tian, N.: Fluid model driven by an *M*/*M*/1/*N* queue with multiple exponential vacations. J. Comput. Inf. Syst. **6**, 1809–1816 (2010)
15. Mao, B., Wang, F., Tian, N.: Fluid model driven by an *M*/*G*/1 queue with multiple exponential vacations. Appl. Math. Comput. **218**, 4041–4048 (2011)
16. Xu, X., Geng, J., Liu, M., Guo, H.: Stationary analysis for fluid model driven by the M/M/c working vacation queue. J. Math. Anal. Appl. **403**, 423–433 (2013)
17. Baek, J.W., Lee, H.W., Lee, S.W., Ahn, S.: A MAP-modulated fluid flow queueing model with multiple vacations. Ann. Oper. Res. **202**, 19–34 (2013)
18. Vijayashree, K.V., Anjuka, A.: Stationary Analysis of an M/M/1 Driven Fluid Queue Subject to Catastrophes and Subsequent Repair. IAENG Int. J. Appl. Math. **43**(4), 238–241 (2013)
19. Ammar, Sherif I.: Analysis of an *M*/*M*/1 driven fluid queue with multiple exponential vacations. Appl. Math. Comput. **227**, 329–334 (2014)
20. Griffiths, J.D., Leonenko, G.M., Williams, J.E.: New generalization of the modified Bessel function and its generating function. Fractional Calculus Appl. Anal. **8**, 267–276 (2005)

An Effective Tool for Optimizing the Number of Test Paths in Data Flow Testing for Anomaly Detection

M. Prabu, D. Narasimhan and S. Raghuram

Abstract Software testing plays an important role in the development process employed in industries. In the field of quality assurance, it is a critical element. Structural oriented testing methods which define test cases on the basis of internal program structure are widely used. We have designed a tool named EFTAD (Effective Tool for Anomaly Detection) based on structural testing, that detects the most effective test paths which actually comprised of data flow anomaly. This tool uses ant colony algorithm for optimizing statically detected paths. This tool minimizes the number of paths to be tested and covers maximum anomalies, thereby ensuring a reliable system. The recursive traversal of the control flow graph along with artificial ants is performed and it provides an efficient set of paths by implementing all def-uses (ADU) strategy of data flow testing technique. The tool also provides a test suite for the evaluation of test results. The test paths are prioritized based on the number of visits and sum value of iteration performed. The unpredictable happenings of the data in a program are noted effectively.

Keywords Structural testing · Data flow anomaly · Ant colony algorithm · Test path · All-Def-Uses

M. Prabu (✉) · D. Narasimhan
Srinivasa Ramanujan Centre, Sastra University, Kumbakonam, Tamil Nadu, India
e-mail: prabu.m@src.sastra.edu

D. Narasimhan
e-mail: narasimhan@src.sastra.edu

S. Raghuram
Tata Consultancy Services, Hyderabad, India
e-mail: meetraghuram92@gmail.com

M. Senthilkumar et al. (eds.), *Computational Intelligence, Cyber Security and Computational Models*, Advances in Intelligent Systems and Computing 412, DOI 10.1007/978-981-10-0251-9_47

1 Introduction

Software testing is the process of covering the undiscovered errors (i.e.) purposely executing a program to find undiscovered errors [1]. As a tester we will gain confidence about the correctness of a program if the testing shows the absence of errors. Software testing is the crucial phase in the entire development process. Techniques in software testing can be classified into functional and structural testing techniques. Globally, to achieve the desired quality in the product, an organization must perform a series of testing activity over the software. Data flow testing is one of its kinds, which comes under structural testing technique category. Data flow testing is centered on variables (data). Software developed today consists of programs and data. Most Programming dialects standards use the idea of variables. Various variables can be utilized together to figure out the estimation of the value of other variables and variables used can get their values from different sources for example, such as input via a keyboard. This expanded level of intricacy can bring about slips inside program references that may be pointed to variables that were not exist or the value of variables may be changed in a surprising and undesired way. The idea of Data Flow testing permits the analyzer to inspect variables all through the system. Most programs developed today will handle enormous amount of data; hence there is a need for the tester to examine the variables in the programs throughout its execution. The concept of data flow testing allows us to ensure that no errors occur within programs that are related to variables such as variables values get changed in an unexpected manner (or) references that may be pointed to variables that were not exist [2]. The complexity of the data flow testing constantly increases with the size of the system and its variables.

Data flow testing is a technique, which comprised of family of strategies that aims to execute sub paths from a point at which variable is defined to a point at which it is referenced. These paths are called as definition-use pairs. Simply, it is called as du-paths or du-chains or du-pairs. Data flow anomalies in the software can be found by data flow testing. Static analysis of the program code will catch most of the anomalies. Dynamic analysis of source code requires execution of different paths to satisfy coverage criteria in data flow testing. Without automation, there is no possibility of executing them at all. Test case design by means of an automated manner is the most challenging task in performing structural testing of software. The automated tool developed by us will generate test paths that can be used for performing data flow testing.

2 Background

2.1 *Ant Colony Optimization*

Several researchers have put enormous amount of effort for using ACO in solving many software testing problems. Ant Colony Optimization [3, 4] is a Meta heuristic approach based on AI technique for solving computational problems of finding optimal paths in the field of software testing. We developed this tool by means of using an algorithm based on ACO technique to generate the optimal paths. Control flow graphs will be used as an input for this algorithm to generate optimal path. The goal of testers is to test all possible paths generated from control flow graphs, but Exhaustive testing is not possible as per Software Engineering principles [5, 6]. The problem here is to find and select optimal paths from the set of possible paths. In solving these kinds of issues, Ant's Behavior described in ACO technique will be very useful. ACO is a Meta heuristic approach which makes of population based searching technique for solving a wide variety of hard combinatorial optimization problems. A concept, known as Stigmergy is applied in ACO which is actually used by ants to communicate with each other through interaction with the environment (i.e.) an indirect communication takes place by placing pheromone trails on the ground and thereby helping other ants to make decisions.

2.2 *Related Work*

Ant Colony Optimization has been widely connected in the field of testing subsequent to 2003 [7]. Doerner and Gutjahr [7] derived test paths for a product framework utilizing a methodology includes ACO and Markov Software utilization model. McMinn and Holcombe [8] applied ACO for the test data generation in evolutionary testing. Li and Lam [9, 10] proposed data generation technique using ACO for performing state-based testing. Bouchachia [11] applied genetic algorithm for the generation of software test data based on condition coverage. Srivastava et al. [12–15] presented an approach for path generation based on control flow graph and test data generation technique using ACO. Suri et al. [16] presented a study on applying ACO to solve the various problems in software testing. References [17–19] used the basic ACO algorithm to generate test cases based on specific coverage criterion. But, detailed experimental and comparative analysis have not been performed in those studies. Srivastava et al. [18, 19] produced the sequences of statement nodes, though they did not focus on real input data generation. Srikanth et al. [20] performed test case optimization using ant bee algorithm and compared ACO with his proposed method. Madhumitha et al. [21] focused on achieving path coverage using genetic algorithm by means of finding feasible paths. Jakkrit et al. [22]

achieved critical paths in performing basis path testing using genetic algorithms. Applying ACO in various software testing process steps made a significant contribution in improving and performing software development activities effectively. Though the concept of applying ACO is a recent field, there is an inference that the large number of publications published and experimentation performed since 2003. But, lot of research and experimentation yet to be achieved in this field.

3 Proposed Approach

Even though most of the industries spend large financial capital on testing, it is the fact that they fall in shortage of time while testing all possible paths and large number of test cases to be executed in it. Our work tries to minimize the total number of paths to be tested, by prioritizing the test paths that which allow the tester to finish the testing activity, far before the stipulated time. We attempted to apply Ant colony optimization (ACO) to solve this issue, especially minimizing the amount of effort needed for testing. The proposed methodology makes use of ACO technique for the generation of optimal paths from the path sets (i.e.) based on covering def-use or du pairs in the program under test. Figure 1 shows the various architectural elements of our tool named "ETFAD".

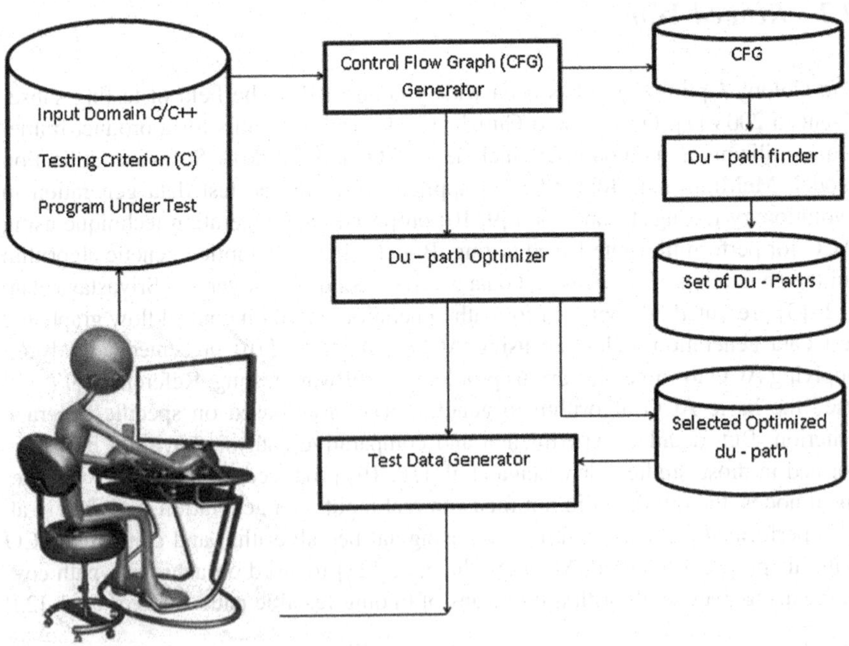

Fig. 1 Architecture diagram of EFTAD

The Input program will get processed to generate the Control Flow Graph. The set of variables involved and decision making nodes in the program are monitored. The appearances of the variables in various parts of the program in combination with its appearance type are processed. The processed data along with the total number of paths are sent for the ant colony optimization technique. The prioritizations of the paths to be followed in order to test are provided. These paths according to the availability of time and work force in the organization can be tested in order. The intermediate data like CFG and variables usage are shown for the clear understanding of the tester. This tool "ETFAD" as indicated in Fig. 1 consists of the following functionalities, the control flow graph generator, DU-path optimizer and test data generator.

3.1 Control Flow Graph Generator

It comprised of 3 sub functions:

3.1.1 Preliminary Input Program Verification

The function checks for the input program to be in proper order. It considers the basic steps for structuring the program like balancing the braces, flower brackets and termination of statements.

3.1.2 CFG Generation

The function forms nodes, with a single statement or set of statements. It links the nodes according to the control flow of the program. The entire data of nodes and links is stored as a matrix in digital form. This matrix when accessed provides the nodes and their statements along with links.

3.1.3 CFG Viewer

The function displays the generated Control Flow Graph using applet viewer.

3.2 DU-Path Optimizer

To find all the paths that have the definition of every variable, to the usage of that variable in the control flow graph, ant colony algorithm is used here to optimize the set of paths found. From the starting node to definition node and usage node to end node, paths are selected randomly. In the selected def-use paths, artificial ants are introduced. These artificial ants one after another iterate the paths, assign and alter the decision making variables for identifying the prioritized paths by means of computing feasible path Set, pheromone trace Set, heuristic Set, visited status set. Each variable has a distinguished purpose for the ant and the path to alter. Here is the generic ant colony algorithm used for optimizing the set of paths after the CFG generation.

3.2.1 A Generic Ant Colony Algorithm

Step 1: Initialization:
Initialize the pheromone trail.
Step 2: Iteration:
For each Ant Repeat

2.1. Solution construction using the current pheromone trail
2.2. Evaluate the solution constructed
2.3. Update the pheromone trail until stopping criteria

We used ant colony algorithm for generating optimized paths and those paths may be used in terms of priority for generating test cases and is as follows.

3.2.2 Ant Colony Algorithm for Path Cover Generation

For each ant "ANT":

Step 1: Follow Step 2 to Step 4 for every du pair

1.1. Select DU: select uncovered def-use pair
1.2. Set start and end node: s_node = def node; e_node = use node;

Step 2: Initialize all parameter
For every branch in CFG

2.1. Set heuristic Value (η): η = 2
2.2. Set pheromone level (τ): τ = 1
2.3. Set visited status (Vs): Vs = 0
2.4. Set Probability level (L): L = 0
2.5. Set α = 1, and β = 1 (desirability and visibility)
2.6. Set count: count = cc (cyclomatic complexity)
2.7. Set key: key = e _node

Step 3: Repeat while count > 0
While (count > 0)

3.1. Perform Evaluation at node "p"
3.2. Initialize: start = p, sum = 0, visit = 0
3.3. Update the track: Update the visited status for the current node "p".
3.4. Evaluate Feasible Set: Evaluate the entire possible path (from current node "p" to all the neighboring nodes).
3.5. Sense the trace: Evaluate the probability (from current node "p" to all non-zero connections in the F(ANT)).
3.6. Move to next node: Select paths (p → q) with maximum probability (Lpq).
3.7. Update the parameter: Update Pheromone: (τpq) = (τpq) α + (ηpq)–β
3.8. Update Heuristic: ηpq = 2*(ηpq)
3.9. Calculate Strength: sum = sum + τpq
strength [count] = sum.
start = next_node
If (start! = end_node): then go to "Step 3.4"
Else
if (visit! = 0)
then discard the path (the redundant path)
3.10. Update count: count = count-1

Step 4: Complete the generated path

4.1. Select path randomly

(i) from the start node to def node.
(ii) from the use node to end node.

4.2. Select another uncovered def-use pair and go to Step 1.

3.3 Test Data Generator

This module generates the test data for a particular efficient test path generated and optimized in the DU-path optimizer module. The Ant Colony Optimization is used for the generation of the test data. Artificial Ants are put at start node of searching model. Ant traverses through the model by visiting several nodes and records the quantity of nodes. If end node does not reached by ant, ant is again set to move to the previous position. If end node is found, recording the path is done and generation of the corresponding data is performed. This allows the tester to generate data and execute the program under test and monitor by recording the execution path.

4 Tool Implementation

The major packages involved in the implementation of the tool are as shown in the Fig. 2 and it clearly shows functionality of the classes involved.

The class Starter in the GUI package contains the interface which includes a File input dialog box which accepts a C/C++ program and provides a button to start generating CFG. The Iface class along with the InfoCarrier class displays the constructed CFG in the CFG viewer interface with the help of java swing package. The DUFinder and Optimizer classes are used to find all the paths that have the definition of every variable to the usage of that variable in the control flow graph and set of paths elicitated was then optimized using Ant Colony Algorithm. Test data generator package is used to generate the test data for a particular test path generated and optimized in the DU-Path optimizer module.

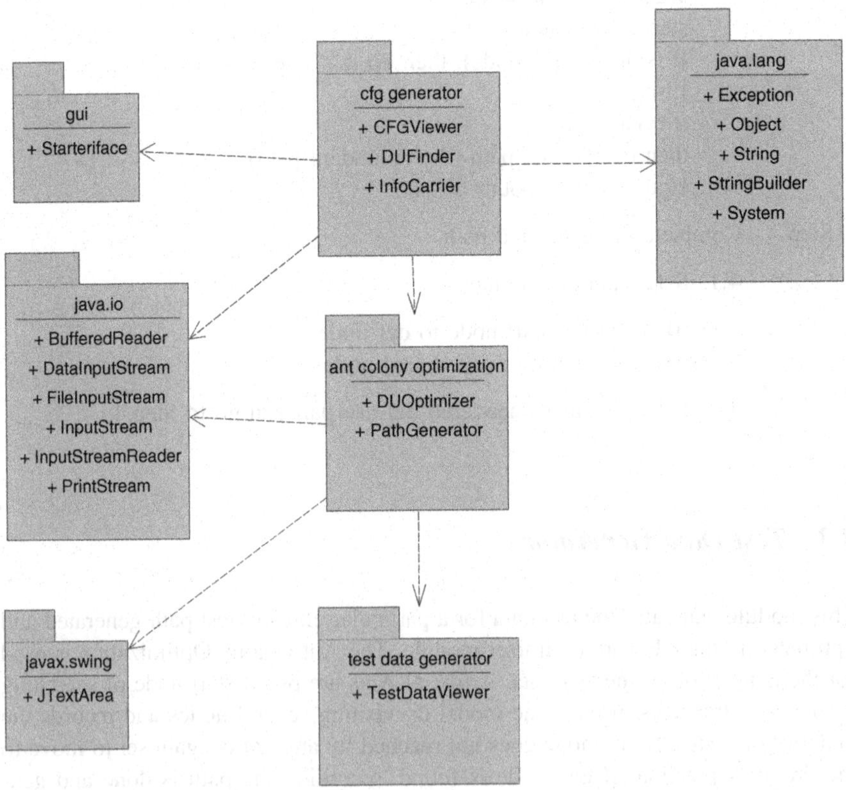

Fig. 2 Package diagram of EFTAD

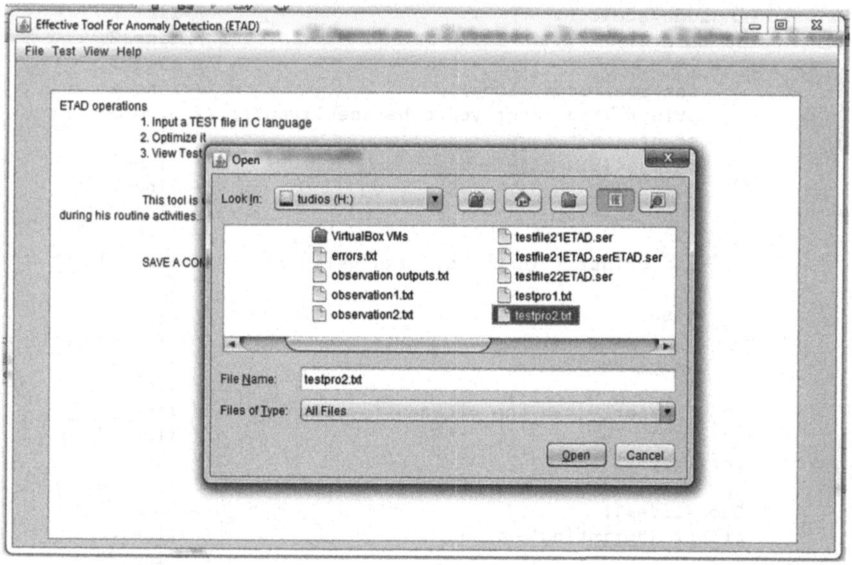

Fig. 3 The tool EFTAD

5 Case Study

Several experiments are conducted using our proposed solution by means of applying our prototyping tool EFTAD on procedural programming unit code level. Figure 3 shows the screenshot of our proposed work and implemented tool EFTAD.

We conducted experiment and compared result with various other approaches using All-Def uses criteria. Although the example we have taken is very simple but there is strong evidence from this one, that this approach will show better result for other complex examples as well. While conducting those experiments, we found that our approach showed better results as to random testing as well as with other genetic algorithm based data flow testing approaches also.

In the input program, the statements are numbered (even blank spaces will have statement number) as indicated in Fig. 4 and control flow graph is generated. We are showing the information about the control flow graph in the form of intermediate data. Then, using those data, paths are generated in the order of decreasing priority (more important to less). The test data are generated for those optimized paths.

The control determiners for the above source code are as follows:

The control flow graph in the form of intermediate data is shown in Fig. 5. Ant colony optimization algorithm applied on the control flow graph which was indicated in the form of intermediate data in Fig. 5 and set of optimized paths was

```
#include<stdio.h>                                          //line 1
int main ()                                                //line 2
{                                                          // line 3
     int age, loop, live;                                  //line 4
     printf ("\n Enter years he/she lived"); // line 5
     scanf ("%d",&age);                                    // line 6
    if (age<18)                                            // line 7
       {                                                   // line 8
         age=45;                                           // line 9
                                                           // line 10
        printf ("\n the citizen valid");                   // line 11
        }                                                  // line 12
     else                                                  // line 13
        {                                                  // line 14
        live=90;                                           // line 15
                                                           // line 16
      printf ("\n the citizen invalid");                   // line 17
      }                                                    // line 18
 for(loop>0)                                               // line 19
 {                                                         // line 20
 loop=live+1;                                              // line 21
 printf ("printing data");                                 // line 22
 }                                                         // line 23
 return 0;                                                 // line 24
 }                                                         / line 25
```

Fig. 4 Source code for evaluating EFTAD

Fig. 5 Control flow graph

The control determiners for the above source code are as follows:
7—if—8—13
13—if—14—19
19—for—20—23
--0—jumps->>1
--1—jumps->>2
--2—jumps->>3
--3—jumps->>4
--4—jumps->>5
--5—jumps->>6
--6—jumps->>7
--7—jumps->>8 on condition false 13
--8—jumps->>9
--9—jumps->>10
--10—jumps->>11
--11—jumps->>12
--12—jumps->>13
--13—jumps->>14 on condition false 19
--14—jumps->>15
--15—jumps->>16
--16—jumps->>17
--17—jumps->>18
--18—jumps->>19
--19—jumps->>20 on condition false 24
--20—jumps->>21
--21—jumps->>22
--22—jumps->>23
--23—jumps->>24
--24—jumps->>25

Table 1 Optimized paths

Priority	Path_id	Decision_logic	Path_followed
1	1	001	1-2-3-4-5-6-7-13-14-15-16-17-18-19-20-21-22-23-24-25
2	0	000	1-2-3-4-5-6-7-8-9-10-11-12-13-14-15-16-17-18-19-20-21-22-23-24-25
3	5	101	1-2-3-4-5-6-7-13-14-15-16-17-18-19-24-25
4	4	100	1-2-3-4-5-6-7-8-9-10-11-12-13-14-15-16-17-18-19-24-25
5	3	011	1-2-3-4-5-6-7-13-19-20-21-22-23-24-25
6	2	010	1-2-3-4-5-6-7-8-9-10-11-12-13-19-20-21-22-23-24-25
7	7	111	1-2-3-4-5-6-7-13-19-24-25
8	6	110	1-2-3-4-5-6-7-8-9-10-11-12-13-19-24-25

generated. Table 1 shows the list of those paths with path id in the order of decreasing priority. The decision logic in the Table 1 indicates the value of the control determiners i.e. predicates truth values. For instance, the decision logic 001 indicates the least significant bit “1” indicates node 7 takes on false side, the intermediate bit “0” indicates node 13 takes on true side and the most significant bit 0 indicates node 19 takes on true side. The nodes 7, 13, and 19 are the control determiners. The test cases can be generated for the optimized paths based on the control determiners. For PATH ID = 1, the test cases might be age = 20 and loop = 1. Similarly the test cases will be generated for all the optimized paths based on the input variable.

6 Experimental Results

We have formulated the following experimental results using our own tool EFTAD by applying proposed technique on unit code level. Sample programs were taken and tested using various testing techniques such as Random testing; Genetic algorithm based testing and using our approach. The results of our technique as shown in Table 4 are arrived and on comparing with other techniques such as random testing and other genetic algorithm which has been shown in Tables 2 and 3, the dataset indicates our approach works well with some programs that we have taken under test. The approach works well when the number of variables is less in number in the program under test (Table 4).

Table 2 Effectiveness of RANDOM testing techniques

Program	Variables used	Random testing	
		Iterations	Coverage (%)
1	2	43	100
2	3	71	100
3	5	121	100

Table 3 Effectiveness of other GA concepts in testing

Program	Variables used	Genetic algorithms	
		Iterations	Coverage (%)
1	2	12	100
2	3	21	100
3	5	51	83

Table 4 Effectiveness of EFTAD

Program	Variables used	Eftad	
		Iterations	Coverage (%)
1	2	4	100
2	**3**	**6**	**100**
3	5	18	100

This approach is effective in terms of reducing the test effort but its efficiency to be determined in our future work. Many people criticized the efficiency of ACO techniques by saying that it will waste computing resources when the program is complex. This approach need to be evaluated with data intensive web or stand alone applications and object oriented systems in future.

7 Conclusion and Future Enhancement

The proposed idea gave methods for effectively reducing the number of paths to be tested. The prioritization of the paths, help the tester to choose the most effective paths that are to be tested first. It provides information that is required to analyze the program data like Control Flow of the program and detailed view of variables and predicative statements. This application as of now is implemented only for finding set of optimized du-paths. This can also be extended to optimize the test data that must be generated to frame and test meaningful test cases. The components build in this tool are easily adaptable and used for extending the tool that provides optimized test data for efficient testing activity. The interface to the tester can be improved by visualizing the Control Flow Graph graphically.. While hovering over the node, the code snippet that the node holds can be made visible. The optimization specified can further be improved by implementing new conceptual functions in Ant Colony. Fine tuning of the algorithm to perform better on finding optimized DU path can be done. During development, we faced lack of existing components as a resource that can be used in our tool. So we made this tool as a project that can be enhanced further and made as a standalone component and released for the reducing the time that took for development of all the basic functions and classes. Conversion of the tool as a plug in and inserting it into development environments after optimizing further provides easier testing functionalities. A separate test sequence can be optioned in the IDEs, making them efficient and effective to different testing strategies. The approach need to be evaluated with the programs where there are large numbers of variables involved. In our future work, this approach can be evaluated using object oriented system.

References

1. Beizer, B.: S/w Testing Techniques. International Thomson Computer press, Boston (1990)
2. Copeland, L.: A Practitioners Guide to Software Test Design. STQE publishing, Massachusetts (2004)
3. Dorigo, M., Stutzle, T.: Ant Colony Optimization. Phi publishers, New York (2005)
4. Ayari, K., Boukitf, S., Antonial, G.: Automatic Mutation Test Input Data Generation via Ant Colony. In: Genetic and Evolutionary Computational Conference, London, pp. 1074–1081 (2007)
5. Mathur, A.P.: Foundation of Software Testing, First Edition. Pearson Education (2007)
6. Sommerville, I.: Software Engineering, Eighth Edition. Pearson Education (2009)
7. Doerner, K., Gutjahr, W.J.: Extracting test sequences from a markov software usage model by ACO. In: Proceedings of GECCO 2003. LNCS, vol. 2724, pp. 2465–2476. Springer-Verlag Berlin Heidelberg (2003)
8. McMinn, F.P., Holcombe, M.: The state problem for evolutionary testing. In: Proceedings of GECCO3. LNCS, vol. 2724, pp. 2488–2500. Springer Verlag (2003)
9. Li, H., Lam, C.P.: Software test data generation using ant colony optimization. In: Transactions on Engineering, Computing and Technology (2005)
10. Li, H., Lam, C.P.: An ant colony optimization approach to test sequence generation for state based software testing. In: Proceedings of the Fifth International Conference on Quality Software, pp. 255–264 (2005)
11. Bouchachia, A.: An immune genetic algorithm for software test data generation. In: Proceedings of 7th International Conference on Hybrid Intelligent Systems (HIS'07), pp. 84–89. IEEE Press (2007)
12. Srivastava, P.R., Baby, K., Raghurama, G.: An approach of optimal path generation using ant colony optimization. In: Proceedings of TENCON, IEEE Press (2009)
13. Srivastava, P.R., Jose, N., Barade, S., Ghosh, D.: Optimized test sequence generation from usage models using ant colony optimization. Int. J. Softw. Eng. Appl. (IJSEA) **2**, 14–28 (2010)
14. Srivastava, P.R., Baby, K.: Automated software testing using metaheuristic technique based on an ant colony optimization. In: International Symposium on Electronic System Design (ISED), pp. 235–240. Bhubaneswar, India (2010)
15. Srivastava, P.R.: Structured testing using ant colony optimization. In: Proceedings of the First International Conference on Intelligent Interactive Technologies and Multimedia (IITM'10), pp. 203–207. ACM Press (2010)
16. Suri, B., Singhal, S.: Literature survey of ant colony optimization in software testing. In: Proceedings of the CSI Sixth International Conference on S/w Engineering, (CONSEG'12), pp. 1–7 (2012)
17. Li, K., Zhang, Z., Liu, W.: Automatic test data generation based on ant colony optimization. In: Proceedings of the Fifth International Conference on Natural Computation (ICNC'09), pp. 216–220 (2009)
18. Srivastava, P.R., Rai, V.K.: An ant colony optimization approach to test sequence generation for control flow based software testing. CCIS31 **12**, 345–346 (2009)

19. Sharma, B., Giridhar, I., Taneja, M., Basia, P., Vadla, S., Srivastava, P.R.: Software coverage: A testing approach through ant colony optimization. In: Proceedings of the Second International Conference on Swarm, Evolutionary and Memetic Computing (SEMCCO'11), pp. 618–625 (2011)
20. Srikanth, A., Kulkarni, N.J., Naveen, K.V., Singh, P., Srivastava, P.R.: Test case optimization using artificial bee colony algorithm. In: CCIS, pp. 570–579. Springer-Verlag Berlin Heidelberg (2011)
21. Panda, M., Mohapatra, D.P.: Generating test data for path coverage based testing using genetic algorithms. In: Advances in Intelligent Systems and Computing, pp. 367–379. Springer-Verlag Berlin Heidelberg (2013)
22. Kaewyotha, J., Songpan, W.: Finding the critical path with loop structure for a basis path testing using genetic algorithm. In: Advances in Intelligent Systems and Computing, pp. 41–52. Springer-Verlag Berlin Heidelberg (2015)

Venus Flytrap Optimization

R. Gowri and R. Rathipriya

Abstract In this paper, we intend to devise a Novel Nature-inspired Meta-Heuristic algorithm named Venus Flytrap Optimization (VFO) suitable for solving various optimization problems. This algorithm is based on the rapid closure behavior of the Venus Flytrap (Dionaea Muscipula) leaves. This is due to the continuous stimulation of the trigger hairs by the fast movement of the prey. The empirical study of the proposed algorithm is done using various test functions.

Keywords Algorithm · Venus flytrap optimization · Metaheuristics · Nature-inspired strategy · Venus flytrap model

1 Introduction

In this modern era, there are numerous meta-heuristic algorithms are devised based on the natural systems to solve optimization problems [1]. Though there are many algorithms present, there prevails the necessity of finding the optimal solution to diverse problems in different types of domains. The optimization algorithms like Ant Colony optimization is based on the stigmergy of the ants, Particle Swarm

---

R. Gowri (✉) · R. Rathipriya
Department of Computer Science, Periyar University, Salem, Tamil Nadu, India
e-mail: gowri.candy@gmail.com

R. Rathipriya
e-mail: rathi_priyar@periyaruniversity.ac.in

M. Senthilkumar et al. (eds.), *Computational Intelligence, Cyber Security and Computational Models*, Advances in Intelligent Systems and Computing 412, DOI 10.1007/978-981-10-0251-9_48

Optimization [2] is inspired based on the flocking behavior of the birds, Firefly optimization [3] is devised based on the mating of fireflies, Cuckoo search [4] is based on the searching behavior of the cuckoo to find optimal nest for laying eggs, and so on. All the modern Meta-Heuristic nature-inspired algorithms [5] are modeled based on the peculiar group behavior or feature of the nature mainly the biological systems.

The Venus Flytrap Optimization is designed based on the prey capturing nature by immediate leaf closure action of the Venus Flytrap plant [6, 7]. From the prey hunting characteristics of the Venus Flytrap plant, a new VFO algorithm is formulated, followed by its implementation. Finally the empirical study of this optimization technique is performed by applying various test functions [8, 9] like De Jong, Rastrigin, Rosenbrock, Griewangk, Schwefel, Ackley, Michalewicz functions.

2 Venus Flytrap Optimization (VFO)

2.1 *Venus Flytrap Behavior*

The botanical name of Venus Flytrap is Dionaea Muscipula shown in Fig. 1a, b [6, 7]. The great scientist Darwin quoted this plant as "one of the most wonderful in the world" [6]. This algorithm is devised on the rapid closure action of its leaves. The closure of Venus Flytrap leaf is about 100 ms one of the fastest movements in the plant kingdom [6]. This trap closure is due to the stimulation of the trigger hairs that present in the two lobes of the leaf by the movements of prey. It is an insectivorous plant [6].

There are various phases in trapping of Venus Flytrap. They are Initial phase, Tightening Phase, Sealing Phase, Re-opening Phase as in Fig. 1d–g [10–13]. There are six triggers present on the surface of two lobes (three in each lobe) shown in Fig. 1c. These trigger hairs are possible indicators of the presence of prey in the trap. The prey may be any small insects or large insects or even small frogs. Whenever the prey moves on the surface of the lobes, may tend to stimulate the trigger hair. If two trigger hairs are stimulated or same trigger hair is stimulated twice within 30 s [10], then it initiates the rapid closure of the trap. The prey will be caught inside the trap. The continuous movement of prey inside the trap causes the tightening of the trap. The trap is then sealed to digest the prey. It will take 5–7 days to digest the prey [11–13]. The trap will be reopened after the digestion is

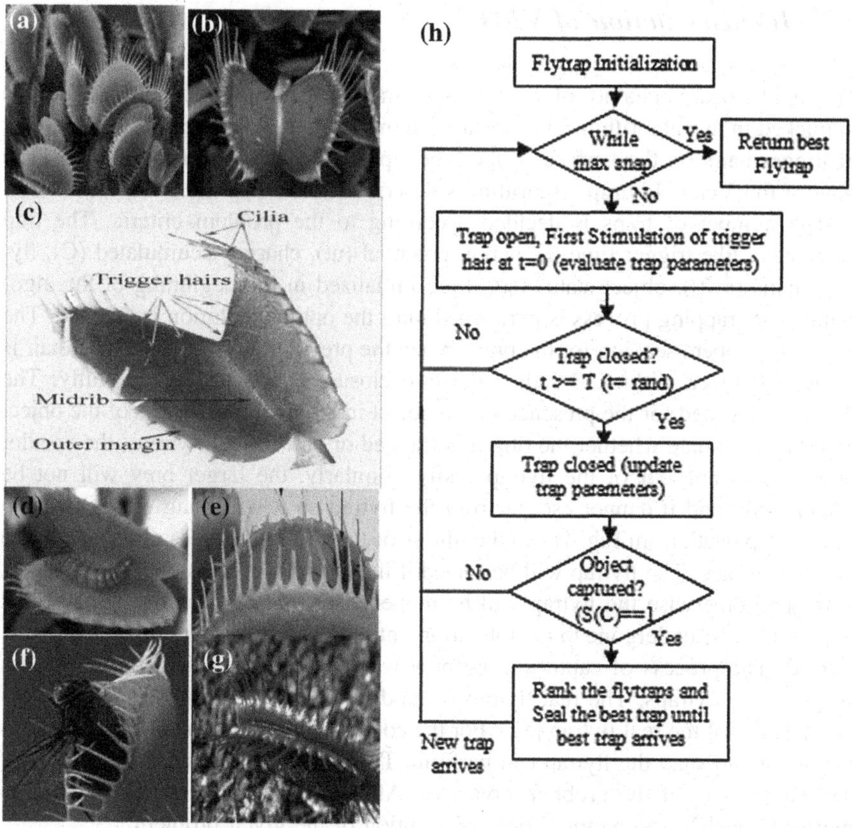

Fig. 1 **a** Venus flytrap (Dionaea Muscipula) plant with many flytraps (leaves), **b** inner view of a flytrap leaf shows trigger hairs on the midrib, **c** parts of Venus flytrap leaf, **d** a worm in Venus flytrap bending trigger hairs, **e** semi closed Venus flytrap with trapped fly using fringed hairs of blade, **f** flytrap with meal hanging out, **g** sealed Venus flytrap in prey digestion, **h** steps involved in VFO

completed. If the prey escapes from the trap then there will be no movements inside the trap so there is no sealing phase and the trap will be opened within 12 h, even if the trap closure is initiated by rain droplets or wind blasts, then there also be no sealing phase and the trap will be reopens.

2.2 Implementation of VFO

The rapid closure behavior of the Venus Flytrap leaves (trap) to capture the prey is mimicked in this algorithm. The closure action is triggered through the stimulations of trigger hairs on the surface of the lobes, performed by the prey (objects). The flow of the Venus Flytrap Algorithm is described in the Fig. 1h.

The number of traps is decided according to the problem criteria. The trap parameters like trigger time (t), action potential (ut), charge accumulated (C), flytrap status (δ(ft)), object status (s(C)) are initialized in the beginning of the algorithm. The trapping process is performed until the optimal solution is attained. The trap is kept open, seeking for the prey. When the prey has arrived, the trigger hair is stimulated twice within 30 s then the trap closure is initiated successfully. The flytrap is checked for the presence of the object in the trap. The fitness of the object is used to evaluate whether the object is trapped or not. In real flytraps, the smaller prey will escape out of the flytrap easily. Similarly, the larger prey will not be caught fully and it cannot escape from the flytrap so it will hang out of the trap causes trap death as in Fig. 1f. So the fitness of the object is necessary to decide the object presence. The flytrap will be sealed if the trap is closed as well as the object is trapped otherwise the flytrap will be reopened for the next capture. The sealed trap will not be undergone to next iteration until better flytrap than the sealed trap is arrived. The process of capturing the prey will be performed until the maximum snaps of the flytraps. The real flytrap is good for 4–6 catches, so this is the maximum snaps of the real flytrap [12]. For the computational flytraps, the max_snap is the maximum snap the flytrap can perform. The max_snap is fixed on the basis of the requirement of the problem objective. After the max_snap the best flytrap is returned which is the required optimal solution of the given problem.

2.3 Parameter Setup

The stimulation of trigger hairs generates the action potential [7, 10] required for the closure of the leaf. The potential generated is dissipated at a particular rate and reaches to zero. The action potential jumped to 0.15 V at 0.001 s and rapidly dissipated to zero after 0.003 s [12]. By recent measurements the action potential ut evoked by the bending of trigger hairs is described using exponential function is given in the Eq. (1).

$$u_t = \begin{cases} 0.15e^{-2000t}, & t \geq 0 \\ 0, & t < 0 \end{cases} \tag{1}$$

where, t is the trigger time point at which the trigger hairs are stimulated, $t < 0$, before the first stimulation, $t = 0$, for the first stimulation and $t > 0$, for the second stimulation. The time between the two successive stimulations will be less than 30 s in order to initiate the flytrap closure. So the time before first stimulation is taken as less than zero, the time at the first stimulation is taken as zero and the time at which the trigger is stimulated for the second time is taken as greater than zero. Thus there is no action potential for the flytrap before the first stimulation. This is shown in the Eq. (1). The charge accumulation may relate to the stepwise accumulation of a bioactive substance, resulting in ion channel activation by the action potential [12–14]. The charge accumulation can be described by the following linear dynamic system

$$\mathrm{C} = -\mathrm{k_c C} + \mathrm{k_a u_t} \tag{2}$$

where, C is the charge accumulated by the lobes for trap shutter, kc is rate of dissipation of charge, ka is rate of accumulation of charge. Two stimulations are required for the trap closure, the accumulated charge of the first stimulation is dissipated at the rate k_c and reaches zero after 30 s. So it is implied that the second stimulation will be occurred within 30 s to attain the maximum charge of 14 μC to shut the flytrap [12]. The k_a is the rate of accumulation of the charge, i.e., the rate at which the charge is builds up based on the action potential attained in the stimulation. For the first stimulation the dissipation part will be zero because the initial charge of the flytrap before first stimulation is zero, it will work for the second stimulation to sum up the previously gained charge with it. This is how the charge is calculated in the real flytraps. In these computational flytraps, the charge is calculated based on the objective function f(x). The next parameter is the flytrap status (δ), used to know the current status of the flytraps. The status of the flytrap will be either 0 or 1 or 2 (open or close or seal).

$$\delta(\mathrm{ft}) = \begin{cases} 2, & S(\mathrm{C}) = 1, \delta(\mathrm{ft}) = 1 \\ 1, & 0 \leq \mathrm{t} \leq \mathrm{T} \\ 0, & \text{otherwise} \end{cases} \tag{3}$$

The flytrap will be initially in the opened the (0) state. If the trap closure is triggered by the prey (object) then the trap will be in closed (1) state. The time point of the first stimulation is taken as zero, here the t represents the second stimulation time. The T is the time threshold between two stimulations (say 30 s). All of the closed traps will not be sealed. The flytrap whose fitness is best will be placed in sealed state. The sealed flytraps will be reopened when the fittest flytrap than it will be achieved.

The various steps involved in the Venus Flytrap Algorithm are shown above. This algorithm has Initial phase, trapping phase and sealing phase. In the initial phase of the algorithm, the objective function of the problem is chosen, the population size (n) is fixed based on the criteria of the problem, the maximum number of snaps (S) each flytrap can perform is chosen, the dimension of the solution space (d) is decided based on the dimension of the input vector. The flytraps in this phase are like new born baby does not perform any trap shutter so far. These traps are ready for capturing preys (objects).

Venus Flytrap Algorithm

```
//Initial phase
Objective function f(x), x = (x1, ..., xd)^S
Initialize a population of flytraps xi (i = 1, 2, ..., n)
While iter< = S
 For i=1 : n all n flytraps                          //trapping phase
        At t=0,                                      //first stimulation
        Evaluate Action Potential ut of flytrap i
        Charge Accumulation C of flytrap i  is determined by f(xi)
        At t=rand()                                  //second stimulation
        if t<=T then
                update ut, C of flytrap i
                evaluate the Object Status
        end if
end for
Rank the flytraps and find the current best      //sealing phase
seal the best flytrap until another best flytrap arrives
end while
Post process results and visualization
```

In the trapping phase, each of the flytraps undergoes two stimulations to trigger flytrap closure. At two time points the trigger hair of each of the flytraps are stimulated, i.e. at time $t = 0$ and at time t between zero and time threshold (T). The first stimulation is at $t = 0$, where the action potential is gained by the flytraps and in turn builds up the charge for snapping [10, 12]. The charge is determined using the objective function (f(x)) of the problem. Then, if the second stimulation occurs within the specified threshold duration, the action potential and charge accumulation will be updated. The status of the object is evaluated, whether the object is present or escaped (1 or 0). The flytrap closure can be performed even by the natural aspects or the object may escape from the trap, so this necessitates the evaluation of the object status. These steps are performed for each flytrap until the maximum number of snaps the flytrap can be performed. In the sealing phase the closed flytraps in each of the snap iteration are ranked based on their fitness. The best flytrap is sealed and will not be iterated further and kept idle until another flytrap with fittest object is arrived.

3 Empirical Study of VFA

3.1 Experimental Setup

There are various benchmark test functions are available to test the performance of optimization algorithms. The novel optimization algorithms can be validated and tested with the help of these test functions. Some of the benchmark test functions [8, 9, 15] used to test this novel Venus Flytrap Algorithm is listed below.

De Jong's first function is essentially a sphere function which is described in $f_0(x)$ as follows.

$$f_0(x) = \sum_{i=1}^{d} x_i^2, \quad x_i \in [-5.12, 5.12]$$

Here d represents dimension of the solution space. The global minima f* = 0 whose x* = (0, 0,…0).

The second test function is Rosenbrock function which is described in the following equation. The d represents dimension of the input vector $x = [x_1, x_2, x_3, \ldots, x_d]$.

$$f_1(x) = \sum_{i=1}^{d-1} \left(100\left(x_{i+1} + x_i^2\right)^2 + (x_i - 1)^2\right), \quad x_i \in [-5.12, 5.12]$$

The third test function is Rastrigin's test function which is described in the following equation. The d represents dimension of the input vector $x = [x_1, x_2, x_3, \ldots, x_d]$.

$$f_2(x) = 10d + \sum_{i=1}^{d} \left[x_i^2 - 10\cos(2\pi x_i)\right], \quad x_i \in [-600, 600]$$

The fourth test function is Griewangk's test function which is described in the following equation. The d represents dimension of the input vector x, x = [x_1, x_2, x_3, …, x_d].

$$f_3(x) = \frac{1}{4000}\sum_{i=1}^{d} x_i^2 - \prod_{i=1}^{d} \cos\left(\frac{x_i}{\sqrt{i}}\right) + 1, \quad x_i \in [-5.12, 5.12]$$

The fifth test function is Schwefel's function which is described in the following equation. The d represents dimension of the input vector x, x = [x_1, x_2, x_3, …, x_d].

$$f_4(x) = \sum_{i=1}^{d} -x_i \sin(\sqrt{x_i}), \quad x_i \in [-500, 500]$$

The sixth test function is Ackley function which is described in the following equation. The d represents dimension of the input vector x, x = [x_1, x_2, x_3, …, x_d].

$$f_5(x) = -20\exp\left[-20\sqrt{\frac{1}{d}\sum_{i=1}^{d} x_i^2}\right] - \exp\left[\frac{1}{d}\sum_{i=1}^{d} \cos 2\pi x_i\right] + (20 + e),$$
$$x_i \in [-32.768, 32.768]$$

The seventh test function is Michalewicz function which is described in the following equation. The d represents dimension of the input vector x, x = [x_1, x_2, x_3, …, x_d].

$$f_6(x) = -\sum_{i=1}^{d} \sin(x_i)\left[\sin\left(\frac{ix_i^2}{\pi}\right)\right]^{2m}, \quad x_i \in [0, \pi]$$

The following Table 1 shows the asymmetric range of values used for initialization of the population for various test functions described above.

Table 1 Asymmetric initialization ranges

Test functions	Asymmetric initialization range
f_0	$(-5.12, 5.12)^d$ (sphere)
f_1	$(-5.12, 5.12)^d$ (Rosenbrock)
f_2	$(-600, 600)^d$ (Rastrigin)
f_3	$(-5.12, 5.12)^d$ (Griewangk)
f_4	$(-500, 500)^d$ (Schwefel)
f_5	$(-32.768, 32.768)^d$ (Ackley)
f_6	$(0, \pi)$ (Michalewicz)

The Tables 2, 3, 4, 5, 6, 7 and 8 shows the experimental results of various test function. The Venus flytrap optimization algorithm is tested with various test functions and their results with four different population sizes such as 20, 40, 80 and 160, three different dimensions taken are 10, 20 and 30, three different maximal snap values used for testing are 1000, 2000 and 3000, and their mean best fitness values are listed. The success rate of all the test functions is 100 %.

From these results it is evidence that the proposed VFO algorithm can perform well in finding the optimal solution at different setup of population size, dimension

Table 2 Mean fitness values for the sphere function

Population size	Dimension	Snaps	Mean best fitness
20	10	1000	0.1437
	20	2000	0.2873
	30	3000	0.4310
40	10	1000	0.1437
	20	2000	0.2873
	30	3000	0.4310
80	10	1000	0.1437
	20	2000	0.2873
	30	3000	0.4310
160	10	1000	0.1437
	20	2000	0.2873
	30	3000	0.4310

Table 3 Mean fitness values for the Rosenbrock function

Population size	Dimension	Snaps	Mean best fitness
20	10	1000	0.2443
	20	2000	0.5156
	30	3000	0.7869
40	10	1000	0.2442
	20	2000	0.5156
	30	3000	0.7869
80	10	1000	0.2443
	20	2000	0.5156
	30	3000	0.7869
160	10	1000	0.2443
	20	2000	0.5156
	30	3000	0.7869

Table 4 Mean fitness values for the Rastrigin's function

Population size	Dimension	Snaps	Mean best fitness
20	10	1000	27.1818
	20	2000	54.3635
	30	3000	81.5452
40	10	1000	27.1818
	20	2000	54.3635
	30	3000	81.5452
80	10	1000	27.1818
	20	2000	54.3635
	30	3000	81.5452
160	10	1000	27.1818
	20	2000	54.3635
	30	3000	81.5452

Table 5 Mean fitness values for the Griewangk's function

Population size	Dimension	Snaps	Mean best fitness
20	10	1000	0.00003650
	20	2000	0.00002027
	30	3000	0.00001440
40	10	1000	0.00003623
	20	2000	0.00002110
	30	3000	0.00001440
80	10	1000	0.00003440
	20	2000	0.00002027
	30	3000	0.00001430
160	10	1000	0.00003643
	20	2000	0.00002120
	30	3000	0.00001447

Table 6 Mean fitness values for the Ackley function

Population size	Dimension	Snaps	Mean best fitness
20	10	1000	2.7183
	20	2000	3.3738
	30	3000	3.2681
40	10	1000	3.1192
	20	2000	3.3738
	30	3000	3.2681
80	10	1000	3.1192
	20	2000	3.3738
	30	3000	3.2681
160	10	1000	3.1192
	20	2000	3.3738
	30	3000	3.2681

Table 7 Mean fitness values for the Michalewicz function

Population size	Dimension	Snaps	Mean best fitness
20	10	1000	−3.6820
	20	2000	−7.3327
	30	3000	−10.9986
40	10	1000	−3.6820
	20	2000	−7.3327
	30	3000	−10.9986
80	10	1000	−3.6820
	20	2000	−7.3327
	30	3000	−10.9986
160	10	1000	−3.6820
	20	2000	−7.3327
	30	3000	−10.9986

and maximum snap values. This algorithm can find optimal solution efficiently under different experimental setup. It will converge to global optimal even with higher population size, maximum snaps and higher dimensional input vector. It is insensitive to the population size, maximum snaps and input dimensionality.

Table 8 Mean fitness values for the Schwefel function

Population size	Dimension	Snaps	Mean best fitness
20	10	1000	−36.1599
	20	2000	−79.6511
	30	3000	−122.6204
40	10	1000	−36.1345
	20	2000	−68.4892
	30	3000	−119.3517
80	10	1000	−35.9879
	20	2000	−69.6511
	30	3000	−121.4789
160	10	1000	−36.5745
	20	2000	−73.9594
	30	3000	−120.4759

4 Conclusion

In this paper, the Venus Flytrap Optimization algorithm has been designed based on the trap closure behavior of the plant Dionaea Muscipula. The implementation of VFO besides the parameter setup has also been discussed in this paper. The Venus flytrap optimization technique has been used to find optimal solution. The test functions were used to show the tendency of Venus Flytrap Optimization algorithm at different setup. This algorithm can perform well even in the higher dimension as well as for maximal iterations. Further performance of this algorithm can be compared with the other optimization algorithms like PSO, Firefly, Cuckoo Search, etc. Further refinements to the various parameters of VFO can also be performed.

References

1. Bonabeau, E., Dorigo, M., Theraulaz, G.: Swarm Intelligence: From Nature to Artificial Systems. Oxford University Press (1999)
2. Kennedy, J., Eberhart, R.: Particle Swarm Optimization. In: Proceedings of IEEE International Conference of Neural Networks, pp. 1942–1948. (1995)
3. Yang, X.-S.: Firefly Algorithm, Stochastic Test Functions and Design Optimisation, pp. 1–12. (2010)
4. Yang, X.-S., Deb, S.: Cuckoo Search via Levy Flights
5. Yang. X.S.: Nature-Inspired Metaheuristic Algorithms. Luniver Press (2008)
6. Darwin, C.: Insectivorous Plants. Murray, London (1875)
7. Hodick, D., Sievers, A.: The action potential of Dionaeamuscipula Ellis. Planta **174**, 8–18 (1988)
8. Chattopadhyay, R.: A study of test functions for optimization algorithms. J. Optim. Theory Appl. **8**, 231–236 (1971)
9. Schoen, F.: A widw class of test functions for global optimization. J. Global Optim. **3**, 133–137 (1993)

10. Volkov, A.G., Adesina, T., Jovanov, E.: Closing of Venus flytrap by electrical stimulation of motor cells. Plant Signal. Behav. **2**, 139–144 (2007)
11. Volkov, A.G., Adesina, T., Jovanov, E.: Closing of Venus flytrap by electrical stimulation of motor cells. Plant Signaling Behav. **2**(3), 139–145 (2007)
12. Yang, R., Lenaghan, S.C., Zhang, M., Xia, L.: A mathematical model on the closing and opening mechanism for Venus flytrap. Plant Signaling Behav. **5**(8), 968–978 (2010)
13. YoelForterre, J.M., Skotheim, J.D., Mahadevan, L.: How the Venus flytrap snaps. Nature **433**, 421–425 (2005)
14. FagerbergWR, H.D.: A quantitative study of tissue dynamics in venus's flytrap Dionaeamuscipula (Droseraceae) II. Trap Reopening. Am. J. Bot. **83**, 836–842 (1996)
15. Shi, Y., Eberhart, R.C.: Empirical Study on Particle Swarm Optimization, pp. 1945–1950. IEEE 1999

Zumkeller Cordial Labeling of Graphs

B.J. Murali, K. Thirusangu and R. Madura Meenakshi

Abstract In this paper we introduce a new graph labeling called Zumkeller cordial labeling of a graph G = (V, E). It can be defined as an injective function f: V → N such that the induced function $f^*: E \rightarrow \{0, 1\}$ defined by $f^*(xy) = f(x)\, f(y)$ is 1 if f(x) f(y) is a Zumkeller number and 0 otherwise with the condition $\left|e_f^*(0) - e_f^*(1)\right| \le 1$. We make use of a technique of generating Zumkeller numbers and the concept of cordiality in the labeling of graphs.

Keywords Graphs · Zumkeller numbers · Zumkeller labeling · Zumkeller cordial labeling

1 Introduction

In the theory of graph labeling, the labels are mathematical objects such as integers, prime numbers, modular integer, elements of a group, etc. The mathematical properties of such objects are used through an evaluating function which assigns a value to a vertex or an edge or a face. The graph labeling has found its origin in the paper [12] by Alex Rosa in the late 1960s. A large number of papers have been devoted to the topic of labeling of graphs, which are updated by Gallian [9]. For all notations and terminologies in graph theory, we follow Harary [10]. Labelled

B.J. Murali (✉)
Research and Development Centre, Bharathiar University, Coimbatore 641 046, Tamil Nadu, India
e-mail: muralibjpgm@gmail.com

K. Thirusangu · R. Madura Meenakshi
Department of Mathematics, SIVET College, Gowrivakkam, Chennai 600 073, Tamil Nadu, India
e-mail: kthirusangu@gmail.com

R. Madura Meenakshi
e-mail: madhurameenakshi@yahoo.co.in

M. Senthilkumar et al. (eds.), *Computational Intelligence, Cyber Security and Computational Models*, Advances in Intelligent Systems and Computing 412, DOI 10.1007/978-981-10-0251-9_49

graphs have wide applications in coding theory, X-ray crystallography, radar, astronomy, circuit design and communication network addressing.

Balamurugan et al. [3] introduced Zumkeller labeling using Zumkeller numbers, which is defined as an injective function $f : V \rightarrow N$ such that the induced function $f^* : E \rightarrow N$ defined by $f^*(xy) = f(x)\, f(y)$ is a Zumkeller number. The concept of Zumkeller labeling, strongly multiplicative Zumkeller labeling and k-Zumkeller labeling of graphs have been introduced and investigated in the literature [1–6]. The concept of cordial labeling was introduced by Cahit [7]. In this paper we introduce the cordiality of edges in the Zumkeller labeling of graphs and call such labeling as Zumkeller cordial labeling of graphs. We discuss the existence of Zumkeller cordial labeling for paths, cycles and star graphs.

2 Preliminaries

In this section we recall the definition of Zumkeller numbers, few properties of Zumkeller numbers [11] and the concept of cordial labeling of a graph [7].

Definition 1 A positive integer n is said to be a Zumkeller number if all the positive factors of n can be partitioned into two disjoint sets so that the sum of the two sets are equal.

We shall call such a partition as Zumkeller partition. For example 20 is a Zumkeller number since we can partition its factors into A = {1, 20} and B = {2, 4, 5, 10} with sum 21. Few Zumkeller numbers are listed below. 6, 12, 20, 24, 28, 30, 40, 42, 48, 54, 56, 60, …, 80, 84, 88, …, 150, 156, 160, …, 220, 222, 224, …, 270, 272, 276, …, 1180, 1182, 1184, 1188, … .

2.1 Properties of Zumkeller Numbers

1. Let the prime factorization of an even Zumkeller number n be $2^k p_1^{k_1} p_2^{k_2} \ldots p_m^{k_m}$. Then at least one of k_i must be an odd number.
2. If n is a Zumkeller number and p is a prime with (n, p) = 1, then np^{ℓ} is a Zumkeller number for any positive integer ℓ.
3. Let n be a Zumkeller number and $p_1^{k_1} p_2^{k_2} \ldots p_m^{k_m}$ be the prime factorization of n. Then for any positive integers $\ell_1, \ell_2, \ldots, \ell_m$, $p_1^{k_1+\ell_1(k_1+1)} p_2^{k_2+\ell_2(k_2+1)} \ldots p_m^{k_m+\ell_m(k_m+1)}$ is a Zumkeller number.
4. For any prime $p \neq 2$ and a positive integer k with $p \leq 2^{k+1}-1$, $2^k p$ is a Zumkeller number.

We use a programming language slightly adopted to C, given by Frank Buss [8] to identify the Zumkeller numbers and Zumkeller partitions.

Definition 2 Let G = (V, E) be a graph. A function $f : V \rightarrow \{0, 1\}$ is said to be a cordial labeling of the graph G if there exists an induced function $f^* : E \rightarrow \{0, 1\}$ such that $f^*(uv) = |f(u) - f(v)|$ satisfies the conditions that the number of zeros and number of ones on the edges differ by at most one and the number of zeros and the number of ones on vertices also differ by at most one. A graph G which admits cordial labeling is called a cordial graph.

3 Main Results

In this section we introduce a new graph labeling called Zumkeller cordial labeling of a graph and we prove the existence of such labeling to path graphs, cycle graphs and star graphs. We adopt the following notations in this paper.

(i) $e_f^*(0)$ is the number of edges of the graph G having label 0 under f^*.
(ii) $e_f^*(1)$ is the number of edges of the graph G having label 1 under f^*.

Definition 3 Let G = (V, E) be a graph. An injective function $f : V \rightarrow N$ is said to be a Zumkeller cordial labeling of the graph G if there exists an induced function $f^* : E \rightarrow \{0, 1\}$ defined by $f^*(xy) = f(x)\, f(y)$ satisfies the following conditions

(i) For every $xy \in E$, $f^*(xy) = \begin{cases} 1 & \text{if } f(x)\, f(y) \text{ is a Zumkeller number} \\ 0 & \text{otherwise.} \end{cases}$

(ii) $\left|e_f^*(0) - e_f^*(1)\right| \leq 1$.

Definition 4 A graph G = (V, E) which admits a Zumkeller cordial labeling is called a Zumkeller cordial graph.

Example 1 A Zumkeller cordial graph is shown in Fig. 1.

Theorem 1 *The path P_n with n vertices admits a Zumkeller cordial labeling.*

Proof Let $V = \{v_i / 1 \leq i \leq n\}$ be the vertex set and $E = \{e_i = v_i\, v_{i+1} / 1 \leq i \leq n - 1\}$ be the edge set of the path P_n.

Define an injective function $f : V \rightarrow N$ such that $f(v_1) = 3$ and for $i \equiv 0 \pmod 2$, $f(v_i) = 2^{\frac{i+2}{2}}$, $f(v_{i+1}) = p_{i+1}$, where p_{i+1} is the smallest prime number such that $p_{i+1} \geq 2^{\frac{i+4}{2}} + 1$ and an induced function $f^* : E \rightarrow \{0, 1\}$ such that

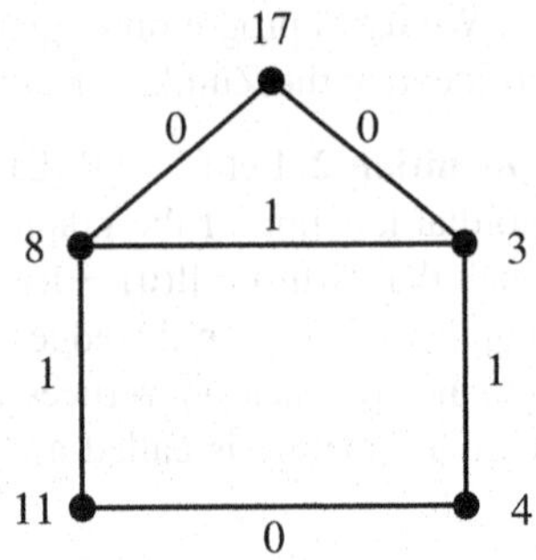

Fig. 1 A Zumkeller cordial graph

$$f^*(e_i) = f^*(v_iv_{i+1}) = \begin{cases} 1 & \text{if } f(v_i)\ f(v_{i+1}) \text{ is a Zumkeller number} \\ 0 & \text{otherwise.} \end{cases}$$

Denote the set of edges whose labels are Zumkeller numbers as E_1 and $E - E_1$ as E_2.

Now we identify the edges which belongs to E_1 and those belongs to E_2 as follows:

(i) $f^*(e_1) = f^*(v_1v_2) = f(v_1)f(v_2) = 3.2^{\frac{2+2}{2}} = 3.2^2 = 12$which is a Zumkeller number and hence $e_1 \in E_1$.

(ii) When $i \equiv 0 \pmod 2$, $f^*(e_i) = f^*(v_iv_{i+1}) = f(v_i)f(v_{i+1}) = 2^{\frac{i+2}{2}}p_{i+1}$ where p_{i+1} is the smallest prime number such that $2^{\frac{i+4}{2}} + 1 \le p_{i+1} \le 2^{\frac{i+6}{2}} - 1$. Therefore $f*(e_i) = 2^{\frac{i+2}{2}}.p_{i+1}$ is not a Zumkeller number since it contradicts the property 4 of Zumkeller numbers.
Hence $e_i \in E_2$ for $i \equiv 0 \pmod 2$.

(iii) When $i \equiv 1 \pmod 2$, $f^*(e_i) = f^*(v_iv_{i+1}) = f(v_i)f(v_{i+1}) = p_i2^{\frac{i+3}{2}}$ where p_i is the smallest prime such that $2^{\frac{i+3}{2}} + 1 \le p_i \le 2^{\frac{i+5}{2}} - 1$
$f^*(e_i) = p_i2^{\frac{i+3}{2}}$, is a Zumkeller number, since it satisfies the property 4 of Zumkeller numbers. Hence $e_i \in E_1$ for $i \equiv 1 \pmod 2$.

Thus we have proved that the labels of the alternate edges from the first edge onwards are Zumkeller numbers.

Now we will prove $\left|e_f^*(0) - e_f^*(1)\right| \le 1$.

Case (i) When $n \equiv 0 \pmod 2$
E_1 has n/2 edges and E_2 has $\frac{n}{2} - 1$ edges. That is $e_f^*(1) = n/2$ and $e_f^*(0) = \frac{n}{2} - 1$.
Therefore $\left|e_f^*(0) - e_f^*(1)\right| \le 1$.

Case (ii) When $n \equiv 1 \pmod 2$
E_1 has $\frac{n-1}{2}$ edges and E_2 has $\frac{n-1}{2}$ edges.

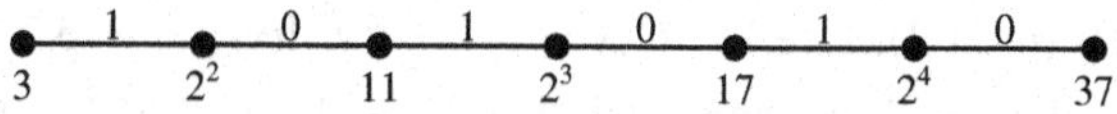

Fig. 2 Zumkeller cordial labeling of P_7

That is

$$e_f^*(1) = \frac{n-1}{2} = \text{and}\, e_f^*(0) = \frac{n-1}{2}.$$

Therefore

$$\left|e_f^*(0) - e_f^*(1)\right| \le 1.$$

Hence the path P_n admits Zumkeller cordial labeling for all n.
The Zumkeller cordial labeling of the path P_7 is shown in the following example.

Example 2 See Fig. 2.

Lemma 1 *The cycle C_n admits Zumkeller cordial labeling when $n \equiv 0 \pmod 2$.*

Proof Let C_n be an even cycle of length n.
Let $V = \{v_i / 1 \le i \le n\}$ be the vertex set and
$E = \{e_i = v_i v_{i+1} / 1 \le i \le n-1\} \cup \{e_n = v_n v_1\}$ be the edge set of the cycle C_n.
Define an injective function $f : V \rightarrow N$ such that
$f(v_1) = 6$, $f(v_2) = 3$ and
for $i \equiv 1 \pmod 2$, $f(v_i) = 2^{\frac{i+1}{2}}$, $f(v_{i+1}) = p_{i+1}$, where p_{i+1} is the smallest prime such that $p_{i+1} \ge 2^{\frac{i+3}{2}} + 1$ and an induced function $f^* : E \rightarrow \{0, 1\}$ such that

$$f^*(e_i) = f^*(v_i v_{i+1}) = \begin{cases} 1 & \text{if } f(v_i)\, f(v_{i+1}) \text{ is a Zumkeller number} \\ 0 & \text{otherwise.} \end{cases}$$

Denote the set of edges whose labels are Zumkeller numbers as E_1 and $E - E_1$ as E_2.

Now we identify the edges which belongs to E_1 and those belongs to E_2 as follows:

(i) $f^*(e_1) = f^*(v_1 v_2) = f(v_1)f(v_2) = 6 \times 3 = 18$which is not a Zumkeller number and hence $e_1 \in E_2$.
(ii) $f^*(e_2) = f^*(v_2 v_3) = f(v_2)f(v_3) = 3.2^{\frac{3+1}{2}} = 3.2^2 = 12$is a Zumkeller number and hence $e_2 \in E_1$.
(iii) When $i \equiv 1 \pmod 2$, $f^*(e_i) = f^*(v_i v_{i+1}) = f(v_i)f(v_{i+1}) = 2^{\frac{i+1}{2}} p_{i+1}$ where p_{i+1} is the smallest prime such that $2^{\frac{i+3}{2}} + 1 \le p_{i+1} \ge 2^{\frac{i+5}{2}} - 1$ $f^*(e_i) = 2^{\frac{i+1}{2}} p_{i+1}$, is not a Zumkeller number and hence $e_i \in E_2$ for $i \equiv 1 \pmod 2$.

(iv) When $i \equiv 0 \pmod 2$, $f^*(e_i) = f^*(v_iv_{i+1}) = f(v_i)f(v_{i+1}) = p_i 2^{\frac{i+2}{2}}$
where p_i is the smallest prime such that $2^{\frac{i+2}{2}} + 1 \le p_i \le 2^{\frac{i+4}{2}} - 1$
$f^*(e_i) = p_i 2^{\frac{i+2}{2}}$, is a Zumkeller number and hence $e_i \in E_1$ for $i \equiv 0 \pmod 2$.

(v) $f^*(e_n) = f^*(v_nv_1) = f(v_n)f(v_1) = p_n 6$where p_n is the smallest prime such that $p_n \ge 2^{\frac{n+2}{2}} + 1$.

From the property 2 of Zumkeller numbers, p_n6 is a Zumkeller number and hence

$e_n \in E_1$. Thus we have proved that the alternate edges starting from first edge onwards are Zumkeller numbers.

Since the cycle C_n has n edges, where n is even, E_1 has n/2 edges and E_2 has n/2 edges, that is $e_f^*(1) = n/2$ and $e_f^*(0) = n/2$.

Therefore $|e_f^*(0) - e_f^*(1)| \le 1$ is satisfied.

Hence the cycle C_n admits Zumkeller cordial labeling when $n \equiv 0 \pmod 2$.

The above lemma is illustrated in the example 3.

Example 3 The Zumkeller cordial labelling of the cycle C_8 is given in Fig. 3.

Lemma 2 *The cycle C_n admits Zumkeller cordial labeling when $n \equiv 1 \pmod 2$.*

Proof Let C_n be an odd cycle of length n. Let $V = \{v_i / 1 \le i \le n\}$ be the vertex and $E = \{e_i = v_iv_{i+1} / 1 \le i \le n-1\} \cup \{e_n = v_nv_1\}$ be the edge set of the cycle C_n. Define a function $f : V \to N$ as in Theorem 1. Define an induced function $f^* : E \to \{0, 1\}$ such that

$$f^*(e_i) = f^*(v_iv_{i+1}) = \begin{cases} 1 & \text{if } f(v_i)\ f(v_{i+1}) \text{ is a Zumkeller number} \\ 0 & \text{otherwise.} \end{cases}$$

Denote the set of edges whose labels are Zumkeller numbers as E_1 and $E - E_1$ as E_2.

From Theorem 1, it is clear that the edges $f^*(e_1) \in E_1$

$f^*(e_i) \in E_2$ for $i \equiv 0 \pmod 2)$

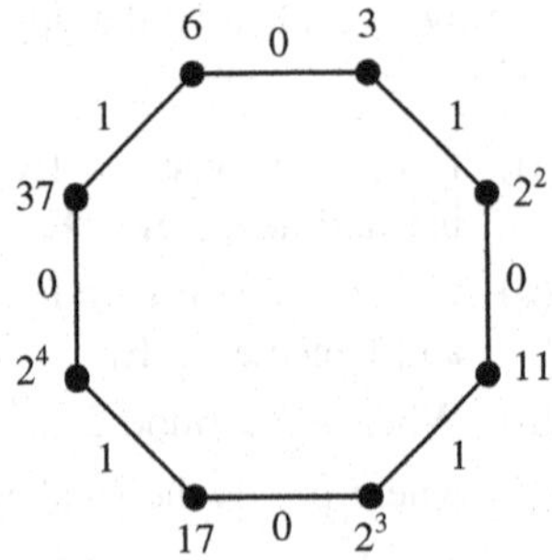

Fig. 3 Zumkeller cordial labelling of C_8

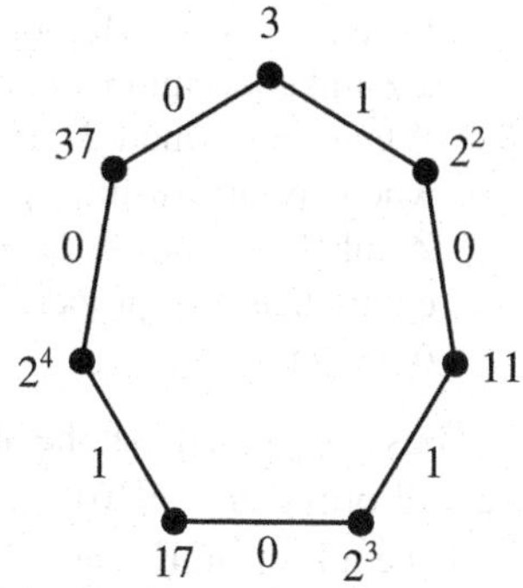

Fig. 4 Zumkeller cordial labelling of C_7

$f^*(e_i) \in E_1$ for $i \equiv 1 \pmod 2$

Consider the edge $f^*(e_n) = f^*(v_n v_1) = f(v_n)f(v_1) = p_n.3$ where $p_n \geq 2^{\frac{n+3}{2}} + 1$ is not a Zumkeller number and hence $e_n \in E_2$.

It is clear that E_1 has $\frac{n-1}{2}$ edges and E_2 has $\frac{n-1}{2} + 1$ edges.

That is, $e_f^*(1) = \frac{n-1}{2}$ and $e_f^*(0) = \frac{n+1}{2}$.

Therefore $\left|e_f^*(0) - e_f^*(1)\right| = \left|\frac{n+1}{2} - \frac{n-1}{2}\right| = 1$. Since the condition $\left|e_f^*(0) - e_f^*(1)\right| \leq 1$ is satisfied, the cycle C_n admits Zumkeller cordial labeling when $n \equiv 1 \pmod 2$.

The above lemma is illustrated by the following example.

Example 4 The Zumkeller cordial labelling of the cycle C_7 is shown in Fig. 4.

Theorem 2 *The cycle C_n admits Zumkeller cordial labeling.*

Proof Proof follows from Lemma 1 and 2.

Theorem 3 *The star graph $K_{1,n}$ admits Zumkeller cordial labeling.*

Proof Let $V = \{v_i / 0 \leq i \leq n\}$ be the vertex set and $E = \{e_i = v_0 v_i / 1 \leq i \leq n\}$ be the edge set of the star $K_{1,n}$. Let v_0 be the central vertex. Define an injective function $f: V \to N$ such that, $f(v_0) = 5$, $f(v_i) = 2^{\frac{i+3}{2}}$ for $i \equiv 1 \pmod 2$, $f(v_i) = p_i$ for $i \equiv 0 \pmod 2$, where p_i are all distinct prime numbers and $p_i \neq 5$ and an induced function $f^*: E \to \{0, 1\}$ such that

$$f^*(e_i) = f^*(v_0 v_i) = \begin{cases} 1 & \text{if } f(v_0)\ f(v_i) \text{ is a Zumkeller number} \\ 0 & \text{otherwise.} \end{cases}$$

Denote the set of edges whose labels are Zumkeller numbers as E_1 and $E - E_1$ as E_2.

Now we identify the edges which belongs to E_1 and those belongs to E_2 as follows

(i) When $i \equiv 1 \pmod 2$, $f^*(e_i) = f^*(v_0 v_i) = f(v_0)f(v_i) = 5 \times 2^{\frac{i+3}{2}}$

Here $5 \leq 2^{\frac{i+5}{2}} - 1$ for all i. By the property 4 of Zumkeller numbers $5 \times 2^{\frac{i+3}{2}}$ is a Zumkeller number and hence $e_i \in E_1\ \forall\ i \equiv 1 \pmod 2$.

(ii) When $i \equiv 0 \pmod 2$, $f^*(e_i) = f^*(v_0 v_i) = f(v_0)f(v_i) = 5 \times p_i$ where p_i are distinct prime numbers $\forall$ i. Here $(5, p_i) = 1$ and 5 is not a Zumkeller number. Therefore $5.p_i$ is also not a Zumkeller number, since this contradicts the property 2 of Zumkeller numbers. Hence $e_i \in E_2\ \forall\ i \equiv 0 \pmod 2$.

Thus it is proved that the alternate edges belongs to E_1 and E_2 respectively. Now we will prove that $|e_f^*(0) - e_f^*(1)| \leq 1$

Since n is even, E_1 has n/2 edges and E_2 has n/2 edges, That is, $e_f^*(0) = n/2$ and $e_f^*(1) = n/2$. Therefore $|e_f^*(0) - e_f^*(1)| = 0$

Case (ii) When $n \equiv 1 \pmod 2$

Since n is odd, E_1 has $\frac{n+1}{2}$ edges and E_2 has $\frac{n-1}{2}$ edges. That is $e_f^*(0) = \frac{n-1}{2}$ and $e_f^*(1) = \frac{n+1}{2}$. Therefore $|e_f^*(0) - e_f^*(1)| = 1$.

In both the cases, the condition $|e_f^*(0) - e_f^*(1)| \leq 1$ is satisfied. Hence the star graph $K_{1,n}$ is a Zumkeller cordial graph for all n.

The Zumkeller cordial labeling of the star graphs $K_{1,6}$ and $K_{1,7}$ are given in the following examples.

Example 5 See Fig. 5.

Example 6 See Fig. 6.

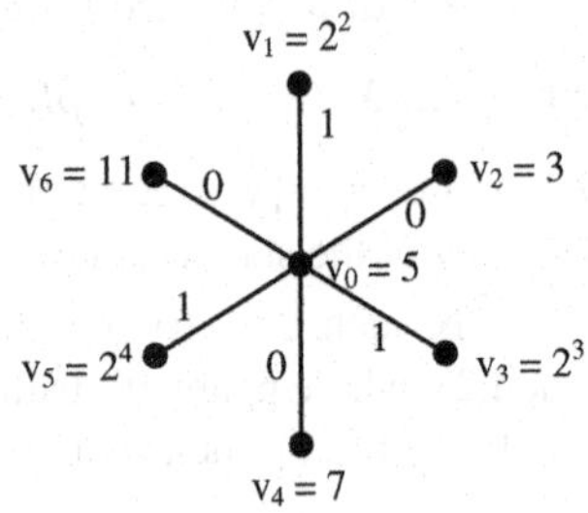

Fig. 5 Zumkeller cordial labelling of $K_{1,6}$

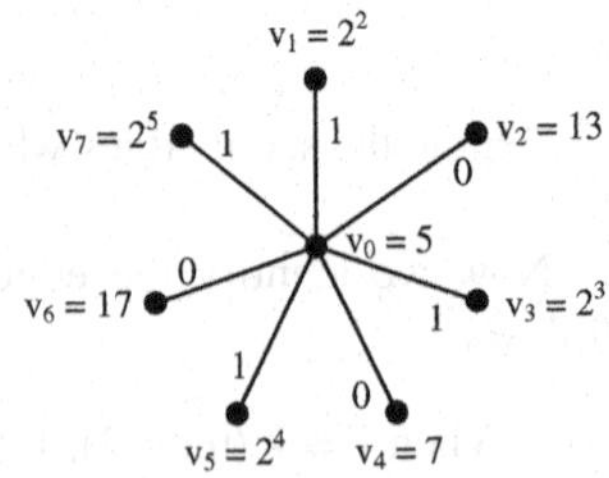

Fig. 6 Zumkeller cordial labelling of $K_{1,7}$

4 Conclusion

In this paper, we have introduced a new graph labeling called Zumkeller cordial labeling of graphs and proved the existence of Zumkeller cordial labeling for graphs such as paths, cycles and star graphs.

References

1. Balamurugan, B.J., Thirusangu, K., Thomas, D.G.: Strongly multiplicative Zumkeller labeling of graphs. In: International Conference on Information and Mathematical Sciences, Elsevier, pp. 349–354 (2013)
2. Balamurugan, B.J., Thirusangu, K., Thomas, D.G.: Strongly multiplicative Zumkeller labeling for acyclic graphs. In: Proceedings of International Conference on Emerging Trends in Science, Engineering, Business and Disaster Management (ICBDM—2014), to appear in IEEE Digital Library
3. Balamurugan, B.J., Thirusangu, K., Thomas, D.G.: Zumkeller labeling of some cycle related graphs. In: Proceedings of International Conferenc on Mathematical Sciences (ICMS—2014), Elsevier, pp. 549–553 (2014)
4. Balamurugan, B.J., Thirusangu, K., Thomas, D.G.: Zumkeller labeling algorithms for complete bipartite graphs and wheel graphs. Adv. Intell. Syst. Comput. Springer **324,** 405–413 (2015)
5. Balamurugan, B.J., Thirusangu, K., Thomas, D.G.: Algorithms for Zumkeller labeling of full binary trees and square grids. Adv. Intell. Syst. Comput. Springer **325**, 183–192 (2015)
6. Balamurugan, B.J., Thirusangu, K., Thomas, D.G.: k-Zumkeller labeling for twig graphs. Electron. Notes Discrete Math. **48**, 119–126 (2015)
7. Cahit, I.: On cordial and 3-equitable labelling of graph. Utilitas Math. **370**, 189–198 (1990)
8. Frank Buss: Zumkeller numbers and partitions. http://groups.google.de/group/de.sci.mathematik/msg/e3fc5afcec2ae540
9. Gallian, J.A.: A dynamic survey of graph labeling. Electronic Journal of Combinatorics. 17 (DS6) (2014)
10. Harary, F.: Graph theory. Addison-Wesley, Reading Mass (1972)
11. Peng, Y., Bhaskara Rao, K.P.S.: On Zumkeller numbers. J. Number Theory. **133**(4), 1135–1155 (2013)
12. Rosa, A.: On certain valuations of the vertices of a graph. In: Gordan, N.B., Dunad (eds.) Theory of graphs. International Symposium, Paris, pp. 349–359 (1966)

Cuckoo Based Resource Allocation for Mobile Cloud Environments

. Durga, S. Mohan, J. Dinesh and A. Aneena

bstract The boom of Mobile Cloud Computing fosters a large volume of smart nobile applications enabling processing and data handlings in the remote cloud ervers. An Efficient allocation of resources to a large number of requests in Mobile Cloud Computing environment is an important aspect that needs special attention in rder to make the environment a highly optimized entity. In this paper, A Cuckoo ased allocation strategy is proposed and the allocation is considered as an optimization problem with the aim of reducing the makespan and the computational ost meeting the deadline constraints, with high resource utilization. The proposed pproach is evaluated using CloudSim framework and the results, indicate that the roposed model provides better quality of service to the mobile cloud customers.

eywords Mobile cloud computing · Cuckoo based resource allocation · esource utilization · Makespan · Tardiness

Introduction

ecent advances in Cloud computing and smartphone technologies lead to the evelopment of integrated Mobile Cloud Computing (MCC) environment, which vercomes obstacles related to the performance such as battery life, computing

Durga (✉) · J. Dinesh · A. Aneena
arunya University, Coimbatore, India
mail: durga.sivan@gmail.com

Dinesh
mail: dineshpeter@karunya.edu

. Aneena
mail: aneenaalex@gmail.com

Mohan
CIS, Al Yamamah University, Riyadh, Kingdom of Saudi Arabia
mail: s.mohan77@gmail.com

. Senthilkumar et al. (eds.), *Computational Intelligence, yber Security and Computational Models*, Advances in Intelligent ystems and Computing 412, DOI 10.1007/978-981-10-0251-9_50

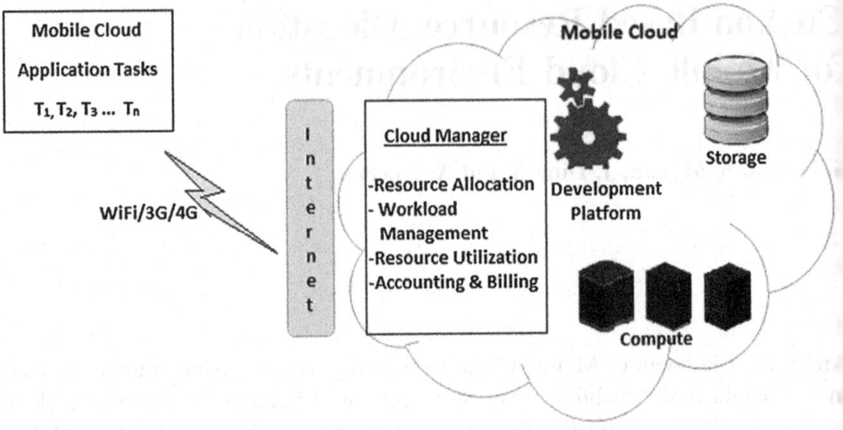

Fig. 1 Fundamental concept of mobile cloud resource provisioning

capabilities, storage and bandwidth etc. [1]. MCC is growing due to the advance ment in wireless technologies like the third generation of mobile communicatior technology (3G), Bluetooth, wireless local area network (WLANs) and worldwide interoperability for microwave access (WIMAX) [2] etc. MCC leads to the optior of accessing applications via users' handset's web browser, from a cloud, where al processing and storage done on a system hosted by a third party providers.

For instance Google's gmail, Google's voice for iphone, skype, social applica tions like facebook, media services such as picasa, youtube, and finally persona financial applications like Mint are some of the famous cloud based mobile applications. In such an era of mobile cloud world, the service providers Quality o Service (QoS) is constrained with many factors such as finding an optimal resource service delay, cost and non-violation of SLA etc. The fundamental concept of MCC Resource Allocation (RA) is shown in Fig. 1. This paper defines a problem of RA as an optimization problem and a cuckoo based allocation approach is proposed to improve the performance of the system.

The rest of the paper is organized as follows. The related work is given in Sect. 2. The proposed Algorithm is introduced in Sect. 3, and the experimenta result is explained in Sect. 4. We conclude in Sect. 5.

2 Related Work

There are number of active researches on mobile cloud resource Management whicl have been proposed in recent times. A review of existing application frameworks i conducted in [3], where the authors discussed various issues, challenges and sug gested the future areas for optimum distributed application processing framework Pandey, Linlin [4] have proposed a Particle Swarm Optimization (PSO) base euristic to schedule applications to cloud resources that takes into account com-utational cost and data transmission cost. In adaptive model [5] an independent uthority is employed to predict and store the resource usages, and the same is used) predict future required resources using ANN. However the processing time and ie communication cost involved with the independent authority is not considered hich ends up in performance deviation. Nishio et al. [6] presented a framework for ervice oriented resource sharing in mobile cloud. They focused on the service itency rather than the service providers management objectives. Ravi et al. [7] ddresses a seamless service provisioning framework in which the application xecution is shared between the cloud and the mobile ad hoc grid in order to reduce ie communication overhead. In [8], author focuses on clonal selection algorithm ased on time cost and energy consumption models in cloud computing environ-ent. In [1], authors claimed that in addition to the customers SLA, other specific ianagement objectives relating to the datacentre infrastructure such as balanced ad, fault tolerance and profit maximization etc. need to be considered. From the bove mentioned state of art approaches it is evident that, in mobile cloud resource location, cost and resource utilization are conflicting objectives that require an fficient optimization approach for optimal non dominated solution.

Proposed Model

.1 *Cuckoo Search via Levy Flights*

uckoo Search Algorithm (CSA) is a new meta-heuristic approach that models the atural behaviour of cuckoos and explore their landscape using a series of straight ight paths punctuated by a sudden 90° turn. The following idealized rules are llowed in [9]:

Each cuckoo lays one egg at a time, and dumps it in a randomly chosen nest.
The best nests with high quality eggs (solutions) will be carried out over to the next generations.
The number of available host nests is fixed, say n, and the

Host can discover an alien egg by a probability p_a [0, 1].
In this case, a host bird can either throw the egg away or abandon the nest, and uild completely new nest. The last assumption is approximated by the fraction p_a of nests are replaced by new nests. The main aim of this algorithm is to use new and otentially better solutions to replace a not so good solution in the nest. For new lution for cuckoo i, a levy flight is performed by the following stochastic equation

$$X_i^{t+1} = X_i^t + \propto \otimes levy(\lambda), \text{ where the step size } \propto\ > 0$$

3.2 System Model

Jobs arrive at unknown intervals and are placed in the queue of scheduled task from where the jobs are assigned to the processor. After a certain interval of time CSA is applied and map the jobs to the corresponding processors. Here we map cuckoo nests as the resources, cuckoo as Cloud resource Manager, cuckoo egg as the newly arrived job and the host's eggs are considered as the jobs in the queue and the characteristics of the eggs are the constraints. If the newly arrived job satisfies the constraints of the jobs in the resources approximately, then the job is chosen for execution in the resource. Else the job is deallocated from that resource and some other optimal resource is chosen for that job [10].

3.3 Mathematical Formulation and Proposed Algorithm

The problem of finding an assignment of suitable resource to the request in terms of reduced cost, delay and makespan is NP-hard. The objective function is to allocate the task to the virtual machine so as to achieve minimum execution time, minimum cost and meet the deadline constraints with optimal utilization of resources. Let E represents the amount of time taken by Task i to be executed in resource j, C denotes the computational cost and $Load_{lt}$ and $Load_{ut}$ denoting the lower and upper threshold respectively. This problem can be expressed as linear programming problem, as depicted below

$$F(x) = Max \sum\nolimits_{i,j=1}^{n} P(X_{ij}) \quad (1)$$

where, $P(X_{ij}) = aU_j - bC_{ij} - cE_{ij}$

Subjected to

Resource Utilization $U_j = Load_{lt} \leq Load \leq Load_{ut}$

C_{ij} is computed as follows,

C_{ij} [task length/processing power of resource] * Resource cost

For Cuckoo Search strategy the required step-size [9] is obtained by

$$stepsize_j = .01 * \left(\frac{u_j}{\vartheta_j}\right)^{\{\frac{1}{\beta}\}} * V - X_{best} \quad (2)$$

where $u = \varphi.randn[d]$, $\vartheta = randn[d]$

where $randn[d]$ generates a random number between [0, 1].

Then V can be generated as: $V = V + stepsize * randn[d]$

The update process of Cuckoo search is defined by:

$$X_{best} \leftarrow f(X_{best}) \leq f(X_i)$$

The algorithmic control parameters of cuckoo search are the scale factor (β) and utation probability value (p_a). In this paper β = 1.50 and p_a = 0.25 have been used in [10, 11].

Pseudo code: CSA based Resource Allocation

Begin
Initialize population randomly
Choose best nest
While (t<maxGeneration) or (stop criterion) /* termination condition */
  Get a cuckoo randomly by Levy Flights
  Perform new nest
  The fitness function
    $F(x) = Max\ \sum_{i,j=1}^{n} P\left(X_{ij}\right)$ /* As per Equation 1 */
  Evaluate its fitness/quality
  Choose a nest among n (say j) randomly
  If ($F_i > F_j$) Replace j by the new solution End if
  A fraction (p) of worse nets are abandoned and new ones are built Keep the best solutions (or nests with quality solutions)
  Rank the solutions and find the current best
End while
Post process results and visualization
End

Performance Evaluation

e proposed algorithm is analyzed using Cloud Sim 3.0 Simulation Tool kit [12] a Windows system with Core i5 CPU at 2.40 GHz and 4 GB RAM. As discussed previous section, the first step in CSA is initializing the population. This is done a vector in which the length of the vector is taken as the number of resources. The algorithm is allowed to run for k cycles (k = 15 in this paper). The proposed A based RA is compared with PSO based RA algorithm by considering the rameters such as the percentage of the deadline met, computational cost) makespan and CPU utilization. Makespan is defined as the total execution time all the tasks.

$$Makespan = \sum_{i=0}^{n} Execution\ time\ (Task_i) \quad (3$$

Tardiness TR is taken to evaluate the percentage of deadline met.

$$TR_i = d_i - f_i \quad (4$$

Total Tardiness is the sum of the tardiness of the each task which did not g executed under the provided deadline. The average tardiness is defined by using th following formulae

$$Avg_{TR} = \frac{\sum_{i=1}^{n} TR_i}{n} \quad (5$$

We assume that the resource cost falls in the range of \$2–\$5 per resources. Th computational cost (C) is calculated as,

$$\mathrm{C} = [\text{task length/processing power of resource}] * \text{Resource cost} \quad (6$$

Based on the comparison and the results from the experiment it is concluded th the proposed approach works better. The impacts of change in no. of tasks c makespan, deadline and cost are illustrated in Figs. 2, 3 and 4. Figure 2 shows th decrease in makespan using CSA in proposed model. Figure 3 shows that th deadline meeting success ratio increases with CSA based RA. The number non-delayed tasks is the total number of tasks whose finishing time was less tha the deadline of the task, i.e., which finished before the deadline given to them. Th expected completion time is calculated as the mean of the completion time for th task at every resource. The analysis of the result suggest that cuckoo search strateg works well minimizing the makespan, computational cost and the percentage deadline met by the strategy is high compared to other meta heuristic algorithm

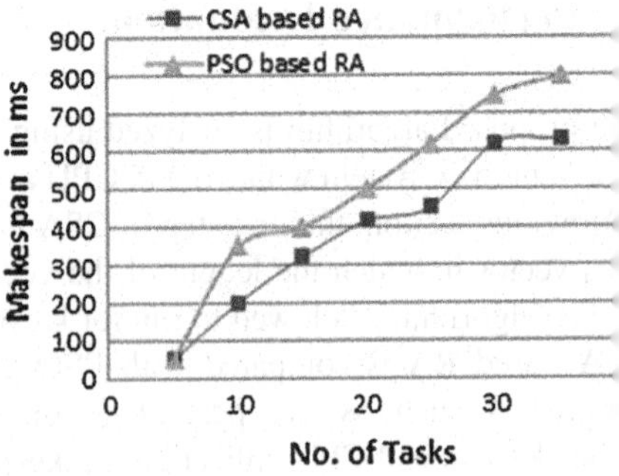

Fig. 2 Influence of change in no. of tasks in makespan

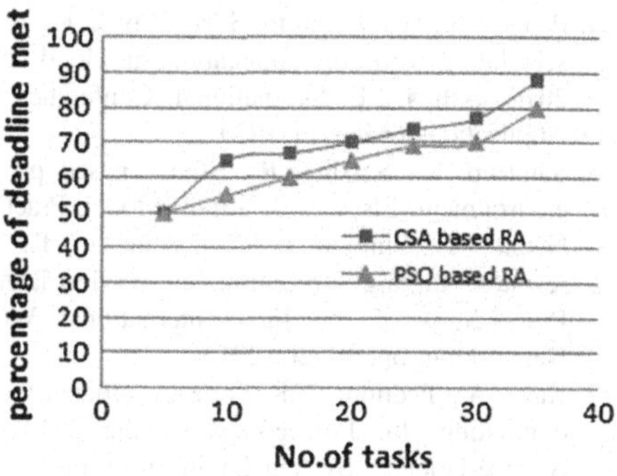

g. 3 Influence of change in). of tasks in makespan

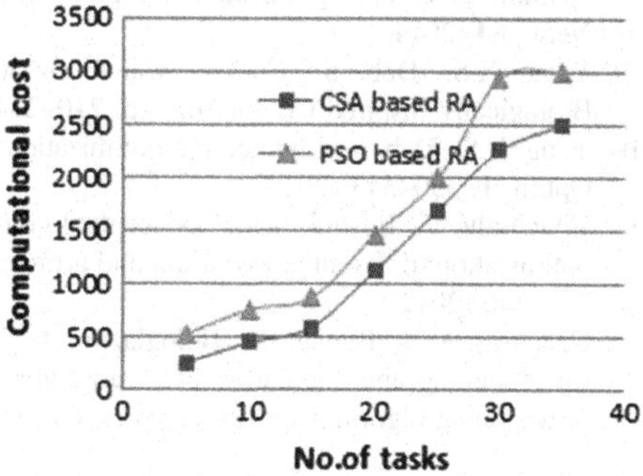

g. 4 Influence of change in . of tasks in computational st

Conclusion and Future Work

this work, a CSA based allocation strategy is presented and aims to minimize the erall execution time and the computational cost, meeting the deadline constraints d such a way meeting the providers QoS. The results indicate the algorithm gnificantly reduces the total executing time, computational cost, and increases the rcentage of meeting the deadline. In Future, experiments will be conducted with fferent optimization strategies such as genetic algorithms and compare their rformance with Cuckoo Search Algorithm.

eferences

. Dinh, H.T., Lee, C., Nyato, D., Wang, P.: A survey of mobile cloud computing: architecture, applications and approaches. Wirel. Commun. Mob. Comput. **13**, 1587–1611 (2013)
. Escalnte, D., Andrew, J.: Cloud services: policy and assessment. Educause Rev. **46** (2011)
. Shiraz, M., Gani, A., Khokhar, R.H., Buyya, R.: A review on distributed application processing frameworks in smart mobile devices for mobile cloud computing. IEEE Commun. Surv. **15**, 1294–1313 (2013)

4. Pandey, S., Wu, L., Guru, S.M., Buyya, R.: A particle swarm optimization-based heuristic f scheduling workflow applications in cloud computing environments. In: Proceedings of th 2010 24th IEEE International Conference on Advanced Information Networking ar Applications (AINA) (2010)
5. Sandeep, K., Sandhu, R.: Matrix based proactive resource provisioning in mobile clou environment. Elsevier. J. Simul. Model. Pract. Theory **50**, 83–95 (2015)
6. Nishio, T., Shinkuma, R., Takahashi, T., Narayan, B.: Service-oriented heterogeneou resource sharing for optimizing service latency in mobile cloud. In: Mobile Cloud '1 Proceedings of the First International Workshop on Mobile Cloud Computing ar Networking, pp. 19–26 (2013)
7. Ravi, A., Peddoju, S.K.: Energy efficient seamless service provisioning in mobile clou computing, In: Proceedings of the 2013 IEEE Seventh International Symposium c Service-Oriented System Engineering. pp. 463–471 (2013)
8. Shu, W., Wang, W., Wang, Y.: A novel energy-efficient resource allocation algorithm bas on immune clonal optimization for green cloud computing. EURASIP J. Wirel. Commu Netw. **64** (2014)
9. Yang, X.S., Deb, S.: Cuckoo search via levy flights. In: World Congress Nature ar Biologically Inspired Computing, pp. 210–214 (2009)
10. Yang, X.S., Deb, S.: Engineering optimization by cuckoo search. Int. J. Math. Model. Nume Optim. **1**, 330–343 (2010)
11. Civicioglu, P., Besdok, E.: A conceptual comparison of the Cuckoo-search, particle swar optimization, differential evolution and artificial bee colony algorithms. Artif. Intell. Rev. **3** 315–346 (2013)
12. Calheiros, R.N., Ranjan, R., Beloglazov, A., De Rose, A.F., Buyya, R.: CloudSim: a tool for modelling and simulation of cloud computing environments and evaluation of resour provisioning algorithms. Softw.: Pract. Experience **41**, 23–50 (2011)

Transient Analysis of an $M/M/c$ Queue Subject to Multiple Exponential Vacation

.V. Vijayashree and B. Janani

bstract In this paper, we consider an $M/M/c$ queueing model subject to multiple xponential vacation wherein arrivals occur according to a Poisson distribution and e c servers provide service according to an exponential distribution. When the stem is empty, all the c servers go on a vacation and the vacation times are sumed to follow exponential distribution. Further arrivals are allowed to join the ueue when servers are in vacation. Explicit analytical expressions for the time ependent probabilities of the number in the system are presented using matrix eometric method.

eywords Transient analysis · Laplace transform · Matrix geometric method · Multiple vacation

Introduction

ueueing system with multiple server is useful and important as it finds a wide nge of applications in computer system, communication networks, production anagement, etc. For example, congestion in vehicular network pose a major threat e to varied reasons like high mobility, short link lifetime and spectrum efficiency. herefore, cognitive radio (CR) plays a predominant role in the effective management of the available spectrum. More recently, Daniel et al. [1] proposed a multi rver queueing model design for CR vehicular traffic and showed its effectiveness terms of reducing the average waiting time for a vehicle or equivalently the

V. Vijayashree (✉)
epartment of Mathematics, College of Engineering, Anna University, Chennai, India
nail: vkviji@annauniv.edu

Janani
epartment of Mathematics, Central Institute of Plastics Engineering and Technology, ennai, India
nail: Jananisrini2009@gmail.com

Senthilkumar et al. (eds.), *Computational Intelligence, ber Security and Computational Models*, Advances in Intelligent stems and Computing 412, DOI 10.1007/978-981-10-0251-9_51 optimal use of the available spectrum resource. Vijayalakshmi and Jyothsna [2 presents the steady state analysis of a renewal input multiple working vacation queue with balking, reneging and heterogeneous servers thereby obtaining th various system performance measures like expected system length, expecte balking rate etc. Lin and Ke [3] discusses the multi-server system with singl working vacation and obtains the stationary probabilities using matrix geometri method. Tian et al. [4] considers an *M*/*M*/*c* queue with multiple vacations an obtains the conditional stochastic decompositions of the stationary queue length an waiting time.

Parthasarathy [5] provided an explicit analytical expression for the tim dependent probabilities of the number in the system at time t in terms of modifie Bessel function of first kind using generating function methodology for a mul server queueing model. Recently, Al-seedy et al. [6] extended the results b introducing the concept of balking and reneging in the multi server queueing mode Using the similar technique as above, Ammar [7] obtains an explicit expression fo the transient system size probabilities for the queue with heterogeneous servers an impatient behavior. In this paper, explicit analytical expressions for the stationar probabilities of the number in the system for an *M*/*M*/*c* queueing model are pre sented using matrix geometric method. However, for the transient analog, th system size probabilities are recursively obtained in the Laplace domain. As special case, when $c = 1$, the transient probabilities in Laplace domain are seen t coincide with the results of an *M*/*M*/*1* queueing model subject to multiple vacatio

2 Model Description

Consider an *M*/*M*/*c* queueing model subject to multiple exponential vacation Arrivals are assumed to follow Poisson distribution with parameter λ. The c serve provide service according to an exponential distribution with parameter μ. Whe the system is empty, all the 'c' servers go on a vacation wherein the vacation time follow exponential distribution with parameter θ. Further, arrivals are allowed t join the queue during the vacation period. Let $N(t)$ denote the number of custome in the system at time t. Define $J(t) = 1$ when the server is busy and $J(t) = 0$ when th server is on vacation at time, t. It is well known that $\{(J(t), N(t)), t \geq 0\}$ is a Marko process with state space $\Omega = \{(0,k) \cup (1,k), k = 1,2, \ldots\}$. The state transitio diagram for the model is given in Fig. 1.

3 Stationary Analysis

This section presents explicit expressions for the stationary probabilities of th above described model. Let π_{jk} denote the stationary probability for the system be in state j with k customers. Define $\pi = [\pi_0, \pi_1, \pi_2, \ldots]$ where $\pi_0 = \pi_{00}$ an

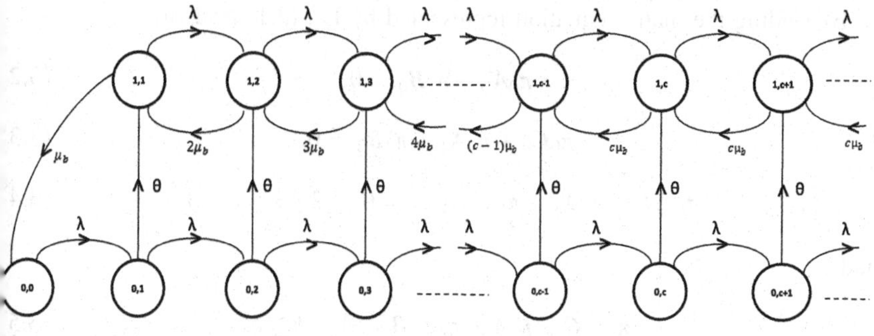

'ig. 1 State transition diagram

$_{0k} = [\pi_{0k}, \pi_{1k}]$ $k \geq 1$, then the system of equations governing the state proba-
ilities under steady state are given by

$$\pi \mathbb{Q} = 0 \tag{3.1}$$

The infinitesimal generator $\mathbb{Q}$ is given by

$$\mathbf{Q} = \begin{bmatrix} A_0 & C_0 & & & & \\ B_0 & A_1 & C & & & \\ & B_2 & A_2 & C & & \\ & & B_2 & A_3 & C & \\ & & & B & A & C \\ & & & & \ddots & \ddots & \ddots \end{bmatrix}$$

'here $A_0 = -\lambda$, $C_0 = (\lambda, 0)$, $B_0 = \begin{pmatrix} 0 \\ \mu \end{pmatrix}$, $B_k = \begin{pmatrix} 0 & 0 \\ 0 & k\mu \end{pmatrix}$ $k = 2, 3, \ldots c-1$,

$= \begin{pmatrix} \lambda & 0 \\ 0 & \lambda \end{pmatrix}$

$_k = \begin{pmatrix} -(\lambda+\theta) & \theta \\ 0 & -(\lambda+k\mu) \end{pmatrix}$ $k = 1, 2 \ldots c-1, B = \begin{pmatrix} 0 & 0 \\ 0 & c\mu \end{pmatrix}$,

$= \begin{pmatrix} -(\lambda+\theta) & \theta \\ 0 & -(\lambda+c\mu) \end{pmatrix}.$

Expanding the matrix equation represented by Eq. (3.1) leads to

$$\pi_0 A_0 + \pi_1 B_0 = 0, \tag{3.2}$$

$$\pi_0 C_0 + \pi_1 A_1 + \pi_2 B_2 = 0, \tag{3.3}$$

$$\pi_{k-1} C + \pi_k A_k + \pi_{k+1} B_{k+1} = 0; \quad 2 \le k \le c-1, \tag{3.4}$$

and

$$\pi_{k-1} C + \pi_k A + \pi_{k+1} B = 0, \quad k \ge c. \tag{3.5}$$

Lemma 1 *If* $\rho = \frac{\lambda}{c\mu} < 1$ *and* $r = \frac{\lambda}{\lambda+\theta}$, *then the matrix quadratic equation*

$$R^2 B + RA + C = 0 \tag{3.6}$$

has the minimal non-negative solution given by

$$R = \begin{pmatrix} r & \frac{\theta r}{c\mu(1-r)} \\ 0 & \rho \end{pmatrix}. \tag{3.7}$$

Proof Since B, A and C are all upper triangular matrices, we assume R has the same structure as $R = \begin{pmatrix} r_{11} & r_{12} \\ 0 & r_{22} \end{pmatrix}$. Substituting R into $R^2 B + RA + C = 0$ leads to

$$\lambda - r_{11}(\lambda+\theta) = 0,$$
$$r_{11}\theta - (\lambda + c\mu) r_{12} + c\mu (r_{11} r_{12} + r_{12} r_{22}) = 0,$$

and

$$\lambda - (\lambda + c\mu) r_{22} + c\mu r_{22}^2 = 0.$$

On solving the above equations, it is seen that $r_{11} = \frac{\lambda}{\lambda+\theta}$, $r_{22} = \frac{\lambda}{c\mu}$, and $r_{12} = \frac{\theta r}{c\mu(1-r)}$. Hence, we obtain $R = \begin{pmatrix} r & \frac{\theta r}{c\mu(1-r)} \\ 0 & \rho \end{pmatrix}$. Observe that for $k = 1, 2, 3, \ldots$

$$R^k = \begin{pmatrix} r^k & \frac{\theta r}{c\mu(1-r)} \sum_{j=0}^{k-1} \rho^{k-1-j} r^j \\ 0 & \rho^k \end{pmatrix}. \tag{3.8}$$

heorem 1 *If $\rho<1$, then the stationary probabilities of the system state are given y*

$$\pi_{0k}=\pi_{00}\left(\frac{\lambda}{\lambda+\theta}\right)^{k};k\geq 1,$$

$$\pi_{1k}=\pi_{00}\left\{\frac{1}{k!}\left(\frac{\lambda}{\mu}\right)^{k}\left(1+\sum_{i=1}^{k-1}i!\left(\frac{r\mu}{\lambda}\right)^{i}\right)\right\};1\leq k\leq c-1,$$

$$\pi_{1k}=\pi_{00}\left\{\frac{1}{c!}\left(\frac{\lambda}{\mu}\right)^{c}\left(1+\sum_{i=1}^{c-1}i!\left(\frac{r\mu}{\lambda}\right)^{i}\right)\right\}+\pi_{00}r^{c}\frac{\theta r}{c\mu(1-r)}\sum_{i=0}^{k-c-1}\rho^{k-c-1-i}r^{i};k\geq c.$$

and π_{00} is found using the normalization condition.

roof On substituting the corresponding matrices in Eq. (3.2), we get, $\lambda\pi_{00}+\mu\pi_{11}=0$, which yields $\pi_{11}=\frac{\lambda}{\mu}\pi_{00}$. Similarly, substituting the corre- onding matrices in Eq. (3.3) yields

$$\pi_{00}\lambda-(\lambda+\theta)\pi_{01}=0,\theta\pi_{01}-(\lambda+\mu)\pi_{11}+2\mu\pi_{12}=0,$$

hich simplifies to $\pi_{01}=r\pi_{00}$ and $\pi_{12}=\pi_{00}\left\{\frac{1}{2!}\left(\frac{\lambda}{\mu}\right)^{2}\left(1+\left(\frac{r\mu}{\lambda}\right)\right)\right\}$. Also, substi- ting the corresponding matrices in Eq. (3.4), we get

$$\lambda\pi_{0k-1}-(\lambda+\theta)\pi_{0k}=0;2\leq k\leq c-1,$$

d

$$\lambda\pi_{1k-1}+\theta\pi_{0k}-(\lambda+k\mu)\pi_{1k}+(k+1)\mu\pi_{1k+1}=0;2\leq k\leq c-1.$$

Therefore, $\pi_{0k}=r^{k}\pi_{00},1\leq k\leq c-1$ and $\pi_{1k}=\pi_{00}\left\{\frac{1}{k!}\left(\frac{\lambda}{\mu}\right)^{k}\left(1+\sum_{i=1}^{k-1}\right.\right.$ $\left.\left.\frac{r\mu}{\lambda})^{i}\right)\right\}$. Assume, $\boldsymbol{\pi}_k=\boldsymbol{\pi}_c\boldsymbol{R}^{k-c}$ for $k\geq c$, then from Eq. (3.5) and using Eq. (3.8), e get $\pi_{0k}=r^{k}\pi_{00};k\geq c$ and $\pi_{1k}=\pi_{00}\left\{\frac{1}{c!}\left(\frac{\lambda}{\mu}\right)^{c}\left(1+\sum_{i=1}^{c-1}i!\left(\frac{r\mu}{\lambda}\right)^{i}\right)\right\}+\pi_{00}r^{c}\frac{\theta r}{c\mu(1-r)}$ $\sum_{i=0}^{k-c-1}\rho^{k-c-1-i}r^{i};k\geq c$. Hence all the stationary probabilities are expressed in rms of π_{00} and π_{00} can be found using the normalization condition.

Transient Analysis

is section provides an analytical expression for the time dependent probabilities the number in the system in Laplace domain, using matrix analytic method. Let $_k(t)=P(J(t)=j,N(t)=k),j=0,1;$ and $k=0,1,2\ldots$ Using standard methods, the system of equations that governs the process $\{(J(t),N(t)),t\geq 0\}$ are given by

$$P'_{00}(t) = -\lambda P_{00}(t) + \mu P_{11}(t),$$
$$P'_{0k}(t) = -(\lambda+\theta)P_{0k}(t) + \lambda P_{0k-1}(t), \quad k\geq 1,$$
$$P'_{11}(t) = -(\lambda+\mu)P_{11}(t) + 2\mu P_{12}(t) + \theta P_{01}(t).$$
$$P'_{1k}(t) = -(\lambda+k\mu)P_{1k}(t) + \lambda P_{1k-1}(t) + (k+1)\mu P_{1k+1}(t) + \theta P_{0k}(t), 2\leq k\leq c-1$$

and

$$P'_{1k}(t) = -(\lambda+c\mu)P_{1k}(t) + \lambda P_{1k-1}(t) + c\mu P_{1k+1}(t) + \theta P_{0k}(t), \quad k\geq c,$$

subject to the condition $P_{00}(0)=1$. Let $\boldsymbol{P}_0(t) = P_{00}(t), \boldsymbol{P}_k(t) = [P_{0k}(t), P_{1k}(t)]$ $k=1,2,\ldots$ Then, the above system of equations can be expressed in the matrix form as

$$\frac{d\boldsymbol{P}(t)}{dt} = \boldsymbol{P}(t)\mathbb{Q},$$

where $\boldsymbol{P}(t) = [\boldsymbol{P}_0(t), \boldsymbol{P}_1(t), \boldsymbol{P}_2(t)\ldots\ldots]$. Let $\hat{P}_{jk}(s)$ denote the Laplace transform of $P_{jk}(t)$ for $j=0,1$ and $k=0,1,2,\ldots$. Taking Laplace Transform of the above equation yields $\hat{\boldsymbol{P}}(s)[\mathbb{Q}-s\boldsymbol{I}] = -\boldsymbol{P}(0)$ which on expansion leads to

$$\hat{\boldsymbol{P}}_0(s)(\boldsymbol{A}_0 - s) + \hat{\boldsymbol{P}}_1(s)\boldsymbol{B}_0 = -1, \tag{4.1}$$

$$\hat{\boldsymbol{P}}_0(s)\boldsymbol{C}_0 + \hat{\boldsymbol{P}}_1(S)(\boldsymbol{A}_1 - s\boldsymbol{I}) + \hat{\boldsymbol{P}}_2(s)\boldsymbol{B}_2 = 0, \tag{4.2}$$

$$\hat{\boldsymbol{P}}_{K-1}(s)\boldsymbol{C} + \hat{\boldsymbol{P}}_K(s)(\boldsymbol{A}_k - s\boldsymbol{I}) + \hat{\boldsymbol{P}}_{k+1}(s)\boldsymbol{B}_{k+1} = 0; \quad 2\leq k\leq c-1, \tag{4.3}$$

and

$$\hat{\boldsymbol{P}}_{k-1}(s)\boldsymbol{C} + \hat{\boldsymbol{P}}_k(s)(\boldsymbol{A} - s\boldsymbol{I}) + \hat{\boldsymbol{P}}_{k+1}(s)\boldsymbol{B} = 0; \quad k\geq c \tag{4.4}$$

where $\boldsymbol{I}$ is the identity matrix of the corresponding order.

Lemma 2 *The quadratic matrix equation*

$$\boldsymbol{R}^2(s)\boldsymbol{B} + \boldsymbol{R}(s)(\boldsymbol{A} - s\boldsymbol{I}) + \boldsymbol{C} = 0 \tag{4.5}$$

has the minimal non-negative solution given by

$$\boldsymbol{R}(s) = \begin{pmatrix} \rho(s) & \beta_c(s) \\ 0 & r_c(s) \end{pmatrix} \tag{4.6}$$

where $\rho(s) = \dfrac{\lambda}{s+\lambda+\theta}, \beta_c(s) = \dfrac{\theta r_c(s)}{s+\lambda+\theta-c\mu r_c(s)},$ *and*

$$r_c(s) = \frac{(s+\lambda+c\mu) - \sqrt{(s+\lambda+c\mu)^2 - 4\lambda c\mu}}{2c\mu}.$$

roof By an analysis similar to the proof of Lemma 1***, we assume $\boldsymbol{R}(s)$ as a pper triangular matrix since $\boldsymbol{B}, \boldsymbol{A} - s\boldsymbol{I}$ and $\boldsymbol{C}$ are all upper triangular matrices. ubstituting $\boldsymbol{R}(s)$ into Eq. (4.5) and upon solving the corresponding equations leads Eq. (4.6). Observe that for $k = 1, 2, 3, \ldots$

$$\boldsymbol{R}^k(s) = \begin{pmatrix} \rho^k(s) & \beta(s) \sum_{i=0}^{k-1} \rho^{k-1-i}(s) r_c^i(s) \\ 0 & r_c^k(s) \end{pmatrix}. \tag{4.7}$$

here $\boldsymbol{R}^1(s) = \boldsymbol{R}(s)$ and $\boldsymbol{R}(0) = \boldsymbol{R}$.

emma 3 *Let* $\boldsymbol{R}_k(s) = \begin{pmatrix} \rho(s) & \lambda\beta_k(s) \\ 0 & r_k(s) \end{pmatrix}, \quad 1 \le k \le c-1,$

where

$$\rho(s) = \frac{\lambda}{s+\lambda+\theta}, \tag{4.8}$$

$$\beta_{c-1}(s) = \frac{\theta + c\mu\beta_c(s)}{(s+\lambda+\theta)(s+\lambda+(c-1)\mu - c\mu r_c(s))}, \tag{4.9}$$

$$\beta_k(s) = \frac{\theta + (k+1)\lambda\mu\beta_{k+1}(s)}{(s+\lambda+\theta)(s+\lambda+k\mu-(k+1)\mu r_{k+1}(s))}, \quad 1 \le k \le c-2, \tag{4.10}$$

and

$$r_k(s) = \frac{\lambda}{s+\lambda+k\mu-(k+1)\mu r_{k+1}(s)}, \quad 1 \le k \le c-1. \tag{4.11}$$

Then $\{\boldsymbol{R}_k(s), 1 \le k \le c-1\}$ *are satisfied by the following recurrence relation*

$$\boldsymbol{C} + \boldsymbol{R}_k(s)(\boldsymbol{A}_k - s\boldsymbol{I}) + \boldsymbol{R}_k(s)\boldsymbol{R}_{k+1}(s)\boldsymbol{B}_{k+1} = 0. \tag{4.12}$$

Proof For $n = c-1$, let $\boldsymbol{R}_{c-1}(s) = \begin{pmatrix} \rho(s) & \lambda\beta_{c-1}(s) \\ 0 & r_{c-1}(s) \end{pmatrix}$. Then the recurrence relation becomes $\boldsymbol{C} + \boldsymbol{R}_{c-1}(s)(\boldsymbol{A}_{c-1} - s\boldsymbol{I}) + \boldsymbol{R}_{c-1}(s)\boldsymbol{R}_c(s)\boldsymbol{B}_c = 0$. Substituting the corresponding matrices and upon simplification yields

$$\lambda - (\lambda + \theta + s)\rho(s) = 0,$$

$$\theta\rho(s) - \lambda\beta_{c-1}(s)(\lambda + (c-1)\mu + s) + c\mu(\lambda\rho(s)\beta_c(s) + \lambda\beta_{c-1}(s)r_c(s) = 0,$$

and

$$\lambda - r_{c-1}(s)(\lambda + (c-1)\mu + s) + c\mu r_c(s) r_{c-1}(s) = 0,$$

which on solving leads to $\rho(s) = \frac{\lambda}{\lambda+\theta+s}$, $\beta_{c-1}(s) = \frac{\theta + c\mu\beta_c(s)}{(\lambda+\theta+s)(\lambda+(c-1)\mu+s-c\mu r_c(s))}$ and $r_{c-1}(s) = \frac{\lambda}{\lambda+(c-1)\mu+s-c\mu r_c(s)}$.

Therefore,

$$\boldsymbol{R}_{c-1}(s) = \begin{pmatrix} \frac{\lambda}{\lambda+\theta+s} & \frac{\theta + c\mu\beta_c(s)}{(\lambda+\theta+s)(\lambda+(c-1)\mu+s-c\mu r_c(s))} \\ 0 & \frac{\lambda}{\lambda+(c-1)\mu+s-c\mu r_c(s)} \end{pmatrix} \tag{4.13}$$

is completely determined. In general, assuming $\boldsymbol{R}_{k+1}(s) = \begin{pmatrix} \rho(s) & \lambda\beta_{k+1}(s) \\ 0 & r_{k+1}(s) \end{pmatrix}$, it can be proved that $\boldsymbol{R}_k(s) = \begin{pmatrix} \rho(s) & \lambda\beta_k(s) \\ 0 & r_k(s) \end{pmatrix}$, satisfies Eq. (4.12). Consider $\boldsymbol{C} + \boldsymbol{R}_k(s)(\boldsymbol{A}_k - s\boldsymbol{I}) + \boldsymbol{R}_k(s)\boldsymbol{R}_{k+1}(s)\boldsymbol{B}_{k+1} = 0$ substituting the corresponding matrices and on simplification leads to

$$\rho(s) = \frac{\lambda}{s+\lambda+\theta}, r_k(s) = \frac{\lambda}{s+\lambda+k\mu-(k+1)\mu r_{k+1}(s)},$$

and

$$\beta_k(s) = \frac{\theta + (k+1)\mu\beta_{k+1}(s)\lambda}{(s+\lambda+\theta)(s+\lambda+k\mu-(k+1)\mu r_{k+1}(s))}.$$

Therefore Eqs. (4.8), (4.9), (4.10) and (4.11) are true for all $n = 1, 2, \ldots, c-1$. Having determined $\boldsymbol{R}_{c-1}(s)$ in Eq. (4.13), from the matrix Eq. (4.12), we get $\boldsymbol{R}_{c-2}(s) = \begin{pmatrix} \rho(s) & \lambda\beta_{c-2}(s) \\ 0 & r_{c-2}(s) \end{pmatrix}$ where

$$\rho(s) = \frac{\lambda}{s+\lambda+\theta}, r_{c-2}(s) = \frac{\lambda}{s+\lambda+(c-2)\mu-(c-1)\mu r_{c-1}(s)},$$

nd

$$\beta_{c-2}(s) = \frac{\theta + (c-1)\mu\beta_{c-1}(s)\lambda}{(s+\lambda+\theta)(s+\lambda+(c-2)\mu-(c-1)\mu r_{c-1}(s))}.$$

Similarly $\boldsymbol{R}_{c-3}(\boldsymbol{s}), \boldsymbol{R}_{c-4}(\boldsymbol{s}), \ldots \boldsymbol{R}_1(\boldsymbol{s})$ can be recursively determined. Hence the roof.

'heorem 2 *The Laplace transform of the transient state probability distribution unctions sequence, $\hat{\boldsymbol{P}}_k(s)$ are satisfied with the following relations.*

$$\hat{\boldsymbol{P}}_k(s) = \hat{\boldsymbol{P}}_{c-1}(s)\boldsymbol{R}^{k-c+1}(s), k \geq c \tag{4.14}$$

$$\hat{\boldsymbol{P}}_k(s) = \hat{\boldsymbol{P}}_{k-1}(s)R_k(s) = R_K^*(s)e_1\hat{P}_0(s), 1 \leq k \leq c-1 \tag{4.15}$$

where $R_K^*(s) = R_K(s)R_{K-1}(s)\ldots R_1(s)$ *for* $k = 1, 2\ldots c-1, e_1 = (1 \quad 0)$ *and* $\hat{P}_0(s) = [s - \boldsymbol{A}_0 - \boldsymbol{e}_1\boldsymbol{R}_1(s)\boldsymbol{B}_0]^{-1}$.

'roof For $k \geq c$, substituting Eq. (4.14) in (4.4) leads to

$$\begin{aligned}\hat{\boldsymbol{P}}_{k-1}(s)\boldsymbol{C} + \hat{\boldsymbol{P}}_k(s)(\boldsymbol{A} - s\boldsymbol{I}) + \hat{\boldsymbol{P}}_{k+1}(s) &= \hat{\boldsymbol{P}}_{c-1}(s)\boldsymbol{R}^{k-c}(s)\left[\boldsymbol{C} + \boldsymbol{R}(s)(\boldsymbol{A} - s\boldsymbol{I}) + \boldsymbol{R}^2(s)\boldsymbol{B}\right] \\ &= 0 \text{ (from Lemma 2)}.\end{aligned}$$

Similarly for $2 \leq k \leq c-1$, substituting Eq. (4.15) in Eq. (4.3) leads to

$$\begin{aligned}\hat{\boldsymbol{P}}_{k-1}(s)\boldsymbol{C} + \hat{\boldsymbol{P}}_k(s)(\boldsymbol{A}_k - s\boldsymbol{I}) + \hat{\boldsymbol{P}}_{k+1}(s)\boldsymbol{B}_{k+1} &= \hat{\boldsymbol{P}}_{k-1}(s)\boldsymbol{R}_{k-2}(s)\boldsymbol{C} + \hat{\boldsymbol{P}}_{k-1}(s)\boldsymbol{R}_k(s)(\boldsymbol{A}_k - s\boldsymbol{I}) + \hat{\boldsymbol{P}}_k(s)\boldsymbol{R}_{k+1}(s)\boldsymbol{B}_{k+1} \\ &= \hat{\boldsymbol{P}}_{k-2}(s)\boldsymbol{R}_{k-1}(s)[\boldsymbol{C} + \boldsymbol{R}_k(s)(\boldsymbol{A}_k - s\boldsymbol{I}) + \boldsymbol{R}_k(s)\boldsymbol{R}_{k+1}(s)\boldsymbol{B}_{k+1}] \\ &= 0 \text{ (from Lemma 3)}.\end{aligned}$$

Also, substituting Eq. (4.15) in Eq. (4.2) and noting that $\boldsymbol{C}_0 = \boldsymbol{e}_1\boldsymbol{C}$ yields

$$\begin{aligned}\hat{\boldsymbol{P}}_0(s)\boldsymbol{C}_0 + \hat{\boldsymbol{P}}_1(s)(\boldsymbol{A}_1 - s\boldsymbol{I}) + \hat{\boldsymbol{P}}_2(s)\boldsymbol{B}_2 &= \hat{\boldsymbol{P}}_0(s)\boldsymbol{e}_1\boldsymbol{C} + \hat{\boldsymbol{P}}_0(s)\boldsymbol{e}_1\boldsymbol{R}_1^*(s)(\boldsymbol{A}_1 - s\boldsymbol{I}) + \hat{\boldsymbol{P}}_0(s)\boldsymbol{e}_1\boldsymbol{R}_2^*(s)\boldsymbol{B}_2 \\ &= \hat{\boldsymbol{P}}_0(s)\boldsymbol{e}_1[\boldsymbol{C} + \boldsymbol{R}_1(s)(\boldsymbol{A}_1 - s\boldsymbol{I}) + \boldsymbol{R}_2(s)\boldsymbol{R}_1(s)\boldsymbol{B}_2] \\ &= 0 \text{ (from Lemma 3 for } k = 1).\end{aligned}$$

Therefore, it is verified that $\hat{\boldsymbol{P}}_n(s)$ expressed by Eqs. (4.14) and (4.15) satisfies ie governing system of differential equations in the Laplace domain as represented y Eqs. (4.2)–(4.4). Hence, from Eq. (4.1), we get

$$\hat{\boldsymbol{P}}_0(s) = [s - \boldsymbol{A}_0 - \boldsymbol{e}_1\boldsymbol{R}_1(s)\boldsymbol{B}_0]^{-1} \tag{4.16}$$

'here $\boldsymbol{A}_0, \boldsymbol{B}_0$ are known and $\boldsymbol{R}_1(s)$ can be recursively determined from Lemma 3. 'hus, the transient state probabilities of the model under consideration are given by

$$\hat{\boldsymbol{P}}_k(s) = \hat{\boldsymbol{P}}_{c-1}(s)\boldsymbol{R}^{k-c+1}(s), k \geq c$$
$$\hat{\boldsymbol{P}}_k(s) = \hat{\boldsymbol{P}}_{k-1}(s)\boldsymbol{R}_k(s) = \boldsymbol{R}_K^*(s)e_1\hat{P}_0(s); 1 \leq k \leq c-1$$

where $\hat{\boldsymbol{P}}_0(s)$ is given by Eq. (4.16), $\boldsymbol{R}^k(s)$ for all k is given by Eq. (4.7) and $\boldsymbol{R}_k^*(s)$ for $k = 1, 2, \ldots$ are recursively determined using Lemma 3.

Remark When $c = 1$, the model reduces to the transient analysis of an *M*/*M*/1 queue subject to multiple exponential vacation. Accordingly, the above solution reduces $\hat{\boldsymbol{P}}_k(s) = \hat{\boldsymbol{P}}_0(s)e_1\boldsymbol{R}^k(s)$, for $k \geq 1$, which on simplification yields

$$\hat{\boldsymbol{P}}_{0k}(s) = \hat{\boldsymbol{P}}_{00}(s)\rho^k(s) \tag{4.17}$$

and

$$\hat{\boldsymbol{P}}_{1k}(s) = \hat{\boldsymbol{P}}_{00}(s)\beta_c(s)\sum_{i=0}^{k-1}\rho^{k-1-i}(s)r_c^i(s) \tag{4.18}$$

for $k \geq 1$. On simplification, Eq. (4.17) reduces to $\hat{\boldsymbol{P}}_{0k}(s) = \hat{\boldsymbol{P}}_{00}(s)\left(\frac{\lambda}{s+\lambda+\theta}\right)^k; k \geq 1$ which is seen to coincide with the equation below (23) obtained by Sudhesh and Raj [8]. When $k = 1$, Eq. (4.18) reduces to $\hat{\boldsymbol{P}}_{1k}(s) = \hat{\boldsymbol{P}}_{00}(s)\beta_c(s) =$ $\hat{\boldsymbol{P}}_{00}(s)\frac{\theta r_c(s)}{(s+\lambda+\theta)\left(1-\frac{c\mu r_c(s)}{s+\lambda+\theta}\right)}$ which on simplification reduces to $\hat{\boldsymbol{P}}_{1k}(s) =$ $\hat{\boldsymbol{P}}_{00}(s)\frac{\theta}{c\mu}\sum_{m=1}^{\infty}\left(\frac{\frac{\lambda}{\beta}}{s+\lambda+\theta}\right)^m\left(\frac{p-\sqrt{p^2-\alpha^2}}{\alpha}\right)^m$ which is seen to coincide with the Eq. (25) obtained by Sudhesh and Raj [8].

5 Numerical Illustrations

This section illustrates the behaviour of time dependent state probabilities of the system during busy and vacation states for varying values of k. Figure 2 depicts the variation of the functional state probabilities against time for $c = 3, \lambda = 1, \mu = 2$ and $\theta = 0.5$. Since the system is assumed to be initially in vacation state, the curves for $P_{1,k}(t)$ (k = 1, 2, 3, 4, 5) begins at 0. The value of $P_{1,k}(t)$ decreases with increase in. It is seen that $P_{1,k}(t)$ increases with time and converges to the corresponding steady state probabilities, $\pi_{1,k}$ as t tends to infinity. The values of $\pi_{1,k}, k = 1, 2, 3, 4, 5$ are depicted in the figure. Figure 3 depicts the variation of probabilities of functional state against time for $= 3, \lambda = 1, \mu = 2$ and $\theta = 0.5$. Since the system is assumed to be initially in vacation state, the curve for $P_{0,0}(t)$ begins at 1. The value of $P_{0,k}(t)$ decreases with 'ig. 2 Variation of robabilities of busy state of gainst time t

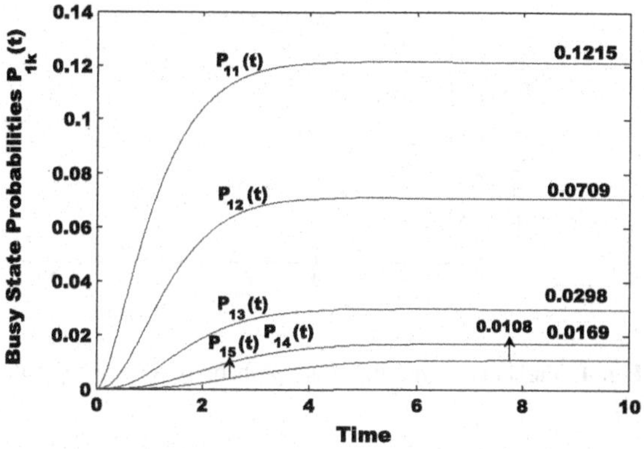

ig. 3 Variation of :obabilities vacation state gainst time t

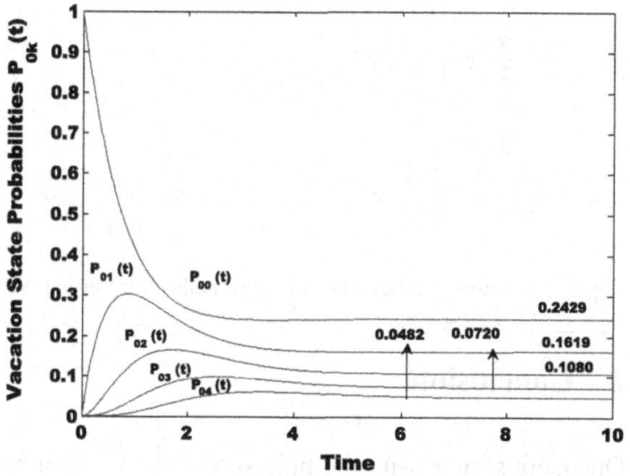

ıcrease in. It is seen that $P_{0,k}(t)(k = 0, 1, 2, 3, 4)$ converges to the corresponding eady state probabilities, $\pi_{0,k}$ as t tends to infinity. The values of $\pi_{0,k}, k =$, 1, 2, 3, 4 are depicted in the figure.

Figures 4 and 5 depicts the variations the probability for the system to be in acation state and functional state respectively against time for varying values of c ıd the same choice of the other parameter values. Mathematically

$$P_0(t) = P(J(t) = 0) = \sum_{k=0}^{\infty} P_{0k}(t) \text{ and } P_1(t) = P(J(t) = 1) = \sum_{k=1}^{\infty} P_{1k}(t)$$

Since the system is assumed to be initially in vacation state, the curve or $P_0(t)$:gins at 1 and that of $P_1(t)$ begins at 0. The value of $P_0(t)$ decreases with increase time and the that of increase with increase in time and converges to the corre- ›onding steady state value. Observe that when $c = 1$, the model reduces to that of ı *M/M/1* queue subject to multiple exponential vacation.

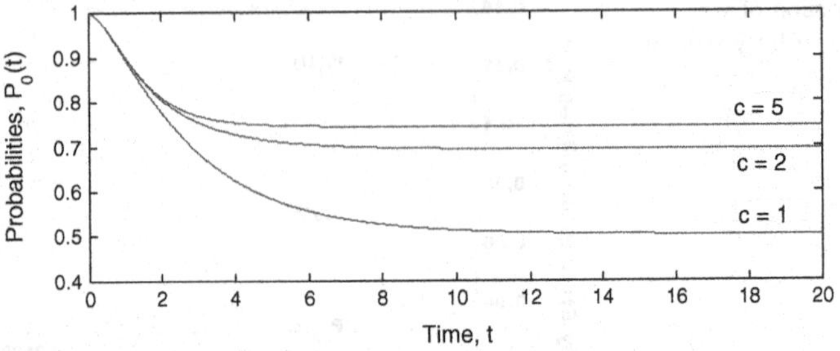

Fig. 4 Variation of vacation state probability for varying values of c

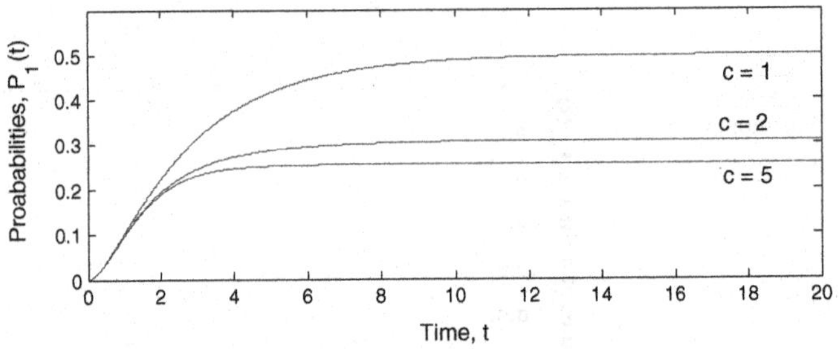

Fig. 5 Variation of functional state probability for varying values of c

6 Conclusion

Queueing system with multiple server plays a vital role owing of their application in computer system, communication network, production management etc.,. I particular, *M/M/c* queue subject to vacation is of a great interest in view of their re time applications. Extensive work on both the steady state and transient analysis c *M/M/c* queueing model and its variations are done by many authors. This paper i the first of its kind to present the transient analysis of multi server model subject t multiple exponential vacation. Explicit solution for the state probabilities are pre sented in the stationary regime, however due to the complexity of the problem, th transient solution are presented in the Laplace domain using matrix analyti method.

References

1. Daniel, A., Paul, A., Ahmed, A.: Queueing model for congnitive radio vehicular network. I International Conference on Platform Technology and Service, pp. 9–10 (2015)
2. Vijayalakshmi, P., Jyothsna, K.: Balking and reneging multiple working vacations queue wi heterogeneous servers. J. Math. Model. Algorithms Oper. Res. **10**(3), 224–236 (2014)

Lin, C.-H., Ke, J.-C.: Multi-server system with single working vacation. Appl. Math. Model. **33**, 2967–2977 (2009)
Tian, N., Li, Q.-L., Gao, J.: Conditional stochastic decompositions in M/M/c queue with server vacations. Commun. Stat. Stochast. Models **15**, 367–377 (1999)
Parthasarathy, P.R., Sharafali, M.: Transient solution to the many-server Poisson queue: a simple approach. J. Appl. Probab. **26**, 584–594 (1989)
Al-Seedy, R.O., El-Sherbiny, A.A., El-Shehawy, S.A., Ammar, S.I.: Transient solution of the M/M/c queue with balking and reneging. Comput. Math Appl. **57**, 1280–1285 (2009)
Ammar, Sherif I.: Transient analysis of a two-heterogeneous servers queue with impatient behavior. J. Egypt. Math. Soc. **22**(1), 90–95 (2014)
Sudhesh, R., Raj, F.: Computational Analysis of Stationary and Transient Distribution of Single Server Queue with Working Vacation. Springer, Berlin, Heidelberg, vol. 1, pp. 480–489 (2014)

ˀractional Filter Based Internal Model ˀontroller for Non Linear Process

ˀ. Anbumalar, C. Barath Kanna and J. Janani

ıbstract Recently, fractional calculus found its applications in the field of process ıutomation. Here the mathematical perspectives of fractional property are utilized to ıontrol the process. This article presents a design of fractional filter based internal ıodel controller (IMC) for the nonlinear process. The process chosen is pH process lays significant role in the industrial applications. Control of pH process is highly ıhallenging due to its severe nonlinear characteristics and process uncertainty as ıchoed in the titration curve. An attempt has been made to design the fractional ılter based IMC controller for pH process based on desired phase margin and cross ıver frequency using MATLAB-SIMULINK. Here Oustaloup approximation ıchnique is used to approximate the Fractional filter. Servo and regulatory perıormance of the process are analyzed and reflects that better control actions are ıchieved by the fractional filter based IMC through simulation and it is found to be ıatisfied.

ıeywords Fractional calculus · Oustaloup approximation technique · Internal ıodel controller

. Anbumalar (✉) · J. Janani
ılectronics and Instrumentation Engineering, B. S. AbdurRahman University,
hennai, Tamil Nadu, India
ımail: anbumalarice@gmail.com

Janani
ımail: janani.govindasamy@gmail.com

. Barath Kanna
ıstrumentation and Control Engineering, National Institute of Technology,
ırichy, Tamil Nadu, India
mail: barath@nitt.edu

I. Senthilkumar et al. (eds.), *Computational Intelligence, ˀyber Security and Computational Models*, Advances in Intelligent ıystems and Computing 412, DOI 10.1007/978-981-10-0251-9_52

1 Introduction

In spite of tremendous development in the field of advanced control schemes in th process industries, use of conventional proportional integral derivativ (PID) controller is remarkable because of the reduced number of parameters to b tuned and also they give satisfactory servo and regulatory performance for a wid range of operating conditions for different processes. The primary function o designed controller is to bring the process variable to set point by utilizing minimun energy optimally. Different tuning methods are available to determine the controlle parameters (proportional, integral and derivative gain) based on the time and th frequency domain specifications as cited in [1]. Now a days, the closed loop performance of the process has been improved by applying the concept of fractiona calculus in controller design. These Fractional-order proportional-integral-derivativ (FOPID) controllers more robust than the integer order (IO) controller due to th fractional powers of the integral and derivative *s* terms. Fractional order calculus i the further development of integer order calculus whose power is of non intege values. Real world processes such as voltage-current relation of a semi-infinite loss RC line, the diffusion of heat into a semi-infinite solid, heating furnace and ga turbine was found to have the fractional characteristics [2]. There are different tunin rules and auto tuning techniques are available for FO PID controller design with th IO process based on frequency domain specifications and QFT techniques t guarantee the system optimum performances [2–9]. The optimization technique such as genetic algorithm and the particle swarm optimization techniques are als used to obtain the FO PID parameters [9, 10]. For FOS, the FO controller parameter are tuned using gain margin and phase margin specifications [11–17]. Advanced F controllers such as fractional order lead compensator [18], self tuning regulator [19 model reference adaptive control [20], adaptive high gain controller [21], and slidin mode controller [22] have been implemented to improve closed loop performanc and robustness. It has been proved that, the stability region of the fractional orde PID controllers is wider than the integer order PID controller [23]. Wider the stability region, then the controller is more robust i.e. it can accommodate more mode uncertainties compared to integer order controller. As further development in th field of fractional control theory, fractional order controller based internal mode control (IMC) has been developed based on the desired band width specification [15]. Design methodology focused on IMC based FO PID controller has bee developed for the IO first order plus time delay process [24] and a class of FO [25, 26] to improve the process stability and robustness. In this paper, an attempt ha been made to design fractional filter based IMC controller for the nonlinear proces Here the process chosen is pH process. Controlling pH is very important in man processes. The excessive non-linearity of the pH process makes control by a conventional linear PID controller difficult.

Process Descriptions

H is a measurement of concentration of hydronium ions ([H_3O^+] commonly bbreviated to [H^+]) in aqueous solution and it has its own unique attributes, the nost distinctive of which is characterized by a neutralization or titration curve. pH the negative logarithm of hydronium ion (or hydrogen ion) concentration, and his results in the familiar 'S' or 'Z' shaped titration curve which defines the teady-state characteristic of a pH process. Because of logarithmic nonlinearity, it is ossible for the gain of a pH process to change by as much as a factor of 10 per pH nit which is given by the following equation.

$$\mathrm{pH} = -\log_{10[\mathrm{H}^+]}$$

The process considered for the control design is a batch process. The acetic acid added with sodium hydroxide to form sodium acetate in a continuously stirred nk as shown in the Fig. 1. Here volume is considered to be constant. Either the out t from the process tank should be at the top of the tank or there should be a level neasurement and control unit to check the level and use an on/off control for the nlet flow. After the volume needed (1000 l) is reached. The content of the tank is nixed well using a stirrer.

Using a pH sensor, the pH value of the solution is measured. Here, base flow rate kept constant and the acid flow-rate is continuously varied in steps. Sodium ydroxide of constant flow-rate is continuously neutralized by Acetic acid which cts as a manipulated variable. The steady-state of process occurs when the con- entrations of hydrogen and hydroxyl ions are equivalent, for different base flow ate steady state pH is obtained. pH of the mixture is measured continuously using H sensor and transmitted to the pH transmitter.

Based on mass balance equation, dynamic equations have been derived to xpress the pH process [27].

$$v\frac{dx_a}{dt} = F_1C_1 - (F_1 + F_2)x_a$$
$$v\frac{dx_b}{dt} = F_2C_2 - (F_1 + F_2)x_b$$

g. 1 pH process

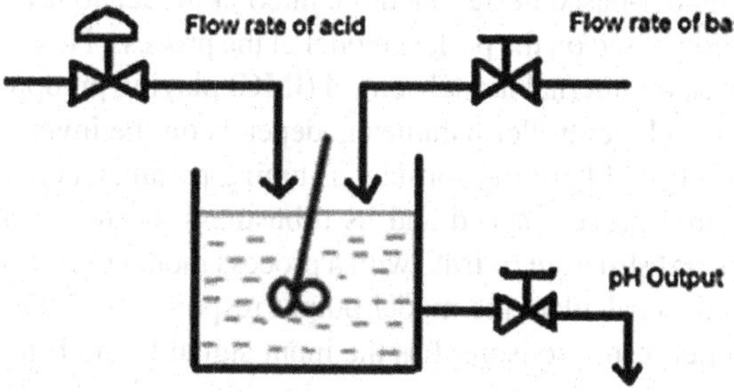

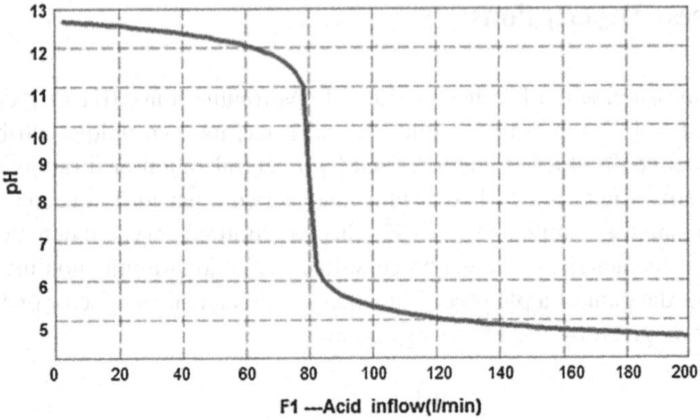

Fig. 2 Titration curve

The pH process considered is acetic acid (weak acid) added with sodium hydroxide (strong base) to form sodium acetate. The general equation for the pH process may be written as follows

$$x_b + 10^{-\mathrm{pH}} - 10^{\mathrm{pH}-14} - \frac{x_a}{1 + 10^{pk_a - \mathrm{pH}}} = 0$$

The titration curve or the I/O characteristic curve should be obtained which gives the perfect dynamics of the process. The titration curve for pH process is shown in Fig. 2, has 'Z' shaped curve due to the neutralization of base with acid flow rate with respect to the process parameters as listed in Table 1. And the pH value stop increasing around 4 because of buffering effect that is caused by the weak acid and its salt. Servo response of pH process with conventional PI controller at different operating regions is shown in Fig. 3. From the response it is clear that, performance of process with PI controller is not found to be satisfactory due to the process nonlinearity.

3 Controller Design

A model-based design method, internal model control (IMC) achieves better control action based on the perfect model of the process. Now a days, design of PID controller based on internal model control (IMC) playing predominant role in the control world in which controller parameters depends on the inverse of process model parameter with IMC filter time constant as tuning parameters tuned based on trade off between desired process speed and its robustness to uncertainties [24]. Figure 4 shows the general structure of IMC with a process model $G_m(s)$ and the controller output 'u' are used to calculate the model output response 'y_m'. Difference of plant response and model response is used as the input signal to the IMC controller $C_{IMC}(s)$.

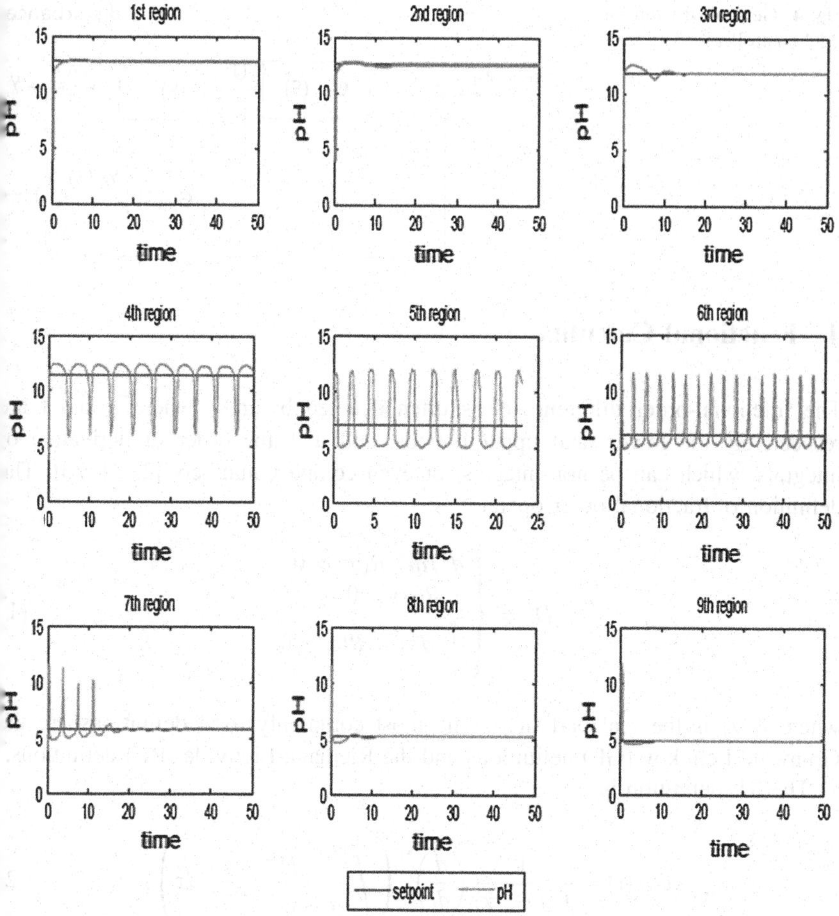

ig. 3 Closed loop response of pH process with PI controller

able 1 Process parameters

Process variables/parameters	Nominal values
Acid dissociation constant (k_a)	$1.8 * 10^{-5}$
Water dissociation constant (K_w)	10^{-14}
Acid inflow rate (l/min)(F_1)	0–200
Base inflow rate (F_2)	512
Acid inflow concentration (mol/l)(C_1)	0.32
Base inflow concentration (mol/l)(C_2)	0.05
Volume of the tank (l^3)	1000

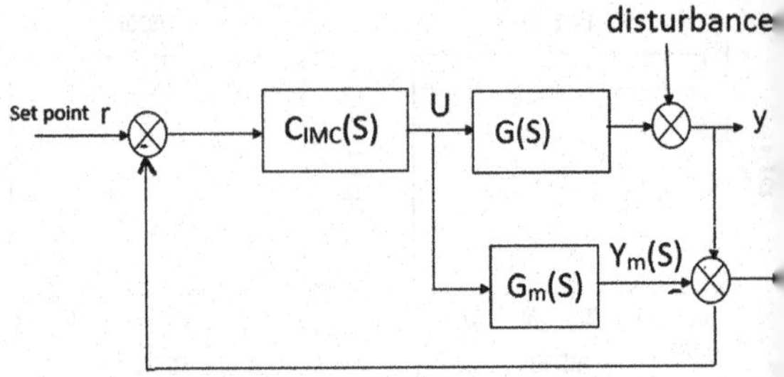

Fig. 4 General structure of IMC controller

4 Fractional Calculus

The fractional-order differentiator is often denoted by aD_t^α, where a and t are respectively the lower- and upper bounds, and α is the order of derivative or integrals, which can be non-integers, or even complex numbers [2, 21–25]. The definition of fractional-order operator is:

$$_aD_t^\alpha = \begin{cases} d^\alpha/dt^\alpha, R(\alpha) > 0 \\ 1, R(\alpha) = 0 \\ \int_a^t (d\tau)^{-\alpha}, R(\alpha) < 0 \end{cases} \quad (1)$$

where $R(\alpha)$ is the real part of α. The most commonly used definitions are the Grunwald-Letnikov (GL) definitions and the Riemann-Liouville (RL) definitions.

The RL definition is

$$_aD_t^\alpha f(t) = \frac{1}{\Gamma(m-\alpha)} \left(\frac{d}{dt}\right)^m \left(\int_a^t \frac{f(\tau)}{(t-\tau)^{1-(m-\alpha)}} d\tau \right) \quad (2)$$

where m − 1 < α < m, and $\Gamma(x)$ is the well-known Euler Gamma function.

The GL definition is

$$_aD_x^\alpha[f(t)] = \lim_{h \to 0} \frac{1}{\Gamma(\alpha)h^\alpha} \sum_{k=0}^{[x-a/h]} \frac{\Gamma(\alpha+1)}{\Gamma(k+1)} f(t-kh) \quad (3)$$

Fractional order system as contradictory to the integer order system represented by powers of non integer values.

$$G(s) = \frac{1}{a_n s^{\beta_n} + a_{n-1} s^{\beta_{n-1}} + \cdots + a_1 s^{\beta 1} + a_0 s^{\beta_0}}$$

where $\beta_{k,} (k = 0, 1, \ldots, n)$ is a non integer numbers.

In other words, fractional order systems better describes the dynamic behavior of ne process expressed as the fractional order differential equations given by the ollowing expressions

$$a_n D_t^{\beta_n} y(t) + a_{n-1} D_t^{\beta_{n-1}} y(t) + \cdots + a_1 D_t^{\beta_1} y(t) + b_0 D_t^{\beta_0} y(t) = u(t)$$

.1 *Fractional Filter Based IMC Controller for Integer Order System*

this section, internal model controller cascaded with filter having fractional haracteristics has been designed for pH process. The primary objective of the esign is to control pH process with desired closed loop specifications based on the ming techniques as reported by Maamar et al. [26] method of designing fractional lter based internal model controller for a class of systems. General procedure to esign of fractional filter based IMC controller is summarized as follows:

To design IMC controller, process model $G_m(s)$, is factorized as non invertible nd invertible part.

$$G_m(s) = G_m^+(s) G_m^-(s)$$

Here $G_m^+(s)$ is the invertible part of process model $G_m(s)$ and $G_m^-(s)$ is the non vertible part of process respectively.

General structure of fractional filter is given by $f(s) = \frac{1}{1+\tau_c s^\alpha}$ $0 < \alpha < 1$

Time constant τ_c and non integer $\boldsymbol{\alpha}$ are selected based on desired phase margin $_m$ and the cross over frequency w_c of the closed loop.

$$\alpha = \frac{\pi - \phi_m}{\pi/2} - 1 \text{ and } \tau_c = \frac{1}{w_c^{\alpha+1}}$$

For the given phase margin and closed loop corner frequency, controller ructure is given by

$$c_{IMC(S)} = \frac{1}{G_m^-(s)} f(s)$$

Feedback controller C(s) is then $C(s) = \frac{c_{IMC(S)}}{1 - c_{IMC(S)} G_m(s)}$

Structure of fractional filter based IMC controller is represented as

$$C(s) = f(s) \cdot k_p \left(1 + \frac{1}{\tau_i s} + \tau_d s\right)$$

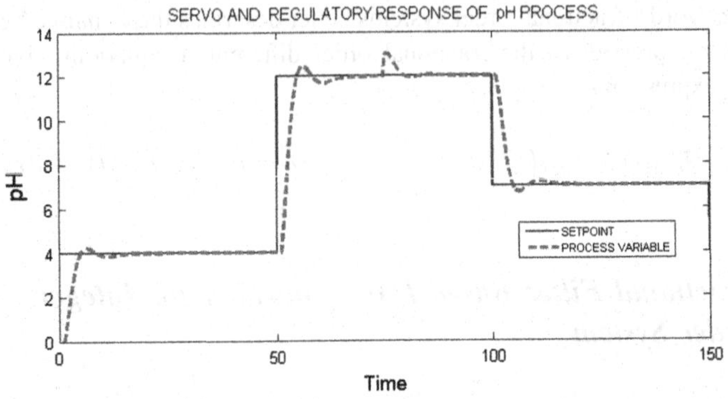

Fig. 5 Servo and regulatory performance of pH process with fractional filter based IMC controller

Table 2 Performance analysis of pH process based on ISE

Integral of squared value of error (ISE)	PI controller	Fractional filter based IMC
τ_c	691	33.25
+10 % change in τ_c	720	34.38
−10 % change in τ_c	705	32.12

For the first order process with time delay, controller structure can be rewritten as

$$C(s) = \frac{1}{1+\frac{2\tau_c s}{\theta}} \frac{2T+\theta}{K\theta}\left(1+\frac{1}{\frac{2T+\theta}{2}s}+\frac{T\theta}{2T+\theta}s\right)$$

where $k_p = \frac{2T+\theta}{K\theta}; \tau_i = \frac{2T+\theta}{2}; \tau_d = \frac{T\theta}{2T+\theta}$

Here a continued fraction expansion method is used to approximate the fractional filter $f(s)$ which is an expression obtained through an iterative process of representing a number as the sum of its integer part and the reciprocal of another number, then writing this other number as the sum of its integer part and another reciprocal, and so on. Figure 5 shows the servo performance of pH process with fractional filter based Internal model controller. From the simulation results, it is clear that the designed fractional filter based IMC controller for particular operating region has an ability to keep track of different set point changes (Fig. 6). Table shows the performance analysis of fractional based IMC controller with PI controller for change in process time constant based on integral of squared value of error. It is observed that fractional filter gives satisfactory servo and regulatory performance.

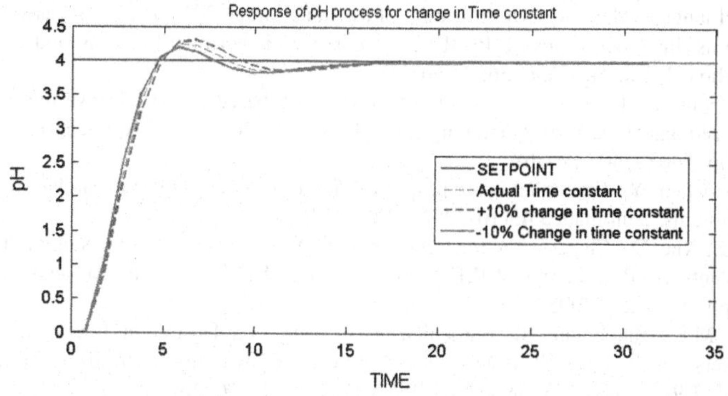

g. 6 Robustness analysis of pH process with fractional filter based IMC controller

Conclusion

this paper fractional filter based internal model controller has been developed for e ph process. Fractional filter is approximated by the continued fraction expansion ethod. It should be noted that designed fractional filter based IMC around the rticular operating point is found to give the satisfactory performances at all other erating points. Further servo and regulatory performance of the non linear pross are analyzed through the simulation results, it has been inferred that fractional ter based internal model control of pH process gives satisfactory performance. rther more real time implementation of controller for pH process can carried out future work.

eferences

. Astrom, K.J., Hagglund, T.: Advanced PID control. ISA (2006)
. Xue, D., Petras, I., Chen, Y.: Fractional order control—a tutorial. In: American Control Conference Hyatt Regency Riverfront, St. Louis, Mo, USA, 10–12 June 2009
. Cervera, Banos, A., Monje, C.A., et al.: Tuning of fractional PID using QFT. In: Proceedings of 32nd IEEE Conference on Industrial Electronics, Paris, IEEE, pp. 5402–5407 (2006)
. Chen, Y., Bhaskaran, T., Xue, D.: Practical tuning rule development for fractional order proportional and derivative controllers. J. Comput. Nonlinear Dyn. pp. 021403.1–021403.8 (2008)
. Li, H., Chen, Y.: A fractional order proportional and derivative (Fopd) controller tuning algorithm. In: Proceedings of 47th IEEE Conference on Control and Decision, Mexico, IEEE, pp. 4059–4063 (2008)
. Maione, G., Lino, P.: New tuning rules for fractional Pi® controllers. Nonlinear Dyn. **49**(1/2), 251–257 (2007)
. Monje, C.A., Vinagre, B.M., Feliu, V., et al.: Tuning and auto tuning of fractional order controllers for industry applications. Control Eng. Pract. **16**(7), 798–812 (2008)
. Valerio, D., Da Costa, J.S.: Tuning of fractional PID controllers with Ziegler-Nichols-type rules. Sig. Process. **86**(10), 2771–2784 (2006)

9. Yi, C., Liang, J., Gang, C.: Optimization of fractional order PID controllers based on geneti algorithms. In: Proceedings of 4th IEEE Conference on Machine Learning and Cybernetics Guangzhou: IEEE, pp. 5686–5689 (2005)
10. Yi, C., Gang, C.: Design of fractional order controllers based on particle swarm optimization In: Proceedings of 1st IEEE Conference on Industrial Electronics and Applications, Singapore IEEE, pp. 5402–5407 (2006)
11. Luo, Y., Chen, Y.: Fractional order proportional derivative controller for a class of fractiona order systems. Automatica **45**(10), 2446–2450 (2009)
12. Zhao, C., Xue, D., Chen, Y.: A fractional order PID tuning algorithm for a class of fractiona order plants. In: Proceedings of IEEE Conference on Mechatronics and Automation, Canada IEEE, pp. 216–221 (2005)
13. Shahri, M.E., Balochian, S., Balochian, H., Zhang, Y.: Design of fractional order PI controllers for time delay systems using differential evolution algorithm. Indian J. Sc Technol. **7**(9), 1307–1315, Sept 2014. ISSN (Print): 0974–6846
14. Zhao, C., Xue, D.: A fractional order pid tuning algorithm for a class of fractional order plant In: IEEE Transactions, July 2005
15. Li, H., Luo, Y., Chen, Y.: A fractional order proportional and derivative (FOPD) motio controller. In: IEEE Transactions on Control Systems Technology (2009)
16. Li, H., QuanChen, Y. Fractional order proportional and derivative (FOPD) controller tunin algorithm. IEEE Transactions on (2008)
17. Zhao, C., Xue, D.: Closed-form solutions to fractional-order linear differential equation Front. Electron. Eng. China 2008:214–217
18. Monje, L.A., Feliu, V.: The fractional order lead compensator. In: Proceedings of 2nd IEE Conference on Computational Cybernetics, Vienna, pp. 347–352 (2004)
19. Maiti, D., Chakraborty, M., Acharya, A. et al.: Design of Fractional order Self Tunin Regulator Using Optimization Algorithm. In: Proceedings of 2nd IEEE Conference o Computer and Information Technology, Singapore, IEEE, pp. 470–475 (2008)
20. Ladaci, S., Charef, A.: Model reference adaptive control with fractional derivative. I Proceedings of International Conference on Tele-Communication Systems, Medic Electronics and Automation, Algeria (2003)
21. Ladaci, S., Loiseau, J.J., Charef, A.: Fractional order adaptive high gain controllers for a cla of linear systems. Commun. Nonlinear Sci. Numer. Simul. **13**(4), 707–714 (2008)
22. Delvari, H., Ghaderi, R., Ranjbar, A., et al.: Fuzzy fractional order sliding mode controller f non linear systems. Commun. Innonlinear Sci. Numer. Simul. **15**(4), 963–978 (2010)
23. Hamamci, S.E., Koksal, M.: Calculation of all stabilizing fractional order PD controllers f integrating time delay systems. Comput. Math Appl. **59**(5), 1621–1629 (2010)
24. Vino Praba, T., Sivakumaran, N., Selvaganesan, N.: Stabilization using fractional order PID controllers for first order time delay system. In: Proceedings of International Conference Advances in Computing Control and Telecommunication Technologies, Kerala
25. Vinopraba, T., Sivakumaran, N., Narayanan, S., Radhakrishnan, T.K.: Design of intern model control based fractional order PID controller. J. Control Theory Appl. **10**(3), 297–3 (2012)
26. Maâmar, B., Rachid, M.: IMC-PID-fractional-order-filter controllers design for integer ord systems. In: ISA Transactions, pp. 1620–1628 (2014)
27. McAvoy, T.J., Hsu, E., Lowenthals, S.: Dynamics of pH in controlled stirred tank reactor. In Eng. Chem. Process. Des. Develop. **11**(1), 68–78 (1972)

\ Novel Approach for Solving Triangular ınd Trapezoidal Intuitionistic Fuzzy ;ames Using Dominance Property ınd Oddment Method

ſI. Joseph Robinson, S. Sheela and A. Sudha Rani

.bstract In this Paper, We examine Intuitionistic Fuzzy Game Theory problem in 'hich cost co-efficients are triangular and trapezoidal intuitionistic fuzzy numbers. ı conventional game theory problem, cost is always certain. This paper develops n approach to solve an intuitionistic fuzzy games where cost is not deterministic umbers but imprecise ones. Here, the elements of the costs (profits) matrix of the ames are triangular and trapezoidal intuitionistic fuzzy numbers. Then its membership and non-membership functions are defined. A ranking technique is used to ompare the intuitionistic fuzzy numbers so that Dominance property in ıtuitionistic Fuzzy Games may be applied and later Dominance property in ıtuitionistic Fuzzy Oddments Method is used to solve the intuitionistic fuzzy ames. Numerical examples show that an intuitionistic fuzzy ranking technique ffers an effective tool for handling an intuitionistic fuzzy games.

.eywords Fuzzy set · Intuitionistic fuzzy set · Triangular intuitionistic fuzzy umber · Trapezoidal intuitionistic fuzzy number · Intuitionistic fuzzy games · ſagnitude of triangular intuitionistic fuzzy number · Magnitude of trapezoidal ıtuitionistic fuzzy number and · Value of the game

Introduction

ı modern era there are lot of situations in the society where there is a conflicting ıterest situation and such situation is handled by game theory. But there are lot of ases where the information given are not in precise manner and in such situation

---

ſ. Joseph Robinson
epartment of Mathematics, T.J.S Engineering College, Peruvoyal 601206, India

Sheela (✉) · A. Sudha Rani
epartment of Mathematics, Gojan School of Business and Technology,
edhills, Chennai 600052, India
mail: sheela2009.s@gmail.com

ſ. Senthilkumar et al. (eds.), *Computational Intelligence,*
yber Security and Computational Models, Advances in Intelligent
ystems and Computing 412, DOI 10.1007/978-981-10-0251-9_53

we apply fuzzy mathematics to get a solution. Fuzziness in matrix games ca appear in so many ways but two classes of fuzziness are very common. These tw classes of fuzzy matrix games are referred as matrix games with Fuzzy goal [1] an matrix games with fuzzy payoff [2]. In such situations, the fuzzy sets introduced by Zadeh [3], is very useful tool in game theory. Various attempts have been made i the literature to study fuzzy game theory. Sakawa and Nishizaki [2] introduce max-min solution procedure for multi-objective fuzzy games. Atanassov [4, 5] firs introduced the concept of IF-set where he explained an element of an IF-set i respect of degree of belongingness, degree of non-belongingness and degree c hesitancy. This degree of hesitancy is nothing but the uncertainty in taking decision by a decision maker (DM).

Atanassov [6] firstly described a game problem using the IF-set. Li [7] studie matrix games with payoffs represented by Atanassov's interval-valued intuitionisti fuzzy sets. Li and Nan [8] studied the matrix games with payoffs of IF-sets. Li an Nan [8] studied matrix game with intuitionistic fuzzy goals. Nayak and Pal [9 studied the bi-matrix games with intuitionistic fuzzy goals. A triangular intuition istic fuzzy number is a special kind of IF-set, and has been used to measure th payoffs in matrix game [10, 11]. The ranking of intuitionistic fuzzy numbers play an important role with finding the solution of matrix games with payoffs of intu itionistic fuzzy numbers. Nan et al. defined the average indices of the membershi degree and the non-membership degree and given a lexicographic ranking metho for triangular intuitionistic fuzzy number. In this paper, We have considered matrix game where the elements of the pay-off matrix are all intuitionistic fuzz numbers.

In this work, we have concentrated on the solution of Two Person Zero Su Games by using Dominance Rule and Oddment method using Triangular an Trapezoidal Intuitionistic Fuzzy Numbers. They are characterized by their simpl formulations and computational efficiency and thus have been used to solve dif ferent problems in Engineering and Management.

This paper is organized as follows: Sect. 2 deals with some basic terminolog and ranking of Triangular and Trapezoidal Intuitionistic fuzzy numbers. Section describes the computational procedure for solving $n \times n$ games using oddment method. Section 4 describes the general rule for Dominance Property, Sect. illustrate the proposed method with numerical examples and followed by th conclusions are given in Sect. 6.

2 Preliminaries

In this section we discuss some basic terminology on Game theory and IFS. Als we discuss about the ranking of Triangular and Trapezoidal Intuitionistic fuzz numbers and some based on ranking of Triangular and Trapezoidal Intuitionisti fuzzy numbers.

Definition 2.1 (*Two Person Zero Sum Games*) If the algebraic sum of gains and losses of all the players is zero in a game then such game is called Zero-Sum Games. Otherwise the game is called a Non-Zero Sum Game.

Definition 2.2 (*Pay-Off Matrix*) Zero Sum Games with 2 Players are called Rectangular Games. In this case the loss (gain) of one player is exactly equal to the gain (loss) of the other. The gains resulting from a Two Person Zero Sum Game can be represented in the matrix form, usually called Pay-off Matrix.

Definition 2.3 [4] Let A be a classical set, $\mu_A(x)$ be a function from A to $[0, 1]$, A fuzzy set A^* with the membership function $\mu_A(x)$ is defined by

$$A^* = \{(x, \mu_A(x)); x \in A \text{ and } \mu_A(x) \in [0, 1]\}$$

Definition 2.4 [6] Let X be a universe of discourse, then an intuitionistic fuzzy set in X is given by a set of ordered triples. $\tilde{A}' = \{<x, \mu_A(x), \vartheta_A(x)> ; x \in X\}$ where $\mu_A, \vartheta_A : X \to [0, 1]$, are functions such that $\leq \mu_A(x) + \vartheta_A(x) \leq 1, \forall x \in X$. For each x the membership $\mu_A(x)$ and $\vartheta_A(x)$ represent the degree of membership and the degree of non-membership of the element $x \in X$ to $A \subset X$ respectively.

Definition 2.5 [6] An intuitionistic fuzzy number $\tilde{A}' = (a_1, a_2, a_3)(a'_1, a_2, a'_3)$ is said to be triangular intuitionistic fuzzy number (TrIFN) if its membership and non-membership functions are respectively given by,

$$\mu_A(x) = \begin{cases} 0 & x < a_1 \\ \frac{x-a_1}{a_2-a_1} & a_1 \leq x \leq a_2 \\ 1 & x = a_2 \\ \frac{a_3-x}{a_3-a_2} & a_2 \leq x \leq a_3 \\ 0 & x > a_3 \end{cases} \qquad \vartheta_A(x) = \begin{cases} 1 & x < a'_1 \\ \frac{a_2-x}{a_2-a'_1} & a'_1 \leq x \leq a_2 \\ 0 & x = a'_1 \\ \frac{x-a_2}{a'_3-a_2} & a_2 \leq x \leq a'_3 \\ 1 & x > a'_3 \end{cases}$$

where $a'_1 \leq a_1 \leq a_2 \leq a_3 \leq a'_3$ and $\mu_A(x), \mu_B(x) \leq 0.5$. For $\mu_A(x) = \vartheta_A(x), \forall x \in R$.

Definition 2.6 [6] An intuitionistic fuzzy number $\tilde{A}' = (a_1, a_2, a_3, a_4)(a'_1, a_2, a_3, a'_4)$ is said to be trapezoidal intuitionistic fuzzy number (TrpIFN) if its membership and non-membership functions are respectively given by,

$$\mu_A(x) = \begin{cases} \frac{x-a_1}{a_2-a_1} & a_1 \leq x \leq a_2 \\ 1 & a_2 \leq x \leq a_3 \\ \frac{a_4-x}{a_4-a_3} & a_3 \leq x \leq a_4 \\ 0 & otherwise \end{cases} \qquad \vartheta_A(x) = \begin{cases} \frac{a_2-x}{a_2-a'_1} & a'_1 \leq x \leq a_2 \\ 0 & a_2 \leq x \leq a_3 \\ \frac{a_3-x}{a_3-a'_4} & a_3 \leq x \leq a'_4 \\ 1 & otherwise \end{cases}$$

where $a'_1 \leq a_1 \leq a_2 \leq a_3 \leq a_4 \leq a'_4$ and $\mu_A(x), \mu_B(x) \leq 0.5$. For $\mu_A(x) = \vartheta_A(x)$, $x \in R$. This TrpIFN is denoted by $(a_1, a_2, a_3, a_4)(a'_1, a_2, a_3, a'_4)$.

Definition 2.7 Let $\tilde{A}' = (a_1, a_2, a_3)(a_1', a_2, a_3')$ be a triangular intuitionistic fuzzy number, then Magnitude of triangular intuitionistic fuzzy number $\tilde{A}'$ is defined as

$$\begin{aligned}|\tilde{A}'| = \mathrm{Mag}[\tilde{A}'] &= \mathrm{Mag}[(a_1, a_2, a_3)(a_1', a_2, a_3')] \\ &= [(-a_3, -a_2, -a_1)(-a_3', -a_2, -a_1')].\end{aligned}$$

Definition 2.8 Let $\tilde{A}' = (a_1, a_2, a_3, a_4)(a_1', a_2, a_3, a_4')$ be a trapezoidal intuitionistic fuzzy number, then Magnitude of trapezoidal intuitionistic fuzzy number $\tilde{A}'$ is defined as,

$$\begin{aligned}|\tilde{A}'| = \mathrm{Mag}[\tilde{A}'] &= \mathrm{Mag}[(a_1, a_2, a_3, a_4)(a_1', a_2, a_3, a_4')] \\ &= [(-a_4, -a_3, -a_2, -a_1)(-a_4', -a_3, -a_2, -a_1')]\end{aligned}$$

Results

i. Let $\tilde{A}' = (a_1, a_2, a_3)(a_1', a_2, a_3')$ be a triangular intuitionistic fuzzy number, then

$$\begin{aligned}&R[(a_1, a_2, a_3)(a_1', a_2, a_3')] = R[(a_3, a_2, a_1)(a_3', a_2, a_1')] \\ &= R[(a_1, a_2, a_3')(a_1', a_2, a_3)] = R[(a_1', a_2, a_3)(a_1, a_2, a_3')]\end{aligned}$$

ii. Let $\tilde{A}' = (a_1, a_2, a_3, a_4)(a_1', a_2, a_3, a_4')$ be a trapezoidal intuitionistic fuzzy number, then

$$\begin{aligned}&R[(a_1, a_2, a_3, a_4)(a_1', a_2, a_3, a_4')] = R[(a_4, a_2, a_3, a_1)(a_4', a_2, a_3, a_1')] \\ &= R[(a_1, a_2, a_3, a_4')(a_1', a_2, a_3, a_4)] = R[(a_1', a_2, a_3, a_4)(a_1, a_2, a_3, a_4')]\end{aligned}$$

Arithmetic Operations

Let $A = (a_1, a_2, a_3, a_4)(a_1', a_2, a_3, a_4')$ and $B = (b_1, b_2, b_3, b_4)(b_1', b_2, b_3, b_4')$ be any two TrpIFNs then the following arithmetic operations as follows:

Addition:

$$A + B = (a_1 + b_1, a_2 + b_2, a_3 + b_3, a_4 + b_4)(a_1' + b_1', a_2 + b_2, a_3 + b_3, a_4' + b_4')$$

Subtraction:

$$A - B = (a_1 - b_4, a_2 - b_2, a_3 - b_3, a_4 - b_1)(a_1' - b_4', a_2 - b_2, a_3 - b_3, a_4' - b_1')$$

Multiplication:

$$A * B = (a_1 * b_1, a_2 * b_2, a_3 * b_3, a_4 * b_4)(a_1' * b_1', a_2 * b_2, a_3 * b_3, a_4' * b_4')$$

Ranking Technique of Intuitionistic Fuzzy Numbers

The Ranking of a triangular intuitionistic fuzzy number $\tilde{A}' = (a_1, a_2, a_3)(a_1', a_2, a_3')$ is defined by

$$R(\tilde{A}') = \frac{1}{3}\left[\frac{(a_3' - a_1')(a_2 - 2a_1' - 2a_3') + (a_3 - a_1)(a_1 + a_2 + a_3) + 3(a_3'^2 - a_1'^2)}{(a_4' - a_1' + a_4 - a_1)}\right]$$

The Ranking of a trapezoidal intuitionistic fuzzy number $\tilde{A}' = (a_1, a_2, a_3, a_4)$ (a_1', a_2, a_3, a_4') is defined by

$$R(\tilde{A}') = \frac{1}{4}\left[\frac{(a_4' - a_1')(a_2 + a_3 - 3a_1' - 3a_4') + (a_4 - a_1)(a_1 + a_2 + a_3 + a_4) + 4(a_4'^2 - a_1'^2)}{(a_4' - a_1' + a_4 - a_1)}\right]$$

The ranking technique is, if $R(A) \leq R(B)$ iff $A \leq B$

3 The Computational Procedure for Solving $n \times n$ Intuitionistic Fuzzy Games Using Oddments Method

In this section, we are discuss about the computational procedure for solving $n \times n$ triangular and trapezoidal intuitionistic fuzzy games using oddments method

Step 1: Let $A = \begin{pmatrix} a_{11} & a_{12} & a_{13} & \cdots & a_{1n} \\ a_{21} & a_{22} & a_{23} & \cdots & a_{2n} \\ . & . & . & \cdots & . \\ . & . & . & \cdots & . \\ . & . & . & \cdots & . \\ a_{n1} & a_{n2} & a_{n3} & \cdots & a_{nn} \end{pmatrix}$ be $n \times n$ TrIIF or TrpIIF pay-off matrix. Obtain a new TrIIF or TrpIIF matrix C and R are as follows:

$$C = \begin{pmatrix} a_{11}-a_{12} & a_{12}-a_{13} & \cdots & a_{1(n-1)}-a_{1n} \\ a_{21}-a_{22} & a_{22}-a_{23} & \cdots & a_{2(n-1)}-a_{2n} \\ \cdot & \cdot & \cdots & \cdot \\ \cdot & \cdot & \cdots & \cdot \\ \cdot & \cdot & \cdots & \cdot \\ a_{n1}-a_{n2} & a_{n2}-a_{n3} & \cdots & a_{n(n-1)}-a_{nn} \end{pmatrix}_{n\times(n-1)}$$

$$R = \begin{pmatrix} a_{11}-a_{21} & a_{12}-a_{22} & \cdots & a_{1(n-1)}-a_{1n} \\ a_{21}-a_{31} & a_{22}-a_{32} & \cdots & a_{2(n-1)}-a_{2n} \\ \cdot & \cdot & \cdots & \cdot \\ \cdot & \cdot & \cdots & \cdot \\ \cdot & \cdot & \cdots & \cdot \\ a_{(n-1)1}-a_{n1} & a_{(n-1)2}-a_{n2} & \cdots & a_{(n-1)(n-1)}-a_{nn} \end{pmatrix}_{(n-1)\times n}$$

Step 2: Compute TrlIF or TrplIF Row oddments C_i where $i = 1, 2, \ldots, n$, by deleting ith row from C.
Compute TrlIF or TrplIF Column oddments R_j where $j = 1, 2, \ldots, n$, by deleting jth column from R.

Step 3: Form the TrlIF or TrplIF Oddments Table

				TrlIF or TrplIF	Row Oddments
$a_{11}-a_{12}$	$a_{12}-a_{13}$	$\cdots$	$a_{1(n-1)}-a_{1n}$	$\lvert C_1\rvert$	$R\{\lvert C_1\rvert\}$
$a_{21}-a_{22}$	$a_{22}-a_{23}$	$\cdots$	$a_{2(n-1)}-a_{2n}$	$\lvert C_2\rvert$	$R\{\lvert C_2\rvert\}$
.	.	$\cdots$	.	.	.
.	.	$\cdots$	.	.	.
.	.	$\cdots$	.	.	.
$a_{n1}-a_{n2}$	$a_{n2}-a_{n3}$	$\cdots$	$a_{n(n-1)}-a_{nn}$	$\lvert C_n\rvert$	$R\{\lvert C_n\rvert\}$
TrlIF or TrplIF					
$\lvert R_1\rvert$	$\lvert R_2\rvert$	$\cdots$	$\lvert R_n\rvert$		
Column Oddments					
$R\{\lvert R_1\rvert\}$	$R\{\lvert R_2\rvert\}$	$\cdots$	$R\{\lvert R_n\rvert\}$		

Step 4: Check $\sum_{i=1}^{n} R\{|C_i|\} = \sum_{j=1}^{n} R\{|R_j|\} = R\left\{\sum_{i=1}^{n} |C_i|\right\} = R\left\{\sum_{j=1}^{n} |R_j|\right\}$
If not, the method fails.

Step 5: Calculate Expected value of the game $V = \frac{\sum_{k=1}^{n} a_{k1} R\{|C_k|\}}{\sum_{i=1}^{n} R\{|C_i|\}}$ or $\frac{\sum_{k=1}^{n} a_{1k} R\{|R_k|\}}{\sum_{j=1}^{n} R\{|R_j|\}}$.

$$\text{Strategy of Row Player} = \left(\frac{R\{|C_1|\}}{R\{\sum_{i=1}^{n} |C_i|\}}, \frac{R\{|C_2|\}}{R\{\sum_{i=1}^{n} |C_i|\}}, \ldots, \frac{R\{|C_n|\}}{R\{\sum_{i=1}^{n} |C_i|\}} \right)$$

$$\text{Strategy of Column Player} = \left(\frac{R\{|R_1|\}}{R\left\{\sum_{j=1}^{n} |R_j|\right\}}, \frac{R\{|R_2|\}}{R\left\{\sum_{j=1}^{n} |R_j|\right\}}, \ldots, \frac{R\{|R_n|\}}{R\left\{\sum_{j=1}^{n} |R_j|\right\}} \right)$$

Dominance Property for Intuitionistic Fuzzy Games

n this section we describes the computational procedure for solving $n \times n$ games sing oddments method.

eneral Rule of Dominance Property for Triangular and Trapezoidal ntuitionistic Fuzzy Games

i. $A = (a_{ij})_{n \times n}$ be Triangular Intuitionistic Fuzzy Payoff Matrix or a Trapezoidal Intuitionistic Fuzzy Payoff Matrix
ii. If $R(a_{k1}) \leq R(a_{r1}), R(a_{k2}) \leq R(a_{r2}), \ldots, R(a_{kn}) \leq R(a_{rn})$, then kth row is dominated by rth row and delete kth row.
iii. If $R(a_{1k}) \geq R(a_{1r}), R(a_{2k}) \geq R(a_{2r}), \ldots, R(a_{nk}) \geq R(a_{nr})$, then kth column is dominated by rth column and delete kth Column.
iv. If some linear combination of some rows dominates ith row, then ith row will be deleted. Similar arguments follows for columns.

Numerical Examples

this section we illustrate the proposed method with numerical examples

xample 5.1 Solve the following Triangular Intuitionistic Fuzzy Game.

	Player B					
ayer A	(3,4,5)(2,4,6)	(1,2,3)(0,2,4)	(−1,0,1)(−2,0,2)	(1,2,3)(0,2,4)	(0,1,2)(−1,1,3)	(0,1,2)(−1,1,3)
	(3,4,5)(2,4,6)	(2,3,4)(1,3,5)	(0,1,2)(−1,1,3)	(2,3,4)(1,3,5)	(1,2,3)(0,2,4)	(1,2,3)(0,2,4)
	(3,4,5)(2,4,6)	(2,3,4)(1,3,5)	(6,7,8)(5,7,9)	(−6,−5,−4)(−7,−5,−3)	(0,1,2)(−1,1,3)	(1,2,3)(0,2,4)
	(3,4,5)(2,4,6)	(2,3,4)(1,3,5)	(3,4,5)(2,4,6)	(−2,−1,0)(−3,−1,1)	(1,2,3)(0,2,4)	(1,2,3)(0,2,4)
	(3,4,5)(2,4,6)	(2,3,4)(1,3,5)	(2,3,4)(1,3,5)	(−3,−2,−1)(−4,−2,0)	(1,2,3)(0,2,4)	(1,2,3)(0,2,4)

Solution:
By Dominance Property, the above Triangular Intuitionistic Fuzzy payoff matrix can be reduced as,

	Player B	
Player A	(0,1,2)(−1,1,3)	(2,3,4)(1,3,5)
	(6,7,8)(5,7,9)	(−6,−5,−4)(−7,−5,−3)

By Triangular Intuitionistic Fuzzy Games Oddments Method,

Row oddments		Column oddments	
(10,12,14)(8,12,16)	12	(6,8,10)(4,8,12)	8
(0,2,4)(−2,2,6)	2	(4,6,8)(2,6,10)	6
(10,14,18)(6,14,22)	**14**	**(10,14,18)(6,14,22)**	**14**

Hence, $S_A = \left(0, \frac{6}{7}, \frac{1}{7}, 0, 0\right)$ $S_B = \left(0, 0, \frac{4}{7}, \frac{3}{7}, 0, 0\right)$ Value of the game $= \frac{13}{7}$

Example 5.2 Solve the following Trapezoidal Intuitionistic Fuzzy Game.

	Player B			
Player A	(0,2,4,6)(−2,2,4,8)	(−1,1,3,5)(−3,1,3,7)	(1,3,5,7)(−1,3,5,9)	(−3,−1,1,3)(−5,−1,1,5)
	(0,2,4,6)(−2,2,4,8)	(1,3,5,7)(−1,3,5,9)	(−1,1,3,5)(−3,1,3,7)	(1,3,5,7)(−1,3,5,9)
	(1,3,5,7)(−1,3,5,9)	(−1,1,3,5)(−3,1,3,7)	(1,3,5,7)(−1,3,5,9)	(−3,−1,1,3)(−5,−1,1,5)
	(−3,−1,1,3)(−5,−1,1,5)	(1,3,5,7)(−1,3,5,9)	(−3,−1,1,3)(−5,−1,1,5)	(5,7,9,11)(3,7,9,13)

Solution
By Dominance Property, the above Trapezoidal Intuitionistic Fuzzy payoff matrix can be reduced as,

	Player B	
Player A	(1,3,5,7)(−1,3,5,9)	(−3,−1,1,3)(−5,−1,1,5)
	(−3,−1,1,3)(−5,−1,1,5)	(5,7,9,11)(3,7,9,13)

By Triangular Intuitionistic Fuzzy Games Oddments Method,

Row oddments		Column oddments	
(2,8,8,14)(−2,8,8,18)	8	(2,8,8,14)(−2,8,8,18)	8
(−2,4,4,10)(−6,4,4,14)	4	(−2,4,4,10)(−6,4,4,14)	4
(0,12,12,24)(−8,12,12,32)	**12**	**(0,12,12,24)(−8,12,12,32)**	**12**

Hence, $S_A = \left(0, 0, \frac{2}{3}, \frac{1}{3}\right)$ $S_B = \left(0, 0, \frac{2}{3}, \frac{1}{3}\right)$ Value of the game $= \frac{8}{3}$

Conclusion

Ve considered the solution of Two Person Zero Sum Games using Triangular and 'rapezoidal Intuitionistic Fuzzy Numbers. Here pay-off is considered as imprecise umbers instead of crisp numbers which takes care of the uncertainty and vague-ess inherent in such problems. We discuss solution of Intuitionistic Fuzzy Games /ith mixed strategies by Dominance Property and convert into a 2 × 2 Intuitionistic ayoff matrix and solution obtained using Intuitionistic Fuzzy Oddment Method. numerical example has illustrated the proposed methods. This theory can be pplied in decision making procedures in areas such as economics, operations esearch, management, war science, online purchasing, etc.

References

1. Nishizaki, I., Sakawa, M.: Equilibrium solutions for multi objective bimatrix games incorporating fuzzy goals. J. Optim. Theory Appl. **86**, 433–457 (1995)
2. Nishizaki, I., Sakawa, M.: Max-min solution of fuzzy multi objective matrix games. Fuzzy Sets Syst. **61**, 265–275 (1994)
3. Zadeh, L.A.: Fuzzy Sets. Inf. Control **8**, 338–352 (1965)
4. Atanassov, K.: Intutionistic fuzzy sets. Fuzzy Sets Syst. **20**, 87–96 (1986)
5. Atanassov, K.: Intuitionistic Fuzzy Sets: Theory and Applications. Physica-Verglag (1999)
6. Atanassov, K.: Ideas for intuitionistic fuzzy equations, inequalities and optimization. Notes on Intuitionistic Fuzzy Sets **1**(1), 17–24 (1995)
7. Li, D.F.: Mathematical-programming approach to matrix games with payoffs represented by Atanassov's interval-valued intuitionistic fuzzy sets. IEEE Trans. Fuzzy Syst. **18**, 1112–1128 (2011)
8. Li, D.F., Nan, J.X.: A nonlinear programming approach to matrix games with payoffs of Atanassov's intuitionistic fuzzy sets. Int. J. Uncertain. Fuzz. Knowl. Based Syst. **17**(4), 585–607 (2009)
9. Nayak, P.K., Pal, M.: Bi-matrix games with intuitionistic fuzzy goals. Iran. J. Fuzzy Syst. **7**, 65–79 (2010)
10. Li, D.F., Nan, J.X., Tang, Z.P., Chen, K.J., Xiang, X.D., Hong, F.X.: A bi-objective programming model for matrix games with payoffs of Atanassov's triangular intuitionistic fuzzy numbers. Iran. J Fuzzy Syst. **9**, 93–110 (2012)
11. Nan, J.X., Li, D.F., Zhang, M.J.: A lexicographic method for matrix games with payoffs of triangular intuitionistic fuzzy numbers. Int. J. Comput. Intell. Syst. **3**, 280–289 (2010)

Author Index

M. Senthilkumar et al. (eds.), *Computational Intelligence, Cyber Security and Computational Models*, Advances in Intelligent Systems and Computing 412, DOI 10.1007/978-981-10-0251-9